WITHDRAWN.

Electrical Principles and Technology for Engineering

To Elizabeth

Electrical Principles and Technology for Engineering

J. O. Bird, BSc(Hons) CEng MIEE FIEIE CMath FIMA FCollP

NEWNES

Newnes
An imprint of Butterworth-Heinemann Ltd
Linacre House, Jordan Hill, Oxford OX2 8DP

A member of the Reed Elsevier plc group

OXFORD LONDON BOSTON
MUNICH NEW DELHI SINGAPORE SYDNEY
TOKYO TORONTO WELLINGTON

First published 1995

British Library Cataloguing in Publication Data
Bird, J.O.
Electrical Principles and Technology for Engineering
I. Title
621.3076

ISBN 0 7506 2196 6

Composition by Scribe Design, Gillingham, Kent
Printed and bound in Great Britain

Contents

Preface

Electrical Principles and Technology for Engineering introduces the principles which describe the operation of d.c. and a.c. circuits, covering both steady and transient states, and applies these principles to three-phase supplies, transformers, d.c. machines and three-phase induction motors.

This textbook provides coverage of two of the advanced GNVQ optional units in Engineering, these being **Electrical Principles** and **Electrical Technology**. However, it can be regarded as a basic textbook in Electrical Principles and Technology for a wider range of courses.

The text is set out in three main parts.

Part 1, involving Chapters 1 to 10, is basic electrical engineering principles; some is revision from the mandatory unit **Science for Engineering**, and some is important additional material which may be needed for assignment/project work.

Part 2, involving Chapters 11 to 15, covers the GNVQ unit **Electrical Principles (Advanced)**.

Part 3, involving Chapters 16 to 19, covers the GNVQ unit **Electrical Technology (Advanced)**.

Each topic considered in the text is presented in a way that assumes in the reader little previous knowledge of that topic. Theory is introduced in each chapter by a reasonably brief outline of essential information, definitions, formulae, procedures, etc. The theory is kept to a minimum, for problem-solving is extensively used to establish and exemplify the theory. It is intended that readers will gain real understanding through seeing problems solved and then through solving similar problems themselves.

Electrical Principles and Technology for Engineering contains some **350 worked problems**, together with **225 multi-choice questions** (with answers at the back of the book). Also included are some **325 short answer questions**, the answers for which can be determined from the preceding material in that particular chapter, and some **500 further questions**, all with answers, in brackets, immediately following each question. Nearly 400 line diagrams further enhance the understanding of the theory. All of the problems – multi-choice, short answer and further questions – mirror practical situations found in electrical engineering. This textbook provides a lead into **Higher Electrical Technology**.

I would like to express my appreciation for the friendly co-operation and helpful advice given by the publishers and to Mrs Elaine Woolley for the excellent typing of the manuscript.

Learning by Example is at the heart of *Electrical Principles and Technology for Engineering.*

JOHN O. BIRD
Highbury College,
Portsmouth

Part 1

Basic Electrical Engineering Principles

1

Units associated with basic electrical quantities

At the end of this chapter you should be able to:

- state the basic SI units
- recognize derived SI units
- understand prefixes denoting multiplication and division
- state the units of charge, force, work and power and perform simple calculations involving these units
- state the units of electrical potential, e.m.f., resistance, conductance, power and energy and perform simple calculations involving these units

1.1 SI units

The system of units used in engineering and science is the Système Internationale d'Unités (International system of units), usually abbreviated to SI units, and is based on the metric system. This was introduced in 1960 and is now adopted by the majority of countries as the official system of measurement.

The basic units in the SI system are listed below with their symbols:

Quantity	Unit and symbols
length	metre, m
mass	kilogram, kg
time	second, s
electric current	ampere, A
thermodynamic temperature	kelvin, K
luminous intensity	candela, cd
amount of substance	mole, mol

Derived SI units use combinations of basic units and there are many of them. Two examples are:

Velocity – metres per second (m/s)

Acceleration – metres per second squared (m/s^2)

SI units may be made larger or smaller by using prefixes which denote multiplication or division by a particular amount. The six most common multiples, with their meaning, are listed below:

Prefix	Name	Meaning	
M	mega	multiply by 1 000 000	(i.e. $\times 10^6$)
k	kilo	multiply by 1000	(i.e. $\times 10^3$)
m	milli	divide by 1000	(i.e. $\times 10^{-3}$)
μ	micro	divide by 1 000 000	(i.e. $\times 10^{-6}$)
n	nano	divide by 1 000 000 000	(i.e. $\times 10^{-9}$)
p	pico	divide by 1 000 000 000 000	(i.e. $\times 10^{-12}$)

1.2 Charge

The **unit of charge** is the **coulomb (C)** where one coulomb is one ampere second (1 coulomb = 6.24×10^{18} electrons). The coulomb is defined as the quantity of electricity which flows past a given point in an electric circuit when a current of one ampere is maintained for one second. Thus,

charge, in coulombs $\boxed{Q = It}$

where I is the current in amperes and t is the time in seconds.

> Problem 1. If a current of 5 A flows for 2 minutes, find the quantity of electricity transferred.

Quantity of electricity $Q = It$ coulombs. $I = 5$ A, $t = 2 \times 60 = 120$ s, hence

$$Q = 5 \times 120 = \mathbf{600\ C}$$

1.3 Force

The **unit of force** is the **newton (N)** where one newton is one kilogram metre per second squared. The newton is defined as the force which, when applied to a mass of one kilogram, gives it an acceleration of one metre per second squared. Thus,

force, in newtons $\boxed{F = ma}$

where m is the mass in kilograms and a is the acceleration in metres per second squared. Gravitational force, or weight, is mg, where $g = 9.81$ m/s^2.

> Problem 2. A mass of 5000 g is accelerated at 2 m/s^2 by a force. Determine the force needed.

Force = mass × acceleration

$$= 5 \text{ kg} \times 2 \text{ m/s}^2 = 10 \frac{\text{kg m}}{\text{s}^2} = \mathbf{10\ N}$$

> Problem 3. Find the force acting vertically downwards on a mass of 200 g attached to a wire.

Mass = 200 g = 0.2 kg, and acceleration due to gravity, $g = 9.81$ m/s^2. Force acting downwards = weight = mass × acceleration

$$= 0.2 \text{ kg} \times 9.81 \text{ m/s}^2$$

$$= \mathbf{1.962\ N}$$

1.4 Work

The **unit of work or energy** is the **joule (J)** where one joule is one newton metre. The joule is defined as the work done or energy transferred when a force of one newton is exerted through a distance of one metre in the direction of the force. Thus

work done on a body, in joules $\boxed{W = Fs}$

where F is the force in newtons and s is the distance in metres moved by the body in the direction of the force. Energy is the capacity for doing work.

1.5 Power

The **unit of power** is the **watt (W)** where one watt is one joule per second. Power is defined as the rate of doing work or transferring energy. Thus,

power in watts $\boxed{P = \dfrac{W}{t}}$

where W is the work done or energy transferred in joules and t is the time in seconds. Thus,

energy, in joules $\boxed{W = Pt}$

> Problem 4. A portable machine requires a force of 200 N to move it. How much work is done if the machine is moved 20 m and what average power is utilized if the movement takes 25 s?

Work done = force × distance = 200 N × 20 m = **4000 Nm** or **4 kJ**.

$$\text{Power} = \frac{\text{work done}}{\text{time taken}} = \frac{4000}{25} \frac{\text{J}}{\text{s}} = \mathbf{160\ W}$$

> Problem 5. A mass of 1000 kg is raised through a height of 10 m in 20 s. What is (a) the work done and (b) the power developed?

(a) Work done = force × distance and force = mass × acceleration, hence, work done = (1000 kg × 9.81 m/s^2) × (10 m)

= 98 100 Nm = **98.1 kN m** or **98.1 kJ**

(b) $\text{Power} = \dfrac{\text{work done}}{\text{time taken}}$

$= \dfrac{98\,100}{20}\ \dfrac{\text{J}}{\text{s}}$ (or W) = 4905 W = **4.904 kW**

1.6 Electrical potential and e.m.f.

The **unit of electric potential** is the **volt (V)** where one volt is one joule per coulomb. One volt is defined as the difference in potential between two points in a conductor which, when carrying a current of one ampere, dissipates a power of one watt, i.e.

$$\text{volts} = \frac{\text{watts}}{\text{amperes}} = \frac{\text{joules/second}}{\text{amperes}}$$
$$= \frac{\text{joules}}{\text{ampere seconds}} = \frac{\text{joules}}{\text{coulombs}}$$

A change in electric potential between two points in an electric circuit is called a **potential difference**. The electromotive force (e.m.f.) provided by a source of energy such as a battery or a generator is measured in volts.

1.7 Resistance and conductance

The **unit of electric resistance** is the **ohm (Ω)** where one ohm is one volt per ampere. It is defined as the resistance between two points in a conductor when a constant electric potential of one volt applied at the two points produces a current flow of one ampere in the conductor. Thus,

resistance, in ohms $\boxed{R = \dfrac{V}{I}}$

where V is the potential difference across the two points in volts and I is the current flowing between the two points in amperes.

The reciprocal of resistance is called **conductance** and is measured in **siemens (S)**.

Thus, conductance, in siemens $\boxed{G = \dfrac{1}{R}}$

where R is the resistance in ohms.

> Problem 6. Find the conductance of a conductor of resistance (a) 10 Ω, (b) 5 kΩ and (c) 100 mΩ.

(a) Conductance $G = \dfrac{1}{R} = \dfrac{1}{10}$ siemen = **0.1 S**

(b) $G = \dfrac{1}{R} = \dfrac{1}{5 \times 10^3}\ \text{S} = 0.2 \times 10^{-3}\ \text{S} = \mathbf{0.2\ mS}$

(c) $G = \dfrac{1}{R} = \dfrac{1}{100 \times 10^{-3}}\ \text{S} = \dfrac{10^3}{100}\ \text{S} = \mathbf{10\ S}$

1.8 Electrical power and energy

When a direct current of I amperes is flowing in an electric circuit and the voltage across the circuit is V volts, then

power, in watts $\boxed{P = VI}$

Electrical energy = Power × time

= VIt joules

Although the unit of energy is the joule, when dealing with large amounts of energy, the unit used is the kilowatt hour (kW h) where

1 kW h = 1000 watt hour

= 1000 × 3600 watt seconds or joules

= 3 600 000 J

> Problem 7. A source e.m.f. of 5 V supplies a current of 3 A for 10 minutes. How much energy is provided in this time?

Energy = power × time and power = voltage × current. Hence,

energy = VIt = 5 × 3 × (10 × 60)

= 9000 W s or J

= **9 kJ**

> Problem 8. An electric heater consumes 1.8 MJ when connected to a 250 V supply for 30 minutes. Find the power rating of the heater and the current taken from the supply.

$$\text{Power} = \frac{\text{energy}}{\text{time}} = \frac{1.8 \times 10^6}{30 \times 60}\ \frac{\text{J}}{\text{s}} \text{ (or W)}$$

$$= 1000 \text{ W}$$

i.e. **power rating of heater = 1 kW**.

$$\text{Power } P = VI, \text{ thus } I = \frac{P}{V} = \frac{1000}{250} = 4 \text{ A}$$

Hence the current taken from the supply is 4 A.

1.9 Summary of terms, units and their symbols

Quantity	Quantity symbol	Unit	Unit symbol
Length	l	metre	m
Mass	m	kilogram	kg
Time	t	second	s
Velocity	v	metres per second	m/s or m s^{-1}
Acceleration	a	metres per second squared	m/s^2 or m s^{-2}
Force	F	newton	N
Electrical charge or quantity	Q	coulomb	C
Electric current	I	ampere	A
Resistance	R	ohm	Ω
Conductance	G	siemen	S
Electromotive force	E	volt	V
Potential difference	V	volt	V
Work	W	joule	J
Energy	E (or W)	joule	J
Power	P	watt	W

1.10 Multi-choice questions on units associated with basic electrical quantities

(Answers on page 283.)

1. A resistance of 50 kΩ has a conductance of (a) 20 S; (b) 0.02 S; (c) 0.02 mS; (d) 20 kS.
2. State which of the following is incorrect: (a) 1 N = 1 kg m/s^2; (b) 1 V = 1 J/C; (c) 30 mA = 0.03 A; (d) 1 J = 1 N/m.
3. The power dissipated by a resistor of 10 Ω when a current of 2 A passes through it is (a) 0.4 W; (b) 20 W; (c) 40 W; (d) 200 W.
4. A mass of 1200 g is accelerated at 200 cm/s^2 by a force. The value of the force required is: (a) 2.4 N; (b) 2400 N; (c) 240 kN; (d) 0.24 N.
5. A charge of 240 C is transferred in 2 minutes. The current flowing is (a) 120 A; (b) 480 A; (c) 2 A; (d) 8 A.
6. A current of 2 A flows for 10 h through a 100 Ω resistor. The energy consumed by the resistor is: (a) 0.5 kW h; (b) 4 kW h; (c) 2 kW h; (d) 0.02 kW h.

1.11 Short answer questions on units associated with basic electrical quantities

1. What does 'SI units' mean?
2. Complete the following: Force = ... × ....
3. What do you understand by the term 'potential difference'?
4. Define electric current in terms of charge and time.
5. Name the units used to measure (a) the quantity of electricity, (b) resistance and (c) conductance.
6. Define the coulomb.
7. Define electrical energy and name its unit.
8. Define electrical power and name its unit.
9. What is electromotive force?
10. Write down a formula for calculating the power in a d.c. circuit.
11. Write down the symbols for the following quantities: (a) electric charge; (b) work; (c) e.m.f.; (d) p.d.
12. State to which units the following abbreviations refer: (a) A; (b) C; (c) J; (d) N; (e) m.

1.12 Further questions on units associated with basic electrical quantities

(Take g = 9.81 m/s^2 where appropriate.)

1. What force is required to give a mass of 20 kg an acceleration of 30 m/s^2? [600 N]
2. Find the accelerating force when a car having a mass of 1.7 Mg increases its speed with a constant acceleration of 3 m/s^2. [5.1 kN]

3. A force of 40 N accelerates a mass at 5 m/s^2. Determine the mass. [8 kg]
4. Determine the force acting downwards on a mass of 1500 g suspended on a string. [14.72 N]
5. A force of 4 N moves an object 200 cm in the direction of the force. What amount of work is done? [8 J]
6. A force of 2.5 kN is required to lift a load. How much work is done if the load is lifted through 500 cm? [12.5 kJ]
7. An electromagnet exerts a force of 12 N and moves a soft-iron armature through a distance of 1.5 cm in 40 ms. Find the power consumed. [4.5 W]
8. A mass of 500 kg is raised to a height of 6 m in 30 s. Find (a) the work done and (b) the power developed. [(a) 29.43 kN m (b) 981 W]
9. What quantity of electricity is carried by 6.24×10^{21} electrons? [1000 C]
10. In what time would a current of 1 A transfer a charge of 30 C? [30 s]
11. A current of 3 A flows for 5 minutes. What charge is transferred? [900 C]
12. How long must a current of 0.1 A flow so as to transfer a charge of 30 C? [5 minutes]
13. Find the conductance of a resistor of resistance (a) 10 Ω (b) 2 kΩ (c) 2 mΩ. [(a) 0.1 S (b) 0.5 mS (c) 500 S]
14. A conductor has a conductance of 50 μs. What is its resistance? [20 kΩ]
15. An e.m.f. of 250 V is connected across a resistance and the current flowing through the resistance is 4 A. What is the power developed? [1 kW]
16. 450 J of energy are converted into heat in 1 minute. What power is dissipated? [7.5 W]
17. A current of 10 A flows through a conductor and 10 W is dissipated. What p.d. exists across the ends of the conductor? [1 V]
18. A battery of e.m.f. 12 V supplies a current of 5 A for 2 minutes. How much energy is supplied in this time? [7.2 kJ]
19. A d.c. electric motor consumes 36 MJ when connected to a 250 V supply for 1 hour. Find the power rating of the motor and the current taken from the supply. [10 kW, 40 A]

2

An introduction to electric circuits

At the end of this chapter you should be able to:

- recognize common electrical circuit diagram symbols
- understand that electric current is the rate of movement of charge and is measured in amperes
- appreciate that the unit of charge is the coulomb
- calculate charge or quantity of electricity Q from $Q = It$
- understand that a potential difference between two points in a circuit is required for current to flow
- appreciate that the unit of p.d. is the volt
- understand that resistance opposes current flow and is measured in ohms
- appreciate what an ammeter, a voltmeter, an ohmmeter, a multimeter and a CRO measure
- distinguish between linear and non-linear devices
- state Ohm's law as $V = IR$ or $I = V/R$ or $R = V/I$
- use Ohm's law in calculations, including multiples and sub-multiples of units
- describe a conductor and an insulator, giving examples of each
- appreciate that electrical power P is given by $P = VI = I^2R = V^2/R$ watts
- calculate electrical power
- define electrical energy and state its unit
- calculate electrical energy
- state the three main effects of an electric current, giving practical examples of each
- explain the importance of fuses in electrical circuits

2.1 Standard symbols for electrical components

Symbols are used for components in electrical circuit diagrams and some of the more common ones are shown in Fig. 2.1.

2.2 Electric current and quantity of electricity

All **atoms** consist of **protons**, **neutrons** and **electrons**. The protons, which have positive electrical charges, and the neutrons, which have

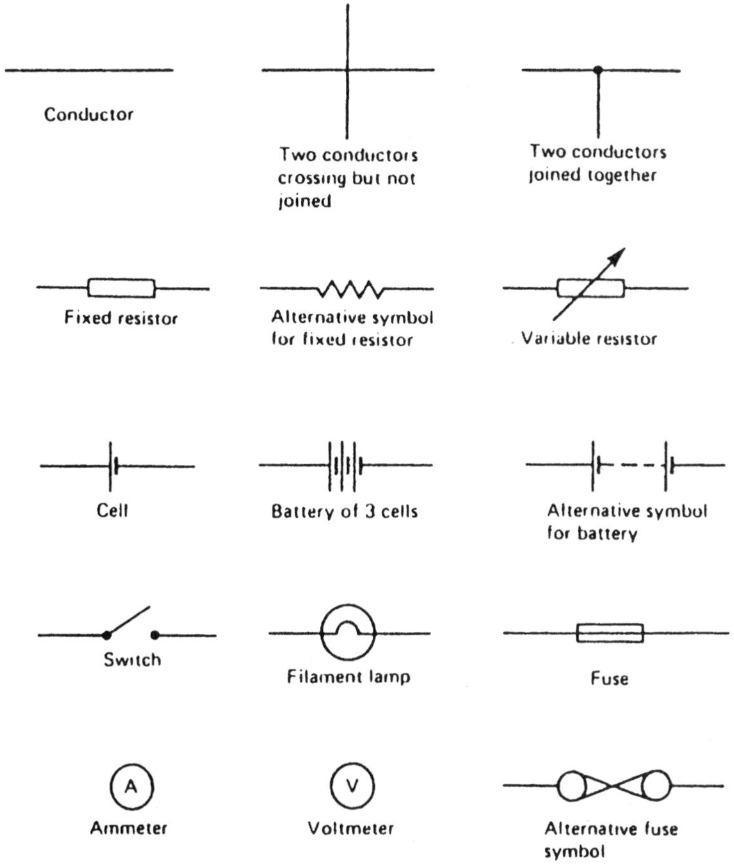

Figure 2.1

no electrical charge, are contained within the **nucleus**. Removed from the nucleus are minute negatively charged particles called electrons. Atoms of different materials differ from one another by having different numbers of protons, neutrons and electrons. An equal number of protons and electrons exist within an atom and it is said to be electrically balanced, as the positive and negative charges cancel each other out. When there are more than two electrons in an atom the electrons are arranged into **shells** at various distances from the nucleus.

All atoms are bound together by powerful forces of attraction existing between the nucleus and its electrons. Electrons in the outer shell of an atom, however, are attracted to their nucleus less powerfully than are electrons whose shells are nearer the nucleus.

It is possible for an atom to lose an electron; the atom, which is now called an **ion**, is not now electrically balanced, but is positively charged and is thus able to attract an electron to itself from another atom. Electrons that move from one atom to another are called free electrons and such random motion can continue indefinitely. However, if an electric pressure or **voltage** is applied across any material there is a tendency for electrons to move in a particular direction. This movement of free electrons, known as **drift**, constitutes an electric current flow.

Thus current is the rate of movement of charge

Conductors are materials that have electrons that are loosely connected to the nucleus and can easily move through the material from one atom to another.

Insulators are materials whose electrons are held firmly to their nucleus.

The unit used to measure the **quantity of electrical charge *Q*** is called the **coulomb C** (where 1 coulomb = 6.24×10^{18} electrons). If the drift of electrons in a conductor takes place at the rate of one coulomb per second the resulting current is said to be a current of one ampere. Thus,

1 ampere = 1 coulomb per second or
1 A = 1 C/s

Hence,

1 coulomb = 1 ampere second or 1 C = 1 A s

Generally, if I is the current in amperes and t the time in seconds during which the current flows, then $I \times t$ represents the quantity of electrical charge in coulombs, i.e. quantity of electrical charge transferred,

$$Q = I \times t \text{ coulombs}$$

Problem 1. What current must flow if 0.24 coulombs is to be transferred in 15 ms?

Since the quantity of electricity, $Q = It$, then

$$I = \frac{Q}{t} = \frac{0.24}{15 \times 10^{-3}} = \frac{0.24 \times 10^3}{15} = \frac{240}{15} = \mathbf{16\ A}$$

Problem 2. If a current of 10 A flows for four minutes, find the quantity of electricity transferred.

Quantity of electricity, $Q = It$ coulombs. $I = 10$ A; $t = 4 \times 60 = 240$ s. Hence

$$Q = 10 \times 240 = \mathbf{2400\ C}$$

2.3 Potential difference and resistance

For a continuous current to flow between two points in a circuit a **potential difference (p.d.)** or **voltage**, V, is required between them; a complete conducting path is necessary to and from the source of electrical energy. The unit of p.d. is the **volt, V**.

Figure 2.2 shows a cell connected across a filament lamp. Current flow, by convention, is considered as flowing from the positive terminal of the cell, around the circuit to the negative terminal.

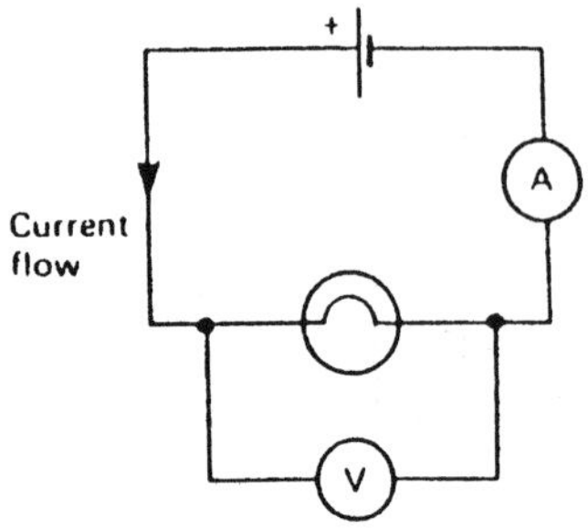

Figure 2.2

The flow of electric current is subject to friction. This friction, or opposition, is called **resistance R** and is the property of a conductor that limits current. The unit of resistance is the **ohm Ω**. 1 ohm is defined as the resistance which will have a current of 1 ampere flowing through it when 1 volt is connected across it, i.e.

$$\textbf{resistance } R = \frac{\textbf{potential difference}}{\textbf{current}}$$

2.4 Basic electrical measuring instruments

An **ammeter** is an instrument used to measure current and must be connected **in series** with the circuit. Figure 2.2 shows an ammeter connected in series with the lamp to measure the current flowing through it. Since all the current in the circuit passes through the ammeter it must have a very **low resistance**.

A **voltmeter** is an instrument used to measure p.d. and must be connected **in parallel** with the part of the circuit whose p.d. is required. In Fig. 2.2, a voltmeter is connected in parallel with the lamp to measure the p.d. across it. To avoid a significant current flowing through it a voltmeter must have a very **high resistance**.

An **ohmmeter** is an instrument for measuring resistance.

A **multimeter**, or universal instrument, may be used to measure voltage, current and resistance. An 'Avometer' is a typical example.

The **cathode ray oscilloscope** (CRO) may be used to observe waveforms and to measure voltages and currents. The display of a CRO involves a spot of light moving across a screen. The amount by which the spot is deflected from

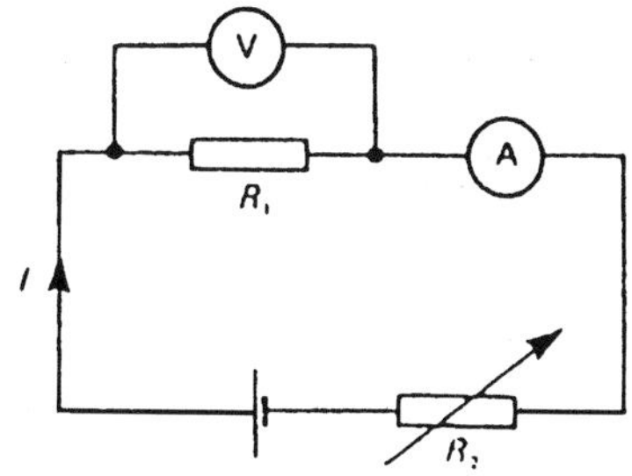

Figure 2.3

its initial position depends on the p.d. applied to the terminals of the CRO and the range selected. The displacement is calibrated in 'volts per cm'. For example, if the spot is deflected 3 cm and the volts/cm switch is on 10 V/cm then the magnitude of the p.d. is 3 cm × 10 V/cm, i.e. 30 V.

(See Chapter 10 for more detail about electrical measuring instruments and measurements.)

2.5 Linear and non-linear devices

Figure 2.3 shows a circuit in which current I can be varied by the variable resistor R_2. For various settings of R_2, the current flowing in resistor R_1, displayed on the ammeter, and the p.d. across R_1, displayed on the voltmeter, are noted and a graph is plotted of p.d. against current. The result is shown in Fig. 2.4(a) where the straight line graph passing through the origin indicates that current is directly proportional to the p.d. Since the gradient, i.e. (p.d.)/(current), is constant, resistance R_1 is constant. A resistor is thus an example of a **linear device**.

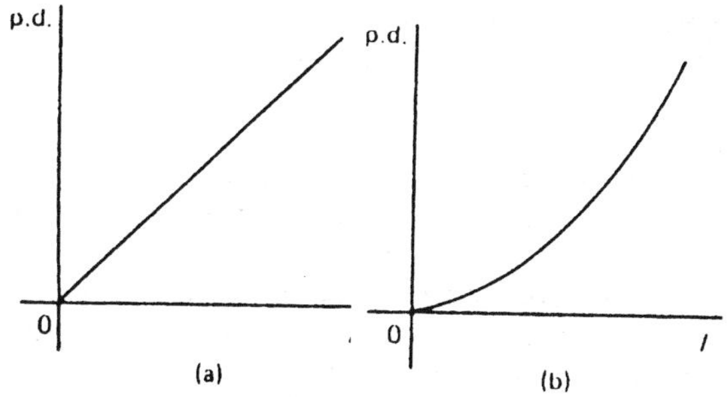

Figure 2.4

If the resistor R_1 in Fig. 2.3 is replaced by a component such as a lamp then the graph shown in Fig. 2.4(b) results when values of p.d. are noted for various current readings. Since the gradient is changing, the lamp is an example of a **non-linear device**.

2.6 Ohm's law

Ohm's law states that the current I flowing in a circuit is directly proportional to the applied voltage V and inversely proportional to the resistance R, provided the temperature remains constant. Thus,

$$I = \frac{V}{R} \text{ or } V = IR \text{ or } R = \frac{V}{I}$$

Problem 3. The current flowing through a resistor is 0.8 A when a p.d. of 20 V is applied. Determine the value of the resistance.

From Ohm's law,

$$\text{resistance } R = \frac{V}{I} = \frac{20}{0.8} = \frac{200}{8} = \mathbf{25\ \Omega}$$

2.7 Multiples and sub-multiples

Currents, voltages and resistances can often be very large or very small. Thus **multiples** and **sub-multiples** of units are often used, as stated in Chapter 1. The most common ones, with an example of each, are listed in Table 2.1.

Problem 4. Determine the p.d. which must be applied to a 2 kΩ resistor in order that a current of 10 mA may flow.

Resistance $R = 2\text{ k}\Omega = 2 \times 10^3 = 2000\ \Omega$.
Current $I = 10\text{ mA} = 10 \times 10^{-3}$ A or $10/10^3$ or (10/1000) A = 0.01 A.
From Ohm's law, potential difference, $V = IR = (0.01)(2000) = \mathbf{20\ V}$.

Problem 5. A coil has a current of 50 mA flowing through it when the applied voltage is 12 V. What is the resistance of the coil?

Table 2.1

Prefix	*Name*	*Meaning*	*Example*
M	mega	multiply by 1 000 000 (i.e. $\times 10^6$)	$2\ \mathrm{M\Omega} = 2\,000\,000$ ohms
k	kilo	multiply by 1000 (i.e. $\times 10^3$)	$10\ \mathrm{kV} = 10\,000$ volts
m	milli	divide by 1000 (i.e. $\times 10^{-3}$)	$25\ \mathrm{mA} = \frac{25}{1000}\ \mathrm{A} = 0.025$ amperes
μ	micro	divide by 1 000 000 (i.e. $\times 10^{-6}$)	$50\ \mu\mathrm{V} = \frac{50}{1\,000\,000}\ \mathrm{V} = 0.000\,05$ volts

$$\text{Resistance, } R = \frac{V}{I} = \frac{12}{50 \times 10^{-3}} = \frac{12 \times 10^3}{50} = \frac{12\,000}{50} = \mathbf{240\ \Omega}$$

Problem 6. A 100 V battery is connected across a resistor and causes a current of 5 mA to flow. Determine the resistance of the resistor. If the voltage is now reduced to 25 V, what will be the new value of the current flowing?

$$\text{Resistance } R = \frac{V}{I} = \frac{100}{5 \times 10^{-3}} = \frac{100 \times 10^3}{5} = 20 \times 10^3 = \mathbf{20\ k\Omega}$$

Current when voltage is reduced to 25 V,

$$I = \frac{V}{R} = \frac{25}{20 \times 10^3} = \frac{25}{20} \times 10^{-3} = \mathbf{1.25\ mA}$$

Problem 7. What is the resistance of a coil which draws a current of (a) 50 mA and (b) 200 μA from a 120 V supply?

(a) $$\text{Resistance } R = \frac{V}{I} = \frac{120}{50 \times 10^{-3}} = \frac{120}{0.05} = \frac{12\,000}{5} = \mathbf{2400\ \Omega} \text{ or } \mathbf{2.4\ k\Omega}$$

(b) $$\text{Resistance } R = \frac{120}{200 \times 10^{-6}} = \frac{120}{0.0002} = \frac{1\,200\,000}{2} = \mathbf{600\,000\ \Omega} \text{ or } \mathbf{600\ k\Omega} \text{ or } \mathbf{0.6\ M\Omega}$$

2.8 Conductors and insulators

A **conductor** is a material having a low resistance which allows electric current to flow in it. All metals are conductors and some examples include copper, aluminium, brass, platinum, silver, gold and also carbon.

An **insulator** is a material having a high resistance which does not allow electric current to flow in it. Some examples of insulators include plastic, rubber, glass, porcelain, air, paper, cork, mica, ceramics and certain oils.

2.9 Electrical power and energy

Electrical power

Power *P* in an electrical circuit is given by the product of potential difference V and current I as stated in Chapter 1. The unit of power is the **watt, W**. Hence

$$\boldsymbol{P = V \times I} \textbf{ watts} \qquad (2.1)$$

From Ohm's law, $V = IR$. Substituting for V in equation (26.1) gives:

$$P = (IR) \times I$$

i.e. $$\boldsymbol{P = I^2R} \textbf{ watts}$$

Also, from Ohm's law, $I = V/R$. Substituting for I in equation (2.1) gives:

$$P = V \times \frac{V}{R}$$

i.e. $$\boldsymbol{P = \frac{V^2}{R}} \textbf{ watts}$$

There are thus three possible formulae which may be used for calculating power.

Problem 8. A 100 W electric light bulb is connected to a 250 V supply. Determine (a) the current flowing in the bulb, and (b) the resistance of the bulb.

Power $P = V \times I$, from which, current $I = P/V$.

(a) Current $I = \dfrac{100}{250} = \dfrac{10}{25} = \dfrac{2}{5} = \mathbf{0.4\ A}$

(b) Resistance $R = \dfrac{V}{I} = \dfrac{250}{0.4} = \dfrac{2500}{4} = \mathbf{625\ \Omega}$

Problem 9. Calculate the power dissipated when a current of 4 mA flows through a resistance of 5 kΩ.

$$\text{Power } P = I^2R = (4 \times 10^{-3})^2(5 \times 10^3)$$
$$= 16 \times 10^{-6} \times 5 \times 10^3 = 80 \times 10^{-3}$$
$$= \mathbf{0.08\ W} \text{ or } \mathbf{80\ mW}$$

Alternatively, since $I = 4 \times 10^{-3}$ and $R = 5 \times 10^3$ then from Ohm's law voltage $V = IR = 4 \times 10^{-3} \times 5 \times 10^3 = 20$ V. Hence

$$\text{Power } P = V \times I = 20 \times 4 \times 10^{-3} = \mathbf{80\ mW}$$

Problem 10. An electric kettle has a resistance of 30 Ω. What current will flow when it is connected to a 240 V supply? Find also the power rating of the kettle.

$$\text{Current, } I = \frac{V}{R} = \frac{240}{30} = 8 \text{ A}$$

Power, $P = VI = 240 \times 8 = 1920$ W $= \mathbf{1.92\ kW}$ = power rating of kettle.

Problem 11. A current of 5 A flows in the winding of an electric motor, the resistance of the winding being 100 Ω. Determine (a) the p.d. across the winding, and (b) the power dissipated by the coil.

(a) Potential difference across winding, $V = IR$ $= 5 \times 100 =$ 500 V.

(b) Power dissipated by coil is given by

$$P = I^2R = 5^2 \times 100$$
$$= \mathbf{2500\ W} \text{ or } \mathbf{2.5\ kW}$$

(Alternatively, $P = V \times I = 500 \times 5 =$ **2500 W** or **2.5 kW.**)

Problem 12. The current/voltage relationship for two resistors A and B is as shown in Fig. 2.5. Determine the value of the resistance of each resistor.

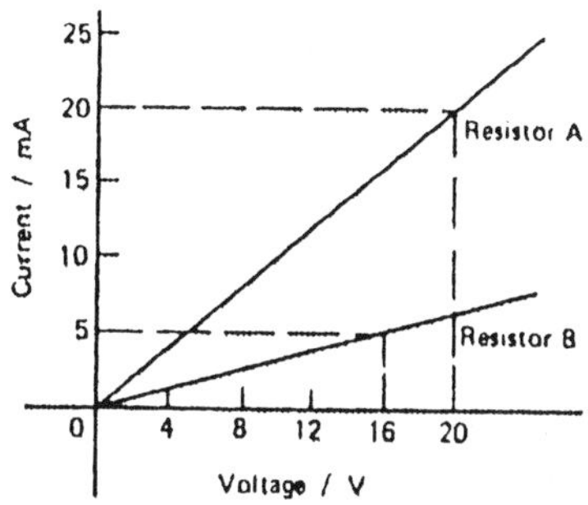

Figure 2.5

For resistor A,

$$R = \frac{V}{I} = \frac{20\text{ A}}{20\text{ mA}} = \frac{20}{0.02} = \frac{2000}{2}$$
$$= \mathbf{1000\ \Omega} \text{ or } \mathbf{1\ k\Omega}$$

For resistor B,

$$R = \frac{V}{I} = \frac{16\text{ V}}{5\text{ mA}} = \frac{16}{0.005} = \frac{16\,000}{5}$$
$$= \mathbf{3200\ \Omega} \text{ or } \mathbf{3.2\ k\Omega}$$

Problem 13. The hot resistance of a 240 V filament lamp is 960 Ω. Find the current taken by the lamp and its power rating.

From Ohm's law,

$$\text{current } I = \frac{V}{R} = \frac{240}{960} = \frac{24}{96} = \mathbf{\frac{1}{4}\ A} \text{ or } \mathbf{0.25\ A}$$

$$\text{Power rating } P = VI = (240)\left(\frac{1}{4}\right) = \mathbf{60\ W}$$

Electrical energy

Electrical energy = power × time

If the power is measured in watts and the time in seconds then the unit of energy is watt-seconds or **joules**. If the power is measured in kilowatts and the time in hours then the unit of energy is **kilowatt-hours**, often called the '**unit of electricity**'. The 'electricity meter' in the home records the number of kilowatt-hours used and is thus an energy meter.

> Problem 14. A 12 V battery is connected across a load having a resistance of 40 Ω. Determine the current flowing in the load, the power consumed and the energy dissipated in 2 minutes.

$$\text{Current } I = \frac{V}{R} = \frac{12}{40} = \mathbf{0.3\ A}$$

Power consumed, $P = VI = (12)(0.3) =$ **3.6 W**.
Energy dissipated = power × time
= (3.6 W)(2 × 60 s) = **432 J** (since 1 J = 1 W s).

> Problem 15. A source of e.m.f. 15 V supplies a current of 2 A for 6 minutes. How much energy is provided in this time?

Energy = power × time, and
power = voltage × current. Hence

$$\text{energy} = VIt = 15 \times 2 \times (6 \times 60)$$

$$= 10\,800 \text{ W s or J}$$

$$= \mathbf{10.8\ kJ}$$

> Problem 16. Electrical equipment in an office takes a current of 13 A from a 240 V supply. Estimate the cost per week of electricity if the equipment is used for 30 hours each week and 1 kW h of energy costs 7p.

Power = VI watts = 240 × 13 = 3120 W = 3.12 kW.
Energy used per week = power × time = (3.12 kW) × (30 h) = 93.6 kW h. Cost at 7p per kW h = 93.6 × 7 = 655.2p. Hence

weekly cost of electricity = £6.55

> Problem 17. An electric heater consumes 3.6 MJ when connected to a 250 V supply for 40 minutes. Find the power rating of the heater and the current taken from the supply.

$$\text{Power} = \frac{\text{energy}}{\text{time}} = \frac{3.6 \times 10^6}{40 \times 60}\ \frac{\text{J}}{\text{s}} \text{ (or W)}$$

$$= 1500 \text{ W}$$

i.e. power rating of heater = **1.5 kW**.
Power $P = VI$, thus

$$I = \frac{P}{V} = \frac{1500}{250} = 6 \text{ A}$$

Hence the current taken from the supply is **6 A**.

> Problem 18. Determine the power dissipated by the element of an electric fire of resistance 20 Ω when a current of 10 A flows through it. If the fire is on for 6 hours determine the energy used and the cost if 1 unit of electricity costs 7p.

Power $P = I^2R = 10^2 \times 20 = 100 \times 20 =$ **2000 W** or **2 kW**.
(Alternatively, from Ohm's law, $V = IR = 10 \times 20 = 200$ V, hence power $P = V \times I = 200 \times 10 = 2000$ W = 2 kW.)
Energy used in 6 hours = power × time = 2 kW × 6 h = **12 kW h**.
1 unit of electricity = 1 kW h. Hence the number of units used is 12. Cost of energy = 12 × 7 = **84p**.

> Problem 19. A business uses two 3 kW fires for an average of 20 hours each per week, and six 150 W lights for 30 hours each per week. If the cost of electricity is 7p per unit, determine the weekly cost of electricity to the business.

Energy = power × time.
Energy used by one 3 kW fire in 20 hours = 3 kW × 20 h = 60 kW h.
Hence weekly energy used by two 3 kW fires = 2 × 60 = 120 kWh.
Energy used by one 150 W light for 30 hours = 150 W × 30 h = 4500 W h = 4.5 kW h.
Hence weekly energy used by six 150 W lamps = 6 × 4.5 = 27 kW h.

Total energy used per week = 120 + 27 = 147 kW h.
1 unit of electricity = 1 kW h of energy, thus weekly cost of energy at 7p per kW h = 7 × 147 = 1029p = **£10.29**.

2.10 Main effects of electric current

The three main effects of an electric current are:

(a) magnetic effect;
(b) chemical effect;
(c) heating effect.

Some practical applications of the effects of an electric current include:

Magnetic effect: bells, relays, motors, generators, transformers, telephones, car ignition and lifting magnets
Chemical effect: primary and secondary cells and electroplating
Heating effect: cookers, water heaters, electric fires, irons, furnaces, kettles and soldering irons

2.11 Fuses

A **fuse** is used to prevent overloading of electrical circuits. The fuse, which is made of material having a low melting point, utilizes the heating effect of an electric current. A fuse is placed in an electrical circuit and if the current becomes too large the fuse wire melts and so breaks the circuit. A circuit diagram symbol for a fuse is shown in Fig. 2.1.

> Problem 20. If 5 A, 10 A and 13 A fuses are available, state which is most appropriate for the following appliances which are both connected to a 240 V supply.
>
> (a) Electric toaster having a power rating of 1 kW
> (b) Electric fire having a power rating of 3 kW

Power $P = VI$, from which,

$$\text{current } I = \frac{P}{V}$$

(a) For the toaster,

$$\text{current } I = \frac{P}{V} = \frac{1000}{240} = \frac{100}{24} = 4\frac{1}{6}\text{ A}$$

Hence a **5 A** fuse is most appropriate.

(b) For the fire,

$$\text{current } I = \frac{P}{V} = \frac{3000}{240} = \frac{300}{24} = 12\frac{1}{2}\text{ A}$$

Hence a **13 A** fuse is most appropriate.

2.12 Multi-choice questions on the introduction to electric circuits

(Answers on page 283.)

1. 60 μs is equivalent to:
   (a) 0.06 s (b) 0.000 06 s (c) 1000 minutes (d) 0.6 s
2. The current which flows when 0.1 coulomb is transferred in 10 ms is:
   (a) 1 A (b) 10 A (c) 10 mA (d) 100 mA
3. The p.d. applied to a 1 kΩ resistance in order that a current of 100 μA may flow is:
   (a) 1 V (b) 100 V (c) 0.1 V (d) 10 V
4. Which of the following formulae for electrical power is incorrect?
   (a) VI (b) V/I (c) I^2R (d) V^2/R
5. The power dissipated by a resistor of 4 Ω when a current of 5 A passes through it is
   (a) 6.25 W (b) 20 W (c) 80 W (d) 100 W
6. Which of the following statements is true?
   (a) Electric current is measured in volts.
   (b) 200 kΩ resistance is equivalent to 0.2 MΩ.
   (c) An ammeter has a low resistance and must be connected in parallel with a circuit.
   (d) An electrical insulator has a high resistance.
7. A current of 3 A flows for 50 h through a 6 Ω resistor. The energy consumed by the resistor is:
   (a) 0.9 kW h (b) 2.7 kW h (c) 9 kW h (d) 27 kW h
8. What must be known in order to calculate the energy used by an electrical appliance?
   (a) voltage and current
   (b) current and time of operation
   (c) power and time of operation
   (d) current and resistance

2.13 Short answer questions on the introduction to electric circuits

1. Draw the preferred symbols for the following components used when drawing electrical circuit diagrams:
   (a) fixed resistor (b) cell (c) filament lamp (d) fuse (e) voltmeter
2. State the unit of (a) current (b) potential difference (c) resistance.
3. State an instrument used to measure (a) current (b) potential difference (c) resistance.
4. What is a multimeter?
5. State Ohm's law.
6. Give one example of (a) a linear device (b) a non-linear device.
7. State the meaning of the following abbreviations of prefixes used with electrical units: (a) k (b) μ (c) m (d) M.
8. What is a conductor? Give four examples.
9. What is an insulator? Give four examples.
10. Complete the following statement:
    An ammeter has a .... resistance and must be connected .... with the load.
11. Complete the following statement:
    A voltmeter has a .... resistance and must be connected .... with the load.
12. State the unit of electrical power. State three formulae used to calculate power.
13. State two units used for electrical energy.
14. State the three main effects of an electric current and give two examples of each.
15. What is the function of a fuse in an electrical circuit?

2.14 Further questions on the introduction to electric circuits

$Q = I \times t$

1. In what time would a current of 10 A transfer a charge of 50 C? [5 s]
2. A current of 6 A flows for 10 minutes. What charge is transferred? [3600 C]
3. How long must a current of 100 mA flow so as to transfer a charge of 50 C? [8 min 20 s]

Ohm's law

4. The current flowing through a heating element is 5 A when a p.d. of 35 V is applied across it. Find the resistance of the element. [7 Ω]

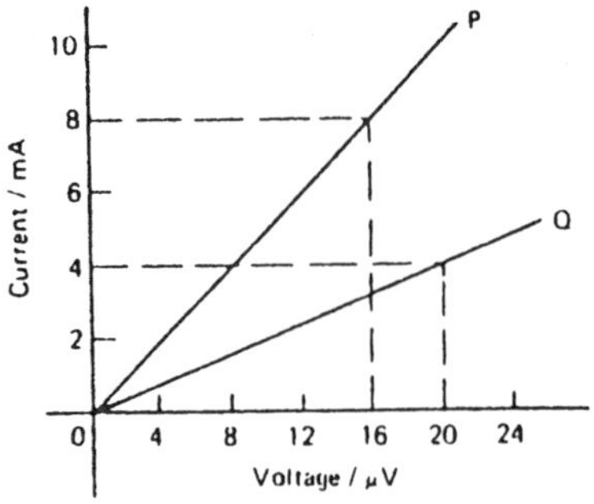

Figure 2.6

5. A 60 W electric light bulb is connected to a 240 V supply. Determine (a) the current flowing in the bulb and (b) the resistance of the bulb. [(a) 0.25 A (b) 960 Ω]
6. Graphs of current against voltage for two resistors P and Q are shown in Fig. 2.6. Determine the value of each resistor. [2 mΩ, 5 mΩ]
7. Determine the p.d. which must be applied to a 5 kΩ resistor such that a current of 6 mA may flow. [30 V]

Power and energy

8. The hot resistance of a 250 V filament lamp is 625 Ω. Determine the current taken by the lamp and its power rating. [0.4 A, 100 W]
9. Determine the resistance of a coil connected to a 150 V supply when a current of (a) 75 mA (b) 300 μA flows through it. [(a) 2 kΩ (b) 0.5 MΩ]
10. Determine the resistance of an electric fire which takes a current of 12 A from a 240 V supply. Find also the power rating of the fire and the energy used in 20 h. [20 Ω, 2.88 kW, 57.6 kW h]
11. Determine the power dissipated when a current of 10 mA flows through an appliance having a resistance of 8 kΩ. [0.8 W]
12. 85.5 J of energy are converted into heat in nine seconds. What power is dissipated? [9.5 W]
13. A current of 4 A flows through a conductor and 10 W is dissipated. What p.d. exists across the ends of the conductor? [2.5 V]
14. Find the power dissipated when:

(a) a current of 5 mA flows through a resistance of 20 kΩ;
(b) a voltage of 400 V is applied across a 120 kΩ resistor;
(c) a voltage applied to a resistor is 10 kV and the current flow is 4 mA.
[(a) 0.5 W (b) 1⅓ W (c) 40 W]

15. A battery of e.m.f. 15 V supplies a current of 2 A for 5 min. How much energy is supplied in this time? [9 kJ]
16. A d.c. electric motor consumes 72 MJ when connected to a 400 V supply for 2 h 30 min. Find the power rating of the motor and the current taken from the supply. [8 kW, 20 A]
17. A p.d. of 500 V is applied across the winding of an electric motor and the resistance of the winding is 50 Ω. Determine the power dissipated by the coil. [5 kW]
18. In a household during a particular week three 2 kW fires are used on average 25 h each and eight 100 W light bulbs are used on average 35 h each. Determine the cost of electricity for the week if 1 unit of electricity costs 7p. [£12.46]
19. Calculate the power dissipated by the element of an electric fire of resistance 30 Ω when a current of 10 A flows in it. If the fire is on for 30 hours in a week determine the energy used. Determine also the weekly cost of energy if electricity costs 7.2p per unit. [3 kW, 90 kW h, £6.48]

Fuses

20. A television set having a power rating of 120 W and an electric lawnmower of power rating 1 kW are both connected to a 240 V supply. If 3 A, 5 A and 10 A fuses are available state which is the most appropriate for each appliance. [3 A, 5 A]

3

Resistance variation

At the end of this chapter you should be able to:

- appreciate that electrical resistance depends on four factors
- appreciate that resistance $R = \rho l/a$, where ρ is the resistivity
- recognize typical values of resistivity and its unit
- perform calculations using $R = \rho l/a$
- define the temperature coefficient of resistance, α
- recognize typical values for α
- perform calculations using $R_\theta = R_0(1 + \alpha\theta)$

3.1 Resistance and resistivity

The resistance of an electrical conductor depends on four factors, these being: (a) the length of the conductor, (b) the cross-sectional area of the conductor, (c) the type of material and (d) the temperature of the material.

Resistance, R, is directly proportional to length, l, of a conductor, i.e. $R \propto l$. Thus, for example, if the length of a piece of wire is doubled, then the resistance is doubled.

Resistance, R, is inversely proportional to cross-sectional area, a, of a conductor, i.e. $R \propto 1/a$. Thus, for example, if the cross-sectional area of a piece of wire is doubled then the resistance is halved.

Since $R \propto 1$ and $R \propto 1/a$ then $R \propto l/a$. By inserting a constant of proportionality into this relationship the type of material used may be taken into account. The constant of proportionality is known as the resistivity of the material and is given the symbol ρ (rho). Thus

resistance $$\boldsymbol{R = \frac{\rho l}{a}} \textbf{ ohms}$$

ρ is measured in ohm metres (Ω m). The value of the resistivity is that resistance of a unit cube of the material measured between opposite faces of the cube.

Resistivity varies with temperature and some typical values of resistivities measured at about room temperature are given below:

Copper 1.7×10^{-8} Ω m (or 0.017 $\mu\Omega$ m)
Aluminium 2.6×10^{-8} Ω m (or 0.026 $\mu\Omega$ m)
Carbon (graphite) 10×10^{-8} Ω m (0.10 $\mu\Omega$ m)
Glass 1×10^{10} Ω m (or 10^4 $\mu\Omega$ m)
Mica 1×10^{13} Ω m (or 10^7 $\mu\Omega$ m)

Note that good conductors of electricity have a low value of resistivity and good insulators have a high value of resistivity.

> Problem 1. The resistance of a 5 m length of wire is 600 Ω. Determine (a) the resistance of an 8 m length of the same wire, and (b) the length of the same wire when the resistance is 420 Ω.

(a) Resistance, R, is directly proportional to length, l, i.e. $R \propto l$. Hence, 600 $\Omega \propto$ 5 m or

$600 = (k)(5)$, where k is the coefficient of proportionality. Hence

$$k = \frac{600}{5} = 120$$

When the length l is 8 m, then

resistance $R = kl = (120)(8) = \mathbf{960\ \Omega}$

(b) When the resistance is 420 Ω, $420 = kl$, from which,

$$\text{length } l = \frac{420}{k} = \frac{420}{120} = \mathbf{3.5\ m}$$

> Problem 2. A piece of wire of cross-sectional area 2 mm^2 has a resistance of 300 Ω. Find (a) the resistance of a wire of the same length and material if the cross-sectional area is 5 mm^2, (b) the cross-sectional area of a wire of the same length and material of resistance 750 Ω.

Resistance R is inversely proportional to cross-sectional area, a, i.e. $R \propto 1/a$. Hence

$$300\ \Omega \propto \frac{1}{2\ \text{mm}^2} \text{ or } 300 = (k)\left(\frac{1}{2}\right)$$

from which, the coefficient of proportionality, $k = 300 \times 2 = 600$.

(a) When the cross-sectional area $a = 5$ mm^2 then

$$R = (k)\left(\frac{1}{5}\right)$$

$$= (600)\left(\frac{1}{5}\right) = \mathbf{120\ \Omega}$$

(Note that resistance has decreased as the cross-sectional area is increased.)

(b) When the resistance is 750 Ω then $750 = (k)(1/a)$, from which

$$\text{cross-sectional area, } a = \frac{k}{750} = \frac{600}{750}$$

$$= \mathbf{0.8\ mm^2}$$

> Problem 3. A wire of length 8 m and cross-sectional area 3 mm^2 has a resistance of 0.16 Ω. If the wire is drawn out until its cross-sectional area is 1 mm^2, determine the resistance of the wire.

Resistance R is directly proportional to length l, and inversely proportional to the cross-sectional area, a, i.e.

$$R \propto \frac{l}{a} \text{ or } R = k\frac{l}{a}$$

where k is the coefficient of proportionality. Since $R = 0.16$, $l = 8$ and $a = 3$, then $0.16 = (k)(8/3)$, from which

$$k = 0.16 \times \frac{3}{8} = 0.06$$

If the cross-sectional area is reduced to $\frac{1}{3}$ of its original area then the length must be tripled to 3×8, i.e. 24 m.

$$\text{New resistance } R = k\frac{l}{a} = 0.06\left(\frac{24}{1}\right)$$

$$= \mathbf{1.44\ \Omega}$$

> Problem 4. Calculate the resistance of a 2 km length of aluminium overhead power cable if the cross-sectional area of the cable is 100 mm^2. Take the resistivity of aluminium to be 0.03×10^{-6} Ω m.

Length $l = 2$ km $= 2000$ m; area, $a = 100$ mm^2 $= 100 \times 10^{-6}$ m^2; resistivity $= \rho = 0.03 \times 10^{-6}$ Ω m.

$$\text{Resistance } R = \frac{\rho l}{a}$$

$$= \frac{(0.03 \times 10^{-6}\ \Omega\ \text{m})(2000\ \text{m})}{(100 \times 10^{-6}\ \text{m}^2)}$$

$$= \frac{0.03 \times 2000}{100}\ \Omega$$

$$= \mathbf{0.6\ \Omega}$$

> Problem 5. Calculate the cross-sectional area, in mm^2, of a piece of copper wire, 40 m in length and having a resistance of 0.25 Ω. Take the resistivity of copper as 0.02×10^{-6} Ω m.

Resistance $R = \rho l/a$ hence cross-sectional area is given by

$$a = \frac{\rho l}{R}$$

$$= \frac{(0.02 \times 10^{-6}\ \Omega\ \text{m})(40\ \text{m})}{0.25\ \Omega}$$

$$= 3.2 \times 10^{-6}\ \text{m}^2$$

$$= (3.2 \times 10^{-6}) \times 10^{6}\ \text{mm}^2$$

$$= \mathbf{3.2\ mm^2}$$

> Problem 6. The resistance of 1.5 km of wire of cross-sectional area 0.17 mm^2 is 150 Ω. Determine the resistivity of the wire.

Resistance, $R = \rho l/a$. Hence

$$\text{resistivity } \rho = \frac{Ra}{l} = \frac{(150\ \Omega)(0.17 \times 10^{-6}\ \text{m}^2)}{(1500\ \text{m})}$$

$$= \mathbf{0.017 \times 10^{-6}\ \Omega\ m} \text{ or } \mathbf{0.017\ \mu\Omega\ m}$$

> Problem 7. Determine the resistance of 1200 m of copper cable having a diameter of 12 mm if the resistivity of copper is 1.7 × 10^{-8} Ω m.

Cross-sectional area of cable, $a = \pi r^2$

$$= \pi \left(\frac{12}{2}\right)^2 = 36\pi\ \text{mm}^2$$

$$= 36\pi \times 10^{-6}\ \text{m}^2$$

$$\text{Resistance } R = \frac{\rho l}{a}$$

$$= \frac{(1.7 \times 10^{-8}\ \Omega\ \text{m})(1200\ \text{m})}{(36\pi \times 10^{-6}\ \text{m}^2)}$$

$$= \frac{1.7 \times 1200 \times 10^6}{10^8 \times 36\pi}\ \Omega$$

$$= \frac{1.7 \times 12}{36\pi}\ \Omega = \mathbf{0.180\ \Omega}$$

3.2 Temperature coefficient of resistance

In general, as the temperature of a material increases, most conductors increase in resistance, insulators decrease in resistance, while the resistance of some special alloys remains almost constant.

The **temperature coefficient of resistance** of a material is the increase in the resistance of a 1 Ω resistor of that material when it is subjected to a rise of temperature of 1°C. The symbol used for the temperature coefficient of resistance is α (alpha). Thus, if some copper wire of resistance 1 Ω is heated through 1°C and its resistance is then measured as 1.0043 Ω then α = 0.0043 Ω/Ω °C for copper. The units are usually expressed only as 'per °C', i.e. α = 0.0043/°C for copper. If the 1 Ω resistor of copper is heated through 100°C then the resistance at 100°C would be 1 + 100 × 0.0043 = 1.43 Ω.

Some typical values of temperature coefficient of resistance measured at 0°C are given below:

Copper	0.0043/°C	Aluminium	0.0038/°C
Nickel	0.0062/°C	Carbon	−0.000 48/°C
Constantan	0	Eureka	0.000 01/°C

(Note that the negative sign for carbon indicates that its resistance falls with increase of temperature.)

If the resistance of a material at 0°C is known the resistance at any other temperature can be determined from:

$$\boldsymbol{R_\theta = R_0(1 + \alpha_0\theta)}$$

where R_0 = resistance at 0°C
R_θ = resistance at temperature θ°C
α_0 = temperature coefficient of resistance at 0°C.

> Problem 8. A coil of copper wire has a resistance of 100 Ω when its temperature is 0°C. Determine its resistance at 100°C if the temperature coefficient of resistance of copper at 0°C is 0.0043/°C.

Resistance $R_\theta = R_0(1 + \alpha_0\theta)$.
Hence resistance at 100°C is given by

$$R_{100} = 100[1 + (0.0043)(100)]$$

$$= 100[1 + 0.43] = 100(1.43) = \mathbf{143\ \Omega}$$

> Problem 9. An aluminium cable has a resistance of 27 Ω at a temperature of 35°C. Determine its resistance at 0°C. Take the temperature coefficient of resistance at 0°C to be 0.0038/°C.

Resistance at θ°C, $R_\theta = R_0\,(1 + \alpha_0\theta)$.
Hence resistance at 0°C is given by

$$R_0 = \frac{R_\theta}{(1 + \alpha_0\theta)} = \frac{27}{[1 + (0.0038)(35)]}$$

$$= \frac{27}{1 + 0.133} = \frac{27}{1.133} = \mathbf{23.83\ \Omega}$$

Problem 10. A carbon resistor has a resistance of 1 kΩ at 0°C. Determine its resistance at 80°C. Assume that the temperature coefficient of resistance for carbon at 0°C is –0.0005.

Resistance at temperature θ°C is given by

$$R_\theta = R_0(1 + \alpha_0\theta)$$

i.e.

$$R_\theta = 1000[1 + (-0.0005)(80)]$$
$$= 1000[1 - 0.040] = 1000(0.96)$$
$$= \mathbf{960\ \Omega}$$

If the resistance of a material at room temperature (approximately 20°C), R_{20}, and the temperature coefficient of resistance at 20°C, α_{20}, are known then the resistance R_θ at temperature θ°C is given by:

$$\boldsymbol{R_\theta = R_{20}[1 + \alpha_{20}(\theta - 20)]}$$

Problem 11. A coil of copper wire has a resistance of 10 Ω at 20°C. If the temperature coefficient of resistance of copper at 20°C is 0.004/°C determine the resistance of the coil when the temperature rises to 100°C.

Resistance at θ°C, $R_\theta = R_{20}[1 + \alpha_{20}(\theta - 20)]$. Hence resistance at 100°C is given by

$$R_{100} = 10[1 + (0.004)(100 - 20)]$$
$$= 10[1 + (0.004)(80)]$$
$$= 10[1 + 0.32]$$
$$= 10(1.32) = \mathbf{13.2\ \Omega}$$

Problem 12. The resistance of a coil of aluminium wire at 18°C is 200 Ω. The temperature of the wire is increased and the resistance rises to 240 Ω. If the temperature coefficient of resistance of aluminium is 0.0039/°C at 18°C determine the temperature to which the coil has risen.

Let the temperature rise to θ°C. Resistance at θ°C, $R_\theta = R_{18}[1 + \alpha_{18}(\theta - 18)]$, i.e.

$$240 = 200[1 + (0.0039)(\theta - 18)]$$
$$240 = 200 + (200)(0.0039)(\theta - 18)$$
$$240 - 200 = 0.78(\theta - 18)$$
$$40 = 0.78(\theta - 18)$$
$$\frac{40}{0.78} = \theta - 18$$

$51.28 = \theta - 18$, from which, $\theta = 51.28 + 18 = 69.28$°C

Hence the temperature of the coil increases to 69.28°C.

If the resistance at 0°C is not known, but is known at some other temperature θ_1, then the resistance at any temperature can be found as follows:

$$R_1 = R_0(1 + \alpha_0\theta_1) \text{ and } R_2 = R_0(1 + \alpha_0\theta_2)$$

Dividing one equation by the other gives:

$$\boldsymbol{\frac{R_1}{R_2} = \frac{1 + \alpha_0\theta_1}{1 + \alpha_0\theta_2}}$$

where R_2 = resistance at temperature θ_2.

Problem 13. Some copper wire has a resistance of 200 Ω at 20°C. A current is passed through the wire and the temperature rises to 90°C. Determine the resistance of the wire at 90°C, correct to the nearest ohm, assuming that the temperature coefficient of resistance is 0.004/°C at 0°C.

$R_{20} = 200\ \Omega$, $\alpha_0 = 0.004$/°C

$$\frac{R_{20}}{R_{90}} = \frac{[1 + \alpha_0(20)]}{[1 + \alpha_0(90)]}$$

Hence

$$R_{90} = \frac{R_{20}[1 + 90\alpha_0]}{[1 + 20\alpha_0]}$$
$$= \frac{200[1 + 90(0.004)]}{[1 + 20(0.004)]}$$
$$= \frac{200[1 + 0.36]}{[1 + 0.08]}$$
$$= \frac{200(1.36)}{(1.08)} = 251.85\ \Omega$$

i.e. **the resistance of the wire at 90°C is 252 Ω.**

3.3 Multi-choice questions on resistance variation

(Answers on page 283.)

1. The unit of resistivity is:
   (a) ohms (b) ohm millimetre
   (c) ohm metre (d) ohm/metre
2. The length of a certain conductor of resistance 100 Ω is doubled and its cross-sectional area is halved. Its new resistance is:
   (a) 100 Ω (b) 200 Ω (c) 50 Ω (d) 400 Ω
3. The resistance of a 2 km length of cable of cross-sectional area 2 mm^2 and resistivity of 2×10^{-8} Ω m is:
   (a) 0.02 Ω (b) 20 Ω (c) 0.02 m Ω
   (d) 200 Ω
4. A piece of graphite has a cross-sectional area of 10 mm^2. If its resistance is 0.1 Ω and its resistivity 10×10^{-8} Ω m, its length is:
   (a) 10 km (b) 10 cm (c) 10 mm (d) 10 m
5. The symbol for the unit of temperature coefficient of resistance is:
   (a) Ω/°C (b) Ω (c) °C (d) Ω/Ω°C
6. A coil of wire has a resistance of 10 Ω at 0°C. If the temperature coefficient of resistance for the wire is 0.004/°C its resistance at 100°C is:
   (a) 0.4 Ω (b) 1.4 Ω (c) 14 Ω (d) 10 Ω
7. A nickel coil has a resistance of 13 Ω at 50°C. If the temperature coefficient of resistance at 0°C is 0.006/°C, the resistance at 0°C is:
   (a) 16.9 Ω (b) 10 Ω (c) 43.3 Ω (d) 0.1 Ω

3.4 Short answer questions on resistance variation

1. Name four factors which can affect the resistance of a conductor.
2. If the length of a piece of wire of constant cross-sectional area is halved, the resistance of the wire is ....
3. If the cross-sectional area of a certain length of cable is trebled, the resistance of the cable is ....
4. What is resistivity? State its unit and the symbol used.
5. Complete the following:
   Good conductors of electricity have a .... value of resistivity and good insulators have a .... value of resistivity.
6. What is meant by the 'temperature coefficient of resistance'? State its units and the symbols used.
7. If the resistance of a metal at 0°C is R_0, R_θ is the resistance at θ°C and α_0 is the temperature coefficient of resistance at 0°C then: R_θ = ...

3.5 Further questions on resistance variation

Resistance and resistivity

1. The resistance of a 2 m length of cable is 2.5 Ω. Determine
   (a) the resistance of a 7 m length of the same cable, and
   (b) the length of the same wire when the resistance is 6.25 Ω.
   [(a) 8.75 Ω (b) 5 m]
2. Some wire of cross-sectional area 1 mm^2 has a resistance of 20 Ω. Determine (a) the resistance of a wire of the same length and material if the cross-sectional area is 4 mm^2, and (b) the cross-sectional area of a wire of the same length and material if the resistance is 32 Ω. [(a) 5 Ω (b) 0.625 mm^2]
3. Some wire of length 5 m and cross-sectional area 2 mm^2 has a resistance of 0.08 Ω. If the wire is drawn out until its cross-sectional area is 1 mm^2, determine the resistance of the wire. [0.32 Ω]
4. Find the resistance of 800 m of copper cable of cross-sectional area 20 mm^2. Take the resistivity of copper as 0.02 μΩ m.
   [0.8 Ω]
5. Calculate the cross-sectional area, in mm^2, of a piece of aluminium wire 100 m long and having a resistance of 2 Ω. Take the resistivity of aluminium as 0.03×10^{-6} Ω m.
   [1.5 mm^2]
6. (a) What does the resistivity of a material mean?
   (b) The resistance of 500 m of wire of cross-sectional area 2.6 mm^2 is 5 Ω. Determine the resistivity of the wire in μΩ m.
   [0.026 μΩ m]
7. Find the resistance of 1 km of copper cable having a diameter of 10 mm if the resistivity of copper is 0.017×10^{-6} Ω m. [0.216 Ω]

Temperature coefficient of resistance

8. A coil of aluminium wire has a resistance of 50 Ω when its temperature is 0°C. Determine

its resistance at 100°C if the temperature coefficient of resistance of aluminium at 0°C is 0.0038/°C. [69 Ω]

9. A copper cable has a resistance of 30 Ω at a temperature of 50°C. Determine its resistance at 0°C. Take the temperature coefficient of resistance of copper at 0°C as 0.0043/°C [24.69 Ω]
10. The temperature coefficient of resistance for carbon at 0°C is −0.000 48/°C. What is the significance of the minus sign? A carbon resistor has a resistance of 500 Ω at 0°C. Determine its resistance at 50°C. [488 Ω]
11. A coil of copper wire has a resistance of 20 Ω at 18°C. If the temperature coefficient of resistance of copper at 18°C is 0.004/°C, determine the resistance of the coil when the temperature rises to 98°C. [26.4 Ω]
12. The resistance of a coil of nickel wire at 20°C is 100 Ω. The temperature of the wire is increased and the resistance rises to 130 Ω. If the temperature coefficient of resistance of nickel is 0.006/°C at 20°C, determine the temperature to which the coil has risen. [70°C]
13. Some aluminium wire has a resistance of 50 Ω at 20°C. The wire is heated to a temperature of 100°C. Determine the resistance of the wire at 100°C, assuming that the temperature coefficient of resistance at 0°C is 0.004/°C. [64.8 Ω]
14. A copper cable is 1.2 km long and has a cross-sectional area of 5 mm^2. Find its resistance at 80°C if at 20°C the resistivity of copper is 0.02×10^{-6} Ω m and its temperature coefficient of resistance is 0.004/°C. [5.952 Ω]

4

Chemical effects of electricity

At the end of this chapter you should be able to:

- understand electrolysis and its applications, including electroplating
- appreciate the purpose and construction of a simple cell
- explain polarization and local action
- explain corrosion and its effects
- define the terms e.m.f., E, and internal resistance, r, of a cell
- perform calculations using $V = E - Ir$
- determine the total e.m.f. and total internal resistance for cells connected in series and in parallel
- distinguish between primary and secondary cells
- explain the construction and practical applications of the Leclanché cell, the mercury cell, the lead–acid cell and the alkaline cell
- list the advantages and disadvantages of alkaline cells over lead–acid cells
- understand the term 'cell capacity' and state its unit

4.1 Introduction

A material must contain **charged particles** to be able to conduct electric current. In **solids**, the current is carried by **electrons**. Copper, lead, aluminium, iron and carbon are some examples of solid conductors. In **liquids** and **gases**, the current is carried by the part of a molecule which has acquired an electric charge, called **ions**. These can possess a positive or negative charge, and examples include hydrogen ion H^+, copper ion Cu^{++} and hydroxyl ion OH^-. Distilled water contains no ions and is a poor conductor of electricity whereas salt water contains ions and is a fairly good conductor of electricity.

4.2 Electrolysis

Electrolysis is the decomposition of a liquid compound by the passage of electric current through it. Practical applications of electrolysis include the electroplating of metals (see section 4.3), the refining of copper and the extraction of aluminium from its ore.

An **electrolyte** is a compound which will undergo electrolysis. Examples include salt water, copper sulphate and sulphuric acid.

The **electrodes** are the two conductors carrying current to the electrolyte. The positive-connected electrode is called the **anode** and the negative-connected electrode the **cathode**.

When two copper wires connected to a battery are placed in a beaker containing a salt water solution, then current will flow through the solution. Air bubbles appear around the wires as the water is changed into hydrogen and oxygen by electrolysis.

4.3 Electroplating

Electroplating uses the principle of electrolysis to apply a thin coat of one metal to another metal. Some practical applications include the tin-plating of steel, silver-plating of nickel alloys and chromium-plating of steel. If two copper electrodes connected to a battery are placed in a beaker containing copper sulphate as the electrolyte it is found that the cathode (i.e. the electrode connected to the negative terminal of the battery) gains copper while the anode loses copper.

4.4 The simple cell

The purpose of an **electric cell** is to convert chemical energy into electrical energy.

A **simple cell** comprises two dissimilar conductors (electrodes) in an electrolyte. Such a cell is shown in Fig. 4.1, comprising copper and zinc electrodes. An electric current is found to flow between the electrodes. Other possible electrode pairs exist, including zinc–lead and zinc–iron. The electrode potential (i.e. the p.d. measured between the electrodes) varies for each pair of metals. By knowing the e.m.f. of each metal with respect to some standard electrode the e.m.f. of any pair of metals may be determined. The standard used is the hydrogen electrode. The **electrochemical series** is a way of listing elements in order of electrical potential, and Table 4.1 shows a number of elements in such a series.

In a simple cell two faults exist – those due to polarization and those due to local action.

Table 4.1 Part of the electrochemical series

potassium
sodium
aluminium
zinc
iron
lead
hydrogen
copper
silver
carbon

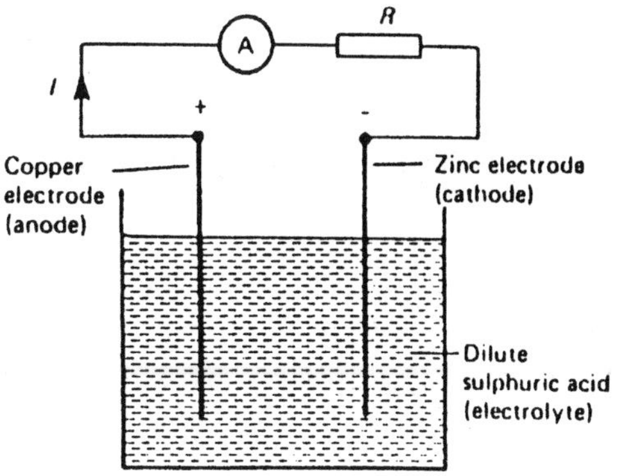

Figure 4.1

Polarization

If the simple cell shown in Fig. 4.1 is left connected for some time, the current I decreases fairly rapidly. This is because of the formation of a film of hydrogen bubbles on the copper anode. This effect is known as the polarization of the cell. The hydrogen prevents full contact between the copper electrode and the electrolyte and this increases the internal resistance of the cell. The effect can be overcome by using a chemical depolarizing agent or depolarizer, such as potassium dichromate which removes the hydrogen bubbles as they form. This allows the cell to deliver a steady current.

Local action

When commercial zinc is placed in dilute sulphuric acid, hydrogen gas is liberated from it and the zinc dissolves. The reason for this is that impurities, such as traces of iron, are present in the zinc which set up small primary cells with the zinc. These small cells are short-circuited by the electrolyte, with the result that localized currents flow causing corrosion. This action is known as 'local action' of the cell. This may be prevented by rubbing a small amount of mercury on the zinc surface, which forms a protective layer on the surface of the electrode.

When two metals are used in a simple cell the electrochemical series may be used to predict the behaviour of the cell:

(i) The metal that is higher in the series acts as the negative electrode, and vice versa. For example, the zinc electrode in the cell shown in Fig. 4.1 is negative and the copper electrode is positive.

(ii) The greater the separation in the series between the two metals the greater is the e.m.f. produced by the cell.

The electrochemical series is representative of the order of reactivity of the metals and their compounds:

(i) The higher metals in the series react more readily with oxygen and vice-versa.
(ii) When two metal electrodes are used in a simple cell the one that is higher in the series tends to dissolve in the electrolyte

4.5 Corrosion

Corrosion is the gradual destruction of a metal in a damp atmosphere by means of simple cell action. In addition to the presence of moisture and air required for rusting, an electrolyte, an anode and a cathode are required for corrosion. Thus, if metals widely spaced in the electrochemical series are used in contact with each other in the presence of an electrolyte, corrosion will occur. For example, if a brass valve is fitted to a heating system made of steel, corrosion will occur.

The **effects of corrosion** include the weakening of structures, the reduction of the life of components and materials, the wastage of materials and the expense of replacement.

Corrosion may be **prevented** by coating with paint, grease, plastic coatings and enamels, or by plating with tin or chromium. Also, iron may be galvanized, i.e. plated with zinc, the layer of zinc helping to prevent the iron from corroding.

4.6 E.m.f. and internal resistance of a cell

The **electromotive force (e.m.f.)**, E, of a cell is the p.d. between its terminals when it is not connected to a load (i.e. the cell is on 'no-load').

The e.m.f. of a cell is measured by using a **high resistance voltmeter** connected in parallel with the cell. The voltmeter must have a high resistance otherwise it will pass current and the cell will not be on no-load. For example, if the resistance of a cell is 1 Ω and that of a voltmeter 1 MΩ then the equivalent resistance of the circuit is 1 MΩ + 1 Ω, i.e. approximately 1 MΩ, hence no current flows and the cell is not loaded.

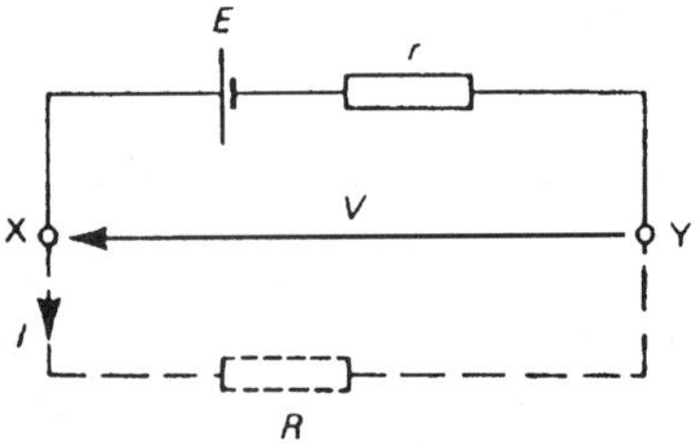

Figure 4.2

The voltage available at the terminals of a cell falls when a load is connected. This is caused by the **internal resistance** of the cell which is the opposition of the material of the cell to the flow of current. The internal resistance acts in series with other resistances in the circuit. Figure 4.2 shows a cell of e.m.f. E volts and internal resistance, r, and XY represents the terminals of the cell.

When a load (shown as resistance R) is not connected, no current flows and the terminal p.d., $V = E$. When R is connected a current I flows which causes a voltage drop in the cell, given by Ir. The p.d. available at the cell terminals is less than the e.m.f. of the cell and is given by:

$$\boldsymbol{V = E - Ir}$$

Thus if a battery of e.m.f. 12 volts and internal resistance 0.1 Ω delivers a current of 100 A, the terminal p.d. is given by

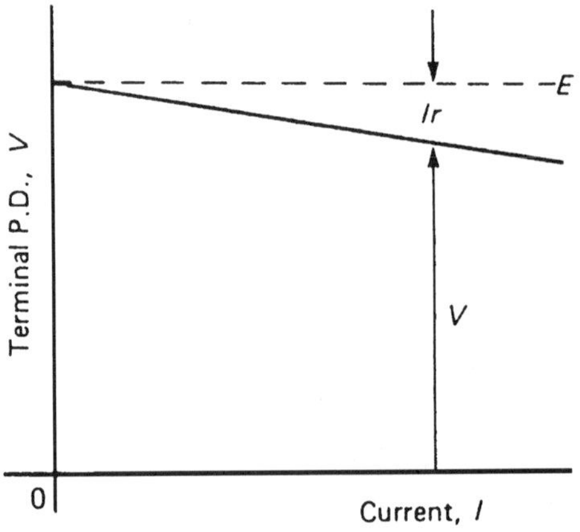

Figure 4.3

$$V = 12 - (100)(0.01)$$
$$= 12 - 1 = 11\text{ V}$$

When different values of potential difference V, across a cell or power supply are measured for different values of current I, a graph may be plotted as shown in Fig. 4.3. Since the e.m.f. E of the cell or power supply is the p.d. across its terminals on no-load (i.e. when $I = 0$), then E is as shown by the broken line.

Since $V = E - Ir$ then the internal resistance may be calculated from

$$r = \frac{E - V}{I}$$

When a current is flowing in the direction shown in Fig. 4.2 the cell is said to be **discharging** ($E > V$).

When a current flows in the opposite direction to that shown in Fig. 4.2 the cell is said to be **charging** ($V > E$).

A battery is a combination of more than one cell. The cells in a battery may be connected in series or in parallel.

(i) For cells connected in series:

> Total e.m.f. = sum of cell's e.m.f.s
> Total internal resistance = sum of cell's internal resistances

(ii) For cells connected in parallel:

> If each cell has the same e.m.f. and internal resistance:
> Total e.m.f. = e.m.f. of one cell
> Total internal resistance of n cells
> $= \frac{1}{n} \times$ internal resistance of one cell

> Problem 1. Eight cells, each with an internal resistance of 0.2 Ω and an e.m.f. of 2.2 V are connected (a) in series, (b) in parallel. Determine the e.m.f. and the internal resistance of the batteries so formed.

(a) When connected in series, total e.m.f.
= sum of cell's e.m.f.
= $2.2 \times 8 = \mathbf{17.6\ V}$
Total internal resistance
= sum of cell's internal resistance
= $0.2 \times 8 = \mathbf{1.6\ \Omega}$

(b) When connected in parallel, total e.m.f.
= e.m.f. of one cell
= **2.2 V**
Total internal resistance of 8 cells

$= \frac{1}{8} \times$ internal resistance of one cell

$= \frac{1}{8} \times 0.2 = \mathbf{0.025\ \Omega}$

> Problem 2. A cell has an internal resistance of 0.02 Ω and an e.m.f. of 2.0 V. Calculate its terminal p.d. if it delivers (a) 5 A, (b) 50 A.

(a) Terminal p.d., $V = E - Ir$ where E = e.m.f. of cell, I = current flowing and r = internal resistance of cell. $E = 2.0$ V, $I = 5$ A and $r = 0.02\ \Omega$, hence

$$V = 2.0 - (5)(0.02) = 2.0 - 0.1 = \mathbf{1.9\ V}$$

(b) When the current is 50 A, terminal p.d. is given by

$$V = E - Ir = 2.0 - 50(0.02)$$

i.e.

$$V = 2.0 - 1.0 = \mathbf{1.0\ V}$$

Thus the terminal p.d. decreases as the current drawn increases.

> Problem 3. The p.d. at the terminals of a battery is 25 V when no-load is connected and 24 V when a load taking 10 A is connected. Determine the internal resistance of the battery.

When no-load is connected the e.m.f. of the battery, E, is equal to the terminal p.d., V, i.e. $E = 25$ V. When current $I = 10$ A and terminal p.d. $V = 24$ V, then $V = E - Ir$, i.e.

$$24 = 25 - (10)r$$

Hence, rearranging, gives

$$10r = 25 - 24 = 1$$

and the internal resistance

$$r = \frac{1}{10} = \mathbf{0.1\ \Omega}$$

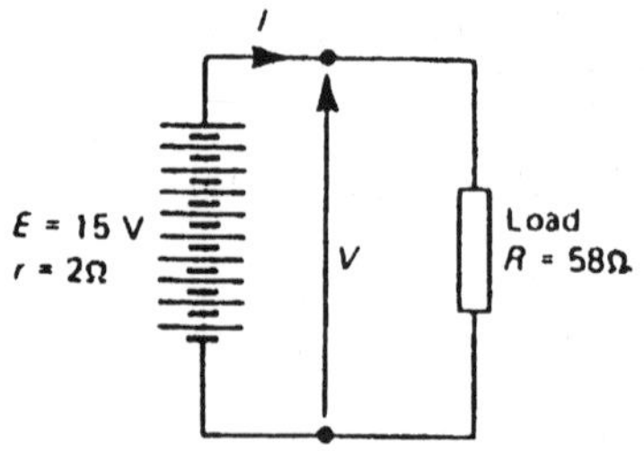

Figure 4.4

> Problem 4. Ten 1.5 V cells, each having an internal resistance of 0.2 Ω, are connected in series to a load of 58 Ω. Determine (a) the current flowing in the circuit and (b) the p.d. at the battery terminals.

(a) For ten cells, battery e.m.f., $E = 10 \times 1.5 = 15$ V, and the total internal resistance, $r = 10 \times 0.2 = 2$ Ω. When connected to a 58 Ω load the circuit is as shown in Fig. 4.4.

$$\text{Current } I = \frac{\text{e.m.f.}}{\text{total resistance}} = \frac{15}{58 + 2}$$

$$= \frac{15}{60} = \mathbf{0.25\ A}$$

(b) p.d. to battery terminals is $V = E - Ir$, i.e.

$V = 15 - (0.25)(2) = \mathbf{14.5\ V}$

4.7 Primary cells

Primary cells cannot be recharged, that is, the conversion of chemical energy to electrical energy is irreversible and the cell cannot be used once the chemicals are exhausted. Examples of primary cells include the Leclanché cell and the mercury cell.

Lechlanché cell

A typical dry Leclanché cell is shown in Fig. 4.5. Such a cell has an e.m.f. of about 1.5 V when new, but this falls rapidly due to polarization if in continuous use. The hydrogen film on the carbon electrode forms faster than can be dissipated by the depolarizer. The Leclanché cell is suitable only for intermittent use, applications including torches, transistor radios, bells, indicator circuits, gas lighters, controlling switch-gear, and so on. The cell is the most commonly used of primary cells, is cheap, requires little maintenance and has a shelf-life of about 2 years.

Mercury cell

A typical mercury cell is shown in Fig. 4.6. Such a cell has an e.m.f. of about 1.3 V which remains constant for a relatively long time. Its main advantages over the Leclanché cell is its smaller size and its long shelf-life. Typical practical

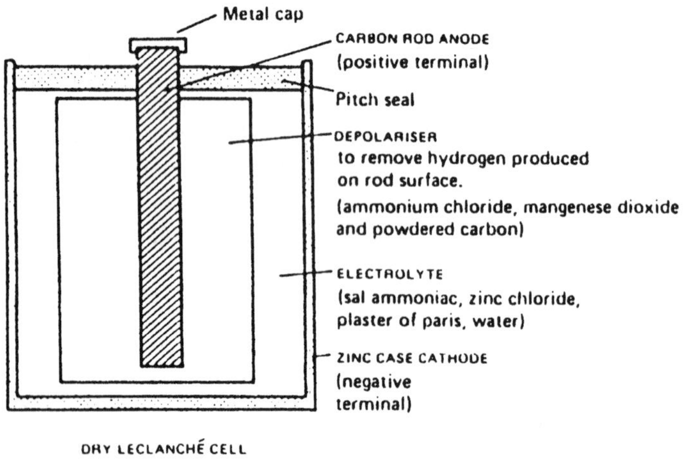

Figure 4.5

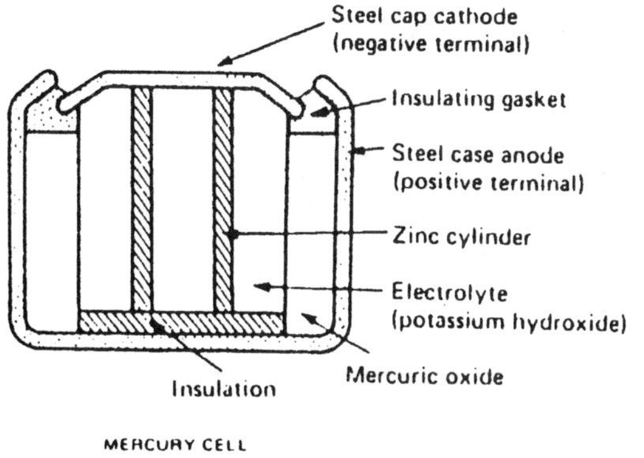

Figure 4.6

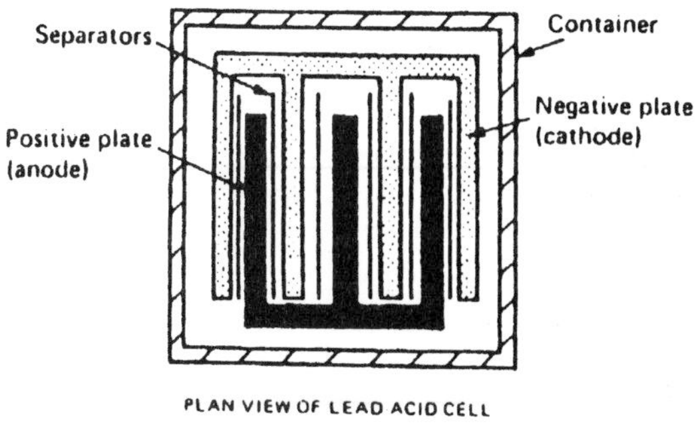

Figure 4.7

applications include hearing aids, medical electronics, cameras and for guided missiles.

4.8 Secondary cells

Secondary cells can be recharged after use, that is, the conversion of chemical energy to electrical energy is reversible and the cell may be used many times. Examples of secondary cells include the lead–acid cell and alkaline cells. Practical applications of such cells include car batteries, telephone circuits and for traction purposes – such as milk delivery vans and fork lift trucks.

Lead–acid cell

A typical lead-acid cell is constructed of:

(i) A container made of glass, ebonite or plastic.
(ii) **Lead plates**
  (a) the negative plate (cathode) consists of spongy lead,
  (b) the positive plate (anode) is formed by pressing lead peroxide into the lead grid.
  The plates are interleaved as shown in the plan view of Fig. 4.7, to increase their effective cross-sectional area and to minimize internal resistance.
(iii) **Separators** made of glass, celluloid or wood.
(iv) An **electrolyte** which is a mixture of sulphuric acid and distilled water.

The relative density (or specific gravity) of a lead–acid cell, which may be measured using a hydrometer, varies between about 1.26 when the cell is fully charged to about 1.19 when discharged. The terminal p.d. of a lead–acid cell is about 2 V.

When a cell supplies current to a load it is said to be **discharging**. During discharge:

(i) the lead peroxide (positive plate) and the spongy lead (negative plate) are converted into lead sulphate, and
(ii) the oxygen in the lead peroxide combines with hydrogen in the electrolyte to form water. The electrolyte is therefore weakened and the relative density falls.

The terminal p.d. of a lead–acid cell when fully discharged is about 1.8 V.

A cell is **charged** by connecting a d.c. supply to its terminals, the positive terminal of the cell being connected to the positive terminal of the supply. The charging current flows in the reverse direction to the discharge current and the chemical action is reversed. During charging:

(i) the lead sulphate on the positive and negative plates is converted back to lead peroxide and lead respectively, and
(ii) the water content of the electrolyte decreases as the oxygen released from the electrolyte combines with the lead of the positive plate. The relative density of the electrolyte thus increases.

The colour of the positive plate when fully charged is dark brown and when discharged is light brown. The colour of the negative plate when fully charged is grey and when discharged is light grey.

Alkaline cell

There are two main types of alkaline cell – the nickel–iron cell and the nickel–cadmium cell. In

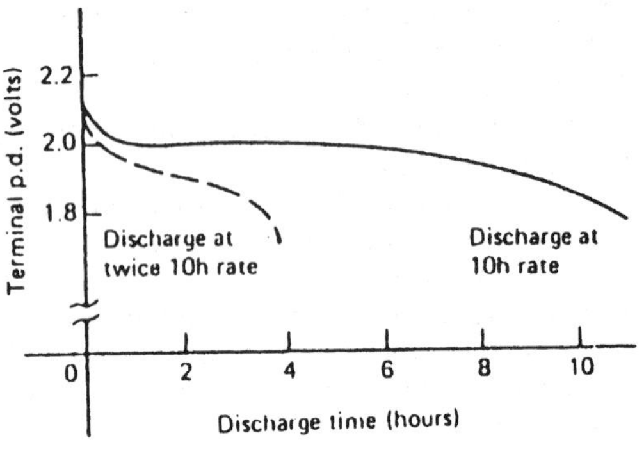

Figure 4.8

both types the positive plate is made of nickel hydroxide enclosed in finely perforated steel tubes, the resistance being reduced by the addition of pure nickel or graphite. The tubes are assembled into nickel-steel plates.

In the nickel–iron cell (sometimes called the Edison cell or nife cell), the negative plate is made of iron oxide, with the resistance being reduced by a little mercuric oxide, the whole being enclosed in perforated steel tubes and assembled in steel plates. In the nickel–cadmium cell the negative plate is made of cadmium. The electrolyte in each type of cell is a solution of potassium hydroxide which does not undergo any chemical change and thus the quantity can be reduced to a minimum. The plates are separated by insulating rods and assembled in steel containers which are then enclosed in a non-metallic crate to insulate the cells from one another. The average discharge p.d. of an alkaline cell is about 1.2 V.

Advantages of an alkaline cell (for example, a nickel–cadmium cell or a nickel–iron cell) over a lead–acid cell include:

(i) More robust construction
(ii) Capable of withstanding heavy charging and discharging currents without damage
(iii) Has a longer life
(iv) For a given capacity is lighter in weight
(v) Can be left indefinitely in any state of charge or discharge without damage
(vi) Is not self-discharging.

Disadvantages of an alkaline cell over a lead–acid cell include:

(i) Is relatively more expensive
(ii) Requires more cells for a given e.m.f.
(iii) Has a higher internal resistance
(iv) Must be kept sealed
(v) Has a lower efficiency.

Alkaline cells may be used in extremes of temperature, in conditions where vibration is experienced or where duties require long idle periods or heavy discharge currents. Practical examples include traction and marine work, lighting in railway carriages, military portable radios and for starting diesel and petrol engines. However, the lead–acid cell is the most common one in practical use.

4.9 Cell capacity

The **capacity** of a cell is measured in ampere-hours (Ah). A fully charged 50 Ah battery rated for 10 h discharge can be discharged at a steady current of 5 A for 10 h, but if the load current is increased to 10 A then the battery is discharged in 3–4 h, since the higher the discharge current, the lower is the effective capacity of the battery. Typical discharge characteristics for a lead–acid cell are shown in Fig. 4.8.

4.10 Multi-choice questions on the chemical effects of electricity

(Answers on page 283.)

1. The terminal p.d. of a cell of e.m.f. 2 V and internal resistance 0.1 Ω when supplying a current of 5 A will be:
   (a) 1.5 V (b) 2 V (c) 1.9 V (d) 2.5 V
2. Five cells, each with an e.m.f. of 2 V and internal resistance 0.5 Ω are connected in series. The resulting battery will have:
   (a) an e.m.f. of 2 V and an internal resistance of 0.5 Ω
   (b) an e.m.f. of 10 V and an internal resistance of 2.5 Ω
   (c) an e.m.f. of 2 V and an internal resistance of 0.1 Ω
   (d) an e.m.f. of 10 V and an internal resistance of 0.1 Ω
3. If the five cells of question 2 are connected in parallel the resulting battery will have:
   (a) an e.m.f. of 2 V and an internal resistance of 0.5 Ω
   (b) an e.m.f. of 10 V and an internal resistance of 2.5 Ω

(c) an e.m.f. of 2 V and an internal resistance of 0.1 Ω
(d) an e.m.f. of 10 V and an internal resistance of 0.1 Ω

4. Which of the following statements is false?
(a) A Leclanché cell is suitable for use in torches.
(b) A nickel–cadmium cell is an example of a primary cell.
(c) When a cell is being charged its terminal p.d. exceeds the cell e.m.f.
(d) A secondary cell may be recharged after use.
5. Which of the following statements is false? When two metal electrodes are used in a simple cell, the one that is higher in the electrochemical series:
(a) tends to dissolve in the electrolyte
(b) is always the negative electrode
(c) reacts most readily with oxygen
(d) acts as the anode
6. Five 2 V cells, each having an internal resistance of 0.2 Ω are connected in series to a load of resistance 14 Ω. The current flowing in the circuit is:
(a) 10 A (b) 1.4 A (c) 1.5 A (d) $\frac{2}{3}$ A
7. For the circuit of question 6, the p.d. at the battery terminals is:
(a) 10 V (b) $9\frac{1}{3}$ V (c) 0 V (d) $10\frac{2}{3}$ V
8. Which of the following statements is true?
(a) The capacity of a cell is measured in volts.
(b) A primary cell converts electrical energy into chemical energy.
(c) Galvanizing iron helps to prevent corrosion.
(d) A positive electrode is termed the cathode.

4.11 Short answer questions on the chemical effects of electricity

1. What is electrolysis?
2. What is an electrolyte?
3. Conduction in electrolytes is due to ....
4. A positive-connected electrode is called the .... and the negative-connected electrode the ....
5. Name two practical applications of electrolysis.
6. The purpose of an electric cell is to convert .... to ....
7. Make a labelled sketch of a simple cell.
8. What is the electrochemical series?
9. What is corrosion?
10. Name two effects of corrosion and state how they may be prevented.
11. What is meant by the e.m.f. of a cell? How may the e.m.f. of a cell be measured?
12. Define internal resistance.
13. If a cell has an e.m.f. of E volts, an internal resistance of r ohms and supplies a current I amperes to a load, the terminal p.d. V volts is given by: V = ...
14. Name the two main types of cells.
15. Explain briefly the difference between primary and secondary cells.
16. Name two types of primary cells.
17. Name two types of secondary cells.
18. State three typical applications of primary cells.
19. State three typical applications of secondary cells.
20. In what unit is the capacity of a cell measured?

4.12 Further questions on the chemical effects of electricity

1. Twelve cells, each with an internal resistance of 0.24 Ω and an e.m.f. of 1.5 V are connected (a) in series, (b) in parallel. Determine the e.m.f. and internal resistance of the batteries so formed.
[(a) 18 V, 2.88 Ω (b) 1.5 V, 0.02 Ω]
2. A piece of chromium and a piece of iron are placed in an electrolyte in a container. A d.c. supply is connected between the pieces of metal, the positive terminal being connected to the chromium. Explain what is likely to happen and why this happens.
3. With reference to conduction in electrolytes, explain briefly how silver plating of nickel alloys is achieved.
4. A cell has an internal resistance of 0.03 Ω and an e.m.f. of 2.2 V. Calculate its terminal p.d. if it delivers (a) 1 A, (b) 20 A, (c) 50 A.
[(a) 2.17 V (b) 1.6 V (c) 0.7 V]
5. The p.d. at the terminals of a battery is 16 V when no-load is connected and 14 V when a load taking 8 A is connected. Determine the internal resistance of the battery.
[0.25 Ω]
6. A battery of e.m.f. 20 V and internal resistance 0.2 Ω supplies a load taking 10 A.

Determine the p.d. at the battery terminals and the resistance of the load. [18 V, 1.8 Ω]

7. Ten 2.2 V cells, each having an internal resistance of 0.1 Ω are connected in series to a load of 21 Ω. Determine (a) the current flowing in the circuit, and (b) the p.d. at the battery terminals. [(a) 1 A (b) 21 V]
8. For the circuits shown in Fig. 4.9 the resistors represent the internal resistance of the batteries. Find, in each case:
   (a) the total e.m.f. across PQ,
   (b) the total equivalent internal resistances of the batteries.
   [(a)(i) 6 V (ii) 2 V (b)(i) 4 Ω (ii) 0.25 Ω]
9. The voltage at the terminals of a battery is 52 V when no load is connected and 48.8 V when a load taking 80 A is connected. Find the internal resistance of the battery. What would be the terminal voltage when a load taking 20 A is connected? [0.04 Ω, 51.2 V]

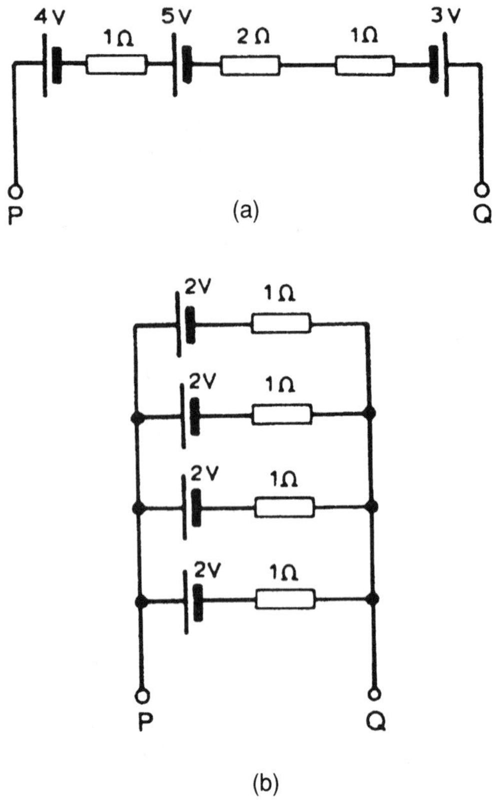

Figure 4.9

10. Define electrolysis and use an example of electrolysis to show the meaning of the terms (a) electrode, (b) electrolyte.
11. The simple cell has two main faults – polarization and local action. Explain these two phenomena.
12. What is corrosion? State its effects and how it may be prevented.
13. Explain the difference between primary and secondary cells. Make a fully labelled sketch of one of each type of cell.
14. Describe the charging and discharging sequence of a simple lead–acid secondary cell.
15. Compare the performances and uses of the Leclanché dry cell and the mercury cell.
16. State three practical applications for each of the following cells:
    (a) dry Leclanché (b) mercury (c) lead–acid (d) alkaline.
17. How may the electrochemical series be used to predict the behaviour of a cell? Also, how is the series representative of the order of reactivity of the metals and their compounds?

5

Series and parallel networks

At the end of this chapter you should be able to:

- calculate unknown voltages, current and resistances in a series circuit
- understand voltage division in a series circuit
- calculate unknown voltages, currents and resistances in a parallel network
- calculate unknown voltages, currents and resistances in series–parallel networks
- understand current division in a two-branch parallel network
- describe the advantages and disadvantages of series and parallel connection of lamps

5.1 Series circuits

Figure 5.1 shows three resistors R_1, R_2 and R_3 connected end to end, i.e. in series, with a battery source of V volts. Since the circuit is closed a current I will flow and the p.d. across each resistor may be determined from the voltmeter readings V_1, V_2 and V_3.

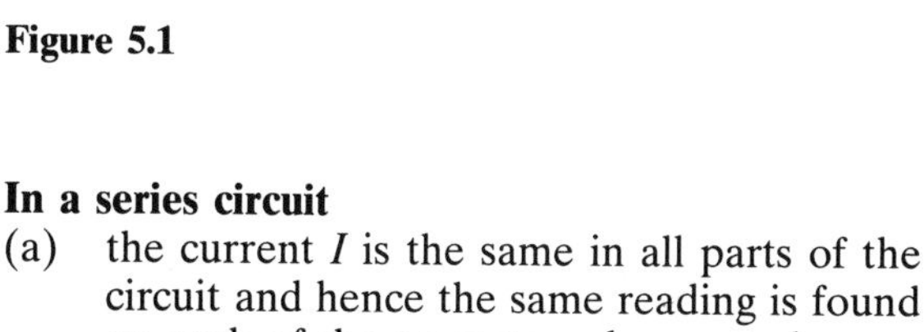

Figure 5.1

In a series circuit

(a) the current I is the same in all parts of the circuit and hence the same reading is found on each of the ammeters shown, and

(b) the sum of the voltages V_1, V_2 and V_3 is equal to the total applied voltage, V, i.e.

$$\boldsymbol{V = V_1 + V_2 + V_3}$$

From Ohm's law:

$$V_1 = IR_1,\ V_2 = IR_2,\ V_3 = IR_3 \text{ and } V = IR$$

where R is the total circuit resistance.
Since $V = V_1 + V_2 + V_3$, then $IR = IR_1 + IR_2 + IR_3$. Dividing throughout by I gives

$$\boldsymbol{R = R_1 + R_2 + R_3}$$

Thus for a series circuit, the total resistance is obtained by adding together the values of the separate resistances.

Problem 1. For the circuit shown in Fig. 5.2, determine (a) the battery voltage V, (b) the total resistance of the circuit, and (c) the values of resistance of resistors R_1, R_2 and R_3, given that the p.d.s across R_1, R_2 and R_3 are 5 V, 2 V and 6 V, respectively.

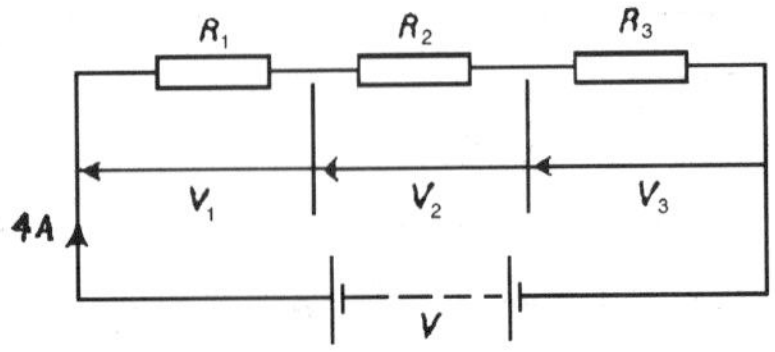

Figure 5.2

(a)

$$\text{Battery voltage } V = V_1 + V_2 + V_3 = 5 + 2 + 6 = \mathbf{13\ V}$$

(b)

$$\text{Total circuit resistance } R = \frac{V}{I} = \frac{13}{4} = \mathbf{3.25\ \Omega}$$

(c)

$$\text{Resistance } R_1 = \frac{V_1}{I} = \frac{5}{4} = \mathbf{1.25\ \Omega}$$

$$\text{Resistance } R_2 = \frac{V_2}{I} = \frac{2}{4} = \mathbf{0.5\ \Omega}$$

$$\text{Resistance } R_3 = \frac{V_3}{I} = \frac{6}{4} = \mathbf{1.5\ \Omega}$$

(Check: $R_1 + R_2 + R_3 = 1.25 + 0.5 + 1.5 = 3.25\ \Omega = R$.)

> Problem 2. For the circuit shown in Fig. 5.3, determine the p.d. across resistor R_3. If the total resistance of the circuit is 100 Ω, determine the current flowing through resistor R_1. Find also the value of resistor R_2.

P.d. across R_3, $V_3 = 25 - 10 - 4 =$ **11 V**.

$$\text{Current } I = \frac{V}{R} = \frac{25}{100} = \mathbf{0.25\ A}$$

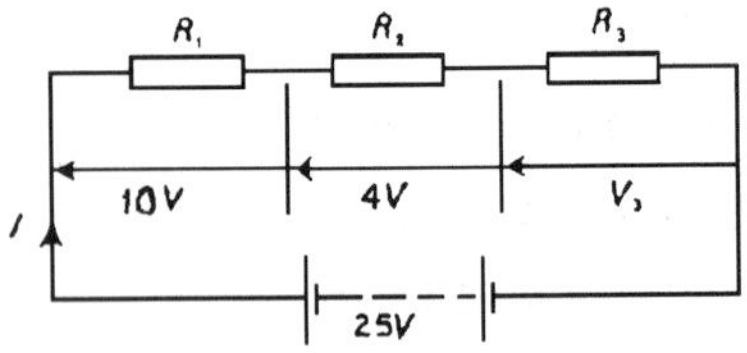

Figure 5.3

which is the current flowing in each resistor.

$$\text{Resistance } R_2 = \frac{V_2}{I} = \frac{4}{0.25} = \mathbf{16\ \Omega}$$

> Problem 3. A 12 V battery is connected in a circuit having three series-connected resistors having resistance of 4 Ω, 9 Ω and 11 Ω. Determine the current flowing through, and the p.d. across the 9 Ω resistor. Find also the power dissipated in the 11 Ω resistor.

The circuit diagram is shown in Fig. 5.4.

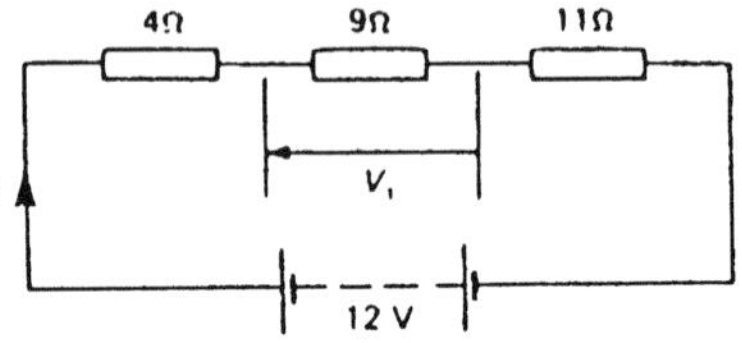

Figure 5.4

Total resistance $R = 4 + 9 + 11 = 24\ \Omega$.

$$\text{Current } I = \frac{V}{R} = \frac{12}{24} = \mathbf{0.5\ A}$$

which is the current in the 9 Ω resistor.
P.d. across the 9 Ω resistor, $V_1 = I \times 9 = 0.5 \times 9$ = **4.5 V**.
Power dissipated in the 11 Ω resistor is given by

$$P = I^2R = 0.5^2(11) = (0.25)(11) = \mathbf{2.75\ W}$$

5.2 Potential divider

The voltage distribution for the circuit shown in Fig. 5.5(a) is given by:

$$V_1 = \left(\frac{R_1}{R_1 + R_2}\right) V$$

$$V_2 = \left(\frac{R_2}{R_1 + R_2}\right) V$$

The circuit shown in Fig. 5.5(b) is often referred to as a **potential divider** circuit. Such a circuit can consist of a number of similar elements in series

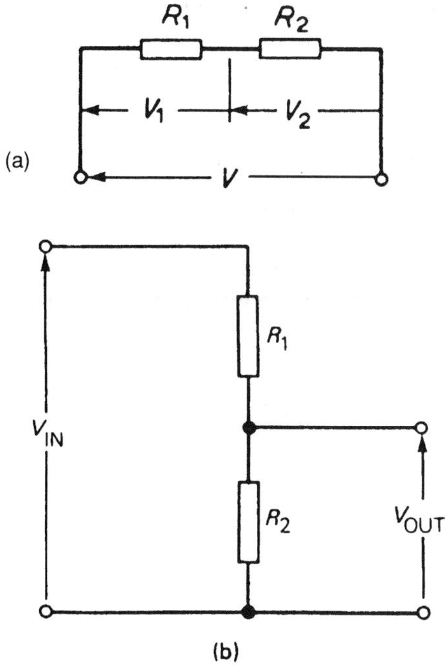

Figure 5.5

connected across a voltage source, voltages being taken from connections between the elements. Frequently the divider consists of two resistors as shown in Fig. 5.5(b), where

$$V_{OUT} = \left(\frac{R_2}{R_1 + R_2}\right) V_{IN}$$

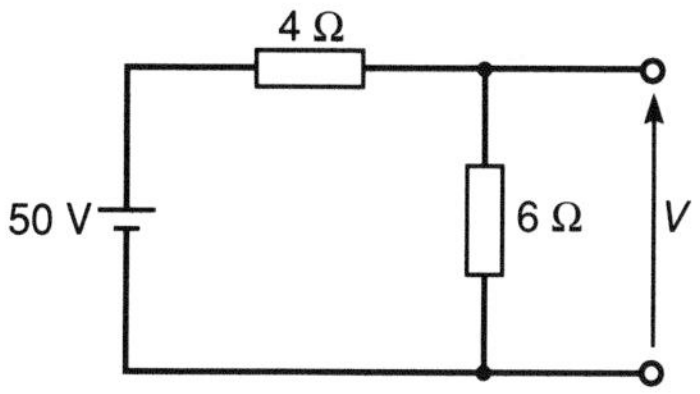

Figure 5.6

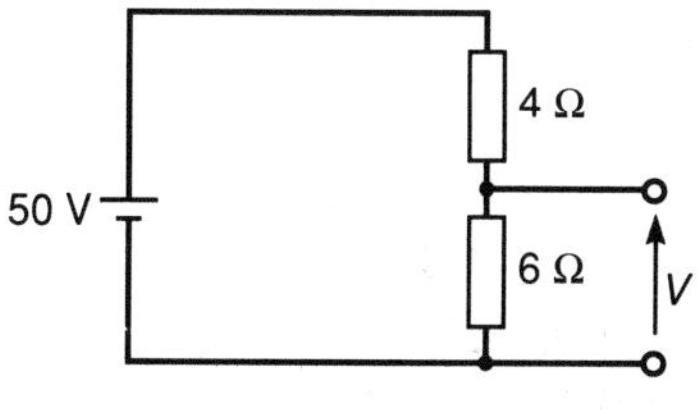

Figure 5.7

> Problem 4. Determine the value of voltage V shown in Fig. 5.6.

Figure 5.6 may be redrawn as shown in Fig. 5.7, and

$$\text{voltage } V = \left(\frac{6}{6 + 4}\right)(50) = \mathbf{30\ V}$$

> Problem 5. Two resistors are connected in series across a 24 V supply and a current of 3 A flows in the circuit. If one of the resistors has a resistance of 2 Ω determine (a) the value of the other resistor, and (b) the p.d. across the 2 Ω resistor. If the circuit is connected for 50 hours, how much energy is used?

The circuit diagram is shown in Fig. 5.8.

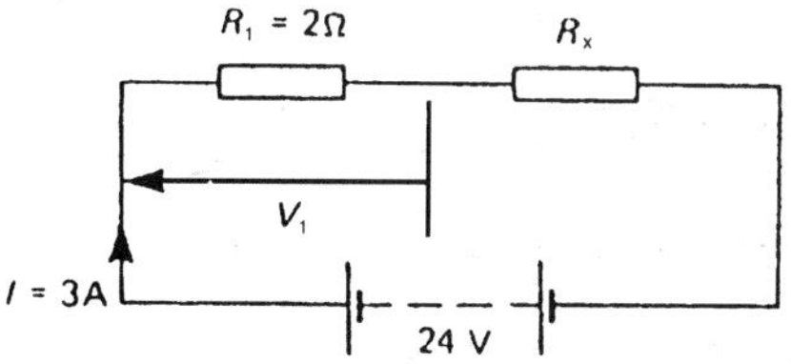

Figure 5.8

(a) Total circuit resistance

$$R = \frac{V}{I} = \frac{24}{3} = 8\ \Omega$$

Value of unknown resistance

$$R_x = 8 - 2 = \mathbf{6\ \Omega}$$

(b) P.d. across 2 Ω resistor

$$V_1 = IR_1 = 3 \times 2 = \mathbf{6\ V}$$

Alternatively, from above,

$$V_1 = \left(\frac{R_1}{R_1 + R_x}\right) V = \left(\frac{2}{2 + 6}\right) 24 = 6\ \text{V}$$

$$\begin{aligned} \text{Energy used} &= \text{power} \times \text{time} \\ &= (V \times I) \times t \\ &= (24 \times 3\ \text{W})(50\ \text{h}) \\ &= 3600\ \text{W h} = \mathbf{3.6\ kW\ h} \end{aligned}$$

5.3 Parallel networks

Figure 5.9 shows three resistors R_1, R_2 and R_3 connected across each other, i.e. in parallel, across a battery source of V volts.

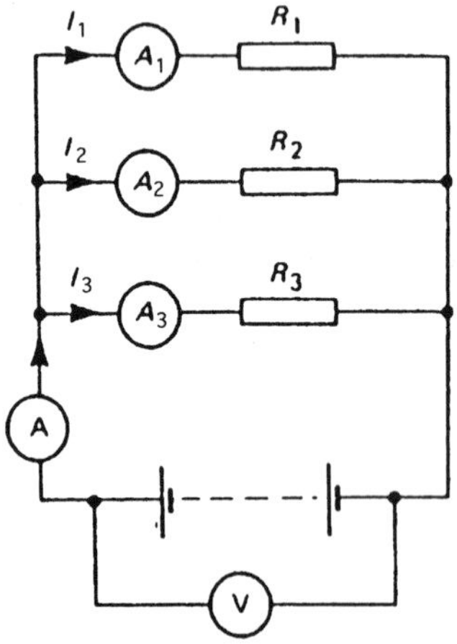

Figure 5.9

In a parallel circuit:

(a) the sum of the currents I_1, I_2 and I_3 is equal to the total circuit current, I, i.e.

$$\boldsymbol{I = I_1 + I_2 + I_3}$$

and

(b) the source p.d., V volts, is the same across each of the resistors.

From Ohm's law:

$$I_1 = \frac{V}{R_1},\ I_2 = \frac{V}{R_2},\ I_3 = \frac{V}{R_3} \text{ and } I = \frac{V}{R}$$

where R is the total circuit resistance. Since

$$I = I_1 + I_2 + I_3$$

then

$$\frac{V}{R} = \frac{V}{R_1} + \frac{V}{R_2} + \frac{V}{R_3}$$

Dividing throughout by V gives

$$\boldsymbol{\frac{1}{R} = \frac{1}{R_1} + \frac{1}{R_2} + \frac{1}{R_3}}$$

This equation must be used when finding the total resistance R of a parallel circuit. For the special case of **two resistors in parallel**

$$\frac{1}{R} = \frac{1}{R_1} + \frac{1}{R_2} = \frac{R_2 + R_1}{R_1 R_2}$$

Hence

$$\boldsymbol{R = \frac{R_1 R_2}{R_1 + R_2}} \quad \left(\text{i.e. } \frac{\text{product}}{\text{sum}}\right)$$

Problem 6. For the circuit shown in Fig. 5.10, determine (a) the reading on the ammeter, and (b) the value of resistor R_2.

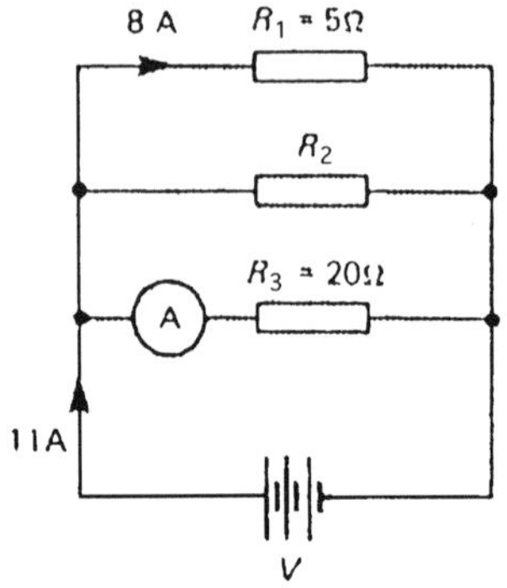

Figure 5.10

P.d. across R_1 is the same as the supply voltage V. Hence supply voltage, $V = 8 \times 5 = 40$ V

(a) Reading on ammeter

$$I = \frac{V}{R_3} = \frac{40}{20} = \mathbf{2\ A}$$

(b) Current flowing through $R_2 = 11 - 8 - 2 = 1$ A, hence

$$R_2 = \frac{V}{I_2} = \frac{40}{1} = \mathbf{40\ \Omega}$$

Problem 7. Two resistors, of resistance 3 Ω and 6 Ω, are connected in parallel across a battery having a voltage of 12 V. Determine (a) the total circuit resistance and (b) the current flowing in the 3 Ω resistor.

The circuit diagram is shown in Fig. 5.11.

(a) The total circuit resistance R is given by

$$\frac{1}{R} = \frac{1}{R_1} + \frac{1}{R_2} = \frac{1}{3} + \frac{1}{6}$$

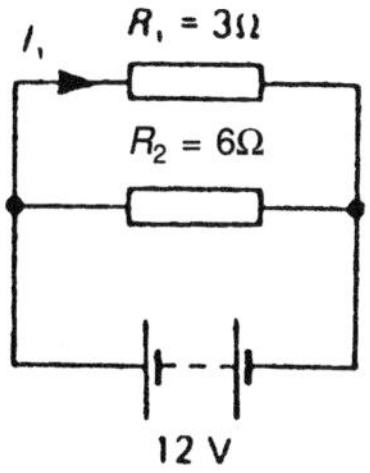

Figure 5.11

$$\frac{1}{R} = \frac{2+1}{6} = \frac{3}{6}$$

Hence, $R = \frac{6}{3} = 2\ \Omega$.

(Alternatively,

$$R = \frac{R_1 R_2}{R_1 + R_2} = \frac{3 \times 6}{3+6} = \frac{18}{9} = 2\ \Omega.)$$

(b) Current in the 3 Ω resistance

$$I_1 = \frac{V}{R_1} = \frac{12}{3} = \mathbf{4\ A}$$

Problem 8. For the circuit shown in Fig. 5.12, find (a) the value of the supply voltage V and (b) the value of current I.

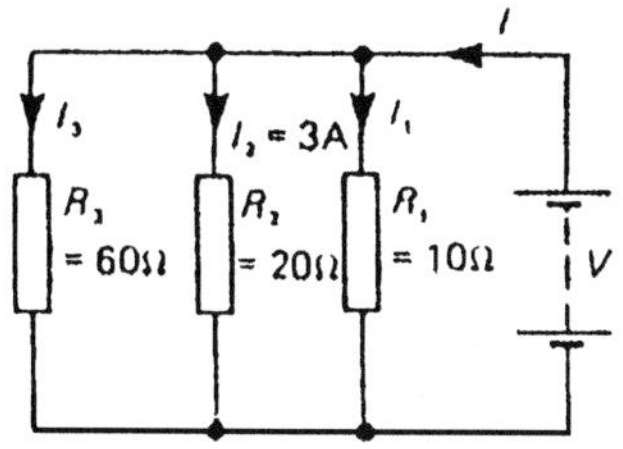

Figure 5.12

(a) P.d. across 20 Ω resistor = $I_2R_2 = 3 \times 20 =$ 60 V, hence supply voltage $\boldsymbol{V} = \mathbf{60\ V}$ since the circuit is connected in parallel.

(b)

$$\text{Current } I_1 = \frac{V}{R_1} = \frac{60}{10} = 6\text{ A};\ I_2 = \mathbf{3\ A}$$

$$I_3 = \frac{V}{R_3} = \frac{60}{60} = 1\text{ A}$$

Current $I = I_1 + I_2 + I_3$ and hence $I = 6 + 3 + 1 =$ **10 A**.

Alternatively

$$\frac{1}{R} = \frac{1}{60} + \frac{1}{20} + \frac{1}{10} = \frac{1+3+6}{60} = \frac{10}{60}$$

Hence total resistance

$$R = \frac{60}{10} = 6\ \Omega, \text{ and}$$

current

$$I = \frac{V}{R} = \frac{60}{6} = \mathbf{10\ A}$$

Problem 9. Given four 1 Ω resistors, state how they must be connected to give an overall resistance of (a) $\frac{1}{4}$ Ω (b) 1 Ω (c) $1\frac{1}{3}$ Ω (d) $2\frac{1}{2}$ Ω, all four resistors being connected in each case.

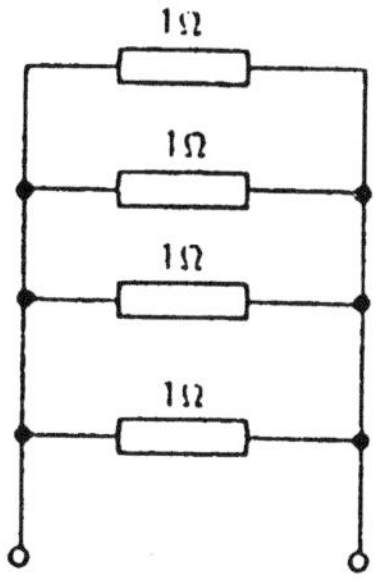

Figure 5.13

(a) **All four in parallel** (see Fig. 5.13), since

$$\frac{1}{R} = \frac{1}{1} + \frac{1}{1} + \frac{1}{1} + \frac{1}{1} = \frac{4}{1}, \text{ i.e.}$$

$$R = \frac{1}{4}\ \Omega$$

(b) **Two in series, in parallel with another two in series** (see Fig. 5.14), since 1 Ω and 1 Ω in series gives 2 Ω, and 2 Ω in parallel with 2 Ω gives:

$$\frac{2 \times 2}{2+2} = \frac{4}{4} = 1\ \Omega$$

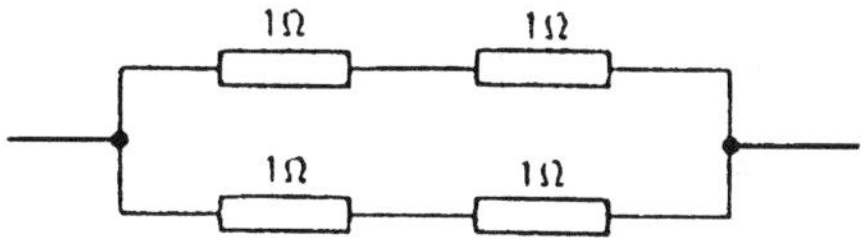

Figure 5.14

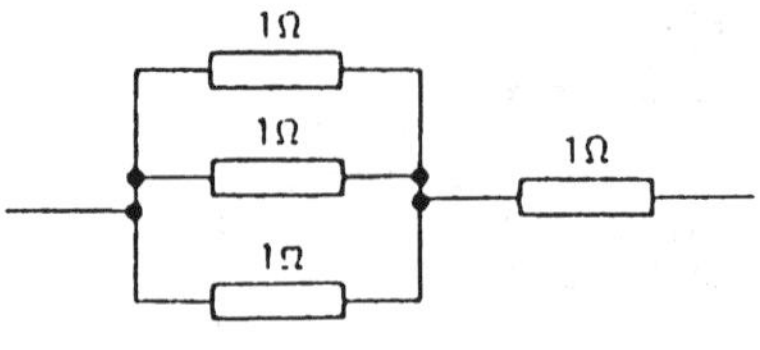

Figure 5.15

(c) **Three in parallel, in series with one** (see Fig. 5.15), since for the three in parallel

$$\frac{1}{R} = \frac{1}{1} + \frac{1}{1} + \frac{1}{1} = \frac{3}{1}$$

i.e. $R = \frac{1}{3}\,\Omega$ and $\frac{1}{3}\,\Omega$ in series with $1\,\Omega$ gives $1\frac{1}{3}\,\Omega$.

(d) **Two in parallel, in series with two in series** (see Fig. 5.16), since for the two in parallel

$$R = \frac{1 \times 1}{1 + 1} = \frac{1}{2}\,\Omega$$

and $\frac{1}{2}\,\Omega$, $1\,\Omega$ and $1\,\Omega$ in series gives $2\frac{1}{2}\,\Omega$.

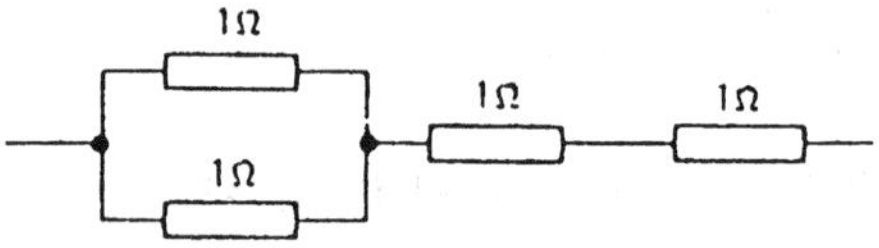

Figure 5.16

Problem 10. Find the equivalent resistance for the circuit shown in Fig. 5.17.

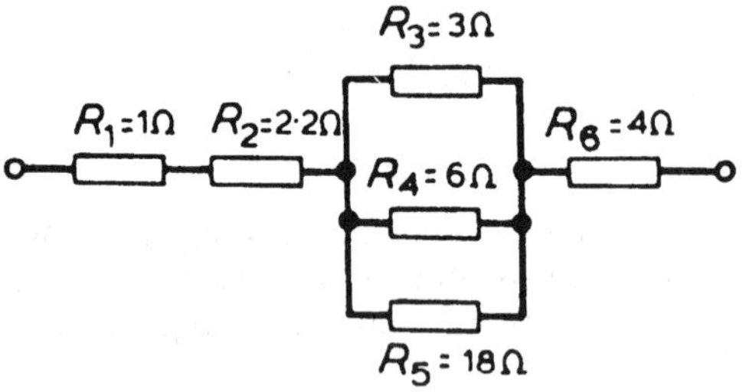

Figure 5.17

R_3, R_4 and R_5 are connected in parallel and their equivalent resistance R is given by:

$$\frac{1}{R} = \frac{1}{3} + \frac{1}{6} + \frac{1}{18} = \frac{6 + 3 + 1}{18} = \frac{10}{18}$$

Hence $R = (18/10) = 1.8\,\Omega$. The circuit is now equivalent to four resistors in series and the equivalent circuit resistance $= 1 + 2.2 + 1.8 + 4 = 9\,\Omega$.

Problem 11. Resistances of $10\,\Omega$, $20\,\Omega$ and $30\,\Omega$ are connected (a) in series and (b) in parallel to a 240 V supply. Calculate the supply current in each case.

(a) The series circuit is shown in Fig. 5.18. The equivalent resistance $R_T = 10\,\Omega + 20\,\Omega + 30\,\Omega = 60\,\Omega$. Supply current

$$I = \frac{V}{R_T} = \frac{240}{60} = \mathbf{4\ A}$$

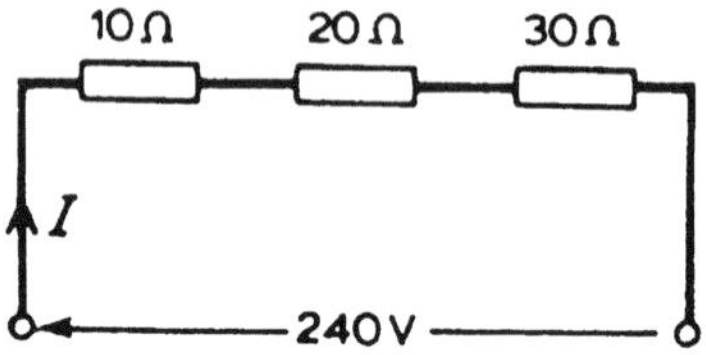

Figure 5.18

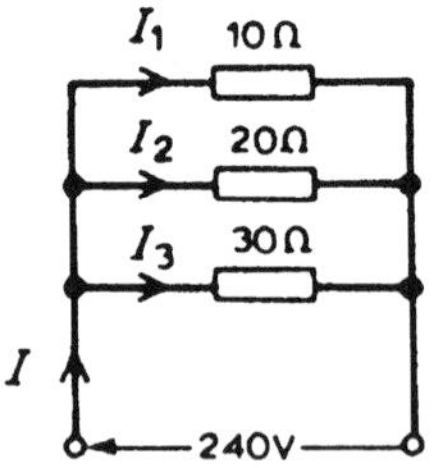

Figure 5.19

(b) The parallel circuit is shown in Fig. 5.19. The equivalent resistance R_T of $10\,\Omega$, $20\,\Omega$ and $30\,\Omega$ resistances connected in parallel is given by:

$$\frac{1}{R_T} = \frac{1}{10} + \frac{1}{20} + \frac{1}{30} = \frac{6 + 3 + 2}{60} = \frac{11}{60}$$

Hence $R_T = (60/11)\,\Omega$. Supply current

$$I = \frac{V}{R_T} = \frac{240}{60/11} = \frac{240 \times 11}{60} = \mathbf{44\ A}$$

(Check:

$$I_1 = \frac{V}{R_1} = \frac{240}{10} = 24\ \text{A};\ I_2 = \frac{V}{R_2} = \frac{240}{20} = 12\ \text{A};$$

$$I_3 = \frac{V}{R_3} = \frac{240}{30} = 8\ \text{A}$$

For a parallel circuit $I = I_1 + I_2 + I_3 = 24 + 12 + 8 =$ **44 A**, as above.)

5.4 Current division

For the circuit shown in Fig. 5.20, the total circuit resistance R_T is given by:

$$R_T = \frac{R_1 R_2}{R_1 + R_2}$$

and

$$V = IR_T = I\left(\frac{R_1 R_2}{R_1 + R_2}\right)$$

$$\text{Current } I_1 = \frac{V}{R_1} = \frac{I}{R_1}\left(\frac{R_1 R_2}{R_1 + R_2}\right)$$

$$= \left(\frac{R_2}{R_1 + R_2}\right)(I)$$

Similarly

$$\text{current } I_2 = \frac{V}{R_2} = \frac{I}{R_2}\left(\frac{R_1 R_2}{R_1 + R_2}\right)$$

$$= \left(\frac{R_1}{R_1 + R_2}\right)(I)$$

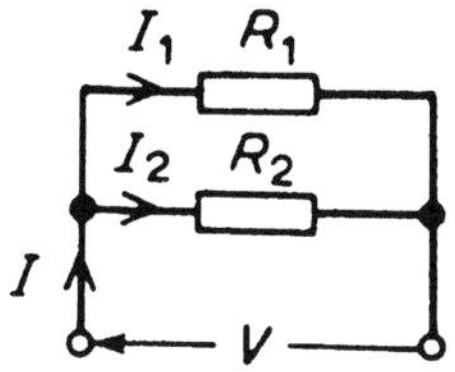

Figure 5.20

Summarizing, with reference to Fig. 5.20

$$\boxed{I_1 = \left(\frac{R_2}{R_1 + R_2}\right)(I)} \quad \text{and} \quad \boxed{I_2 = \left(\frac{R_1}{R_1 + R_2}\right)(I)}$$

Problem 12. For the series–parallel arrangement shown in Fig. 5.21, find (a) the supply current, (b) the current flowing through each resistor and (c) the p.d. across each resistor.

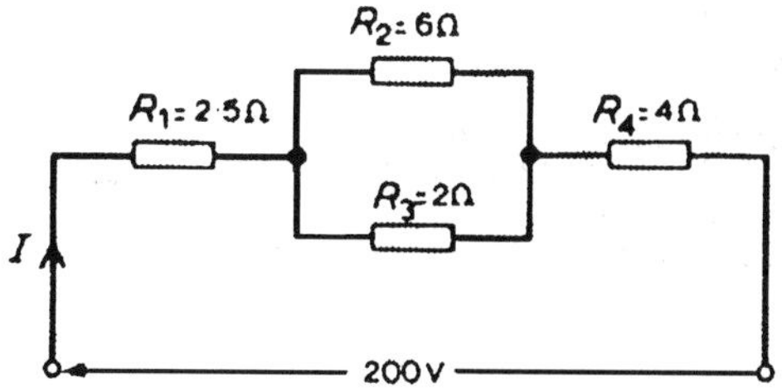

Figure 5.21

(a) The equivalent resistance R_x of R_2 and R_3 in parallel is:

$$R_x = \frac{6 \times 2}{6 + 2} = \frac{12}{8} = 1.5\ \Omega$$

The equivalent resistance R_T of R_1, R_x and R_4 in series is:

$$R_T = 2.5 + 1.5 + 4 = 8\ \Omega$$

Supply current

$$I = \frac{V}{R_T} = \frac{200}{8} = \mathbf{25\ A}$$

(b) The current flowing through R_1 and R_4 is 25 A.
The current flowing through R_2 is

$$\left(\frac{R_3}{R_2 + R_3}\right) I = \left(\frac{2}{6 + 2}\right) 25$$

$$= \mathbf{6.25\ A}$$

The current flowing through R_3 is

$$\left(\frac{R_3}{R_2 + R_3}\right) I = \left(\frac{6}{6 + 2}\right) 25$$

$$= \mathbf{18.75\ A}$$

(Note that the currents flowing through R_2 and R_3 must add up to the total current flowing into the parallel arrangement, i.e. 25 A.)

(c) The equivalent circuit of Fig. 5.21 is shown in Fig. 5.22.

p.d. across R_1, i.e. $V_1 = IR_1 = (25)(2.5)$ = **62.5 V**

p.d. across R_x, i.e. $V_x = IR_x = (25)(1.5) =$ **37.5 V**

p.d. across R_4, i.e. $V_4 = IR_4 = (25)(4) =$ **100 V**

Hence the p.d. across R_2 = p.d. across R_3 = 37.5 V.

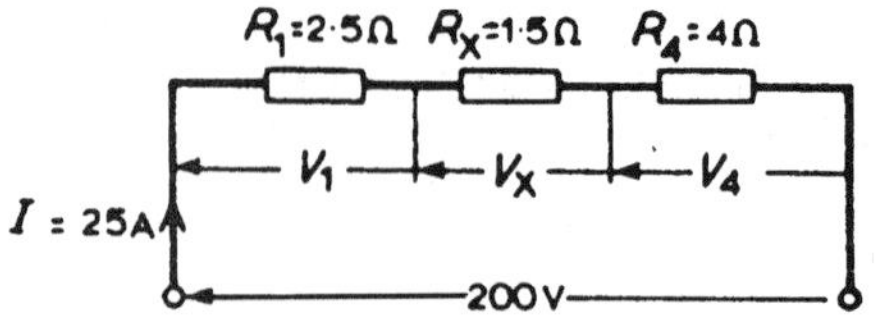

Figure 5.22

Problem 13. For the circuit shown in Fig. 5.23 calculate (a) the value of resistor R_x such that the total power dissipated in the circuit is 2.5 kW, (b) the current flowing in each of the four resistors.

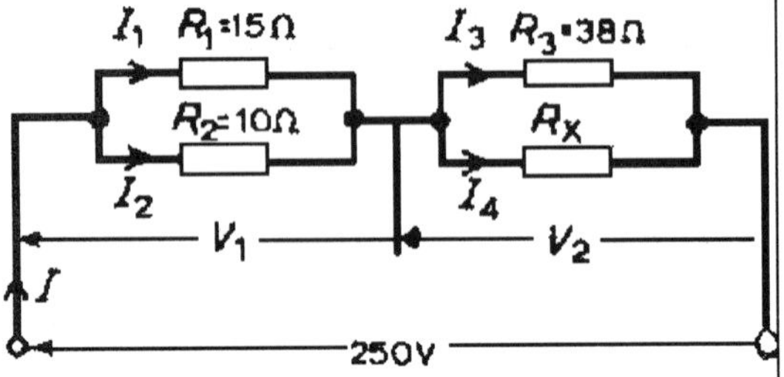

Figure 5.23

(a) Power dissipated $P = VI$ watts, hence $2500 = (250)(I)$, i.e.

$$I = \frac{2500}{250} = 10 \text{ A}$$

From Ohm's law

$$R_T = \frac{V}{I} = \frac{250}{10} = 25\ \Omega$$

where R_T is the equivalent circuit resistance. The equivalent resistance of R_1 and R_2 in parallel is

$$\frac{15 \times 10}{15 + 10} = \frac{150}{25} = 6\ \Omega$$

The equivalent resistance of resistors R_3 and R_x in parallel is equal to 25 Ω – 6 Ω, i.e. 19 Ω.

There are three methods whereby R_x can be determined.

Method 1

The voltage $V_1 = IR$, where R is 6 Ω, from above, i.e. $V_1 = (10)(6) = 60$ V. Hence

$$V_2 = 250 \text{ V} - 60 \text{ V} = 190 \text{ V}$$

= p.d. across R_3

= p.d. across R_x

$$I_3 = \frac{V_2}{R_3} = \frac{190}{38} = 5 \text{ A}$$

Thus $I_4 = 5$ A also, since $I = 10$ A. Thus

$$R_x = \frac{V_2}{I_4} = \frac{190}{5} = 38\ \Omega$$

Method 2

Since the equivalent resistance of R_3 and R_x in parallel is 19 Ω,

$$19 = \frac{38R_x}{38 + R_x} \left(\text{i.e. } \frac{\text{product}}{\text{sum}}\right)$$

Hence

$$19(38 + R_x) = 38R_x$$

$$722 + 19R_x = 38R_x$$

$$722 = 38R_x - 19R_x = 19R_x$$

Thus

$$R_x = \frac{722}{19} = 38\ \Omega$$

Method 3

When two resistors having the same value are connected in parallel the equivalent resistance is always half the value of one of the resistors. Thus, in this case, since $R_T = 19\ \Omega$ and $R_3 = 38\ \Omega$, then $R_x = 38\ \Omega$ could have been deduced on sight.

(b) Current

$$I_1 = \left(\frac{R_2}{R_1 + R_2}\right) I = \left(\frac{10}{15 + 10}\right) 10$$

$$= \left(\frac{2}{5}\right) 10 = \mathbf{4\ A}$$

Current

$$I_2 = \left(\frac{R_1}{R_1 + R_2}\right) I = \left(\frac{15}{15 + 10}\right) 10$$

$$= \left(\frac{3}{5}\right) 10 = \mathbf{6\ A}$$

From part (a), method 1, $I_3 = I_4 = 5$ A.

Problem 14. For the arrangement shown in Fig. 5.24, find the current I_x.

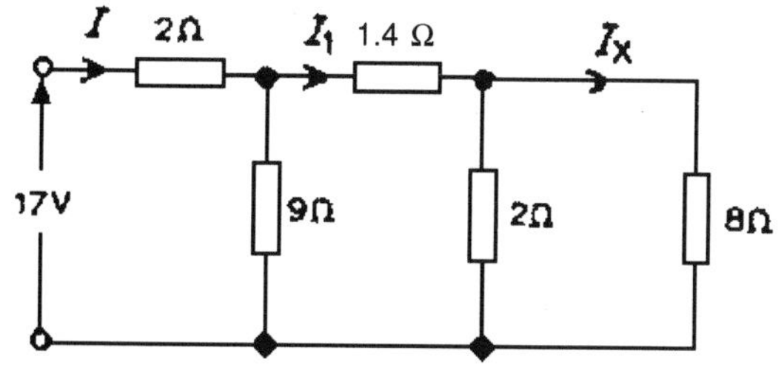

Figure 5.24

Commencing at the right-hand side of the arrangement shown in Fig. 5.24, the circuit is gradually reduced in stages as shown in Fig. 5.25(a)–(d). From Fig. 5.25(d)

$$I = \frac{17}{4.25} = 4 \text{ A}$$

From Fig. 5.25(b)

$$I_1 = \left(\frac{9}{9+3}\right) I = \left(\frac{9}{12}\right) 4 = 3 \text{ A}$$

From Fig. 5.24

$$I_x = \left(\frac{2}{2+8}\right) I_1 = \left(\frac{2}{10}\right) 3 = \mathbf{0.6 \text{ A}}$$

5.5 Wiring lamps in series and in parallel

Series connection

Figure 5.26 shows three lamps, each rated at 240 V, connected in series across a 240 V supply.

(i) Each lamp has only (240/3) V, i.e. 80 V across it and thus each lamp glows dimly.
(ii) If another lamp of similar rating is added in series with the other three lamps then each lamp now has (240/4) V, i.e. 60 V across it and each now glows even more dimly.

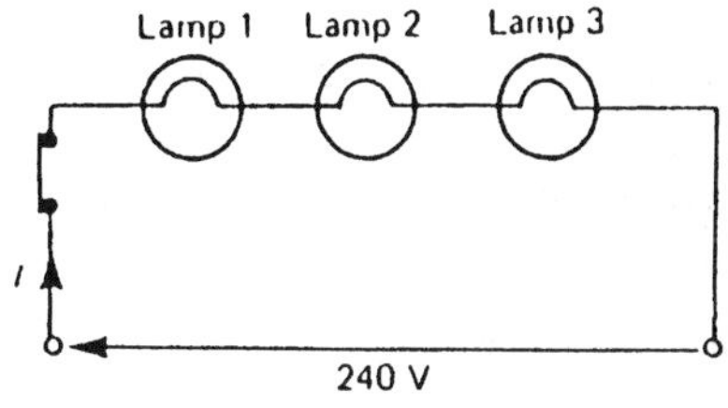

Figure 5.26

(iii) If a lamp is removed from the circuit or if a lamp develops a fault (i.e. an open circuit) or

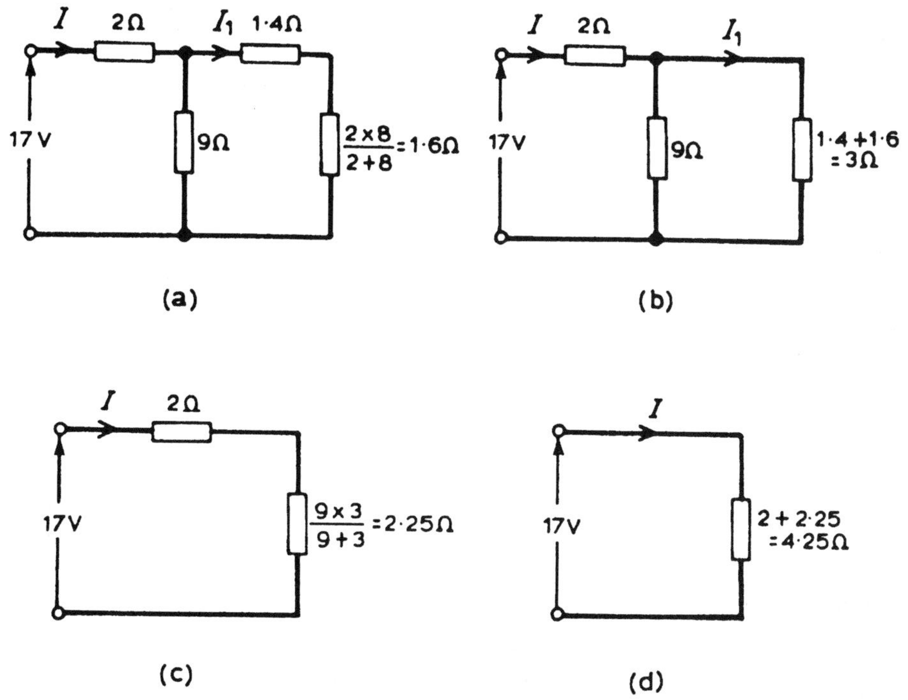

Figure 5.25

if the switch is opened then the circuit is broken, no current flows, and the remaining lamps will not light up.

(iv) Less cable is required for a series connection than for a parallel one.

The series connection of lamps is usually limited to decorative lighting such as for Christmas tree lights.

Parallel connection

Fig. 5.27 shows three similar lamps, each rated at 240 V, connected in parallel across a 240 V supply.

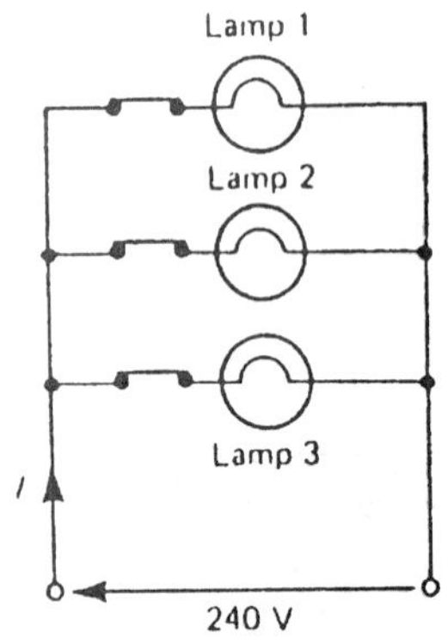

Figure 5.27

(i) Each lamp has 240 V across it and thus each will glow brilliantly at their rated voltage.
(ii) If any lamp is removed from the circuit or develops a fault (open circuit) or a switch is opened, the remaining lamps are unaffected.
(iii) The addition of further similar lamps in parallel does not affect the brightness of the other lamps.
(iv) More cable is required for parallel connection than for a series one.

The parallel connection of lamps is the most widely used in electrical installations.

> Problem 15. If three identical lamps are connected in parallel and the combined resistance is 150 Ω, find the resistance of one lamp.

Let the resistance of one lamp be R, then

$$\frac{1}{150} = \frac{1}{R} + \frac{1}{R} + \frac{1}{R} = \frac{3}{R}$$

from which, $R = 3 \times 150 =$ **450 Ω**.

> Problem 16. Three identical lamps A, B and C are connected in series across a 150 V supply. State (a) the voltage across each lamp, and (b) the effect of lamp C failing.

(a) Since each lamp is identical and they are connected in series there is (150/3) V, i.e. 50 V across each.
(b) If lamp C fails, i.e. open circuits, no current will flow and **lamps A and B will not operate**.

5.6 Multi-choice questions on series and parallel networks

(Answers on page 283.)

1. If two 4 Ω resistors are connected in series the effective resistance of the circuit is:
(a) 8 Ω (b) 4 Ω (c) 2 Ω (d) 1 Ω
2. If two 4 Ω resistors are connected in parallel the effective resistance of the circuit is:
(a) 8 Ω (b) 4 Ω (c) 2 Ω (d) 1 Ω
3. With the switch in Fig. 5.28 closed, the ammeter reading will indicate:
(a) $1\frac{2}{3}$ A (b) 75 A (c) $\frac{1}{3}$ A (d) 3 A

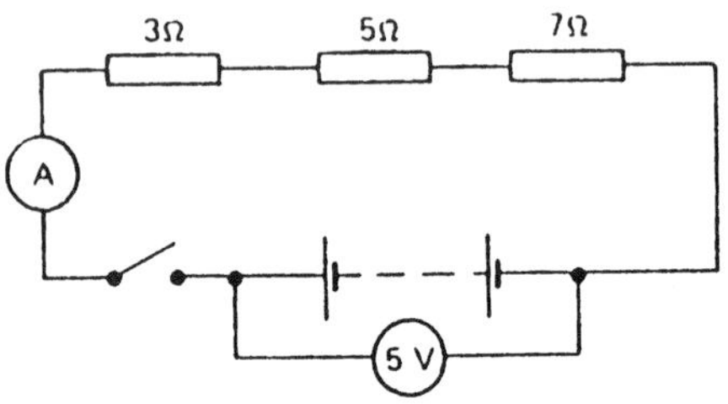

Figure 5.28

4. The effect of connecting an additional parallel load to an electrical supply source is to increase the
(a) resistance of load
(b) voltage of the source
(c) current taken from the source
(d) p.d. across the load
5. The equivalent resistance when a resistor of

$\frac{1}{3}\,\Omega$ is connected in parallel with a $\frac{1}{4}\,\Omega$ resistor is
(a) 1/7 Ω (b) 7 Ω

6. With the switch in Fig. 5.29 closed the ammeter reading will indicate:
(a) 108 A (b) $\frac{1}{3}$ A (c) 3 A (d) $4\frac{3}{5}$ A

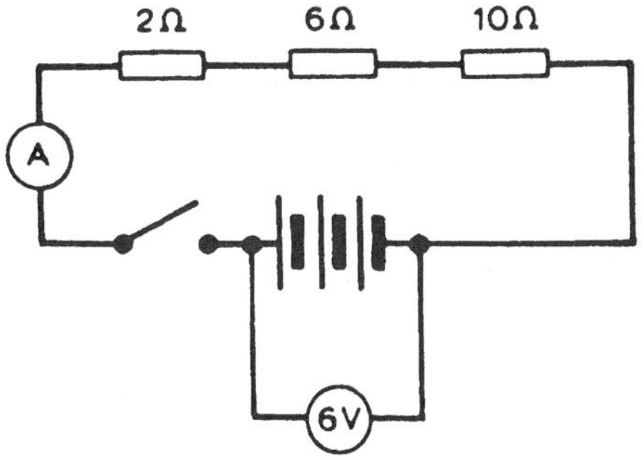

Figure 5.29

7. A 6 Ω resistor is connected in parallel with the three resistors of Fig. 5.29. With the switch closed the ammeter reading will indicate:
(a) $\frac{3}{4}$ A (b) 4 A (c) $\frac{1}{4}$ A (d) $1\frac{1}{3}$ A
8. A 10 Ω resistor is connected in parallel with a 15 Ω resistor and the combination is connected in series with a 12 Ω resistor. The equivalent resistance of the circuit is:
(a) 37 Ω (b) 18 Ω (c) 27 Ω (d) 4 Ω

5.7 Short answer questions on series and parallel networks

1. Name three characteristics of a series circuit.
2. Show that for three resistors R_1, R_2 and R_3 connected in series the equivalent resistance R is given by $R = R_1 + R_2 + R_3$.
3. Name three characteristics of a parallel circuit.
4. Show that for three resistors R_1, R_2 and R_3 connected in parallel the equivalent resistance R is given by

$$\frac{1}{R} = \frac{1}{R_1} + \frac{1}{R_2} + \frac{1}{R_3}$$

5. Explain the potential divider circuit.
6. Compare the merits of wiring lamps in (a) series, (b) parallel.

5.8 Further questions on series and parallel networks

1. The p.d.s measured across three resistors connected in series are 5 V, 7 V and 10 V, and the supply current is 2 A. Determine (a) the supply voltage, (b) the total circuit resistance and (c) the values of the three resistors.
[(a) 22 V (b) 11 Ω (c) 2.5 Ω, 3.5 Ω, 5 Ω]
2. For the circuit shown in Fig. 5.30, determine the value of V_1. If the total circuit resistance is 36 Ω determine the supply current and the value of the resistors R_1, R_2 and R_3.
[10 V, 0.5 A, 20 Ω, 10 Ω, 6 Ω]

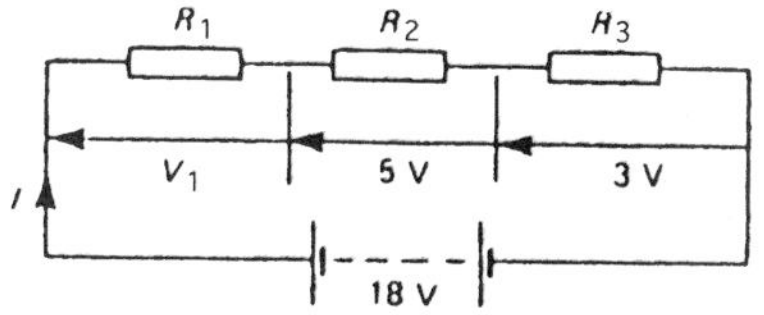

Figure 5.30

3. When the switch in the circuit in Fig. 5.31 is closed the reading on voltmeter 1 is 30 V and that on voltmeter 2 is 10 V. Determine the reading on the ammeter and the value of resistor R_x. [4 A, 2.5 Ω]

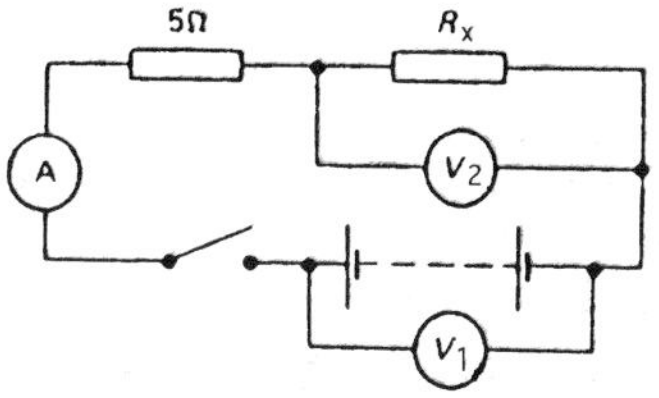

Figure 5.31

4. Two resistors are connected in series across an 18 V supply and a current of 5 A flows. If one of the resistors has a value of 2.4 Ω determine (a) the value of the other resistor and (b) the p.d. across the 2.4 Ω resistor.
[(a) 1.2 Ω (b) 12 V]
5. Resistances of 4 Ω and 12 Ω are connected in parallel across a 9 V battery. Determine (a) the equivalent circuit resistance, (b) the supply current, and (c) the current in each resistor.
[(a) 3 Ω (b) 3 A (c) $2\frac{1}{4}$ A, $\frac{3}{4}$ A]

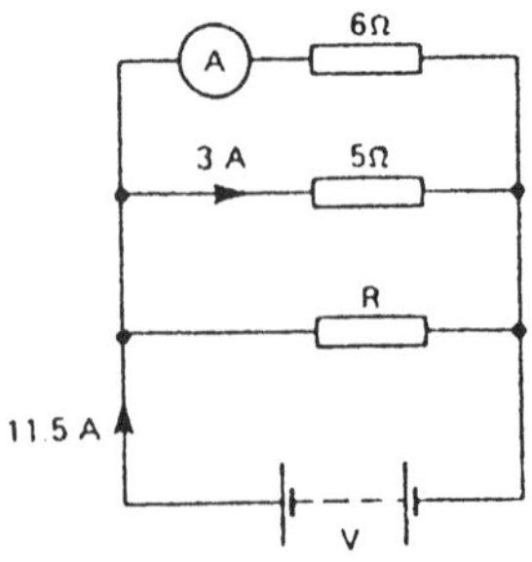

Figure 5.32

6. For the circuit shown in Fig. 5.32 determine (a) the reading on the ammeter, and (b) the value of resistor R. [2.5 A, 2.5 Ω]
7. Find the equivalent resistance when the following resistances are connected (a) in series, (b) in parallel:
(i) 3 Ω and 2 Ω (ii) 20 kΩ and 40 kΩ (iii) 4 Ω, 8 Ω and 16 Ω (iv) 800 Ω, 4 kΩ and 1500 Ω.
[(a)(i) 5 Ω (ii) 60 kΩ (iii) 28 Ω (iv) 6.3 kΩ (b)(i) 1.2 Ω (ii) $13\frac{1}{3}$ kΩ (iii) $2\frac{2}{7}$ Ω (iv) 461.5 Ω]
8. Find the total resistance between terminals A and B of the circuit shown in Fig. 5.33(a). [8 Ω]

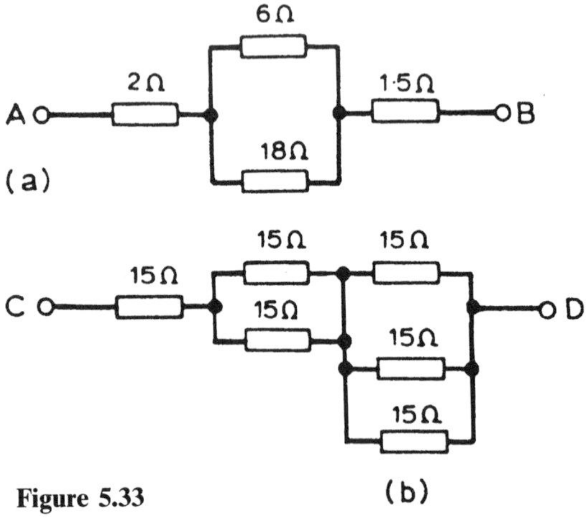

Figure 5.33

9. Find the equivalent resistance between terminals C and D of the circuit shown in Fig. 5.33(b). [27.5 Ω]
10. Resistors of 20 Ω, 20 Ω and 30 Ω are connected in parallel. What resistance must be added in series with the combination to obtain a total resistance of 10 Ω. If the complete circuit expends a power of 0.36 kW, find the total current flowing. [2.5 Ω, 6 A]
11. (a) Calculate the current flowing in the 30 Ω resistor shown in Fig. 5.34.
(b) What additional value of resistance would have to be placed in parallel with the 20 Ω and 30 Ω resistors to change the supply current to 8 A, the supply voltage remaining constant.
[(a) 1.6 A (b) 6 Ω]

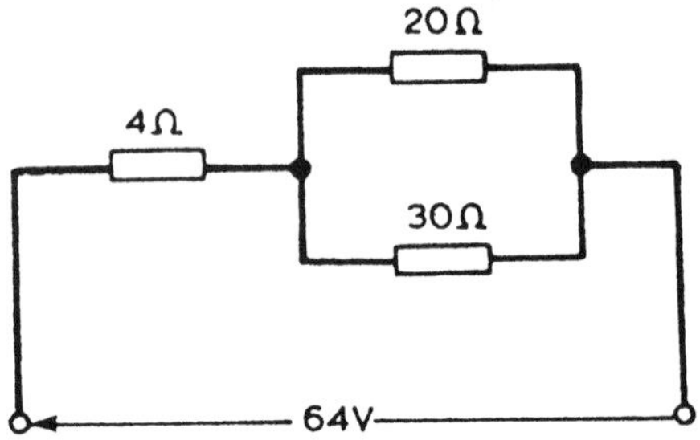

Figure 5.34

12. For the circuit shown in Fig. 5.35, find (a) V_1, (b) V_2, without calculating the current flowing. [(a) 30 V (b) 42 V]

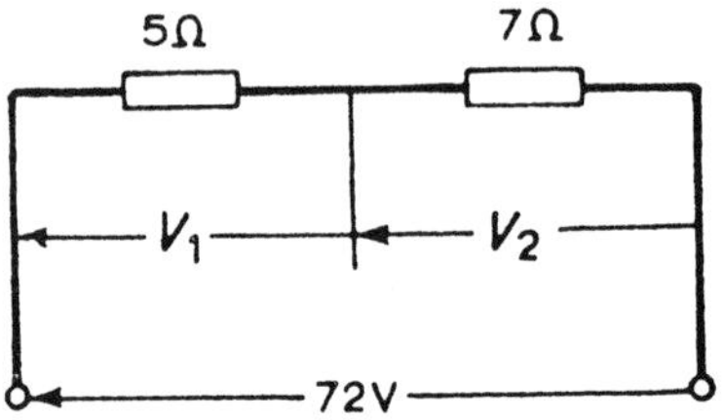

Figure 5.35

13. Determine the currents and voltages indicated in the circuit shown in Fig. 5.36.
[$I_1 = 5$ A, $I_2 = 2.5$ A, $I_3 = 1\frac{2}{3}$ A, $I_4 = \frac{5}{6}$ A, $I_5 = 3$ A, $I_6 = 2$ A, $V_1 = 20$ V, $V_2 = 5$ V, $V_3 = 6$ V]
14. Find the current I in Fig. 5.37. [1.8 A]
15. If four identical lamps are connected in parallel and the combined resistance is 100 Ω, find the resistance of one lamp [400 Ω]

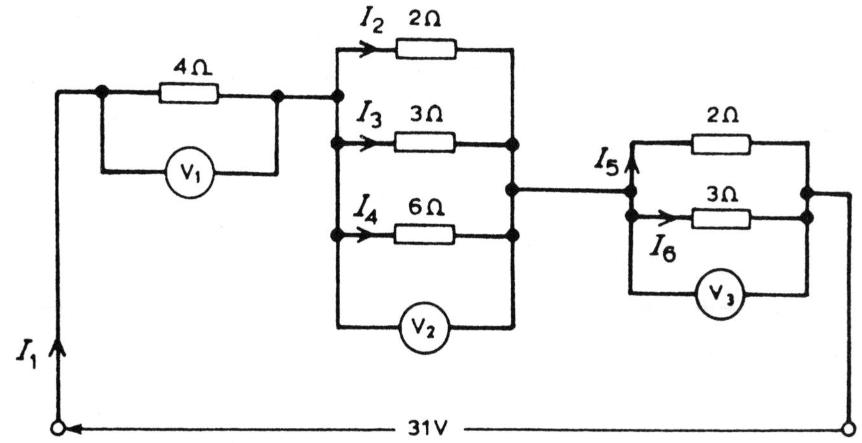

Figure 5.36

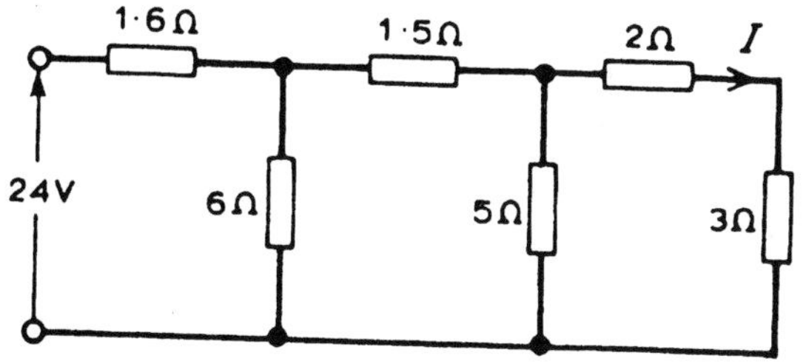

Figure 5.37

16. Three identical filament lamps are connected (a) in series, (b) in parallel across a 210 V supply. State for each connection the p.d. across each lamp. [(a) 70 V (b) 210 V]

6

Capacitors and capacitance

At the end of this chapter you should be able to:

- describe an electrostatic field
- define electric field strength E and state its unit
- define capacitance and state its unit
- describe a capacitor and draw the circuit diagram symbol
- perform simple calculations involving $C = Q/V$ and $Q = It$
- define electric flux density D and state its unit
- define permittivity, distinguishing between ϵ_0, ϵ_r and ϵ
- perform simple calculations involving $D = Q/A$, $E = V/d$ and $D/E = \epsilon_0\epsilon_r$
- understand that for a parallel plate capacitor, $C = \epsilon_0\epsilon_r A(n - 1)/d$
- perform calculations involving capacitors connected in parallel and in series
- define dielectric strength and state its unit
- state that the energy stored in a capacitor is given by $W = \frac{1}{2}CV^2$ joules
- describe practical types of capacitor
- understand the precautions needed when discharging capacitors

6.1 Electrostatic field

Figure 6.1 represents two parallel metal plates, A and B, charged to different potentials. If an electron that has a negative charge is placed between the plates, a force will act on the electron tending to push it away from the negative plate B towards the positive plate A. Similarly, a positive charge would be acted on by a force tending to move it toward the negative plate. Any region such as that shown between the plates in Fig. 6.1, in which an electric charge experiences a force, is called an **electrostatic field**. The direction of the field is defined as that of the force acting on a positive charge placed in the field. In Fig. 6.1, for example, the direction of the force is from the positive plate to the negative plate. Such a field may be represented in magnitude and direction by **lines of electric force** drawn between the charged surfaces. The closeness of the lines is an indication of the field strength. Whenever a p.d. is established between two points, an electric field will always exist. Figure 6.2(a) shows a typical field pattern for an isolated point charge, and Fig. 6.2(b) shows the field pattern for adjacent charges of opposite polarity. Electric lines of force (often called electric flux lines) are continuous and start and finish on point charges. Also, the lines cannot cross each other. When a charged body is placed close to an uncharged body, an induced charge of opposite sign appears on the surface of the uncharged body. This is because lines of force from the charged body terminate on its surface.

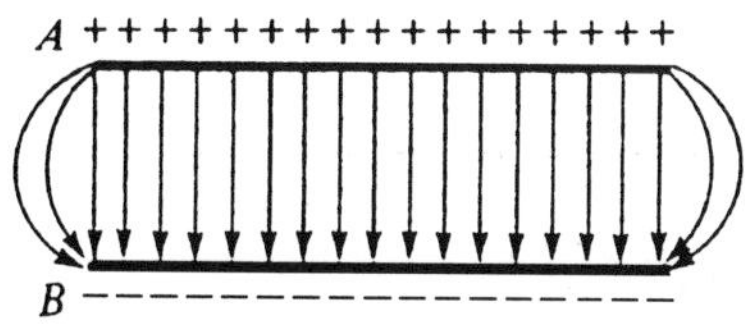

Figure 6.1 Electrostatic field

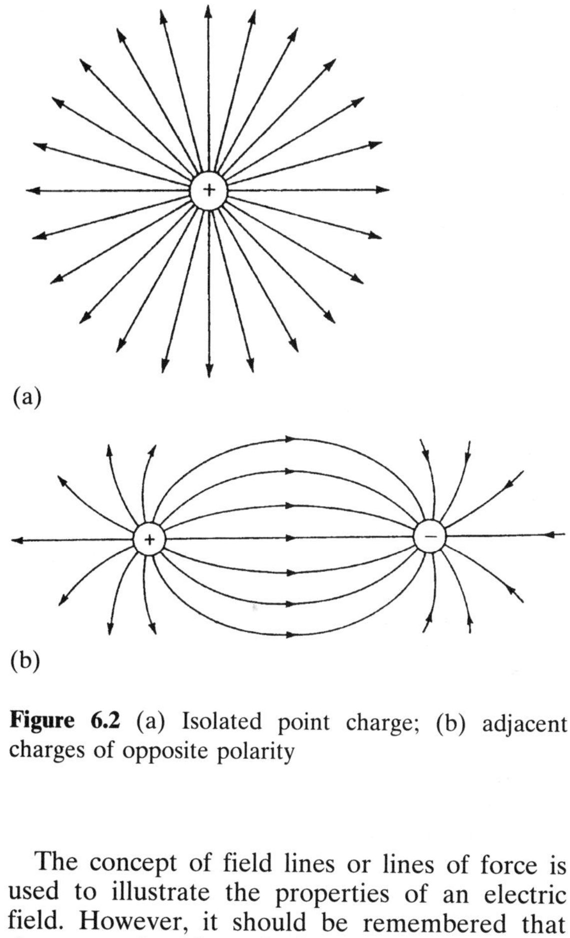

Figure 6.2 (a) Isolated point charge; (b) adjacent charges of opposite polarity

The concept of field lines or lines of force is used to illustrate the properties of an electric field. However, it should be remembered that they are only aids to the imagination.

6.2 Electric field strength

Figure 6.3 shows two parallel conducting plates separated from each other by air. They are connected to opposite terminals of a battery of voltage V volts. There is therefore an electric field in the space between the plates. If the plates are close together, the electric lines of force will be straight and parallel and equally spaced,

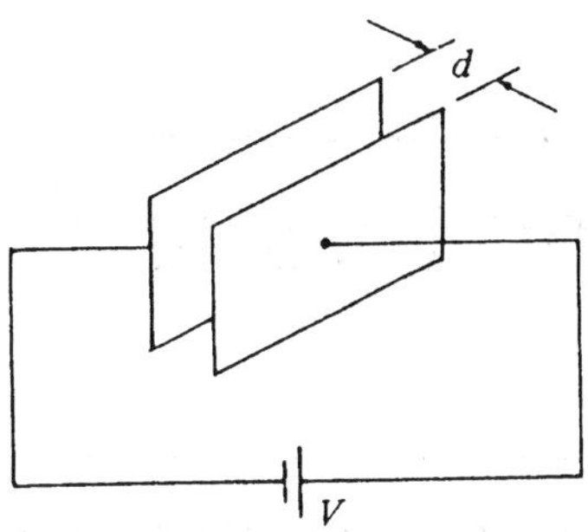

Figure 6.3

except near the edge where fringing will occur (see Fig. 6.1). Over the area in which there is negligible fringing,

Electric field strength, $E = \dfrac{V}{d}$ **volts/metre**

where d is the distance between the plates. Electric field strength is also called **potential gradient**.

6.3 Capacitance

Static electric fields arise from electric charges, electric field lines beginning and ending on electric charges. Thus the presence of the field indicates the presence of equal positive and negative electric charges on the two plates of Fig. 6.3. Let the charge be $+Q$ coulombs on one plate and $-Q$ coulombs on the other. The property of this pair of plates which determines how much charge corresponds to a given p.d. between the plates is called their capacitance:

capacitance $C = \dfrac{Q}{V}$

The **unit of capacitance** is the **farad**, **F** (or more usually $\mu F = 10^{-6}$ F or $pF = 10^{-12}$ F), which is defined as the capacitance when a p.d. of one volt appears across the plates when charged with one coulomb.

6.4 Capacitors

Every system of electrical conductors possesses capacitance. For example, there is capacitance between the conductors of overhead transmission lines and also between the wires of a telephone

Figure 6.4

cable. In these examples the capacitance is undesirable but has to be accepted, minimized or compensated for. There are other situations where capacitance is a desirable property.

Devices specially constructed to possess capacitance are called **capacitors** (or **condensers**, as they used to be called). In its simplest form a capacitor consists of two plates which are separated by an insulating material known as a **dielectric**. A capacitor has the ability to store a quantity of static electricity.

The symbols for a fixed capacitor and a variable capacitor used in electrical circuit diagrams are shown in Fig. 6.4.

The **charge Q** stored in a capacitor is given by:

$$\boldsymbol{Q = I \times t} \textbf{ coulombs}$$

where I is the current in amperes and t the time in seconds.

> Problem 1. (a) Determine the p.d. across a 4 μF capacitor when charged with 5 mC. (b) Find the charge on a 50 pF capacitor when the voltage applied to it is 2 kV.

(a) $C = 4\,\mu\text{F} = 4 \times 10^{-6}\,\text{F}$; $Q = 5\,\text{mC} = 5 \times 10^{-3}\,\text{C}$. Since

$$C = \frac{Q}{V} \text{ then } V = \frac{Q}{C} = \frac{5 \times 10^{-3}}{4 \times 10^{-6}}$$

$$= \frac{5 \times 10^{6}}{4 \times 10^{3}} = \frac{5000}{4}$$

Hence p.d. = 1250 V or 1.25 kV.

(b) $C = 50\,\text{pF} = 50 \times 10^{-12}\,\text{F}$; $V = 2\,\text{kV} = 2000\,\text{V}$. $Q = CV$

$$= 50 \times 10^{-12} \times 2000 = \frac{5 \times 2}{10^{8}} = 0.1 \times 10^{-6}$$

Hence charge = 0.1 μC.

> Problem 2. A direct current of 4 A flows into a previously uncharged 20 μF capacitor for 3 ms. Determine the p.d. between the plates.

$I = 4\,\text{A}$; $C = 20\,\mu\text{F} = 20 \times 10^{-6}\,\text{F}$; $t = 3\,\text{ms} = 3 \times 10^{-3}\,\text{s}$. $Q = It = 4 \times 3 \times 10^{-3}\,\text{C}$.

$$V = \frac{Q}{C} = \frac{4 \times 3 \times 10^{-3}}{20 \times 10^{-6}} = \frac{12 \times 10^{6}}{20 \times 10^{-3}}$$

$$= 0.6 \times 10^{3} = 600\,\text{V}$$

Hence, the p.d. between the plates is 600 V.

> Problem 3. A 5 μF capacitor is charged so that the p.d. between its plates is 800 V. Calculate how long the capacitor can provide an average discharge current of 2 mA.

$C = 5\,\mu\text{F} = 5 \times 10^{-6}\,\text{F}$; $V = 800\,\text{V}$; $I = 2\,\text{mA} = 2 \times 10^{-3}\,\text{A}$. $Q = CV = 5 \times 10^{-6} \times 800 = 4 \times 10^{-3}\,\text{C}$. Also, $Q = It$. Thus

$$t = \frac{Q}{I} = \frac{4 \times 10^{-3}}{2 \times 10^{-3}} = 2\,\text{s}$$

Hence the capacitor can provide an average discharge current of 2 mA for 2 s.

6.5 Electric flux density

Unit flux is defined as emanating from a positive charge of 1 coulomb. Thus electric flux ψ is measured in coulombs, and for a charge of Q coulombs, the flux $\psi = Q$ coulombs.

Electric flux density D is the amount of flux passing through a defined area A that is perpendicular to the direction of the flux:

electric flux density,

$$\boldsymbol{D = \frac{Q}{A}} \textbf{ coulombs/metre}^2$$

Electric flux density is also called **charge density**, σ.

6.6 Permittivity

At any point in an electric field, the electric field strength E maintains the electric flux and produces a particular value of electric flux density D at that point. For a field established in **vacuum** (or for practical purposes in air), the ratio D/E is a constant ϵ_0, i.e.

$$\frac{\boldsymbol{D}}{\boldsymbol{E}} = \epsilon_0$$

where ϵ_0 is called the **permittivity of free space** or the free space constant. The value of ϵ_0 is 8.85×10^{-12} F/m.

When an insulating medium, such as mica, paper, plastic or ceramic, is introduced into the region of an electric field the ratio of D/E is modified:

$$\boldsymbol{\frac{D}{E} = \epsilon_0\epsilon_r}$$

where ϵ_r, the **relative permittivity** of the insulating material, indicates its insulating power compared with that of vacuum:

relative permittivity,

$$\epsilon_r = \frac{\textbf{flux density in material}}{\textbf{flux density in vacuum}}$$

ϵ_r has no unit. Typical values of ϵ_r include air, 1.00; polythene, 2.3; mica, 3–7; glass, 5–10; water, 80; ceramics, 6–1000.

The product $\epsilon_0\epsilon_r$ is called the **absolute permittivity**, ϵ, i.e.

$$\boldsymbol{\epsilon = \epsilon_0\epsilon_r}$$

The insulating medium separating charged surfaces is called a **dielectric**. Compared with conductors, dielectric materials have very high resistivities. They are therefore used to separate conductors at different potentials, such as capacitor plates or electric power lines.

Problem 4. Two parallel rectangular plates measuring 20 cm by 40 cm carry an electric charge of 0.2 μC. Calculate the electric flux density. If the plates are spaced 5 mm apart and the voltage between them is 0.25 kV determine the electric field strength.

Charge Q = 0.2 μC = 0.2×10^{-6} C; area A = 20 cm × 40 cm = 800 cm^2 = 800×10^{-4} m^2.
Electric flux density is given by

$$D = \frac{Q}{A} = \frac{0.2 \times 10^{-6}}{800 \times 10^{-4}} = \frac{0.2 \times 10^{4}}{800 \times 10^{6}}$$

$$= \frac{2000}{800} \times 10^{-6}$$

$$= \mathbf{2.5\ \mu C/m^2}$$

Voltage V = 0.25 kV = 250 V; plate spacing, d = 5 mm = 5×10^{-3} m.
Electric field strength is given by

$$E = \frac{V}{d} = \frac{250}{5 \times 10^{-3}} = \mathbf{50\ kV/m}$$

Problem 5. The flux density between two plates separated by mica of relative permittivity 5 is 2 μC/m^2. Find the voltage gradient between the plates.

Flux density D = 2 μC/m^2 = 2×10^{-6} C/m^2; ϵ_0 = 8.85×10^{-12} F/m; ϵ_r = 5. $(D/E) = \epsilon_0\epsilon_r$, hence

$$\text{voltage gradient } E = \frac{D}{\epsilon_0\epsilon_r}$$

$$= \frac{2 \times 10^{-6}}{8.85 \times 10^{-12} \times 5} \text{ V/m} = \mathbf{45.2\ kV/m}$$

Problem 6. Two parallel plates having a p.d. of 200 V between them are spaced 0.8 mm apart. What is the electric field strength? Find also the flux density when the dielectric between the plates is (a) air, and (b) polythene of relative permittivity 2.3.

Electric field strength $E = (V/d) = (200/(0.8 \times 10^{-3}))$ = 250 kV/m.

(a) For air: $\epsilon_r = 1$; $(D/E) = \epsilon_0\epsilon_r$, hence

electric flux density D

$$= E\epsilon_0\epsilon_r$$

$$= 250 \times 10^3 \times 8.85 \times 10^{-12} \times 1 \text{ C/m}^2$$

$$= \mathbf{2.213\ \mu C/m^2}$$

(b) For polythene, $\epsilon_r = 2.3$. Electric flux density is given by

$$D = E\epsilon_0\epsilon_r = 250 \times 10^3 \times 8.85 \times 10^{-12} \times 2.3 \text{ C/m}^2$$

$$= \mathbf{5.089\ \mu C/m^2}$$

6.7 The parallel-plate capacitor

For a parallel-plate capacitor, as shown in Fig. 6.5(a), experiments show that capacitance C is proportional to the area A of a plate, inversely proportional to the plate spacing d (i.e. the

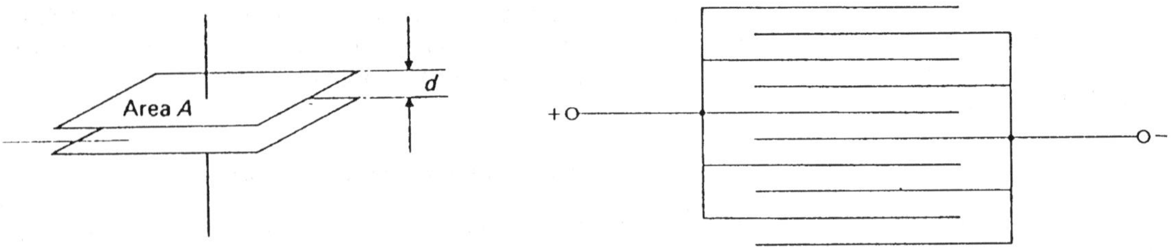

(a) (b)

Figure 6.5

dielectric thickness) and depends on the nature of the dielectric:

capacitance $\boxed{C = \dfrac{\epsilon_0 \epsilon_r A}{d} \text{ farads}}$

where ϵ_0 = 8.85×10^{-12} F/m (constant)
ϵ_r = relative permittivity
A = area of one of the plates, in m^2, and
d = thickness of dielectric in m

Another method used to increase the capacitance is to interleave several plates as shown in Fig. 6.5(b). Ten plates are shown, forming nine capacitors with a capacitance nine times that of one pair of plates.

If such an arrangement has n plates then capacitance $C \propto (n - 1)$. Thus

capacitance $\boxed{C = \dfrac{\epsilon_0 \epsilon_r A(n-1)}{d} \text{ farads}}$

> Problem 7. (a) A ceramic capacitor has an effective plate area of 4 cm² separated by 0.1 mm of ceramic of relative permittivity 100. Calculate the capacitance of the capacitor in picofarads. (b) If the capacitor in part (a) is given a charge of 1.2 μC what will be the p.d. between the plates?

(a) Area $A = 4\text{ cm}^2 = 4 \times 10^{-4}\text{ m}^2$; $d = 0.1$ mm $= 0.1 \times 10^{-3}$ m; $\epsilon_0 = 8.85 \times 10^{-12}$ F/m; $\epsilon_r = 100$. Capacitance is given by

$$C = \frac{\epsilon_0 \epsilon_r A}{d} \text{ farads}$$

$$= \frac{8.85 \times 10^{-12} \times 100 \times 4 \times 10^{-4}}{0.1 \times 10^{-3}} \text{ F}$$

$$= \frac{8.85 \times 4}{10^{10}} \text{ F} = \frac{8.85 \times 4 \times 10^{12}}{10^{10}} \text{ pF}$$

$$= \mathbf{3540\ pF}$$

(b) $Q = CV$ thus $V = \dfrac{Q}{C} = \dfrac{1.2 \times 10^{-6}}{3540 \times 10^{-12}}$ V

$$= \mathbf{339\ V}$$

> Problem 8. A waxed paper capacitor has two parallel plates, each of effective area 800 cm². If the capacitance of the capacitor is 4425 pF determine the effective thickness of the paper if its relative permittivity is 2.5.

$A = 800\text{ cm}^2 = 800 \times 10^{-4}\text{ m}^2 = 0.08\text{ m}^2$; $C = 4425\text{ pF} = 4425 \times 10^{-12}$ F; $\epsilon_0 = 8.85 \times 10^{-12}$ F/m; $\epsilon_r = 2.5$. Since $C = \epsilon_0 \epsilon_r A/d$ then $d = \epsilon_0 \epsilon_r A/C$, hence

$$d = \frac{8.85 \times 10^{-12} \times 2.5 \times 0.08}{4425 \times 10^{-12}} = 0.0004 \text{ m}$$

Hence the thickness of the paper is 0.4 mm.

> Problem 9. A parallel-plate capacitor has 19 interleaved plates each 75 mm by 75 mm separated by mica sheets 0.2 mm thick. Assuming the relative permittivity of the mica is 5, calculate the capacitance of the capacitor.

$n = 19$; $n - 1 = 18$; $A = 75 \times 75 = 5625\text{ mm}^2 = 5625 \times 10^{-6}\text{ m}^2$; $\epsilon_r = 5$; $\epsilon_0 = 8.85 \times 10^{-12}$ F/m; $d = 0.2\text{ mm} = 0.2 \times 10^{-3}$ m. Capacitance is given by

$$C = \frac{\epsilon_0 \epsilon_r A(n-1)}{d}$$

$$= \frac{8.85 \times 10^{-12} \times 5 \times 5625 \times 10^{-6} \times 18}{0.2 \times 10^{-3}} \text{ F}$$

= **0.0224 μF** or **22.4 nF**

6.8 Capacitors connected in parallel and series

(a) Capacitors connected in parallel

Figure 6.6 shows three capacitors, C_1, C_2 and C_3, connected in parallel with a supply voltage V applied across the arrangement.

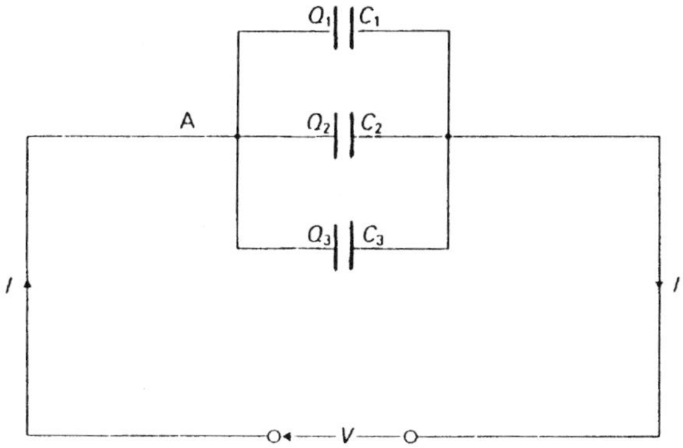

Total charge, $Q_T = Q_1 + Q_2 + Q_3$

Figure 6.6

When the charging current I reaches point A it divides, some flowing into C_1, some flowing into C_2 and some into C_3. Hence the total charge Q_T $(= I \times t)$ is divided between the three capacitors. The capacitors each store a charge and these are shown as Q_1, Q_2 and Q_3, respectively. Hence

$$Q_T = Q_1 + Q_2 + Q_3$$

But $Q_T = CV$, $Q_1 = C_1V$, $Q_2 = C_2V$ and $Q_3 = C_3V$; therefore $CV = C_1V + C_2V + C_3V$ where C is the total equivalent circuit capacitance, i.e.

$$\mathbf{C = C_1 + C_2 + C_3}$$

It follows that for n parallel-connected capacitors,

$$\boxed{C = C_1 + C_2 + C_3 \ldots\ldots + C_n}$$

i.e. the equivalent capacitance of a group of parallel-connected capacitors is the sum of the capacitances of the individual capacitors. (Note that this formula is similar to that used for **resistors** connected in **series**.)

(b) Capacitors connected in series

Figure 6.7 shows three capacitors, C_1, C_2 and C_3, connected in series across a supply voltage V. Let the p.d. across the individual capacitors be V_1, V_2 and V_3, respectively, as shown.

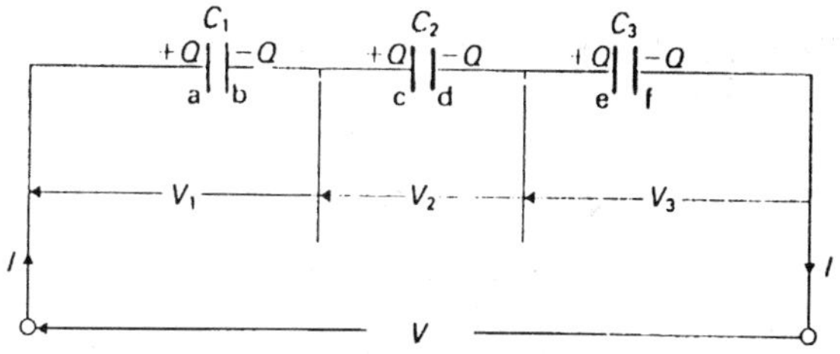

Charge on each capacitor = Q

Figure 6.7

Let the charge on plate 'a' of capacitor C_1 be $+Q$ coulombs. This induces an equal but opposite charge of $-Q$ coulombs on plate 'b'. The conductor between plates 'b' and 'c' is electrically isolated from the rest of the circuit so that an equal but opposite charge of $+Q$ coulombs must appear on plate 'c', which, in turn, induces an equal and opposite charge of $-Q$ coulombs on plate 'd', and so on.

Hence when capacitors are connected in series the charge on each is the same. In a series circuit

$$V = V_1 + V_2 + V_3$$

Since $V = \frac{Q}{C}$ then $\frac{Q}{C} = \frac{Q}{C_1} + \frac{Q}{C_2} + \frac{Q}{C_3}$

where C is the total equivalent circuit capacitance, i.e.

$$\mathbf{\frac{1}{C} = \frac{1}{C_1} + \frac{1}{C_2} + \frac{1}{C_3}}$$

It follows that for n series-connected capacitors:

$$\boxed{\frac{1}{C} = \frac{1}{C_1} + \frac{1}{C_2} + \frac{1}{C_3} + \ldots + \frac{1}{C_n}}$$

i.e. for series-connected capacitors, the reciprocal of the equivalent capacitance is equal to the sum of the reciprocals of the individual capacitances. (Note that this formula is similar to that used for **resistors** connected in **parallel**.)

For the special case of two capacitors in series:

$$\frac{1}{C} = \frac{1}{C_1} + \frac{1}{C_2} = \frac{C_2 + C_1}{C_1 C_2}$$

Hence

$$\boldsymbol{C = \frac{C_1 C_2}{C_1 + C_2}} \quad \left(\text{i.e. } \frac{\text{product}}{\text{sum}}\right)$$

> Problem 10. Calculate the equivalent capacitance of two capacitors of 6 μF and 4 μF connected (a) in parallel and (b) in series.

(a) In parallel, equivalent capacitance $C = C_1 + C_2 = 6\ \mu\text{F} + 4\ \mu\text{F} =$ **10 μF**.

(b) In series, equivalent capacitance C is given by

$$C = \frac{C_1 C_2}{C_1 + C_2}$$

This formula is used for the special case of two capacitors in series. Thus

$$C = \frac{6 \times 4}{6 + 4} = \frac{24}{10} = \mathbf{2.4\ \mu F}$$

> Problem 11. What capacitance must be connected in series with a 30 μF capacitor for the equivalent capacitance to be 12 μF?

Let $C = 12\ \mu\text{F}$ (the equivalent capacitance), $C_1 = 30\ \mu\text{F}$ and C_2 be the unknown capacitance. For two capacitors in series

$$\frac{1}{C} = \frac{1}{C_1} + \frac{1}{C_2}$$

Hence

$$\frac{1}{C_2} = \frac{1}{C} - \frac{1}{C_1} = \frac{C_1 - C}{CC_1}$$

$$C_2 = \frac{CC_1}{C_1 - C} = \frac{12 \times 30}{30 - 12} = \frac{360}{18} = \mathbf{20\ \mu F}$$

> Problem 12. Capacitances of 1 μF, 3 μF, 5 μF and 6 μF are connected in parallel to a direct voltage supply of 100 V. Determine (a) the equivalent circuit capacitance, (b) the total charge and (c) the charge on each capacitor.

(a) The equivalent capacitance C for four capacitors in parallel is given by:

$C = C_1 + C_2 + C_3 + C_4$

i.e.

$C = 1 + 3 + 5 + 6 =$ **15 μF**

(b) Total charge $Q_T = CV$ where C is the equivalent circuit capacitance, i.e.

$Q_T = 15 \times 10^{-6} \times 100 = 1.5 \times 10^{-3}$ C
$=$ **1.5 mC**

(c) The charge on the 1 μF capacitor $Q_1 = C_1 V = 1 \times 10^{-6} \times 100 =$ **0.1 mC**.
The charge on the 3 μF capacitor $Q_2 = C_2 V = 3 \times 10^{-6} \times 100 =$ **0.3 mC**.
The charge on the 5 μF capacitor $Q_3 = C_3 V = 5 \times 10^{-6} \times 100 =$ **0.5 mC**.
The charge on the 6 μF capacitor $Q_4 = C_4 V = 6 \times 10^{-6} \times 100 =$ **0.6 mC**.
(Check: In a parallel circuit $Q_T = Q_1 + Q_2 + Q_3 + Q_4$;

$Q_1 + Q_2 + Q_3 + Q_4 = 0.1 + 0.3 + 0.5 + 0.6$
$= 1.5$ mC $= Q_T$.)

> Problem 13. Capacitances of 3 μF, 6 μF and 12 μF are connected in series across a 350 V supply. Calculate (a) the equivalent circuit capacitance, (b) the charge on each capacitor, and (c) the p.d. across each capacitor.

The circuit diagram is shown in Fig. 6.8.

(a) The equivalent circuit capacitance C for three capacitors in series is given by

$$\frac{1}{C} = \frac{1}{C_1} + \frac{1}{C_2} + \frac{1}{C_3}$$

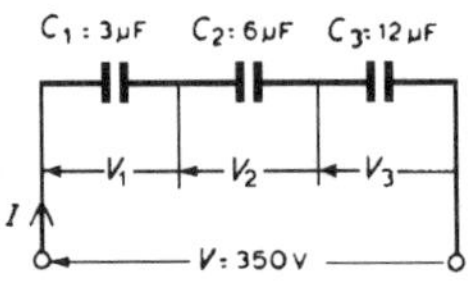

Figure 6.8

i.e.

$$\frac{1}{C} = \frac{1}{3} + \frac{1}{6} + \frac{1}{12} = \frac{4+2+1}{12} = \frac{7}{12}$$

Hence the equivalent circuit capacitance

$$C = \frac{12}{7} = 1\frac{5}{7}\ \mu F$$

(b) Total charge $Q_T = CV$, hence

$$Q_T = \frac{12}{7} \times 10^{-6} \times 350 = 600\ \mu C \text{ or } 0.6\ mC$$

Since the capacitors are connected in series 0.6 mC is the charge on each of them.

(c) The voltage across the 3 μF capacitor is given by

$$V_1 = \frac{Q}{C_1} = \frac{0.6 \times 10^{-3}}{3 \times 10^{-6}} = 200\ V$$

The voltage across the 6 μF capacitor is given by

$$V_2 = \frac{Q}{C_2} = \frac{0.6 \times 10^{-3}}{6 \times 10^{-6}} = 100\ V$$

The voltage across the 12 μF capacitor is given by

$$V_3 = \frac{Q}{C_3} = \frac{0.6 \times 10^{-3}}{12 \times 10^{-6}} = 50\ V$$

(Check: In a series circuit $V = V_1 + V_2 + V_3$;

$V_1 + V_2 + V_3 = 200 + 100 + 50 = 350\ V =$ supply voltage.)

In practice, capacitors are rarely connected in series unless they are of the same capacitance. The reason for this can be seen from the above problem where the lowest valued capacitor (i.e. 3 μF) has the highest p.d. across it (i.e. 200 V) which means that if all the capacitors have an identical construction they must all be rated at the highest voltage.

Problem 14. For the arrangement shown in Fig. 6.9 find (a) the equivalent capacitance of the circuit, (b) the voltage across QR, and (c) the charge on each capacitor.

(a) 2 μF in parallel with 3 μF gives an equivalent capacitance of 2 μF + 3 μF = 5 μF. The circuit is now as shown in Fig. 6.10. The

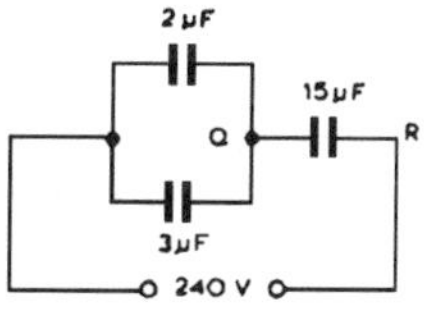

Figure 6.9

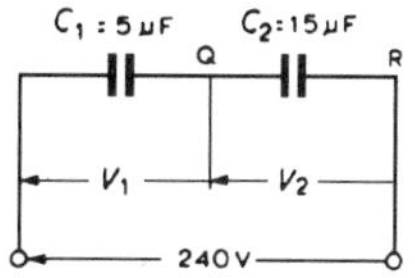

Figure 6.10

equivalent capacitance of 5 μF in series with 15 μF is given by

$$\frac{5 \times 15}{5 + 15}\ \mu F = \frac{75}{20} \text{ or } \mathbf{3.75\ \mu F}$$

(b) The charge on each of the capacitors shown in Fig. 6.10 will be the same since they are connected in series. Let this charge be Q coulombs. Then

$$Q = C_1V_1 = C_2V_2$$

i.e. $5V_1 = 15V_2$

$$V_1 = 3V_2 \qquad (6.1)$$

Also $V_1 + V_2 = 240$ V; hence $3V_2 + V_2 = 240$ V from (6.1), thus

$V_2 = 60$ V and $V_1 = 180$ V

Hence the voltage across QR is 60 V.

(c) The charge on the 15 μF capacitor is $C_2V_2 = 15 \times 10^{-6} \times 60 =$ **0.9 mC**.
The charge on the 2 μF capacitor is $2 \times 10^{-6} \times 180 =$ **0.36 mC**.
The charge on the 3 μF capacitor is $3 \times 10^{-6} \times 180 =$ **0.54 mC**.

6.9 Dielectric strength

The maximum amount of field strength that a dielectric can withstand is called the dielectric strength of the material.

Dielectric strength $\boxed{E_m = \dfrac{V_m}{d}}$

Problem 15. A capacitor is to be constructed so that its capacitance is 0.2 μF and to take a p.d. of 1.25 kV across its terminals. The dielectric is to be mica which, after allowing a safety factor of 2, has a dielectric strength of 50 MV/m. Find (a) the thickness of the mica needed, and (b) the area of a plate assuming a two-plate construction. (Assume ϵ_r for mica to be 6.)

(a) Dielectric strength, $E = V/d$, i.e.

$$d = \frac{V}{E} = \frac{1.25 \times 10^3}{50 \times 10^6} \text{ m} = \mathbf{0.025\ m}$$

(b) Capacitance, $C = \epsilon_0 \epsilon_r A/d$, hence

$$\text{area } A = \frac{Cd}{\epsilon_0 \epsilon_r}$$

$$= \frac{0.2 \times 10^{-6} \times 0.025 \times 10^{-3}}{8.85 \times 10^{-12} \times 6} \text{ m}^2$$

$$= 0.094\,16 \text{ m}^2 = \mathbf{941.6\ cm^2}$$

6.10 Energy stored

The energy, W, stored by a capacitor is given by

$$\mathbf{W = \frac{1}{2} CV^2 \text{ joules}}$$

Problem 16. (a) Determine the energy stored in a 3 μF capacitor when charged to 400 V. (b) Find also the average power developed if this energy is dissipated in a time of 10 μs.

(a) Energy stored $W = \frac{1}{2}CV^2$ joules

$$= \frac{1}{2} \times 3 \times 10^{-6} \times 400^2 = \frac{3}{2} \times 16 \times 10^{-2}$$

$$= 0.24 \text{ J}$$

(b) $\text{Power} = \dfrac{\text{Energy}}{\text{time}} = \dfrac{0.24}{10 \times 10^{-6}}$ watts = **24 kW**

Problem 17. A 12 μF capacitor is required to store 4 J of energy. Find the p.d. to which the capacitor must be charged.

Energy stored $W = \frac{1}{2}CV^2$ hence $V^2 = 2W/C$ and

$$V = \sqrt{\left(\frac{2W}{C}\right)} = \sqrt{\left(\frac{2 \times 4}{12 \times 10^{-6}}\right)} = \sqrt{\left(\frac{2 \times 10^6}{3}\right)}$$

$$= \mathbf{816.5\ V}$$

Problem 18. A capacitor is charged with 10 mC. If the energy stored is 1.2 J find (a) the voltage and (b) the capacitance.

Energy stored $W = \frac{1}{2}CV^2$ and $C = Q/V$, hence

$$W = \frac{1}{2}\left(\frac{Q}{V}\right)V^2 = \frac{1}{2} QV$$

from which, $V = 2W/Q$
$Q = 10 \text{ mC} = 10 \times 10^{-3}$ C and $W = 1.2$ J.

(a) Voltage $V = \dfrac{2W}{Q} = \dfrac{2 \times 1.2}{10 \times 10^{-3}} =$ **0.24 kV** or **240 V**

(b) Capacitance $C = \dfrac{Q}{V} = \dfrac{10 \times 10^{-3}}{240}$ F

$$= \frac{10 \times 10^6}{240 \times 10^3} \text{ μF} = \mathbf{41.67\ μF}$$

6.11 Practical types of capacitor

Practical types of capacitor are characterized by the material used for their dielectric. The main types include: variable air, mica, paper, ceramic, plastic and electrolytic.

1. **Variable air capacitors.** These usually consist of two sets of metal plates (such as aluminium) one fixed, the other variable. The set of moving plates rotate on a spindle as shown by the end view of Fig. 6.11. As the moving plates are rotated through half a revolution, the meshing, and therefore the capacitance, varies from a minimum to a maximum value. Variable air capacitors are used in radio and electronic circuits where very low losses are required, or where a

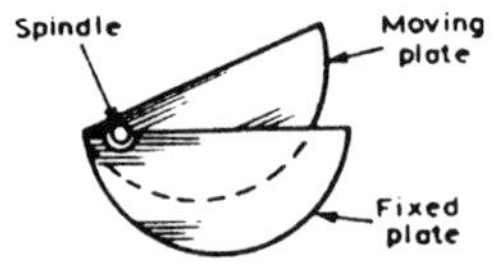

Figure 6.11

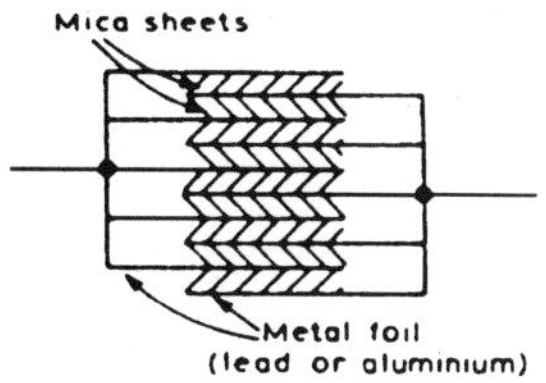

Figure 6.12

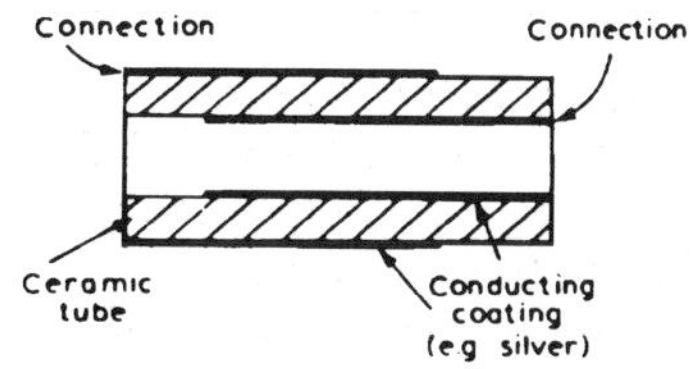

Figure 6.14

variable capacitance is needed. The maximum value of such capacitors is between 500 pF and 1000 pF.

2. **Mica capacitors.** A typical older type construction is shown in Fig. 6.12. Usually the whole capacitor is impregnated with wax and placed in a bakelite case. Mica is easily obtained in thin sheets and is a good insulator. However, mica is expensive and is not used in capacitors above about 0.1 μF. A modified form of mica capacitor is the silvered mica type. The mica is coated on both sides with a thin layer of silver which forms the plates. Capacitance is stable and less likely to change with age. Such capacitors have a constant capacitance with change of temperature, a high working voltage rating and a long service life and are used in high frequency circuits with fixed values of capacitance up to about 1000 pF.
3. **Paper capacitors.** A typical paper capacitor is shown in Fig. 6.13 where the length of the roll corresponds to the capacitance required. The whole is usually impregnated with oil or wax to exclude moisture, and then placed in a plastic or aluminium container for protection. Paper capacitors are made in various working voltages up to about 10 μF. Disadvantages of paper capacitors include variation in capacitance with temperature change and a shorter service life than most other types of capacitor.

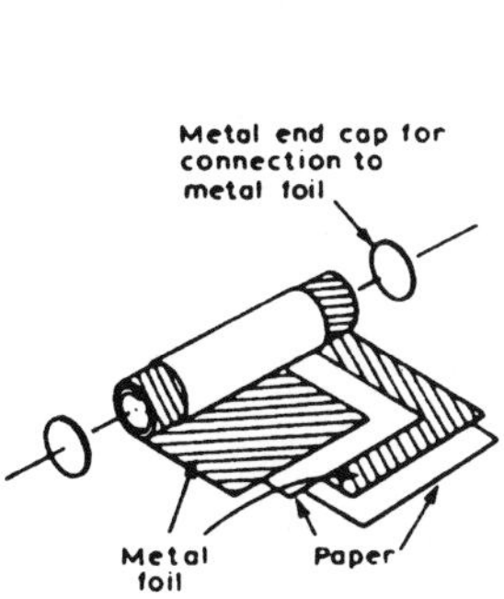

Figure 6.13

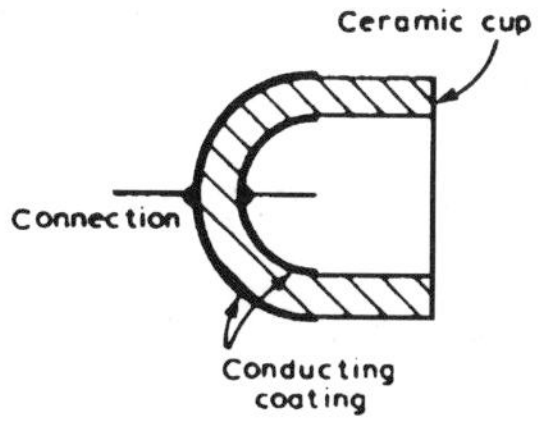

Figure 6.15

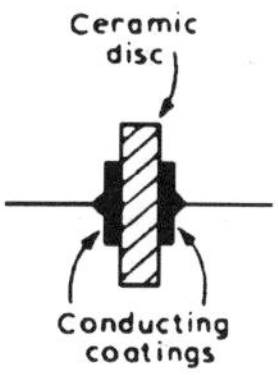

Figure 6.16

4. **Ceramic capacitors.** These are made in various forms, each type of construction depending on the value of capacitance required. For high values, a tube of ceramic material is used as shown in the cross-section of Fig. 6.14. For smaller values the cup construction is used as shown in Fig. 6.15, and for still smaller values the disc construction shown in Fig. 6.16 is used. Certain ceramic materials have a very high permittivity and this enables capacitors of high capacitance to be made which are of small physical size with a high working voltage rating. Ceramic capacitors are available in the range 1 pF to 0.1 μF and may be used in high-frequency electronic circuits subject to a wide range of temperatures.
5. **Plastic capacitors.** Some plastic materials such as polystyrene and Teflon can be used as dielectrics. Construction is similar to the

paper capacitor but using a plastic film instead of paper. Plastic capacitors operate well under conditions of high temperature, provide a precise value of capacitance, a very long service life and high reliability.

6. **Electrolytic capacitors.** Construction is similar to the paper capacitor with aluminium foil used for the plates and with a thick absorbent material, such as paper, impregnated with an electrolyte (ammonium borate), separating the plates. The finished capacitor is usually assembled in an aluminium container and hermetically sealed. Its operation depends on the formation of a thin aluminium oxide layer on the positive plate by electrolytic action when a suitable direct potential is maintained between the plates. This oxide layer is very thin and forms the dielectric. (The absorbent paper between the plates is a conductor and does not act as a dielectric.) Such capacitors must always be used on d.c. and must be connected with the correct polarity; if this is not done the capacitor will be destroyed since the oxide layer will be destroyed. Electrolytic capacitors are manufactured with working voltage from 6 V to 500 V, although accuracy is generally not very high. These capacitors possess a much larger capacitance than other types of capacitors of similar dimensions due to the oxide film being only a few microns thick. The fact that they can be used only on d.c. supplies limits their usefulness.

6.12 Discharging capacitors

When a capacitor has been disconnected from the supply it may still be charged and it may retain this charge for some considerable time. Thus precautions must be taken to ensure that the capacitor is automatically discharged after the supply is switched off. This is done by connecting a high value resistor across the capacitor terminals.

6.13 Multi-choice questions on capacitors and capacitance

(Answers on page 283.)

1. Electrostatics is a branch of electricity concerned with
   (a) energy flowing across a gap between conductors;
   (b) charges at rest;
   (c) charges in motion;
   (d) energy in the form of charges.
2. The capacitance of a capacitor is the ratio
   (a) charge to p.d. between plates;
   (b) p.d. between plates to plate spacing;
   (c) p.d. between plates to thickness of dielectric;
   (d) p.d. between plates to charge.
3. The p.d. across a 10 μF capacitor to charge it with 10 mC is
   (a) 100 V; (b) 1 kV; (c) 1 V; (d) 10 V.
4. The charge on a 10 pF capacitor when the voltage applied to it is 10 kV is
   (a) 100 μC; (b) 0.1 C; (c) 0.1 μC; (d) 0.01 μC.
5. Four 2 μF capacitors are connected in parallel. The equivalent capacitance is (a) 8 μF; (b) 0.5 μF.
6. Four 2 μF capacitors are connected in series. The equivalent capacitance is: (a) 8 μF; (b) 0.5 μF.
7. State which of the following is false. The capacitance of a capacitor
   (a) is proportional to the cross-sectional area of the plates;
   (b) is proportional to the distance between the plates;
   (c) depends on the number of plates;
   (d) is proportional to the relative permittivity of the dielectric.
8. State which of the following statements is false.
   (a) An air capacitor is normally a variable type.
   (b) A paper capacitor generally has a shorter service life than most other types of capacitor.
   (c) An electrolytic capacitor must be used only on a.c. supplies.
   (d) Plastic capacitors generally operate satisfactorily under conditions of high temperature.
9. The energy stored in a 10 μF capacitor when charged to 500 V is
   (a) 1.25 mJ; (b) 0.025 μJ; (c) 1.25 J; (d) 1.25 C.
10. The capacitance of a variable air capacitor is at maximum when
    (a) the movable plates half overlap the fixed plates;
    (b) the movable plates are most widely separated from the fixed plates;
    (c) both sets of plates are exactly meshed;
    (d) the movable plates are closer to one side of the fixed plate than to the other.

6.14 Short answer questions on capacitors and capacitance

1. Explain the term 'electrostatics'.
2. Complete the statements:
   Like charges ....; unlike charges ....
3. How can an 'electric field' be established between two parallel metal plates?
4. What is capacitance?
5. State the unit of capacitance.
6. Complete the statement:

   Capacitance $= \frac{\cdots\cdots}{\cdots\cdots}$

7. Complete the statements:

   (a) 1 μF = ... F;

   (b) 1 pF = ... F.

8. Complete the statement:

   Electric field strength $E = \frac{\cdots\cdots}{\cdots\cdots}$

9. Complete the statement:

   Electric flux density $D = \frac{\cdots\cdots}{\cdots\cdots}$

10. Draw the electrical circuit diagram symbol for a capacitor.
11. Name two practical examples where capacitance is present, although undesirable.
12. The insulating material separating the plates of a capacitor is called the ....
13 10 volts applied to a capacitor results in a charge of 5 coulombs. What is the capacitance of the capacitor?
14. Three 3 μF capacitors are connected in parallel. The equivalent capacitance is .... .
15. Three 3 μF capacitors are connected in series. The equivalent capacitance is ..... .
16. State an advantage of series-connected capacitors.
17. Name three factors upon which capacitance depends.
18. What does 'relative permittivity' mean?
19. Define 'permittivity of free space'.
20. Name five types of capacitor commonly used.
21. Sketch a typical rolled paper capacitor.
22. Explain briefly the construction of a variable air capacitor.
23. State three advantages and one disadvantage of mica capacitors.
24. Name two disadvantages of paper capacitors.
25. Between what values of capacitance are ceramic capacitors normally available?
26. What main advantages do plastic capacitors possess?
27. Explain briefly the construction of an electrolytic capacitor.
28. What is the main disadvantage of electrolytic capacitors?
29. Name an important advantage of electrolytic capacitors.
30. What safety precautions should be taken when a capacitor is disconnected from a supply?
31. What is meant by the 'dielectric strength' of a material?
32. State the formula used to determine the energy stored by a capacitor.

6.15 Further questions on capacitors and capacitance

(Where appropriate take ϵ_0 as 8.85×10^{-12} F/m.)

Charge and capacitance

1. Find the charge on a 10 μF capacitor when the applied voltage is 250 V. [2.5 mC]
2. Determine the voltage across a 1000 pF capacitor to charge it with 2 μC. [2 kV]
3. The charge on the plates of a capacitor is 6 mC when the potential between them is 2.4 kV. Determine the capacitance of the capacitor. [2.5 μF]
4. For how long must a charging current of 2 A be fed to a 5 μF capacitor to raise the p.d. between its plates by 500 V? [1.25 ms]
5. A direct current of 10 A flows into a previously uncharged 5 μF capacitor for 1 ms. Determine the p.d. between the plates. [2 kV]
6. A 16 μF capacitor is charged at a constant current of 4 A for 2 min. Calculate the final p.d. across the capacitor and the corresponding charge in coulombs. [30 V, 480 μC]
7. A steady current of 10 A flows into a previously uncharged capacitor for 1.5 ms when the p.d. between the plates is 2 kV. Find the capacitance of the capacitor. [7.5 μF]

Electric field strength, electric flux density and permittivity

8. A capacitor uses a dielectric 0.04 mm thick and operates at 30 V. What is the electric field strength across the dielectric at this voltage? [750 kV/m]

9. A two-plate capacitor has a charge of 25 C. If the effective area of each plate is 5 cm^2 find the electric flux density of the electric field. [50 kC/m^2]
10. A charge of 1.5 μC is carried on two parallel rectangular plates each measuring 60 mm by 80 mm. Calculate the electric flux density. If the plates are spaced 10 mm apart and the voltage between them is 0.5 kV determine the electric field strength. [312.5 $\mu C/m^2$, 50 kV/m]
11. Two parallel plates are separated by a dielectric and charged with 10 μC. Given that the area of each plate is 50 cm^2, calculate the electric flux density in the dielectric separating the plates. [2 mC/m^2]
12. The electric flux density between two plates separated by polystyrene of relative permittivity 2.5 is 5 $\mu C/m^2$. Find the voltage gradient between the plates. [226 kV/m]
13. Two parallel plates having a p.d. of 250 V between them are spaced 1 mm apart. Determine the electric field strength. Find also the electric flux density when the dielectric between the plates is (a) air, and (b) mica of relative permittivity 5. [250 kV/m (a) 2.213 $\mu C/m^2$ (b) 11.063 $\mu C/m^2$]

Parallel-plate capacitor

14. A capacitor consists of two parallel plates each of area 0.01 m^2, spaced 0.1 mm in air. Calculate the capacitance in picofarads. [885 pF]
15. A waxed paper capacitor has two parallel plates, each of effective area 0.2 m^2. If the capacitance is 4000 pF determine the effective thickness of the paper if its relative permittivity is 2. [0.885 mm]
16. Calculate the capacitance of a parallel-plate capacitor having 5 plates, each 30 mm by 20 mm and separated by a dielectric 0.75 mm thick having a relative permittivity of 2.3. [65.14 pF]
17. How many plates has a parallel-plate capacitor having a capacitance of 5 nF, if each plate is 40 mm square and each dielectric is 0.102 mm thick with a relative permittivity of 6. [7]
18. A parallel-plate capacitor is made from 25 plates, each 70 mm by 120 mm interleaved with mica of relative permittivity 5. If the capacitance of the capacitor is 3000 pF determine the thickness of the mica sheet. [2.97 mm]
19. A capacitor is constructed with parallel plates and has a value of 50 pF. What would be the capacitance of the capacitor if the plate area is doubled and the plate spacing is halved? [200 pF]
20. The capacitance of a parallel-plate capacitor is 1000 pF. It has 19 plates, each 50 mm by 30 mm separated by a dielectric of thickness 0.40 mm. Determine the relative permittivity of the dielectric. [1.67]
21. The charge on the square plates of a multiplate capacitor is 80 μC when the potential between them is 5 kV. If the capacitor has 25 plates separated by a dielectric of thickness 0.102 mm and relative permittivity 4.8, determine the width of a plate. [40 mm]
22. A capacitor is to be constructed so that its capacitance is 4250 pF and to operate at a p.d. of 100 V across its terminals. The dielectric is to be polythene ($\epsilon_r = 2.3$) which, after allowing a safety factor, has a dielectric strength of 20 MV/m. Find (a) the thickness of the polythene needed, and (b) the area of a plate. [(a) 0.005 mm (b) 10.44 cm^2]

Capacitors in parallel and in series

23. Capacitors of 2 μF and 6 μF are connected (a) in parallel, and (b) in series. Determine the equivalent capacitance in each case. [(a) 8 μF (b) 1.5 μF]
24. Find the capacitance to be connected in series with a 10 μF capacitor for the equivalent capacitance to be 6 μF. [15 μF]
25. What value of capacitance would be obtained if capacitors of 0.15 μF and 0.10 μF are connected (a) in series, and (b) in parallel. [(a) 0.06 μF (b) 0.25 μF]
26. Two 6 μF capacitors are connected in series with one having a capacitance of 12 μF. Find the total equivalent circuit capacitance. What capacitance must be added in series to obtain a capacitance of 1.2 μF? [2.4 μF, 2.4 μF]
27. Determine the equivalent capacitance when the following capacitors are connected (a) in parallel, and (b) in series:
(i) 2 μF, 4 μF and 8 μF;
(ii) 0.02 μF, 0.05 μF and 0.10 μF;
(iii) 50 pF and 450 pF;
(iv) 0.01 μF and 200 pF.
[(a)(i) 14 μF (ii) 0.17 μF (iii) 500 pF (iv) 0.0102 μF
(b)(i) $1\frac{1}{7}$ μF (ii) 0.0125 μF (iii) 45 pF (iv) 196.1 pF]

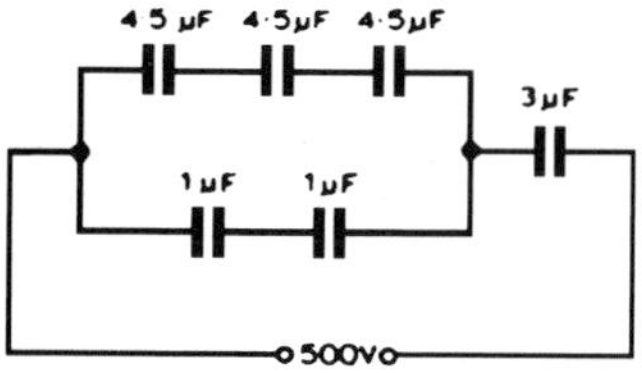

Figure 6.17

28. For the arrangement shown in Fig. 6.17 find (a) the equivalent circuit capacitance, and (b) the voltage across a 4.5 μF capacitor. [(a) 1.2 μF (b) 100 V]
29. Three 12 μF capacitors are connected in series across a 750 V supply. Calculate (a) the equivalent capacitance, (b) the charge on each capacitor, and (c) the p.d. across each capacitor. [(a) 4 μF (b) 3 mC (c) 250 V]
30. If two capacitors having capacitances of 3 μF and 5 μF respectively are connected in series across a 240 V supply determine (a) the p.d. across each capacitor, and (b) the charge on each capacitor. [(a) 150 V, 90 V (b) 0.45 mC on each]
31. In Fig. 6.18 capacitors P, Q and R are identical and the total equivalent capacitance of the circuit is 3 μF. Determine the values of P, Q and R. [4.2 μF each]

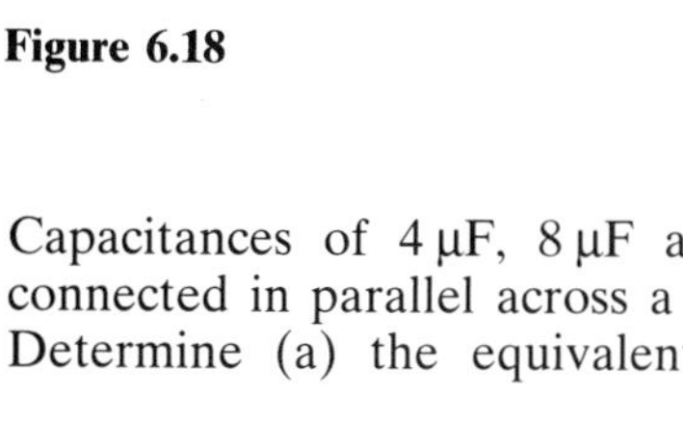

Figure 6.18

32. Capacitances of 4 μF, 8 μF and 16 μF are connected in parallel across a 200 V supply. Determine (a) the equivalent capacitance, (b) the total charge, and (c) the charge on each capacitor. [(a) 28 μF (b) 5.6 mC (c) 0.8 mC, 1.6 mC, 3.2 mC]
33. A circuit consists of two capacitors P and Q in parallel, connected in series with another capacitor R. The capacitances of P, Q and R are 4 μF, 12 μF and 8 μF, respectively. When the circuit is connected across a 300 V d.c. supply find (a) the total capacitance of the circuit, (b) the p.d. across each capacitor, and (c) the charge on each capacitor. [(a) $5\frac{1}{3}$ μF (b) 100 V across P, 100 V across Q, 200 V across R (c) 0.4 mC on P, 1.2 mC on Q, 1.6 mC on R]

Energy stored

34. When a capacitor is connected across a 200 V supply the charge is 4 μC. Find (a) the capacitance, and (b) the energy stored. [(a) 0.02 μF (b) 0.4 mJ]
35. Find the energy stored in a 10 μF capacitor when charged to 2 kV. [20 J]
36. A 3300 pF capacitor is required to store 0.5 mJ of energy. Find the p.d. to which the capacitor must be charged. [550 V]
37. A capacitor is charged with 8 mC. If the energy stored is 0.4 J find (a) the voltage, and (b) the capacitance. [(a) 100 V (b) 80 μF]
38. A capacitor, consisting of two metal plates each of area 50 cm^2 and spaced 0.2 mm apart in air, is connected across a 120 V supply. Calculate (a) the energy stored, (b) the electric flux density, and (c) the potential gradient. [(a) 1.593 μJ (b) 5.31 $\mu C/m^2$ (c) 6000 kV/m]
39. A bakelite capacitor is to be constructed to have a capacitance of 0.04 μF and to have a steady working potential of 1 kV maximum. Allowing a safe value of field stress of 25 MV/m find (a) the thickness of bakelite required, (b) the area of plate required if the relative permittivity of bakelite is 5, (c) the maximum energy stored by the capacitor, and (d) the average power developed if this energy is dissipated in a time of 20 μs. [(a) 0.04 mm (b) 361.6 cm^2 (c) 0.02 J (d) 1 kW]

7

Magnetic circuits

At the end of this chapter you should be able to:

- describe the magnetic field around a permanent magnet
- state the laws of magnetic attraction and repulsion for two magnets in close proximity
- define magnetic flux, Φ, and magnetic flux density, B, and state their units
- perform simple calculations involving $B = \Phi/A$
- define magnetomotive force, F_m, and magnetic field strength, H, and state their units
- perform simple calculations involving $F_m = NI$ and $H = NI/l$
- define permeability, distinguishing between μ_0, μ_r and μ
- understand the B–H curves for different magnetic materials
- appreciate typical values of μ_r
- perform calculations involving $B = \mu_0\mu_r H$
- define reluctance, S, and state its units
- perform calculations involving

  $$S = \frac{\text{m.m.f.}}{\phi} = \frac{l}{\mu_0\mu_r A}$$
- perform calculations on composite series magnetic circuits
- compare electrical and magnetic quantities
- appreciate how a hysteresis loop is obtained and that hysteresis loss is proportional to its area

7.1 Magnetic fields

A **permanent magnet** is a piece of ferromagnetic material (such as iron, nickel or cobalt) which has properties of attracting other pieces of these materials. A permanent magnet will position itself in a north and south direction when freely suspended. The north-seeking end of the magnet is called the **north pole, N**, and the south-seeking end the **south pole, S.**

The area around a magnet is called the **magnetic field** and it is in this area that the effects of the **magnetic force** produced by the magnet can be detected. A magnetic field cannot be seen, felt, smelt or heard and therefore is difficult to represent. Michael Faraday suggested that the magnetic field could be represented pictorially, by imagining the field to consist of **lines of magnetic flux**, which enables investigation of the distribution and density of the field to be carried out.

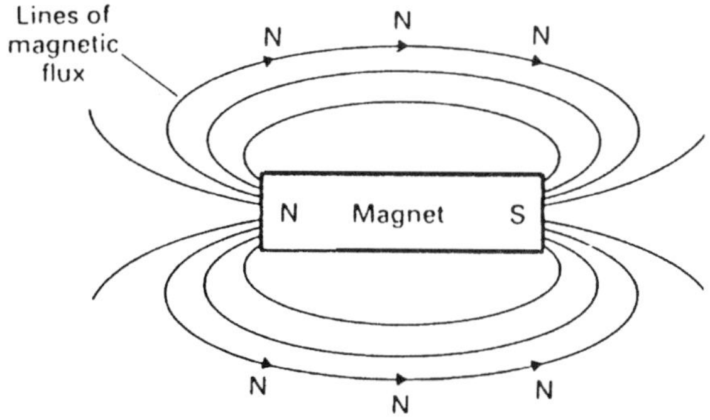

Figure 7.1

The distribution of a magnetic field can be investigated by using some iron fillings. A bar magnet is placed on a flat surface covered by, say, cardboard, upon which is sprinkled some iron filings. If the cardboard is gently tapped the filings will assume a pattern similar to that shown in Fig. 7.1. If a number of magnets of different strength are used, it is found that the stronger the field the closer are the lines of magnetic flux and vice-versa. Thus a magnetic field has the property of exerting a force, demonstrated in this case by causing the iron filings to move into the pattern shown. The strength of the magnetic field decreases as we move away from the magnet. It should be realized, of course, that the magnetic field is three-dimensional in its effect, and not acting in one plane as appears to be the case in this experiment.

If a compass is placed in the magnetic field in various positions, the direction of the lines of flux may be determined by noting the direction of the compass pointer. The direction of a magnetic field at any point is taken as that in which the north-seeking pole of a compass needle points when suspended in the field. The direction of a line of flux is from the north pole to the south pole on the outside of the magnet and is then assumed to continue through the magnet back to the point at which it emerged at the north pole. Thus such lines of flux always form complete closed loops or paths, they never intersect and always have a definite direction. The laws of magnetic attraction and repulsion can be demonstrated by using two bar magnets. In Fig. 7.2(a), **with unlike poles adjacent, attraction takes place.**

Lines of flux are imagined to contract and the magnets try to pull together. The magnetic field

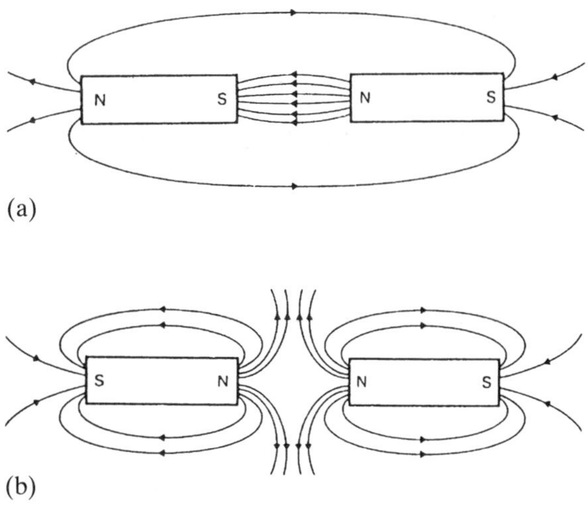

Figure 7.2

is strongest in between the two magnets, shown by the lines of flux being close together. In Fig. 7.2(b), **with similar poles adjacent (i.e. two north poles), repulsion occurs**, i.e. the two north poles try to push each other apart, since magnetic flux lines running side by side in the same direction repel.

7.2 Magnetic flux and flux density

Magnetic flux is the amount of magnetic field (or the number of lines of force) produced by a magnetic source. The symbol for magnetic flux is Φ (Greek letter 'phi'). The unit of magnetic flux is the **weber, Wb**.

Magnetic flux density is the amount of flux passing through a defined area that is perpendicular to the direction of the flux:

$$\textbf{Magnetic flux density} = \frac{\textbf{magnetic flux}}{\textbf{area}}$$

The symbol for magnetic flux density is B. The unit of magnetic flux density is the **tesla, T**, where $1\ T = 1\ Wb/m^2$

Hence

$$B = \frac{\Phi}{A} \text{ tesla}$$

where A (m^2) is the area.

Problem 1. A magnetic pole face has a rectangular section having dimensions 200 mm by 100 mm. If the total flux emerging from the pole is 150 μWb, calculate the flux density.

Flux Φ = 150 μWb = 150×10^{-6} Wb. Cross-sectional area $A = 200 \times 100 = 20\,000\ \text{mm}^2 = 20\,000 \times 10^{-6}\ \text{m}^2$.

$$\text{Flux density } B = \frac{\Phi}{A} = \frac{150 \times 10^{-6}}{20\,000 \times 10^{-6}}$$

$$= \mathbf{0.0075\ T} \text{ or } \mathbf{7.5\ mT}$$

Problem 2. The maximum working flux density of a lifting electromagnet is 1.8 T and the effective area of a pole face is circular in cross-section. If the total magnetic flux produced is 353 mWb, determine the radius of the pole face.

Flux density B = 1.8 T; flux Φ = 353 mWb = 353×10^{-3} Wb. Since $B = \Phi/A$, cross-sectional area $A = \Phi/B$

$$= \frac{353 \times 10^{-3}}{1.8}\ \text{m}^2 = 0.1961\ \text{m}^2$$

The pole face is circular, hence area $= \pi r^2$, where r is the radius; i.e. $\pi r^2 = 0.1961$, from which $r^2 = 0.1961/\pi$ and radius $r = \sqrt{(0.1961/\pi)} = 0.250$ m, i.e. the radius of the pole face is **250 mm**.

7.3 Magnetomotive force and magnetic field strength

Magnetomotive force (m.m.f.) is the cause of the existence of a magnetic flux in a magnetic circuit,

m.m.f. $\quad F_m = NI$ **amperes**

where N is the number of conductors (or turns) and I is the current in amperes. The unit of m.m.f. is sometimes expressed as 'ampere-turns'. However since 'turns' have no dimensions, the SI unit of m.m.f. is the ampere.

Magnetic field strength (or **magnetizing force**),

$H = NI/l$ **ampere per metre**

where l is the mean length of the flux path in metres. Thus **m.m.f. =** $NI = Hl$ ***amperes***.

Problem 3. A magnetizing force of 8000 A/m is applied to a circular magnetic circuit of mean diameter 30 cm by passing a current through a coil wound on the circuit. If the coil is uniformly wound around the circuit and has 750 turns, find the current in the coil.

H = 8000 A/m; $l = \pi d = \pi \times 30 \times 10^{-2}$ m; N = 750 turns; since $H = NI/l$, then

$$I = \frac{Hl}{N} = \frac{8000 \times \pi \times 30 \times 10^{-2}}{750}$$

Thus, current I = 10.05 A.

7.4 Permeability and *B–H* curves

For air, or any non-magnetic medium, the ratio of magnetic flux density to magnetizing force is a constant, i.e. B/H = a constant. This constant is μ_0, the **permeability of free space** (or the magnetic space constant) and is equal to $4\pi \times 10^{-7}$ H/m, i.e. **for air, or any non-magnetic medium**, the ratio

$$B/H = \mu_0$$

(Although all non-magnetic materials, including air, exhibit slight magnetic properties, these can effectively be neglected.)

For all media other than free space,

$$B/H = \mu_0\mu_r$$

where μ_r is the **relative permeability**, and is defined as

$$\mu_r = \frac{\textbf{flux density in material}}{\textbf{flux density in a vacuum}}$$

μ_r varies with the type of magnetic material and, since it is a ratio of flux densities, it has no unit. From its definition, μ_r for a vacuum is 1. $\mu_0\mu_r = \mu$, called the **absolute permeability**.

By plotting measured values of flux density B against magnetic field strength H, a **magnetization curve** (or ***B–H* curve**) is produced. For non-

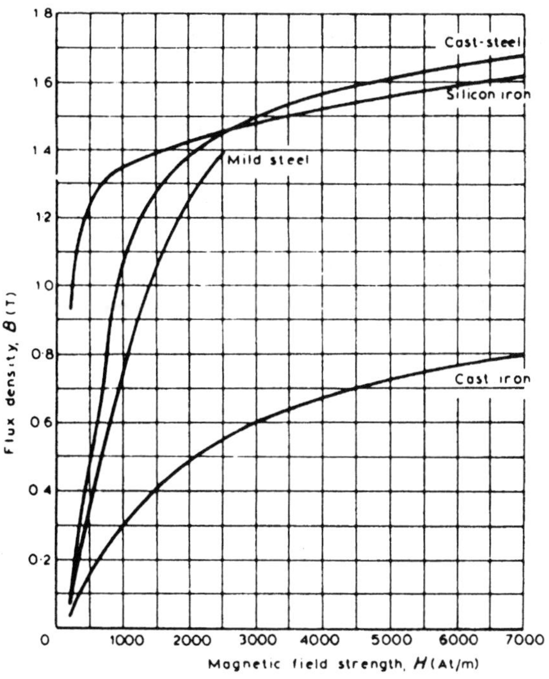

Figure 7.3 *B–H* curves for four materials

magnetic materials this is a straight line. Typical curves for four magnetic materials are shown in Fig. 7.3.

The **relative permeability** of a ferromagnetic material is proportional to the slope of the *B–H* curve and thus varies with the magnetic field strength. The approximate range of values of relative permeability μ_r for some common magnetic materials are:

Cast iron	$\mu_r = 100–250$
Silicon iron	$\mu_r = 1000–5000$
Mumetal	$\mu_r = 200–5000$
Mild steel	$\mu_r = 200–800$
Cast steel	$\mu_r = 300–900$
Stalloy	$\mu_r = 500–6000$

Problem 4. A flux density of 1.2 T is produced in a piece of cast steel by a magnetizing force of 1250 A/m. Find the relative permeability of the steel under these conditions.

For a magnetic material:

$$B = \mu_0\mu_r H$$

i.e. $$u_r = \frac{B}{\mu_0 H} = \frac{1.2}{(4\pi \times 10^{-7})(1250)} = \mathbf{764}$$

Problem 5. Determine the magnetic field strength and the m.m.f. required to produce a flux density of 0.25 T in an air gap of length 12 mm.

For air, $B = \mu_0 H$ (since $\mu_r = 1$).

Magnetic field strength

$$H = \frac{B}{\mu_0} = \frac{0.25}{4\pi \times 10^{-7}} = \mathbf{198\,940\ A/m}$$

$$\text{m.m.f.} = Hl = 198\,940 \times 12 \times 10^{-3} = \mathbf{2387\ A}$$

Problem 6. A coil of 300 turns is wound uniformly on a ring of non-magnetic material. The ring has a mean circumference of 40 cm and a uniform cross-sectional area of 4 cm². If the current in the coil is 5 A, calculate (a) the magnetic field strength, (b) the flux density and (c) the total magnetic flux in the ring.

(a) Magnetic field strength,

$$H = \frac{NI}{l} = \frac{300 \times 5}{40 \times 10^{-2}} = \mathbf{3750\ A/m}$$

(b) For a non-magnetic material, $\mu_r = 1$, thus flux density $B = \mu_0 H$, i.e.

$$B = 4\pi \times 10^{-7} \times 3750 = \mathbf{4.712\ mT}$$

(c) Flux $\Phi = BA = (4.712 \times 10^{-3})(4 \times 10^{-4}) =$ **1.885 μWb**

Problem 7. An iron ring of mean diameter 10 cm is uniformly wound with 2000 turns of wire. When a current of 0.25 A is passed through the coil a flux density of 0.4 T is set up in the iron. Find (a) the magnetizing force and (b) the relative permeability of the iron under these conditions.

$l = \pi d = \pi \times 10\text{ cm} = \pi \times 10 \times 10^{-2}\text{ m}$; $N = 2000$ turns; $I = 0.25$ A; $B = 0.4$ T.

(a) $H = \dfrac{NI}{l} = \dfrac{2000 \times 0.25}{\pi \times 10 \times 10^{-2}} = \dfrac{5000}{\pi} = \mathbf{1592\ A/m}$

(b) $B = \mu_0\mu_r H$. Hence

$$\mu_r = \frac{B}{\mu_0 H} = \frac{0.4}{(4\pi \times 10^{-7})(1592)} = \mathbf{200}$$

> Problem 8. A uniform ring of cast iron has a cross-sectional area of 10 cm² and a mean circumference of 20 cm. Determine the m.m.f. necessary to produce a flux of 0.3 mWb in the ring. The magnetization curve for cast iron is shown on page 63.

$A = 10\ \text{cm}^2 = 10 \times 10^{-4}\ \text{m}^2$; $l = 20\ \text{cm} = 0.2\ \text{m}$; $\Phi = 0.3 \times 10^{-3}$ Wb. Flux density,

$$B = \frac{\Phi}{A} = \frac{0.3 \times 10^{-3}}{10 \times 10^{-4}} = 0.3\ \text{T}$$

From the magnetization curve for cast iron on page 63, when $B = 0.3$ T, $H = 1000$ A/m, hence

$$\text{m.m.f.} = Hl = 1000 \times 0.2 = \mathbf{200\ A}$$

A tabular method could have been used in this problem. Such a solution is shown in Table 7.1.

> Problem 9. From the magnetization curve for cast iron, shown on page 63, derive the curve of μ_r against H.

$B = \mu_0\mu_r H$, hence

$$\mu_r = \frac{B}{\mu_0 H} = \frac{1}{\mu_0} \times \frac{B}{H} = \frac{10^7}{4\pi} \times \frac{B}{H}$$

A number of coordinates are selected from the *B–H* curve and μ_r is calculated for each as shown in Table 7.2.

μ_r is plotted against H as shown in Fig. 7.4. The curve demonstrates the change that occurs in the relative permeability as the magnetizing force increases.

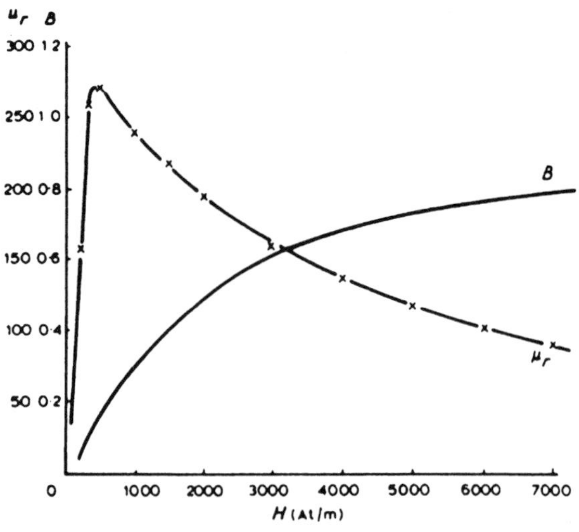

Figure 7.4 *B–H* and μ_r–*H* curves for cast iron

7.5 Reluctance

Reluctance *S* (or R_M) is the 'magnetic resistance' of a magnetic circuit to the presence of magnetic flux:

$$\text{Reluctance } S = \frac{F_M}{\Phi} = \frac{NI}{\Phi} = \frac{Hl}{BA} = \frac{l}{(B/H)A} = \frac{l}{\mu_0\mu_r A}$$

The unit of reluctance is 1/H (or H^{-1}) or A/Wb.

Ferromagnetic materials have a low reluctance and can be used as **magnetic screens** to prevent magnetic fields affecting materials within the screen.

> Problem 10. Determine the reluctance of a piece of mumetal of length 150 mm, cross-sectional area 1800 mm² when the relative permeability is 4000. Find also the absolute permeability of the mumetal.

$$\text{Reluctance } S = \frac{l}{\mu_0\mu_r A}$$

Table 7.1

Part of circuit	Material	Φ (Wb)	A (m²)	$B = \dfrac{\Phi}{A}$ (T)	H from graph	l (m)	m.m.f. = Hl (A)
Ring	Cast iron	0.3×10^{-3}	10×10^{-4}	0.3	1000	0.2	200

Table 7.2

B (T)	0.04	0.13	0.17	0.30	0.41	0.49	0.60	0.68	0.73	0.76	0.79
H (A/m)	200	400	500	1000	1500	2000	3000	4000	5000	6000	7000
$u_r = \dfrac{10^7}{4\pi} \times \dfrac{B}{H}$	159	259	271	239	218	195	159	135	116	101	90

$$= \frac{150 \times 10^{-3}}{(4\pi \times 10^{-7})(4000)(1800 \times 10^{-6})}$$

$= \mathbf{16\,580/H}$

Absolute permeability, $\mu = \mu_0\mu_r = 4\pi \times 10^{-7}$ (4000) = $\mathbf{5.027 \times 10^{-3}}$ **H/m.**

Problem 11. A mild steel ring has a radius of 50 mm and a cross-sectional area of 400 mm^2. A current of 0.5 A flows in a coil wound uniformly around the ring and the flux produced is 0.1 mWb. If the relative permeability at this value of current is 200 find (a) the reluctance of the mild steel and (b) the number of turns on the coil.

$l = 2\pi r = 2 \times \pi \times 50 \times 10^{-3}$ m; $A = 400 \times 10^{-6}$ m^2; $I = 0.5$ A; $\Phi = 0.1 \times 10^{-3}$ Wb; $\mu_r = 200$.

(a) Reluctance $S = \dfrac{l}{\mu_0\mu_r A}$

$$= \frac{2 \times \pi \times 50 \times 10^{-3}}{(4\pi \times 10^{-7})(200)(400 \times 10^{-6})}$$

$= \mathbf{3.125 \times 10^6}$/H

(b) S = m.m.f./Φ, i.e. m.m.f. = $S\Phi$, so that $NI = S\Phi$, and hence

$$N = \frac{S\Phi}{I} = \frac{3.125 \times 10^6 \times 0.1 \times 10^{-3}}{0.5}$$

= 625 turns.

7.6 Composite series magnetic circuits

For a series magnetic circuit having n parts, the **total reluctance S** is given by:

$$S = S_1 + S_2 + ... + S_n$$

(This is similar to resistors connected in series in an electrical circuit.)

Problem 12. A closed magnetic circuit of cast steel contains a 6 cm long path of cross-sectional area 1 cm^2 and a 2 cm path of cross-sectional area 0.5 cm^2. A coil of 200 turns is wound around the 6 cm length of the circuit and a current of 0.4 A flows. Determine the flux density in the 2 cm path, if the relative permeability of the cast steel is 750.

For the 6 cm long path:

Reluctance $S_1 = \dfrac{l_1}{\mu_0\mu_r A_1}$

$$= \frac{6 \times 10^{-2}}{(4\pi \times 10^{-7})(750)(1 \times 10^{-4})}$$

$= 6.366 \times 10^5$/H

For the 2 cm long path:

Reluctance $S_2 = \dfrac{l_2}{\mu_0\mu_r A_2}$

$$= \frac{2 \times 10^{-2}}{(4\pi \times 10^{-7})(750)(0.5 \times 10^{-4})}$$

$= 4.244 \times 10^5$/H

Total circuit reluctance $S = S_1 + S_2 = (6.366 + 4.244) \times 10^5 = 10.61 \times 10^5$ /H. S = m.m.f./Φ, i.e. Φ = m.m.f./S = NI/S

$$= \frac{200 \times 0.4}{10.61 \times 10^5} = 7.54 \times 10^{-5} \text{ Wb}$$

Flux density in the 2 cm path,

$$B = \frac{\Phi}{A} = \frac{7.54 \times 10^{-5}}{0.5 \times 10^{-4}} = \mathbf{1.51\ T}$$

Problem 13. A silicon iron ring of cross-sectional area 5 cm² has a radial air gap of 2 mm cut into it. If the mean length of the silicon iron path is 40 cm, calculate the magnetomotive force to produce a flux of 0.7 mWb. The magnetization curve for silicon is shown on page 63.

There are two parts to the circuit – the silicon iron and the air gap. The total m.m.f. will be the sum of the m.m.f.s of each part.

For the silicon iron

$$B = \frac{\Phi}{A} = \frac{0.7 \times 10^{-3}}{5 \times 10^{-4}} = 1.4\ \text{T}$$

From the B–H curve for silicon iron on page 63, when $B = 1.4$ T, $H = 1650$ A/m. Hence the m.m.f. for the iron path $= Hl = 1650 \times 0.4 =$ **660 A**.

For the air gap

The flux density will be the same in the air gap as in the iron, i.e. 1.4 T. (This assumes no leakage or fringing occurring.)

For air,

$$H = \frac{B}{\mu_0} = \frac{1.4}{4\pi \times 10^{-7}} = 1\,114\,000\ \text{A/m}$$

Hence the m.m.f. for the air gap $= Hl = 1\,114\,000 \times 2 \times 10^{-3} = 2228$ A. Total m.m.f. to produce a flux of 0.6 mWb $= 660 + 2228 =$ **2888 A**.

A tabular method could have been used as shown in Table 7.3.

Problem 14. Figure 7.5 shows a ring formed with two different materials – cast steel and mild steel. The dimensions are:

	Mean length	Cross-sectional area
Mild steel	400 mm	500 mm²
Cast steel	300 mm	312.5 mm²

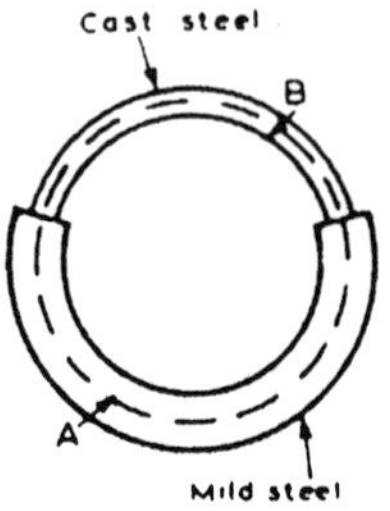

Figure 7.5

Find the total m.m.f. required to cause a flux of 500 μWb in the magnetic circuit. Determine also the total circuit reluctance.

A tabular solution is shown in Table 7.4.

Total circuit reluctance,

$$S = \frac{\text{m.m.f.}}{\Phi} = \frac{2000}{500 \times 10^{-6}} = \mathbf{4 \times 10^6/H}$$

Table 7.3

Part of circuit	Material	Φ (Wb)	A (m²)	B (T)	H (A/m)	l (m)	m.m.f. = Hl (A)
Ring	Silicon iron	0.7×10^{-3}	5×10^{-4}	1.4	1650 (from graph)	0.4	660
Air gap	Air	0.7×10^{-3}	5×10^{-4}	1.4	$\frac{1.4}{4\pi \times 10^{-7}}$ = 1 114 000	2×10^{-3}	2228
							Total: **2888 A**

Table 7.4

Part of circuit	Material	Φ (Wb)	A (m^2)	B (T) (= Φ/A)	H (A/m) (from graphs, p. 63)	l (m)	m.m.f. = Hl (A)
A	Mild steel	500×10^{-6}	500×10^{-6}	1.0	1400	400×10^{-3}	560
B	Cast steel	500×10^{-6}	312.5×10^{-6}	1.6	4800	300×10^{-3}	1440

Total: **2000 A**

Problem 15. A section through a magnetic circuit of uniform cross-sectional area 2 cm^2 is shown in Fig. 7.6. The cast steel core has a mean length of 25 cm. The air gap is 1 mm wide and the coil has 5000 turns. The B–H curve for cast steel is shown on page 63. Determine the current in the coil to produce a flux density of 0.80 T in the air gap, assuming that all the flux passes through both parts of the magnetic circuit.

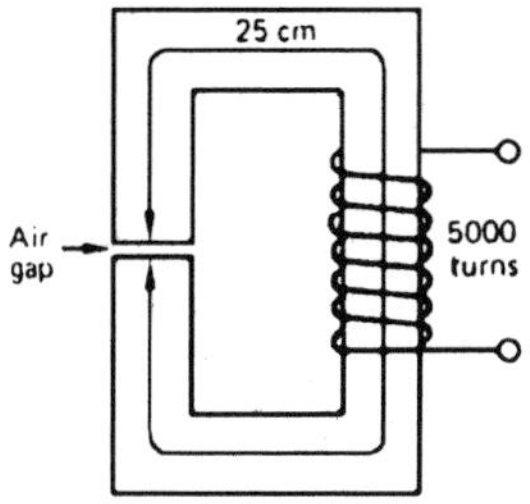

Figure 7.6

For the cast steel core, when $B = 0.80$ T, $H = 750$ A/m (from page 63).
Reluctance of core $S_1 = l_1/\mu_0\mu_r A_1$ and, since $B = \mu_0\mu_r H$, then $\mu_r = B/\mu_0 H$. Thus

$$S_1 = \frac{l_1}{\mu_0\left(\dfrac{B}{\mu_0 H}\right)A} = \frac{l_1 H}{BA}$$

$$= \frac{(25 \times 10^{-2})(750)}{(0.8)(2 \times 10^{-4})}$$

$$= 1\,172\,000/\text{H}$$

For the air gap, reluctance,

$$S_2 = \frac{l_2}{\mu_0\mu_r A_2} = \frac{l_2}{\mu_0 A_2} \text{ (since } \mu_r = 1 \text{ for air)}$$

$$= \frac{1 \times 10^{-3}}{(4\pi \times 10^{-7})(2 \times 10^{-4})} = 3\,979\,000/\text{H}$$

Total circuit reluctance,

$$S = S_1 + S_2 = 1\,172\,000 + 3\,979\,000$$

$$= 5\,151\,000/\text{H}$$

Flux $\Phi = BA = 0.80 \times 2 \times 10^{-4} = 1.6 \times 10^{-4}$ Wb; S = m.m.f./Φ, thus m.m.f. = $S\Phi$. Hence $NI = S\Phi$ and

$$\text{current } I = \frac{S\Phi}{N} = \frac{(5\,151\,000)(1.6 \times 10^{-4})}{5000}$$

$$= \mathbf{0.165\ A}$$

7.7 Comparison between electrical and magnetic quantities

Electrical circuit		Magnetic circuit	
e.m.f. E	(V)	m.m.f. F_m	(A)
current I	(A)	flux Φ	(Wb)
resistance R	(Ω)	reluctance S	(H^{-1})
$I = \dfrac{E}{R}$		$\Phi = \dfrac{\text{m.m.f.}}{S}$	
$R = \dfrac{\rho l}{A}$		$S = \dfrac{l}{\mu_0\mu_r A}$	

7.8 Hysteresis and hysteresis loss

Hysteresis loop

Let a ferromagnetic material which is completely demagnetized, i.e. one in which $B = H = 0$, be subjected to increasing values of magnetic field strength H and the corresponding flux density B measured. The resulting relationship between B and H is shown by the curve $0ab$ in Fig. 7.7. At a particular value of H, shown as $0y$, it becomes difficult to increase the flux density any further. The material is said to be saturated. Thus **by** is the **saturation flux density**.

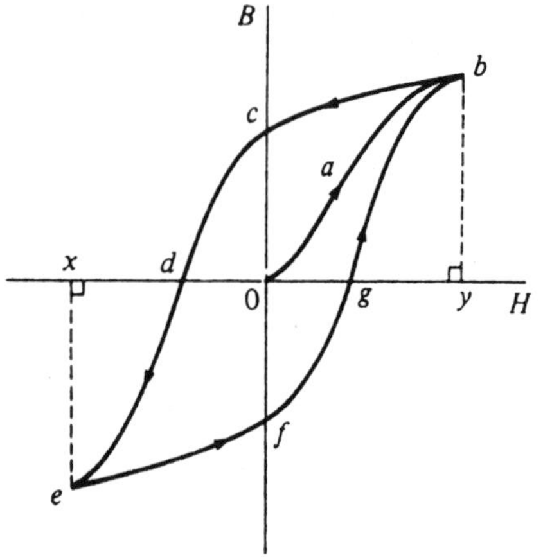

Figure 7.7

If the value of H is now reduced it is found that the flux density follows curve **bc**. When H is reduced to zero, flux remains in the iron. This **remanent flux density** or **remanence** is shown as 0c in Fig. 7.7. When H is increased in the opposite direction, the flux density decreases until, at a value shown as 0d, the flux density has been reduced to zero. The magnetic field strength 0d required to remove the residual magnetism, i.e. reduce B to zero, is called the **coercive force**.

Further increase of H in the reverse direction causes the flux density to increase in the reverse direction until saturation is reached, as shown by curve **de**. If H is varied backwards from 0x to 0y, the flux density follows the curve **efgb**, similar to curve **bcde**. It is seen from Fig. 7.7 that the flux density changes lag behind the changes in the magnetic field strength. This effect is called **hysteresis**. The closed figure **bcdefgb** is called the **hysteresis loop** (or the B/H loop).

Hysteresis loss

A disturbance in the alignment of the domains (i.e. groups of atoms) of a ferromagnetic material causes energy to be expended in taking it through a cycle of magnetization. This energy appears as heat in the specimen and is called the **hysteresis loss**.

The energy loss associated with hysteresis is proportional to the area of the hysteresis loop.

The area of a hysteresis loop varies with the type of material. The area, and thus the energy loss, is much greater for hard materials than for soft materials.

Figure 7.8 shows typical hysteresis loops for (a) hard steel, which has a high remanence 0c and a large coercivity 0d, (b) soft steel, which has a large remanence and small coercivity and (c) ferrite, this being a ceramic-like magnetic substance made from oxides of iron, nickel, cobalt, magnesium, aluminium and manganese. The hysteresis of ferrite is very small.

For AC-excited devices the hysteresis loop is repeated every cycle of alternating current. Thus a hysteresis loop with a large area (as with hard steel) is often unsuitable since the energy loss would be considerable.

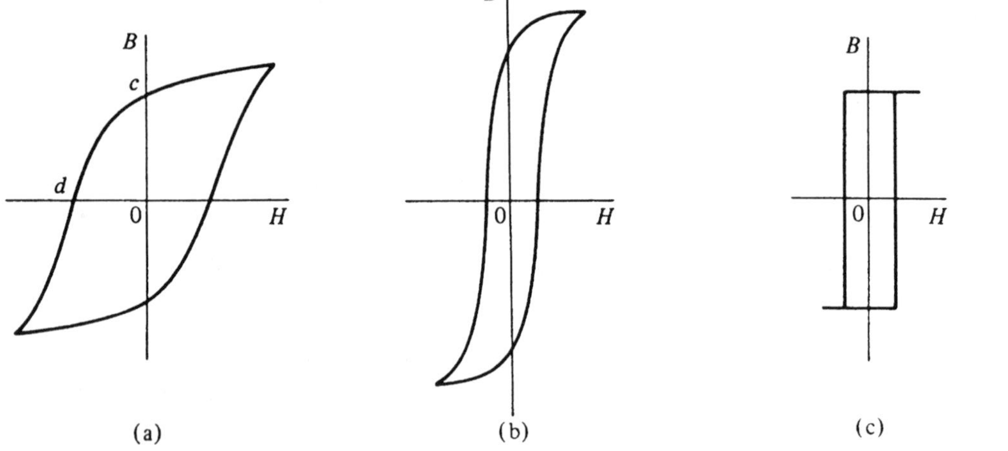

Figure 7.8

Silicon steel has a narrow hysteresis loop, and thus small hysteresis loss, and is suitable for transformer cores and rotating machine armatures.

7.9 Multi-choice questions on magnetic circuits

(Answers on page 283.)

1. The unit of magnetic flux density is the:
   (a) weber; (b) weber per metre;
   (c) ampere per metre; (d) tesla.
2. The total flux in the core of an electrical machine is 20 mWb and its flux density is 1 T. The cross-sectional area of the core is
   (a) 0.05 m^2; (b) 0.02 m^2; (c) 20 m^2;
   (d) 50 m^2.

A coil of 100 turns is wound uniformly on a wooden ring. The ring has a mean circumference of 1 m and a uniform cross-sectional area of 10 cm^2. The current in the coil is 1 A.

In Problems 3 to 7 select the correct answer for each of the required quantities from the following list:

(a) 40π mT (b) 100 A (c) $4\pi \times 10^{-10}$ H
(d) 100 A (e) 0.01 A/m (f) 40π μT (g) 40π mWb
(h) $\dfrac{2.5}{\pi} \times 10^9$/H (i) 100 A/m (j) 0.04π μWb.

3. Magnetomotive force
4. Magnetic field strength
5. Magnetic flux density
6. Magnetic flux
7. Reluctance
8. Which of the following statements is false?
   (a) For non-magnetic materials reluctance is high.
   (b) Energy loss due to hysteresis is greater for harder magnetic materials than for softer magnetic materials.
   (c) The remanence of a ferrous material is measured in ampere/metre.
   (d) Absolute permeability is measured in henrys per metre.
9. The current flowing in a 500 turn coil wound on an iron ring is 4 A. The reluctance of the circuit is 2×10^6/H. The flux produced is
   (a) 1 Wb; (b) 1000 Wb; (c) 1 mWb;
   (d) 62.5 μWb.
10. A comparison can be made between magnetic and electrical quantities. From the following list, match the magnetic quantities with their equivalent electrical quantities.
    (a) Current (b) reluctance (c) e.m.f.
    (d) flux (e) m.m.f. (f) resistance

7.10 Short answer questions on magnetic circuits

1. What is a permanent magnet?
2. Sketch the pattern of the magnetic field associated with a bar magnet. Mark the direction of the field.
3. Define magnetic flux.
4. The symbol for magnetic flux is .... and the unit of flux is the ....
5. Define magnetic flux density.
6. The symbol for magnetic flux density is .... and the unit of flux density is ....
7. The symbol for m.m.f. is .... and the unit of m.m.f. is the ....
8. Another name for the magnetizing force is ....; its symbol is .... and its unit is ....
9. Complete the statement:
   For magnetic materials
   $$\frac{\text{flux density}}{\text{magnetic field strength}} = \ldots\ldots\ldots\ldots$$
10. What is absolute permeability?
11. The value of the permeability of free space is ....
12. What is a magnetization curve?
13. The symbol for reluctance is .... and the unit of reluctance is ....
14. Make a comparison between magnetic and electrical quantities.
15. What is hysteresis?
16. Draw a typical hysteresis loop and identify (a) saturation flux density, (b) remanence, and (c) coercive force.
17. State the units of (a) remanence, (b) coercive force.
18. How is magnetic screening achieved?
19. Complete the statement: magnetic materials have a .... reluctance; non-magnetic materials have a .... reluctance.
20. What loss is associated with hysteresis?

7.11 Further questions on magnetic circuits

(Where appropriate, assume $\mu_0 = 4\pi \times 10^{-7}$ H/m.)

Magnetic circuit quantities

1. What is the flux density in a magnetic field of cross-sectional area 20 cm^2 having a flux of 3 mWb? [1.5 T]
2. Determine the total flux emerging from a magnetic pole face having dimensions 5 cm by 6 cm, if the flux density is 0.9 T. [2.7 mWb]
3. The maximum working flux density of a lifting electromagnet is 1.9 T and the effective area of a pole face is circular in cross-section. If the total magnetic flux produced is 611 mWb determine the radius of the pole face. [32 cm]
4. An electromagnet of square cross-section produces a flux density of 0.45 T. If the magnetic flux is 720 μWb find the dimensions of the electromagnet cross-section. [4 cm by 4 cm]
5. Find the magnetic field strength and the magnetomotive force needed to produce a flux density of 0.33 T in an air gap of length 15 mm. [(a) 262 600 At/m (b) 3939 A]
6. An air gap between two pole pieces is 20 mm in length and the area of the flux path across the gap is 5 cm^2. If the flux required in the air gap is 0.75 mWb find the m.m.f. necessary. [23 870 A]
7. Find the magnetic field strength applied to a magnetic circuit of mean length 50 cm when a coil of 400 turns is applied to it carrying a current of 1.2 A. [960 A/m]
8. A solenoid 20 cm long is wound with 500 turns of wire. Find the current required to establish a magnetizing force of 2500 A/m inside the solenoid. [1 A]
9. A magnetic field strength of 5000 A/m is applied to a circular magnetic circuit of mean diameter 250 mm. If the coil has 500 turns find the current in the coil. [7.85 A]
10. (a) Determine the flux density produced in an air-cored solenoid due to a uniform magnetic field strength of 8000 A/m.
    (b) Iron having a relative permeability 150 at 8000 A/m is inserted into the solenoid of part (a). Find the flux density now in the solenoid.
    [(a) 10.05 mT (b) 1.508 T]
11. Find the relative permeability of a material if the absolute permeability is 4.084×10^{-4} H/m. [325]
12. Find the relative permeability of a piece of silicon iron if a flux density of 1.3 T is produced by a magnetic field strength of 700 A/m. [1478]
13. Part of a magnetic circuit is made from steel of length 120 mm, cross-sectional area 15 cm^2 and relative permeability 800. Calculate (a) the reluctance, and (b) the absolute permeability of the steel.
    [(a) 79 580 /H (b) 1 mH/m]
14. A steel ring of mean diameter 120 mm is uniformly wound with 1500 turns of wire. When a current of 0.30 A is passed through the coil a flux density of 1.5 T is set up in the steel. Find the relative permeability of the steel under these conditions. [1000]
15. A mild steel closed magnetic circuit has a mean length of 75 mm and a cross-sectional area of 320.2 mm^2. A current of 0.40 A flows in a coil wound uniformly around the circuit and the flux produced is 200 μWb. If the relative permeability of the steel at this value of current is 400 find (a) the reluctance of the material and (b) the number of turns of the coil. [(a) 466 000/H (b) 233]
16. A uniform ring of cast steel has a cross-sectional area of 5 cm^2 and a mean circumference of 15 cm. Find the current required in a coil of 1200 turns wound on the ring to produce a flux of 0.8 mWb. (Use the magnetization curve for cast steel shown on page 63.) [0.60 A]
17. (a) A uniform mild steel ring has a diameter of 50 mm and a cross-sectional area of 1 cm^2. Determine the m.m.f. necessary to produce a flux of 50 μWb in the ring. Use the *B–H* curve for mild steel shown on page 63.)
    (b) If a coil of 440 turns is wound uniformly around the ring in part (a) what current would be required to produce the flux?
    [(a) 110 A (b) 0.25 A]
18. From the magnetization curve for mild steel shown on page 63, derive the curve of relative permeability against magnetic field strength. From your graph determine (a) the value of μ_r when the magnetic field strength is 1200 A/m, and (b) the value of the magnetic field strength when μ_r is 500.
    [(a) 590–600 (b) 2000]

Composite series magnetic circuits

19. A magnetic circuit of cross-sectional area 0.4 cm^2 consists of one part 3 cm long of

material having relative permeability 1200, and a second part 2 cm long of material having relative permeability 750. With a 100 turn coil carrying 2 A, find the value of flux existing in the circuit. [0.195 mWb]

20. (a) A cast steel ring has a cross-sectional area of 600 mm^2 and a radius of 25 mm. Determine the m.m.f. necessary to establish a flux of 0.8 mWb in the ring. Use the *B–H* curve for cast steel shown on page 63.
    (b) If a radial air gap 1.5 mm wide is cut in the ring of part (a) find the m.m.f. now necessary to maintain the same flux in the ring. [(a) 267 A (b) 1859 A]

21. A closed magnetic circuit made of silicon iron consists of a 40 mm long path of cross-sectional area 90 mm^2 and a 15 mm long path of cross-sectional area 70 mm^2. A coil of 50 turns is wound around the 40 mm length of the circuit and a current of 0.39 A flows. Find the flux density in the 15 mm length path if the relative permeability of the silicon iron at this value of magnetizing force is 3000. [1.59 T]

22. For the magnetic circuit shown in Fig. 7.9 find the current *I* in the coil needed to produce a flux of 0.45 mWb in the air gap. The silicon iron magnetic circuit has a uniform cross-sectional area of 3 cm^2 and its magnetization curve is as shown on page 63. [0.83 A]

23. A ring forming a magnetic circuit is made from two materials; one part is mild steel of mean length 25 cm and cross-sectional area 4 cm^2, and the remainder is cast iron of mean length 20 cm and cross-sectional area 7.5 cm^2. Use a tabular approach to determine the total m.m.f. required to cause a flux of 0.30 mWb in the magnetic circuit. Find also the total reluctance of the circuit. Use the magnetization curves shown on page 63. [540 A, 18×10^5/H]

24. Figure 7.10 shows the magnetic circuit of a relay. When each of the air gaps are 1.5 mm wide find the m.m.f. required to produce a flux density of 0.75 T in the air gaps. Use the *B–H* curves shown on page 63. [2990 A]

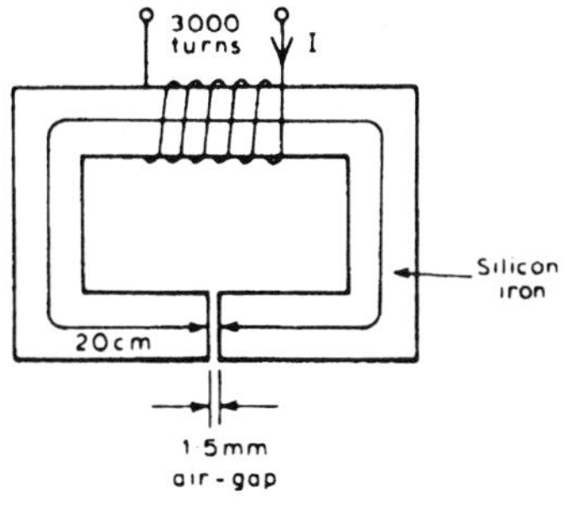

Figure 7.9

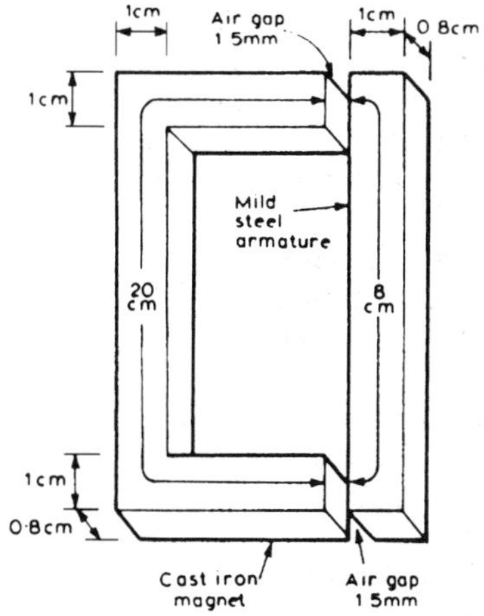

Figure 7.10

8

Electromagnetism

At the end of this chapter you should be able to:

- understand that magnetic fields are produced by electric currents
- apply the screw rule to determine direction of magnetic field
- recognize that the magnetic field around a solenoid is similar to a magnet
- apply the screw rule or grip rule to a solenoid to determine magnetic field direction
- recognize and describe practical applications of an electromagnet, i.e. electric bell, relay, lifting magnet, telephone receiver
- appreciate factors upon which the force F on a current-carrying conductor depends
- perform calculations using $F = BIl$ and $F = BIl \sin \theta$
- recognize that a loudspeaker is a practical application of force F
- use Fleming's left-hand rule to predetermine direction of force in a current-carrying conductor
- describe the principle of operation of a simple d.c. motor
- describe the principle of operation and construction of a moving-coil instrument
- appreciate the force F on a charge in a magnetic field is given by $F = QvB$
- perform calculations using $F = QvB$

8.1 Magnetic field due to an electric current

Magnetic fields can be set up not only by permanent magnets, as shown in Chapter 7, but also by electric currents.

Let a piece of wire be arranged to pass vertically through a horizontal sheet of cardboard, on which are placed some iron filings, as shown in Fig. 8.1(a). If a current is now passed through the wire, then the iron filings will form a definite circular field pattern with the wire at the centre, when the cardboard is gently tapped. By placing a compass in different positions the lines of flux are seen to have a definite direction as shown in Fig. 8.1(b).

If the current direction is reversed, the direction of the lines of flux is also reversed. The effect on both the iron filings and the compass needle disappears when the current is switched off. The magnetic field is thus produced by the electric current. The magnetic flux produced has the same properties as the flux produced by a permanent magnet. If the current is increased the strength of the field increases and, as for the permanent magnet, the field strength decreases as we move away from the current-carrying conductor.

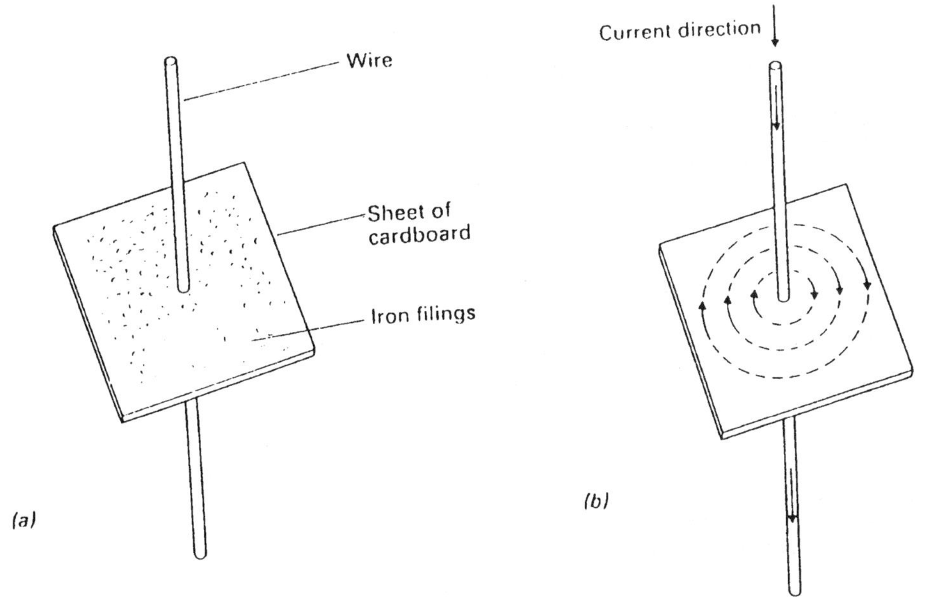

Figure 8.1

In Fig. 8.1, the effect of only a small part of the magnetic field is shown. If the whole length of the conductor is similarly investigated it is found that the magnetic field round a straight conductor is in the form of concentric cylinders as shown in Fig. 8.2, the field direction depending on the direction of the current flow.

When dealing with magnetic fields formed by electric current it is usual to portray the effect as shown in Fig. 8.3. The convention adopted is:

(i) Current flowing away from the viewer, i.e. into the paper, is indicated by ⊕. This may be thought of as the feathered end of the shaft of an arrow. See Fig. 8.3(a).

(ii) Current flowing towards the viewer, i.e. out of the paper, is indicated by ⊙. This may be thought of as the point of an arrow. See Fig. 8.3(b).

The direction of the magnetic lines of flux is best remembered by the **screw rule**. This states that:

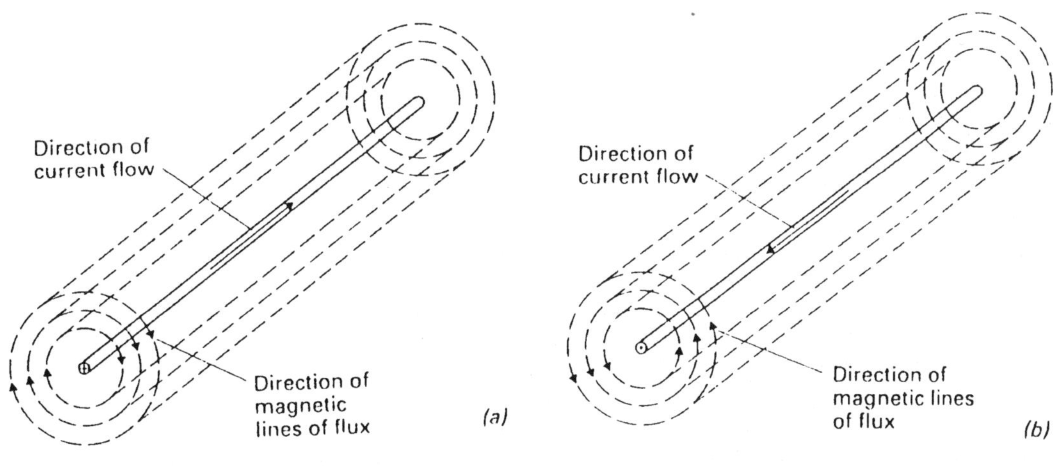

Figure 8.2

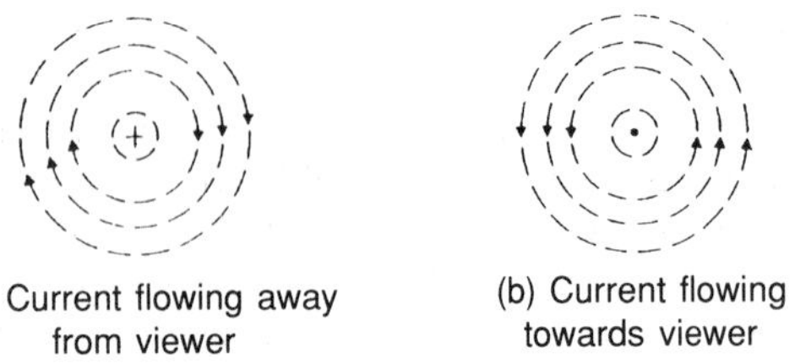

Figure 8.3

If a normal right-hand thread screw is screwed along the conductor in the direction of the current, the direction of rotation of the screw is in the direction of the magnetic field.

For example, with current flowing away from the viewer (Fig. 8.3(a)) a right-hand thread screw driven into the paper has to be rotated clockwise. Hence the direction of the magnetic field is clockwise.

A magnetic field set up by a long coil, or **solenoid**, is shown in Fig. 8.4(a) and is seen to be similar to that of a bar magnet. If the solenoid is wound on an iron bar, as shown in Fig. 8.4(b), an even stronger magnetic field is produced, the iron becoming magnetized and behaving like a permanent magnet. The direction of the magnetic field produced by the current I in the solenoid may be found by either of two methods, i.e. the screw rule or the grip rule.

(a) **The screw rule** states that if a normal right-hand thread screw is placed along the axis of the solenoid and is screwed in the direction of the current it moves in the direction of the magnetic field **inside** the solenoid. The direction of the magnetic field **inside** the solenoid is from south to north. Thus in Figs. 8.4(a) and (b) the north pole is to the right.

(b) **The grip rule** states that if the coil is gripped with the **right** hand, with the fingers pointing in the direction of the current, then the thumb, outstretched parallel to the axis of the solenoid, points in the direction of the magnetic field **inside** the solenoid.

Problem 1. Figure 8.5 shows a coil of wire wound on an iron core connected to a battery. Sketch the magnetic field pattern associated with the current-carrying coil and determine the polarity of the field.

The magnetic field associated with the solenoid in Fig. 8.5 is similar to the field associated with a bar magnet and is as shown in Fig. 8.6. The polarity of the field is determined either by the screw rule or by the grip rule. Thus the north pole is at the bottom and the south pole at the top.

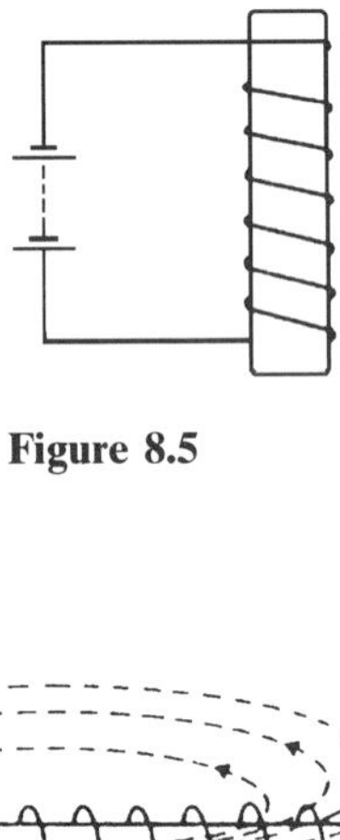

Figure 8.5

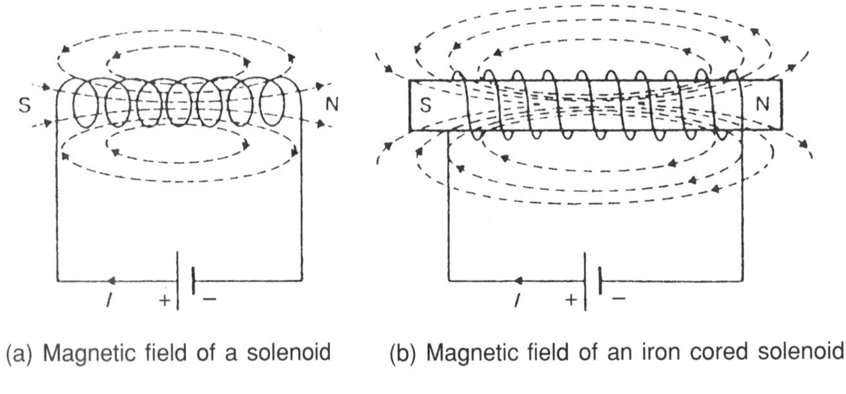

(a) Magnetic field of a solenoid (b) Magnetic field of an iron cored solenoid

Figure 8.4

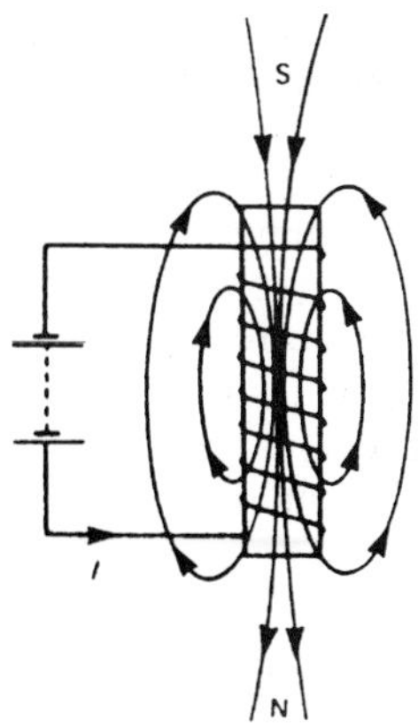

Figure 8.6

8.2 Electromagnets

The solenoid is very important in electromagnetic theory since the magnetic field inside the solenoid is practically uniform for a particular current, and is also versatile, inasmuch that a variation of the current can alter the strength of the magnetic field. An electromagnet, based on the solenoid, provides the basis of many items of electrical equipment, examples of which include electric bells, relays, lifting magnets and telephone receivers.

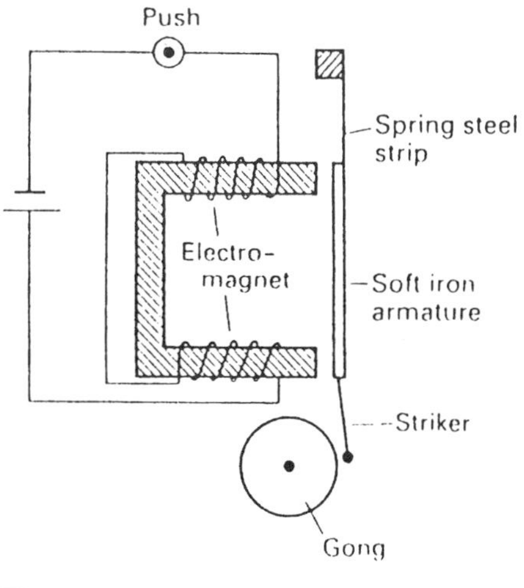

Figure 8.7

Electric bell

There are various types of electric bell, including the single-stroke bell, the trembler bell, the buzzer and a continuously ringing bell, but all depend on the attraction exerted by an electromagnet on a soft iron armature. A typical single-stroke bell circuit is shown in Fig. 8.7. When the push button is operated a current passes through the coil. Since the iron-cored coil is energized the soft iron armature is attracted to the electromagnet. The armature also carries a striker which hits the gong. When the circuit is broken the coil becomes demagnetized and the spring steel strip pulls the armature back to its original position. The striker will only operate when the push is operated.

Relay

A relay is similar to an electric bell except that contacts are opened or closed by operation instead of a gong being struck. A typical simple relay is shown in Fig. 8.8, which consists of a coil wound on a soft iron core. When the coil is energized the hinged soft iron armature is attracted to the electromagnet and pushes against two fixed contacts so that they are connected together, thus closing some other electrical circuit.

Lifting magnet

Lifting magnets, incorporating large electromagnets, are used in iron and steel works for lifting scrap metal. A typical robust lifting magnet, capable of exerting large attractive forces, is

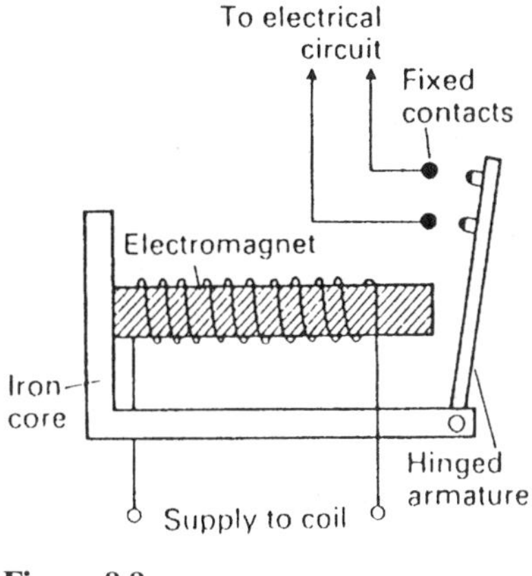

Figure 8.8

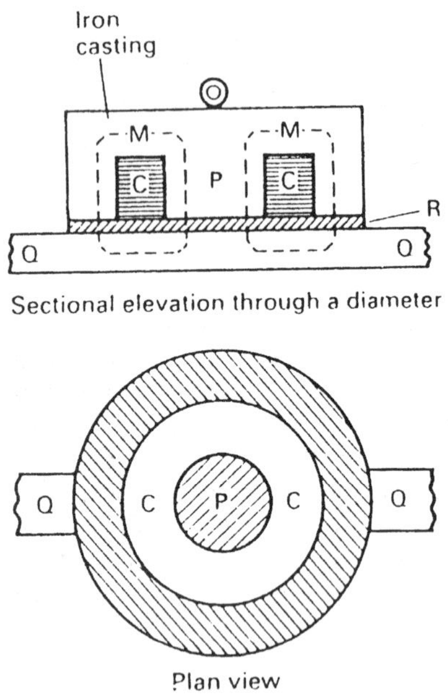

Figure 8.9

shown in the elevation and plan view of Fig. 8.9 where a coil, C, is wound round a central core, P, of the iron casting. Over the face of the electromagnet is placed a protective non-magnetic sheet of material, R. The load, Q, which must be of magnetic material, is lifted when the coils are energized, the magnetic flux paths, M, being shown by the broken lines.

Telephone receiver

Whereas a transmitter or microphone changes sound waves into corresponding electrical signals,

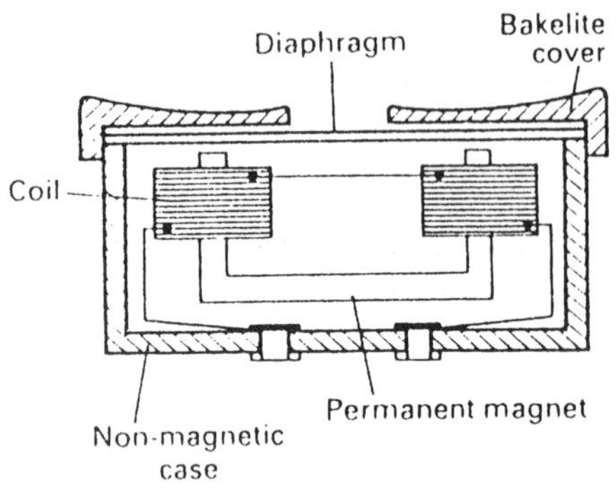

Figure 8.10

a telephone receiver converts the electrical waves back into sound waves. A typical telephone receiver is shown in Fig. 8.10 and consists of a permanent magnet with coils wound on its poles. A thin, flexible diaphragm of magnetic material is held in position near to the magnetic poles but not touching them. Variation in current from the transmitter varies the magnetic field and the diaphragm consequently vibrates. The vibration produces sound variations corresponding to those transmitted.

8.3 Force on a current-carrying conductor

If a current-carrying conductor is placed in a magnetic field produced by permanent magnets, then the fields due to the current-carrying conductor and the permanent magnets interact and cause a force to be exerted on the conductor. The force on the current-carrying conductor in a magnetic field depends upon:

(a) the flux density of the field, B teslas;
(b) the strength of the current, I amperes;
(c) the length of the conductor perpendicular to the magnetic field, l metres; and
(d) the directions of the field and the current.

When the magnetic field, the current and the conductor are mutually at right angles then:

Force $F = BIl$ newtons

When the conductor and the field are at an angle $\theta°$ to each other then:

Force $F = BIl \sin\theta$ newtons

Since when the magnetic field, current and conductor are mutually at right angles, $F = BIl$, the magnetic flux density B may be defined by $B = F/Il$, i.e. the flux density is 1 T if the force exerted on 1 m of a conductor when the conductor carries a current of 1 A is 1 N.

Loudspeaker

A simple application of the above force is the moving-coil loudspeaker. The loudspeaker is used to convert electrical signals into sound waves. Figure 8.11 shows a typical loudspeaker having a magnetic circuit comprising a permanent

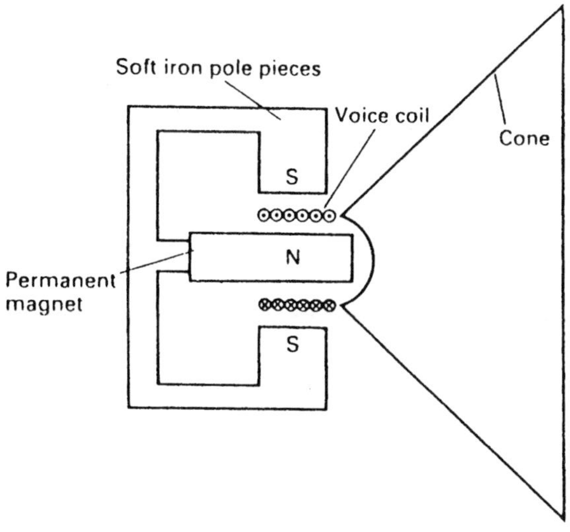

Figure 8.11

magnet and soft iron pole pieces so that a strong magnetic field is available in the short cylindrical air gap. A moving coil, called the voice or speech coil, is suspended from the end of a paper or plastic cone so that it lies in the gap. When an electric current flows through the coil it produces a force which tends to move the cone backwards and forwards according to the direction of the current. The cone acts as a piston, transferring this force to the air, and producing the required sound waves.

Problem 2. A conductor carries a current of 20 A and is at right angles to a magnetic field having a flux density of 0.9 T. If the length of the conductor in the field is 30 cm, calculate the force acting on the conductor. Determine also the value of the force if the conductor is inclined at an angle of 30° to the direction of the field.

$B = 0.9$ T; $I = 20$ A; $l = 30$ cm $= 0.30$ m. Force $F = BIl = (0.9)(20)(0.30)$ newtons when the conductor is at right angles to the field, as shown in Fig. 8.12(a), i.e. $\boldsymbol{F = 5.4}$ **N**.

When the conductor is inclined at 30° to the field, as shown in Fig. 8.12(b), then force $F = BIl \sin\theta$, i.e.

$$F = (0.9)(20)(0.30) \sin 30°$$

$$\boldsymbol{F = 2.7\ \mathrm{N}}$$

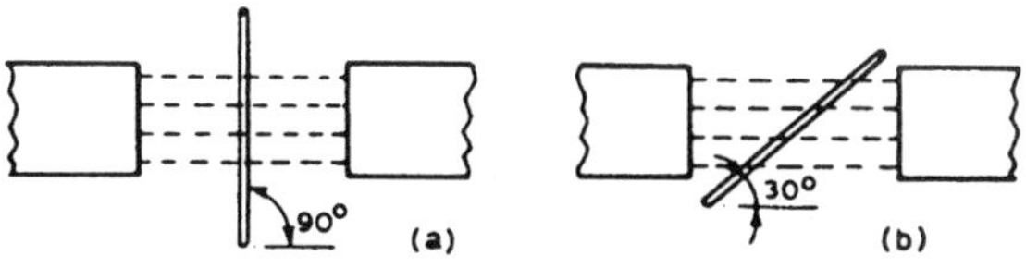

Figure 8.12

If the current-carrying conductor shown in Fig. 8.3(a) is placed in the magnetic field shown in Fig. 8.13(a), then the two fields interact and cause a force to be exerted on the conductor as shown in Fig. 8.13(b). The field is strengthened above the conductor and weakened below, thus tending to move the conductor downwards. This is the basic principle of operation of the electric motor (see section 8.4) and the moving-coil instrument (see section 8.5).

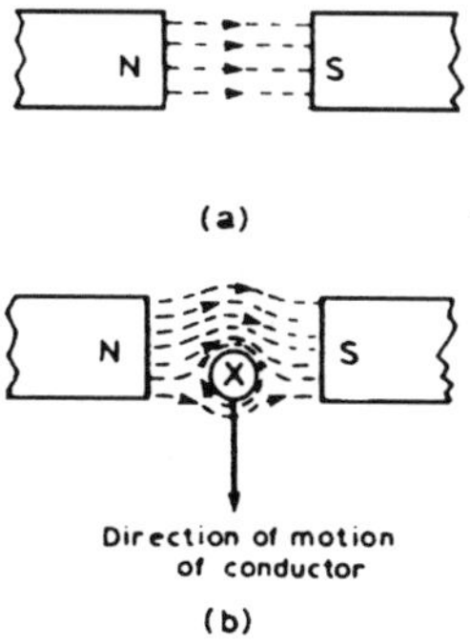

Figure 8.13

The direction of the force exerted on a conductor can be predetermined by using **Fleming's left-hand rule** (often called the motor rule) which states:

Let the thumb, first finger and second finger of the left hand be extended such that they are all at right angles to each other [as shown in Fig. 8.14]. If the first finger points in the direction of the magnetic field, the second finger points in the direction of the current, then the thumb will point in the direction of the motion of the conductor.

Summarizing:

First finger – Field
SeCond finger – Current
ThuMb – Motion

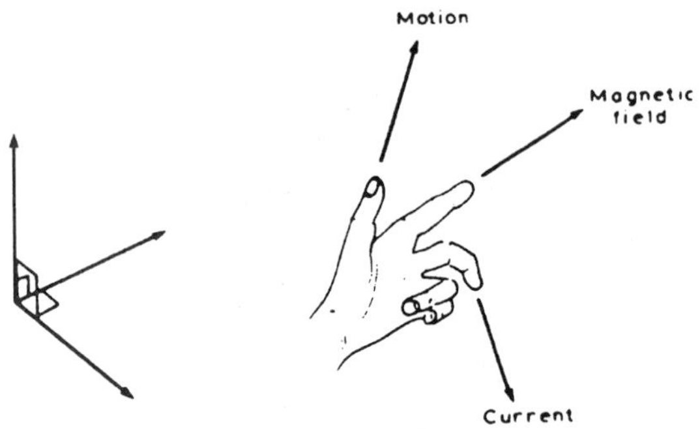

Figure 8.14

> Problem 3. Determine the current required in a 400 mm length of conductor of an electric motor when the conductor is situated at right angles to a magnetic field of flux density 1.2 T, if a force of 1.92 N is to be exerted on the conductor. If the conductor is vertical, the current flowing downwards and the direction of the magnetic field is from left to right, what is the direction of the force?

Force = 1.92 N; l = 400 mm = 0.40 m; B = 1.2 T. Since $F = BIl$, $I = F/Bl$, hence

$$\text{current } I = \frac{1.92}{(1.2)(0.4)} = \mathbf{4\ A}$$

If the current flows downwards, the direction of its magnetic field due to the current alone will be clockwise when viewed from above. The lines of flux will reinforce (i.e. strengthen) the main magnetic field at the back of the conductor and will be in opposition in the front (i.e. weaken the field). **Hence the force on the conductor will be from back to front (i.e. towards the viewer).** This direction may also have been deduced using Fleming's left-hand rule.

> Problem 4. A conductor 350 mm long carries a current of 10 A and is at right angles to a magnetic field lying between two circular pole faces each of radius 60 mm. If the total flux between the pole faces is 0.5 mWb, calculate the magnitude of the force exerted on the conductor.

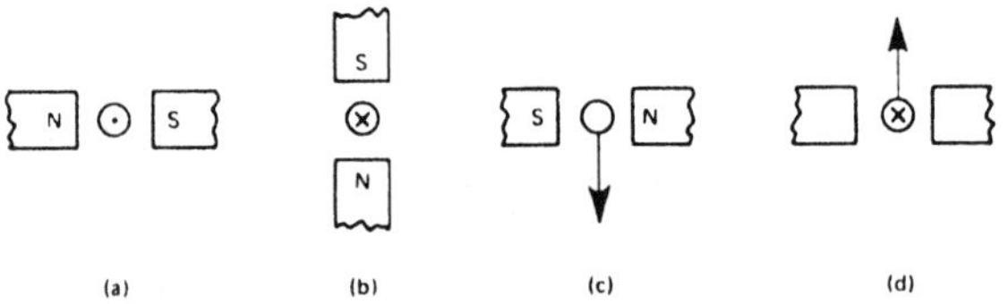

Figure 8.15

l = 350 mm = 0.35 m; I = 10 A; area of pole face, $A = \pi r^2 = \pi(0.06)^2\ \text{m}^2$; Φ = 0.5 mWb = 0.5 × 10^{-3} Wb. Force $F = BIl$, and $B = \Phi/A$, hence

$$\text{force} = \left(\frac{\Phi}{A}\right)Il$$

$$= \frac{(0.5 \times 10^{-3})}{\pi(0.06)^2}(10)(0.35) \text{ newtons}$$

i.e. **force = 0.155 N.**

> Problem 5. With reference to Fig. 8.15 determine (a) the direction of the force on the conductor in Fig. 8.15(a), (b) the direction of the force on the conductor in Fig. 8.15(b), (c) the direction of the current in Fig. 8.15(c), (d) the polarity of the magnetic system in Fig. 8.15(d).

(a) The direction of the main magnetic field is from north to south, i.e. left to right. The

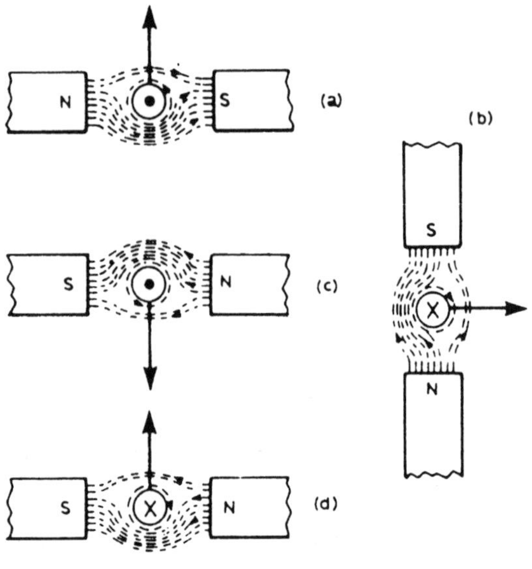

Figure 8.16

current is flowing towards the viewer, and using the screw rule, the direction of the field is anticlockwise. Hence either by Fleming's left-hand rule, or by sketching the interacting magnetic field as shown in Fig. 8.16(a), the direction of the force on the conductor is seen to be upward.

(b) Using a similar method to part (a) it is seen that the force on the conductor is to the right, see Fig. 8.16(b).

(c) Using Fleming's left-hand rule, or by sketching as in Fig. 8.16(c), it is seen that the current is toward the viewer, i.e. out of the paper.

(d) Similar to part (c), the polarity of the magnetic system is as shown in Fig. 8.16(d).

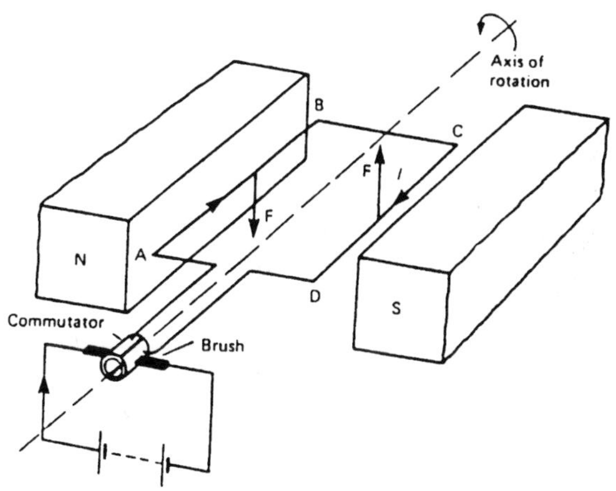

Figure 8.17

Problem 6. A coil is wound on a rectangular former of width 24 mm and length 30 mm. The former is pivoted about an axis passing through the middle of the two shorter sides and is placed in a uniform magnetic field of flux density 0.8 T, the axis being perpendicular to the field. If the coil carries a current of 50 mA, determine the force on each coil side (a) for a single-turn coil, (b) for a coil wound with 300 turns.

(a) Flux density $B = 0.8$ T; length of conductor lying at right angles to field $l = 30$ mm $= 30 \times 10^{-3}$ m; current $I = 50$ mA $= 50 \times 10^{-3}$ A. For a single-turn coil, force on each coil side $F = BIl = 0.8 \times 50 \times 10^{-3} \times 30 \times 10^{-3}$ $=$ $\mathbf{1.2 \times 10^{-3}}$ **N**, or **0.0012 N**.

(b) When there are 300 turns on the coil there are effectively 300 parallel conductors each carrying a current of 50 mA. Thus the total force produced by the current is 300 times that for a single-turn coil. Hence force on coil side, $F = 300\ BIl = 300 \times 0.0012 =$ **0.36 N**.

8.4 Principle of operation of a simple d.c. motor

A rectangular coil which is free to rotate about a fixed axis is shown placed inside a magnetic field produced by permanent magnets in Fig. 8.17. A direct current is fed into the coil via carbon brushes bearing on a commutator, which consists of a metal ring split into two halves separated by insulation. When current flows in the coil a magnetic field is set up around the coil which interacts with the magnetic field produced by the magnets. This causes a force F to be exerted on the current-carrying conductor which, by Fleming's left-hand rule, is downwards between points A and B and upwards between C and D for the current direction shown. This causes a torque and the coil rotates anticlockwise. When the coil has turned through 90° from the position shown in Fig. 8.17 the brushes connected to the positive and negative terminals of the supply make contact with different halves of the commutator ring, thus reversing the direction of the current flow in the conductor. If the current is not reversed and the coil rotates past this position the forces acting on it change direction and it rotates in the opposite direction thus never making more than half a revolution. The current direction is reversed every time the coil swings through the vertical position and thus the coil rotates anticlockwise for as long as the current flows. This is the principle of operation of a d.c. motor which is thus a device that takes in electrical energy and converts it into mechanical energy.

8.5 Principle of operation of a moving-coil instrument

A moving-coil instrument operates on the motor principle. When a conductor carrying current is placed in a magnetic field, a force F is exerted on the conductor, given by $F = BIl$. If the flux

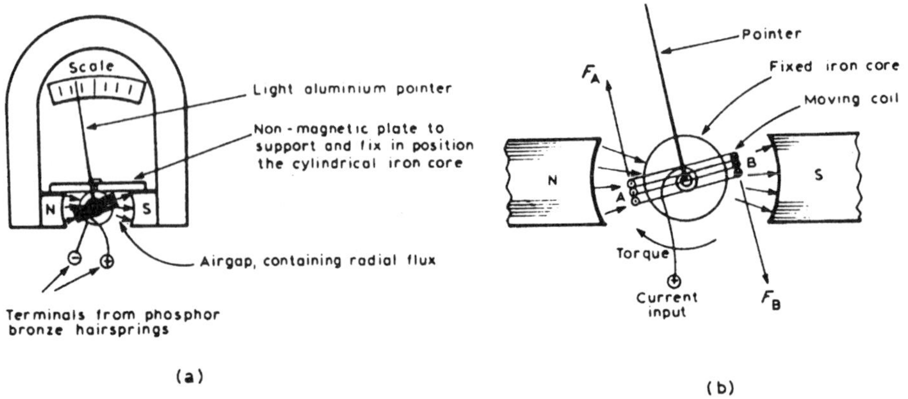

Figure 8.18

density B is made constant (by using permanent magnets) and the conductor is a fixed length (say, a coil) then the force will depend only on the current flowing in the conductor.

In a moving-coil instrument a coil is placed centrally in the gap between shaped pole pieces as shown by the front elevation in Fig. 8.18(a). (The air gap is kept as small as possible, although for clarity it it shown exaggerated in Fig. 8.18.) The coil is supported by steel pivots, resting in jewel bearings, on a cylindrical iron core. Current is led into and out of the coil by two phosphor bronze spiral hairsprings which are wound in opposite directions to minimize the effect of temperature change and to limit the coil swing (i.e. to **control** the movement) and return the movement to zero position when no current flows. Current flowing in the coil produces forces as shown in Fig. 8.18(b), the directions being obtained by Fleming's left-hand rule. The two forces, F_A and F_B, produce a torque which will move the coil in a clockwise direction, i.e. move the pointer from left to right. Since force is proportional to current the scale is linear.

When the aluminium frame, on which the coil is wound, is rotated between the poles of the magnet, small currents (called eddy currents) are induced into the frame, and this provides automatically the necessary **damping** of the system due to the reluctance of the former to move within the magnetic field.

The moving-coil instrument will measure only direct current or voltage and the terminals are marked positive and negative to ensure that the current passes through the coil in the correct direction to deflect the pointer 'up the scale'.

The range of this sensitive instrument is extended by using shunts and multipliers (see Chapter 10).

8.7 Force on a charge

When a charge of Q coulombs is moving at a velocity of v m/s in a magnetic field of flux density B teslas, the charge moving perpendicular to the field, then the magnitude of the force F exerted on the charge is given by:

$$\mathbf{F = QvB \text{ newtons}}$$

Problem 17. An electron in a television tube has a charge of 1.6×10^{-19} N coulombs and travels at 3×10^7 m/s perpendicular to a field of flux density 18.5 μT. Determine the force exerted on the electron in the field.

From above, force $F = QvB$ newtons, where Q = charge in coulombs = 1.6×10^{-19} C; v = velocity of charge = 3×10^7 m/s; and B = flux density = 18.5×10^{-6} T. Hence

force on electron F

$$= 1.6 \times 10^{-19} \times 3 \times 10^7 \times 18.5 \times 10^{-6}$$

$$= 1.6 \times 3 \times 18.5 \times 10^{-18}$$

$$= 88.8 \times 10^{-18} = \mathbf{8.88 \times 10^{-17}\ N}$$

8.7 Multi-choice questions on electromagnetism

(Answers on page 283.)

1. A conductor carries a current of 10 A at right angles to a magnetic field having a flux density of 500 mT. If the length of the conductor in the field is 20 cm, the force on the conductor is:
   (a) 100 kN; (b) 1 kN; (c) 100 N; (d) 1 N.
2. If a conductor is horizontal, the current flowing from left to right and the direction of the surrounding magnetic field is from above to below, the force exerted on the conductor is:
   (a) from left to right;
   (b) from below to above;
   (c) away from the viewer;
   (d) towards the viewer.
3. For the current-carrying conductor lying in the magnetic field shown in Fig. 8.19(a), the direction of the force on the conductor is:
   (a) to the left; (b) upwards;
   (c) to the right; (d) downwards

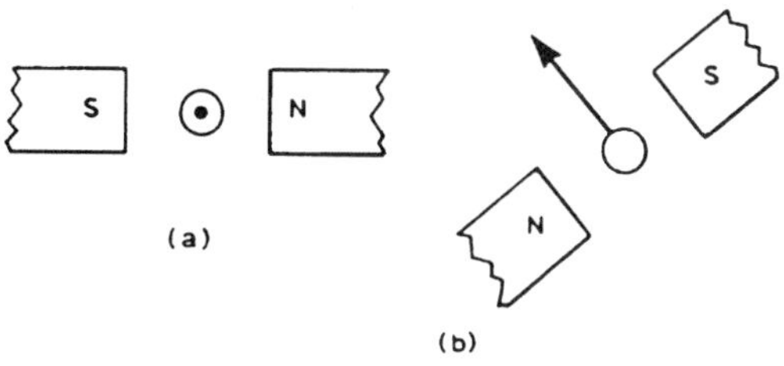

Figure 8.19

4. For the current-carrying conductor lying in the magnetic field shown in Fig. 8.19(b) the direction of the current in the conductor is (a) towards the viewer; (b) away from the viewer.
5. Figure 8.20 shows a rectangular coil of wire placed in a magnetic field and free to rotate about axis AB. If current flows into the coil at C, the coil will:
   (a) commence to rotate anticlockwise;
   (b) commence to rotate clockwise;
   (c) remain in the vertical position;
   (d) experience a force towards the north pole.
6. The force on an electron travelling at 10^7 m/s in a magnetic field of density 10 μT is 1.6×10^{-17} N. The electron has a charge of:

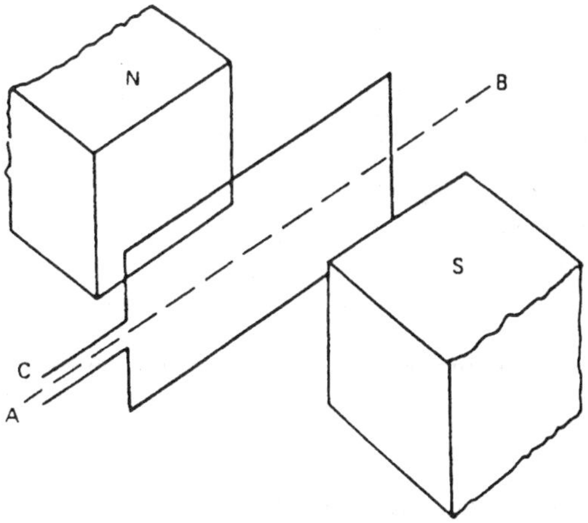

Figure 8.20

   (a) 1.6×10^{-28} C; (b) 1.6×10^{-15} C;
   (c) 1.6×10^{-19} C; (d) 1.6×10^{-25} C.

8.8 Short answer questions on electromagnetism

1. The direction of the magnetic field around a current-carrying conductor may be remembered using the .... rule.
2. Sketch the magnetic field pattern associated with a solenoid connected to a battery and wound on an iron bar. Show the direction of the field.
3. Name three applications of electromagnetism.
4. State what happens when a current-carrying conductor is placed in a magnetic field between two magnets.
5. The force on a current-carrying conductor in a magnetic field depends on four factors. Name them.
6. The direction of the force on a conductor in a magnetic field may be predetermined using Fleming's .... rule.
7. State three applications of the force on a current-carrying conductor.

8.9 Further questions on electromagnetism

1. A conductor carries a current of 70 A at right angles to a magnetic field having a flux

density of 1.5 T. If the length of the conductor in the field is 200 mm calculate the force acting on the conductor. What is the force when the conductor and field are at an angle of 45°? [21.0 N, 14.8 N]

2. Calculate the current required in a 240 mm length of conductor of a d.c. motor when the conductor is situated at right angles to the magnetic field of flux density 1.25 T, if a force of 1.20 N is to be exerted on the conductor. [4.0 A]
3. A conductor 30 cm long is situated at right angles to a magnetic field. Calculate the strength of the magnetic field if a current of 15 A in the conductor produces a force on it of 3.6 N. [0.80 T]
4. A conductor 300 mm long carries a current of 13 A and is at right angles to a magnetic field between two circular pole faces, each of diameter 80 mm. If the total flux between the pole faces is 0.75 mWb calculate the force exerted on the conductor. [0.582 N]
5. (a) A 400 mm length of conductor carrying a current of 25 A is situated at right angles to a magnetic field between two poles of an electric motor. The poles have a circular cross-section. If the force exerted on the conductor is 80 N and the total flux between the pole faces is 1.27 mWb, determine the diameter of a pole face.
   (b) If the conductor in part (a) is vertical, the current flowing downwards and the direction of the magnetic field is from left to right, what is the direction of the 80 N force?
   [(a) 14.2 mm (b) towards the viewer]
6. A coil is wound uniformly on a former having a width of 18 mm and a length of 25 mm. The former is pivoted about an axis passing through the middle of the two shorter sides and is placed in a uniform magnetic field of flux density 0.75 T, the axis being perpendicular to the field. If the coil carries a current of 120 mA, determine the force exerted on each coil side, (a) for a single-turn coil, (b) for a coil wound with 400 turns.
   [(a) 2.25×10^{-3} N (b) 0.9 N]
7. Figure 8.21 shows a simplified diagram of a section through the coil of a moving-coil instrument. For the direction of current flow shown in the coil determine the direction that the pointer will move. [To the right]

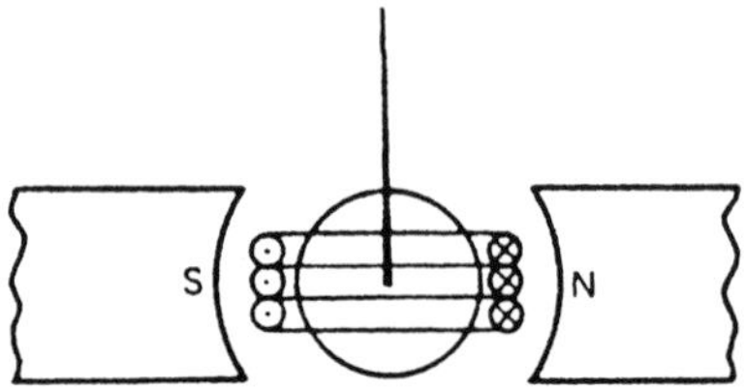

Figure 8.21

8. Explain, with the aid of a sketch, the action of a simplified d.c. motor.
9. Sketch and label the movement of a moving-coil instrument. Briefly explain the principle of operation of such an instrument.
10. Calculate the force exerted on a charge of 2×10^{-18} C travelling at 2×10^{6} m/s perpendicular to a field of density 2×10^{-7} T.
   [8×10^{-19} N]
11. Determine the speed of a 10^{-19} C charge travelling perpendicular to a field of flux density 10^{-7} T, if the force on the charge is 10^{-20} N. [10^{6} m/s]

9

Electromagnetic induction

At the end of this chapter you should be able to:

- understand how an e.m.f. may be induced in a conductor
- state Faraday's laws of electromagnetic induction
- state Lenz's law
- use Fleming's right-hand rule for relative directions
- appreciate that the induced e.m.f., $E = Blv$ or $E = Blv \sin \theta$
- calculate induced e.m.f. given B, l, v and θ and determine relative directions
- define inductance L and state its unit
- define mutual inductance
- appreciate that

  $$\text{e.m.f. } E = N\left(\frac{\Delta\Phi}{t}\right) = L\left(\frac{\Delta I}{t}\right)$$
- calculate induced e.m.f. given N, t, L, change of flux or change of current
- appreciate factors which affect the inductance of an inductor
- draw the circuit diagram symbols for inductors
- calculate the energy stored in an inductor using $W = \frac{1}{2}LI^2$ joules
- calculate inductance L of a coil, given $L = N\Phi/I$
- calculate mutual inductance using $E_2 = M(\Delta I_1/t)$

9.1 Introduction to electromagnetic induction

When a conductor is moved across a magnetic field so as to cut through the lines of force (or flux), an electromotive force (e.m.f.) is produced in the conductor. If the conductor forms part of a closed circuit then the e.m.f. produced causes an electric current to flow round the circuit. Hence an e.m.f. (and thus current) is 'induced' in the conductor as a result of its movement across the magnetic field. This effect is known as **electromagnetic induction**.

Figure 9.1(a) shows a coil of wire connected to a centre-zero galvanometer, which is a sensitive ammeter with the zero-current position in the centre of the scale.

(a) When the magnet is moved at constant speed towards the coil (Fig. 9.1(a)), a deflection is noted on the galvanometer showing that a current has been produced in the coil.
(b) When the magnet is moved at the same speed as in (a) but away from the coil the same deflection is noted but is in the opposite direction (see Fig. 9.1(b)).
(c) When the magnet is held stationary, even within the coil, no deflection is recorded.

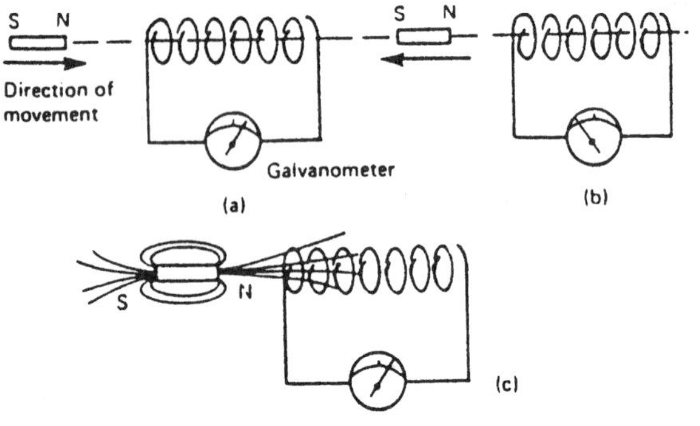

Figure 9.1

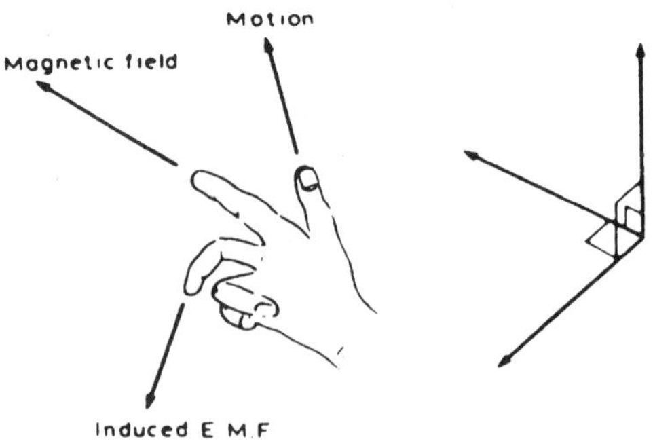

Figure 9.2

(d) When the coil is moved at the same speed as in (a) and the magnet held stationary the same galvanometer deflection is noted.
(e) When the relative speed is, say, doubled, the galvanometer deflection is doubled.
(f) When a stronger magnet is used, a greater galvanometer deflection is noted.
(g) When the number of turns of wire of the coil is increased, a greater galvanometer deflection is noted.

Figure 9.1(c) shows the magnetic field associated with the magnet. As the magnet is moved towards the coil, the magnetic flux of the magnet moves across, or cuts, the coil. **It is the relative movement of the magnetic flux and the coil that causes an e.m.f., and thus current, to be induced in the coil.** This effect is known as electromagnetic induction. The laws of electromagnetic induction stated in section 9.2 evolved from experiments such as those described above.

9.2 Laws of electromagnetic induction

Faraday's laws of electromagnetic induction state:

(i) An induced e.m.f. is set up whenever the magnetic field linking that circuit changes.
(ii) The magnitude of the induced e.m.f. in any circuit is proportional to the rate of change of the magnetic flux linking the circuit.

Lenz's law states:

The direction of an induced e.m.f. is always such that it tends to set up a current opposing the motion or the change of flux responsible for inducing that e.m.f.

An alternative method to Lenz's law of determining relative directions is given by **Fleming's Right-hand rule** (often called the geneRator rule) which states:

Let the thumb, first finger and second finger of the right hand be extended such that they are all at right angles to each other [as shown in Fig. 9.2]. If the first finger points in the direction of the magnetic field, the thumb points in the direction of motion of the conductor relative to the magnetic field, then the second finger will point in the direction of the induced e.m.f.

Summarizing:

First finger – Field
ThuMb – Motion
SEcond finger – E.m.f.

In a generator, conductors forming an electric circuit are made to move through a magnetic field. By Faraday's law, an e.m.f. is induced in the conductors and thus a source of e.m.f. is created. A generator converts mechanical energy into electrical energy. (The action of a simple a.c.

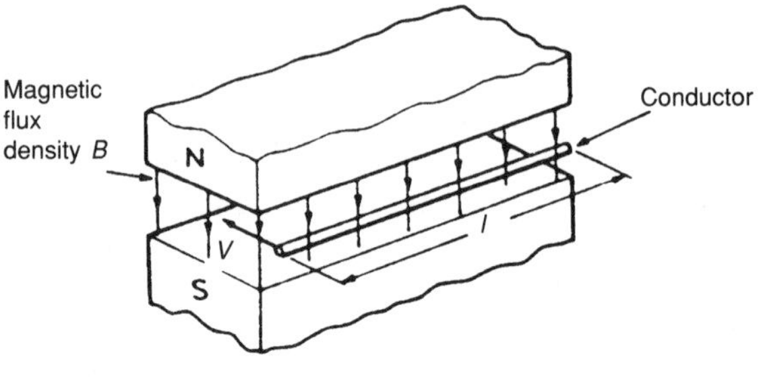

Figure 9.3

generator is described in Chapter 12.) The induced e.m.f. E set up between the ends of the conductor shown in Fig. 9.3 is given by

$$\boldsymbol{E = Blv} \textbf{ volts}$$

where B, the flux density, is measured in teslas, l, the length of conductor in the magnetic field, is measured in metres, and v, the conductor velocity, is measured in metres per second.

If the conductor moves at an angle θ°C to the magnetic field (instead of at 90° as assumed above) then

$$\boldsymbol{E = Blv \sin \theta}$$

Problem 1. A conductor 300 mm long moves at a uniform speed of 4 m/s at right angles to a uniform magnetic field of flux density 1.25 T. Determine the current flowing in the conductor when (a) its ends are open-circuited, (b) its ends are connected to a load of 20 Ω resistance.

When a conductor moves in a magnetic field it will have an e.m.f. induced in it but this e.m.f. can only produce a current if there is a closed circuit. Induced e.m.f. $E = Blv = (1.25)(300/1000)(4) = 1.5$ V.

(a) If the ends of the conductor are open-circuited no current will flow even though 1.5 V has been induced.

(b) From Ohm's law,

$$I = \frac{E}{R} = \frac{1.5}{20} = \mathbf{0.075\ A} \text{ or } \mathbf{75\ mA}$$

Problem 2. At what velocity must a conductor 75 mm long cut a magnetic field of flux density 0.6 T if an e.m.f. of 9 V is to be induced in it? Assume the conductor, the field and the direction of motion are mutually perpendicular.

Induced e.m.f. $E = Blv$, hence velocity $v = E/Bl$, i.e.

$$v = \frac{9}{(0.6)(75 \times 10^{-3})} = \frac{9 \times 10^3}{0.6 \times 75} = \mathbf{200\ m/s}$$

Problem 3. A conductor moves with a velocity of 15 m/s at an angle of (a) 90°, (b) 60° and (c) 30° to a magnetic field produced between two square-faced poles of side length 2 cm. If the flux leaving a pole face is 5 μWb, find the magnitude of the induced e.m.f. in each case.

$v = 15$ m/s; length of conductor in magnetic field, $l = 2$ cm $= 0.02$ m; $A = 2 \times 20\ \text{cm}^2 = 4 \times 10^{-4}\ \text{m}^2$, $\Phi = 5 \times 10^{-6}$ Wb.

(a) $E_{90} = Blv \sin 90°$

$$= \frac{\Phi}{A} lv = \frac{(5 \times 10^{-6})}{(4 \times 10^{-6})}(0.02)(15) = \mathbf{3.75\ mV}$$

(b) $E_{60} = Blv \sin 60°$

$$= E_{90} \sin 60° = 3.75 \sin 60° = \mathbf{3.25\ mV}$$

(c) $E_{30} = Blv \sin 30°$

$$= E_{90} \sin 30° = 3.75 \sin 30° = \mathbf{1.875\ mV}$$

Problem 4. The wing span of a metal aeroplane is 36 m. If the aeroplane is flying at 400 km/h, determine the e.m.f. induced between its wing tips. Assume the vertical component of the earth's magnetic field is 40 μT.

Induced e.m.f. across wing tips, $E = Blv$. $B = 40\ \mu\text{t} = 40 \times 10^{-6}$ T; $l = 36$ m.

$$v = 400 \frac{\text{km}}{\text{h}} \times 1000 \frac{\text{m}}{\text{km}} \times \frac{1\ \text{h}}{60 \times 60\ \text{s}}$$

$$= \frac{(400)(1000)}{3600} = \frac{4000}{36}\ \text{m/s}$$

Hence $E = (40 \times 10^{-6})(36)(4000/36) = \mathbf{0.16\ V}$.

Problem 5. The diagram shown in Fig. 9.4 represents the generation of e.m.f.s. Determine (i) the direction in which the conductor has to be moved in Fig. 9.4(a), (ii) the direction of the induced e.m.f. in Fig. 9.4(b), (iii) the polarity of the magnetic system in Fig. 9.4(c).

The direction of the e.m.f., and thus the current due to the e.m.f., may be obtained by either Lenz's law or Fleming's Right-hand rule (i.e. GeneRator rule).

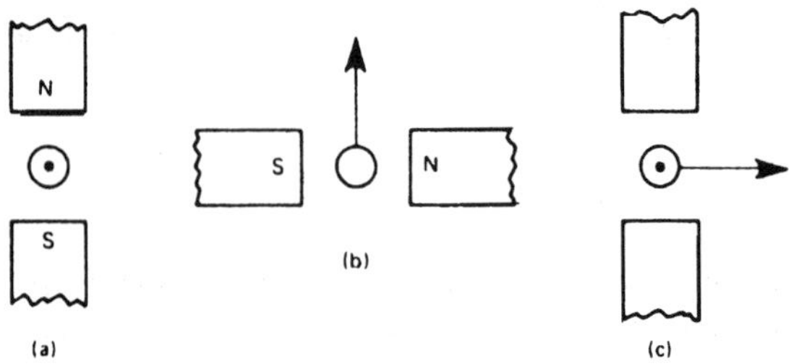

Figure 9.4

(i) Using Lenz's law: The field due to the magnet and the field due to the current-carrying conductor are shown in Fig. 9.5(a) and are seen to reinforce to the left of the conductor. Hence the force on the conductor is to the right. However, Lenz's law states that the direction of the induced e.m.f. is always such as to oppose the effect producing it. **Thus the conductor will have to be moved to the left.**

(ii) Using Fleming's right-hand rule:
First finger – Field, i.e. N → S, or right to left;
ThuMb – Motion, i.e. upwards;
SEcond finger – E.m.f., i.e. **towards the viewer or out of the paper**, as shown in Fig. 9.5(b).

(iii) The polarity of the magnetic system of Fig. 9.4(c) is shown in Fig. 9.5(c) and is obtained using Fleming's right-hand rule.

9.3 Inductance

Inductance is the name given to the property of a circuit whereby there is an e.m.f. induced into the circuit by the change of flux linkages produced by a current change.

When the e.m.f. is induced in the same circuit as that in which the current is changing, the property is called **self-inductance**, ***L***. When the e.m.f. is induced in a circuit by a change of flux due to current changing in an adjacent circuit, the property is called **mutual inductance**, ***M***. The unit of inductance is the **henry**, **H**.

A circuit has an inductance of one henry when an e.m.f. of one volt is induced in it by a current changing at the rate of one ampere per second.

Induced e.m.f. in a coil of N turns,

$$\boldsymbol{E = N\left(\frac{\Delta\Phi}{t}\right)} \textbf{ volts}$$

where $\Delta\Phi$ is the change in flux, in webers, and t is the time taken for the flux to change, in seconds.
Induced e.m.f. in a coil of inductance L henrys,

$$\boldsymbol{E = L\left(\frac{\Delta I}{t}\right)} \textbf{ volts}$$

where ΔI is the change in current, in amperes, and t is the time taken for the current to change, in seconds.

The above equations for e.m.f. E are often stated as $E = -N(\Delta\Phi/t)$ and $E = -L(\Delta I/t)$, the minus sign reminding us of its direction (given by Henry's law). In the following problems, the direction of the e.m.f. is, however, assumed.

Problem 6. Determine the e.m.f. induced in a coil of 200 turns when there is a change of flux of 25 mWb linking with it in 50 ms.

$$\text{Induced e.m.f. } E = N\left(\frac{\Delta\Phi}{t}\right)$$

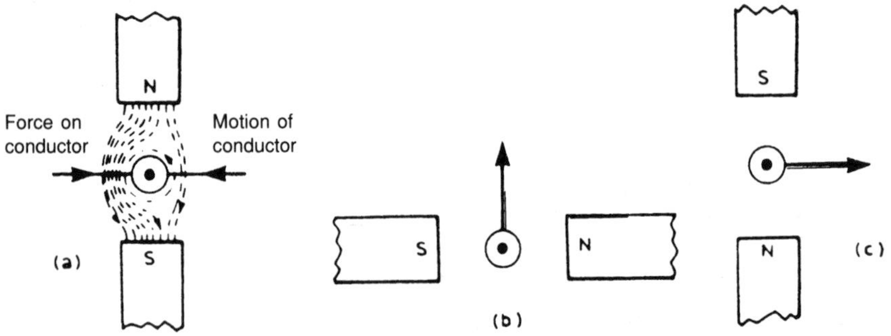

Figure 9.5

$$= (200)\left(\frac{25 \times 10^{-3}}{50 \times 10^{-3}}\right) = \mathbf{100\ volts}$$

> Problem 7. A flux of 400 μWb passing through a 150 turn coil is reversed in 40 ms. Find the average e.m.f. induced.

Since the flux reverses, the flux changes from +400 μWb to –400 μWb, a total change of flux of 800 μWb.

$$\text{Induced e.m.f. } E = N\left(\frac{\Delta\Phi}{t}\right)$$

$$= (150)\left(\frac{800 \times 10^{-6}}{40 \times 10^{-3}}\right) = \frac{800 \times 150 \times 10^3}{40 \times 10^6}$$

Hence the average e.m.f. induced, E = **3 volts.**

> Problem 8. Calculate the e.m.f. induced in a coil of inductance 12 H by a current changing at the rate of 4 A/s.

$$\text{Induced e.m.f. } E = L\left(\frac{\Delta I}{t}\right) = (12)\,(4)$$

$$= \mathbf{48\ volts}$$

> Problem 9. An e.m.f. of 1.5 kV is induced in a coil when a current of 4 A collapses uniformly to zero in 8 ms. Determine the inductance of the coil.

Change in current, $\Delta I = (4 - 0) = 4$ A; $t = 8$ ms $= 8 \times 10^{-3}$ s;

$$\frac{\Delta I}{t} = \frac{4}{8 \times 10^{-3}} = \frac{4000}{8} = 500 \text{ A/s}$$

E = 1.5 kV = 1500 V. Since $E = L(\Delta I/t)$,

$$L = \frac{E}{(\Delta I/t)} = \frac{1500}{500} = \mathbf{3\ H}$$

> Problem 10. An average e.m.f. of 40 V is induced in a coil of inductance 150 mH when a current of 6 A is reversed. Calculate the time taken for the current to reverse.

E = 40 V, L = 150 mH = 0.15 H; change in current, $I = 6 - (-6) = 12$ A (since the current is reversed). Since $E = L(\Delta I/t)$,

$$\text{time } t = \frac{L(\Delta I)}{E} = \frac{(0.15)(12)}{40} = \mathbf{0.045}\text{ s or }\mathbf{45\ ms}$$

9.4 Inductors

A component called an inductor is used when the property of inductance is required in a circuit. The basic form of an inductor is simply a coil of wire. Factors which affect the inductance of an inductor include:

(i) the number of turns of wire – the more turns the higher the inductance;
(ii) the cross-sectional area of the coil of wire – the greater the cross-sectional area the higher the inductance;
(iii) the presence of a magnetic core – when the coil is wound on an iron core the same current sets up a more concentrated magnetic field and the inductance is increased;
(iv) the way the turns are arranged – a short thick coil of wire has a higher inductance than a long thin one.

Two examples of practical inductors are shown in Fig. 9.6, and the standard electrical circuit diagram symbols for air-cored and iron-cored inductors are shown in Fig. 9.7.

An iron-cored inductor is often called a **choke** since, when used in a.c. circuits, it has a choking effect, limiting the current flowing through it.

Inductance is often undesirable in a circuit. To

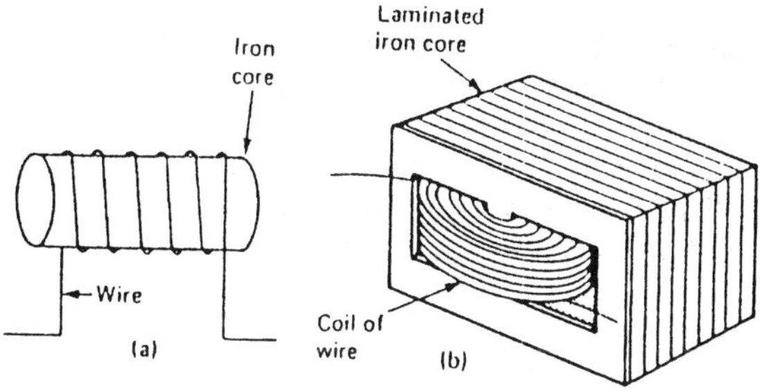

Figure 9.6

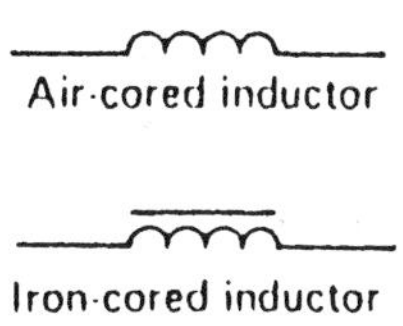

Figure 9.7

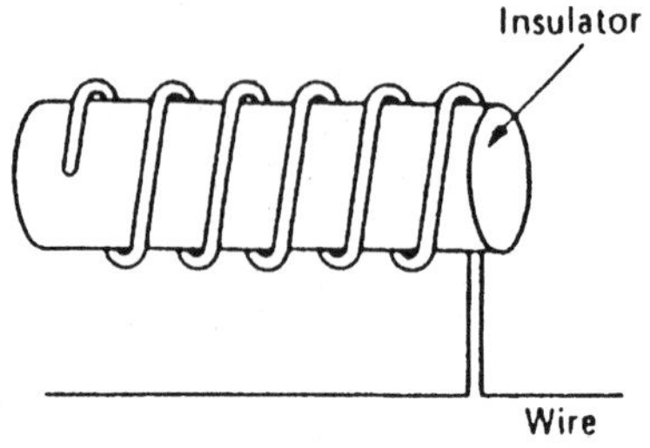

Figure 9.8

reduce inductance to a minimum the wire may be bent back on itself, as shown in Fig. 9.8, so that the magnetizing effect of one conductor is neutralized by that of the adjacent conductor. The wire may be coiled around an insulator, as shown, without increasing the inductance. Standard resistors may be non-inductively wound in this manner.

9.5 Energy stored

An inductor possesses an ability to store energy. The energy stored, W, in the magnetic field of an inductor is given by:

$$W = \tfrac{1}{2} LI^2 \text{ joules}$$

Problem 11. An 8 H inductor has a current of 3 A flowing through it. How much energy is stored in the magnetic field of the inductor?

$$\text{Energy stored, } W = \frac{1}{2} LI^2 = \frac{1}{2}(8)(3)^2$$

$$= \mathbf{36\ joules}$$

9.6 Inductance of a coil

If a current changing from 0 to I amperes, produces a flux change from 0 to Φ webers, then $\Delta I = I$ and $\Delta\Phi = \Phi$. Then, from section 9.3, induced e.m.f. $E = N\Phi/t = LI/t$, from which

$$\textbf{inductance of coil, } L = \frac{N\Phi}{I} \textbf{ henrys}$$

Problem 12. Calculate the coil inductance when a current of 4 A in a coil of 800 turns produces a flux of 5 mWb linking with the coil.

For a coil,

$$\text{inductance } L = \frac{N\Phi}{I} = \frac{(800)(5 \times 10^{-3})}{4} = \mathbf{1\ H}$$

Problem 13. A flux of 25 mWb links with a 1500 turn coil when a current of 3 A passes through the coil. Calculate (a) the inductance of the coil, (b) the energy stored in the magnetic field, and (c) the average e.m.f. induced if the current falls to zero in 150 ms.

(a) **Inductance,** $L = \frac{N\Phi}{I}$

$$= \frac{(1500)(25 \times 10^{-3})}{3} = \mathbf{12.5\ H}$$

(b) **Energy stored in field,** $W = \frac{1}{2} LI^2$

$$= \frac{1}{2}(12.5)(3)^2 = \mathbf{56.25\ J}$$

(c) Induced e.m.f., $E = N\left(\frac{\Delta\Phi}{t}\right)$

$$= (1500)\left(\frac{25 \times 10^{-3}}{150 \times 10^{-3}}\right) = \mathbf{250\ V}$$

Problem 14. When a current of 1.5 A flows in a coil the flux linking with the coil is 90 μWb. If the coil inductance is 0.60 H calculate the number of turns of the coil.

For a coil, $L = N\Phi/I$. Thus

$$N = \frac{LI}{\Phi} = \frac{(0.6)(1.5)}{90 \times 10^{-6}} = \mathbf{10\,000\ turns}$$

Problem 15. When carrying a current of 3 A, a coil of 750 turns has a flux of 12 mWb linking with it. Calculate the coil inductance and the e.m.f. induced in the coil when the current collapses to zero in 18 ms.

Coil inductance, $L = \frac{N\Phi}{I}$

$= \frac{(750)(12 \times 10^{-3})}{3} = \mathbf{3\ H}$

Induced e.m.f. $E = L\left(\frac{\Delta I}{t}\right)$

$= 3\left(\frac{3-0}{18 \times 10^{-3}}\right) = \mathbf{500\ V}$

(Alternatively

$$E = N\left(\frac{\Delta\Phi}{t}\right) = (750)\left(\frac{12 \times 10^{-3}}{18 \times 10^{-3}}\right) = \mathbf{500\ V})$$

9.7 Mutual inductance

Mutually induced e.m.f. in the second coil, $E_2 = M(\Delta I_1/t)$ volts, where M is the mutual inductance between two coils in henrys, ΔI_1 is the change in current in the first coil, in amperes, and t is the time the current takes to change in the first coil, in seconds.

The phenomenon of mutual inductance is used in **transformers** (see Chapter 17).

> Problem 16. Calculate the mutual inductance between two coils when a current changing at 200 A/s in one coil induces an e.m.f. of 1.5 V in the other.

Induced e.m.f. $E_2 = M(\Delta I_1/t)$, or 1.5 = M(200); thus mutual inductance,

$$M = \frac{1.5}{200} = \mathbf{0.0075\ H} \text{ or } \mathbf{7.5\ mH}$$

> Problem 17. The mutual inductance between two coils is 18 mH. Calculate the steady rate of change of current in one coil to induce an e.m.f. of 0.72 V in the other.

Induced e.m.f., $E_2 = M(\Delta I_1/t)$; hence rate of change of current,

$$\left(\frac{\Delta I_1}{t}\right) = \frac{E_2}{M} = \frac{0.72}{0.018} = \mathbf{40\ A/s}$$

> Problem 18. Two coils have a mutual inductance of 0.2 H. If the current in one coil is changed from 10 A to 4 A in 10 ms, calculate (a) the average induced e.m.f. in the second coil, (b) the change of flux linked with the second coil if it is wound with 500 turns.

(a) Induced e.m.f.

$$E_2 = M\left(\frac{\Delta I_1}{t}\right) = (0.2)\left(\frac{10-4}{10 \times 10^{-3}}\right) = \mathbf{120\ V}$$

(b) Induced e.m.f.

$$E = N\left(\frac{\Delta\Phi}{t}\right), \text{ hence } \Delta\Phi = \frac{Et}{N}$$

Thus the change of flux,

$$\Delta\Phi = \frac{120(10 \times 10^{-3})}{500} = \mathbf{24\ mWb}$$

9.8 Multi-choice questions on electromagnetic induction

(Answers on page 283.)

1. A current changing at a rate of 5 A/s in a coil of inductance 5 H induces an e.m.f. of:
   (a) 25 V in the same direction as the applied voltage;
   (b) 1 V in the same direction as the applied voltage;
   (c) 25 V in the opposite direction to the applied voltage;
   (d) 1 V in the opposite direction to the applied voltage.
2. A bar magnet is moved at a steady speed of 1.0 m/s towards a coil of wire which is connected to a centre-zero galvanometer. The magnet is now withdrawn along the same path at 0.5 m/s. The deflection of the galvanometer is in the:
   (a) same direction as previously, with the magnitude of the deflection doubled;
   (b) opposite direction as previously, with the magnitude of the deflection halved;
   (c) same direction as previously, with the magnitude of the deflection halved;
   (d) opposite direction as previously, with the magnitude of the deflection doubled.

3. When a magnetic flux of 10 Wb links with a circuit of 20 turns in 2 s, the induced e.m.f. is:
(a) 1 V; (b) 4 V; (c) 100 V; (d) 400 V.
4. A current of 10 A in a coil of 1000 turns produces a flux of 10 mWb linking with the coil. The coil inductance is:
(a) 10^6 H; (b) 1 H; (c) 1 H; (d) 1 mH.
5. An e.m.f. of 1 V is induced in a conductor moving at 10 cm/s in a magnetic field of 0.5 T. The effective length of the conductor in the magnetic field is:
(a) 20 cm; (b) 5 m; (c) 20 m; (d) 50 m.
6. Which of the following statements is false?
(a) Fleming's left-hand rule or Lenz's law may be used to determine the direction of an induced e.m.f.
(b) An induced e.m.f. is set up whenever the magnetic field linking that circuit changes.
(c) The direction of an induced e.m.f. is always such as to oppose the effect producing it.
(d) The induced e.m.f. in any circuit is proportional to the rate of change of the magnetic flux linking the circuit.
7. The effect of inductance occurs in an electrical circuit when:
(a) the resistance is changing;
(b) the flux is changing;
(c) the current is changing.
8. Which of the following statements is false? The inductance of an inductor increases:
(a) with a short, thick coil;
(b) when wound on an iron core;
(c) as the number of turns increases;
(d) as the cross-sectional area of the coil decreases.
9. The mutual inductance between two coils, when a current changing at 20 A/s in one coil induces an e.m.f. of 10 mV in the other, is:
(a) 0.5 H; (b) 200 mH; (c) 0.5 mH; (d) 2 H.

9.9 Short answer questions on electromagnetic induction

1. What is electromagnetic induction?
2. State Faraday's laws of electromagnetic induction.
3. State Lenz's law.
4. Explain briefly the principle of the generator.
5. The direction of an induced e.m.f. in a generator may be determined using Fleming's .... rule.
6. The e.m.f. E induced in a moving conductor may be calculated using the formula $E = Blv$. Name the quantities represented and their units.
7. What is self-inductance? State its symbol.
8. State and define the unit of inductance.
9. When a circuit has an inductance L and the current changes at a rate of $(\Delta I/t)$ then the induced e.m.f. E is given by E = ... volts.
10. If a current of I amperes flowing in a coil of N turns produces a flux of Φ webers, the coil inductance L is given by L = ... henrys.
11. The energy W stored by an inductor is given by W = ... joules.
12. What is mutual inductance? State its symbol.
13. The mutual inductance between two coils is M. The e.m.f. E_2 induced in one coil by the current changing at $(\Delta I_1/t)$ in the other is given by E_2 = ... volts.

9.10 Further questions on electromagnetic induction

Induced e.m.f.

1. A conductor of length 15 cm is moved at 750 mm/s at right angles to a uniform flux density of 1.2 T. Determine the e.m.f. induced in the conductor. [0.135 V]
2. Find the speed that a conductor of length 120 mm must be moved at right angles to a magnetic field of flux density 0.6 T to induce in it an e.m.f. of 1.8 V. [25 m/s]
3. A 25 cm long conductor moves at a uniform speed of 8 m/s through a uniform magnetic field of flux density 1.2 T. Determine the current flowing in the conductor when (a) its ends are open-circuited, (b) its ends are connected to a load of 15 ohms resistance. [(a) 0 (b) 0.16 A]
4. A straight conductor 500 mm long is moved with constant velocity at right angles both to its length and to a uniform magnetic field. Given that the e.m.f. induced in the conductor is 2.5 V and the velocity is 5 m/s, calculate the flux density of the magnetic field. If the conductor forms part of a closed circuit of total resistance 5 ohms, calculate the force on the conductor. [1 T, 0.25 N]
5. A car is travelling at 80 km/h. Assuming the back axle of the car is 1.76 m in length and the vertical component of the earth's

magnetic field is 40 μT, find the e.m.f. generated in the axle due to motion. [1.56 mV]

6. A conductor moves with a velocity of 20 m/s at an angle of (a) 90° (b) 45° (c) 30°, to a magnetic field produced between two square-faced poles of side length 2.5 cm. If the flux on the pole face is 60 mWb, find the magnitude of the induced e.m.f. in each case. [(a) 48 V (b) 33.9 V (c) 24 V]
7. Find the e.m.f. induced in a coil of 200 turns when there is a change of flux of 30 mWb linking with it in 40 ms. [150 V]
8. An e.m.f. of 25 V is induced in a coil of 300 turns when the flux linking with it changes by 12 mWb. Find the time, in milliseconds, in which the flux makes the change. [144 ms]
9. An ignition coil having 10 000 turns has an e.m.f. of 8 kV induced in it. What rate of change of flux is required for this to happen? [0.8 Wb/s]
10. A flux of 0.35 mWb passing through a 125 turn coil is reversed in 25 ms. Find the average e.m.f. induced. [3.5 V]
11. Calculate the e.m.f. induced in a coil of inductance 6 H by a current changing at a rate of 15 A/s. [90 V]

Energy stored

12. An inductor of 20 H has a current of 2.5 A flowing in it. Find the energy stored in the magnetic field of the inductor. [62.5 J]
13. Calculate the value of the energy stored when a current of 30 mA is flowing in a coil of inductance 400 mH. [0.18 mJ]
14. The energy stored in the magnetic field of an inductor is 80 J when the current flowing in the inductor is 2 A. Calculate the inductance of the coil. [40 H]

Inductance of a coil

15. A flux of 30 mWb links with a 1200 turn coil when a current of 5 A is passing through the coil. Calculate (a) the inductance of the coil, (b) the energy stored in the magnetic field, and (c) the average e.m.f. induced if the current is reduced to zero in 0.20 s. [(a) 7.2 H (b) 90 J (c) 180 V]
16. An e.m.f. of 2 kV is induced in a coil when a current of 5 A collapses uniformly to zero in 10 ms. Determine the inductance of the coil. [4 H]
17. An average e.m.f. of 50 V is induced in a coil of inductance 160 mH when a current of 7.5 A is reversed. Calculate the time taken for the current to reverse. [40 ms]
18. A coil of 2500 turns has a flux of 10 mWb linking with it when carrying a current of 2 A. Calculate the coil inductance and the e.m.f. induced in the coil when the current collapses to zero in 20 ms. [12.5 H, 1.25 kV]
19. Calculate the coil inductance when a current of 5 A in a coil of 1000 turns produces a flux of 8 mWb linking with the coil. [1.6 H]
20. A coil is wound with 600 turns and has a self-inductance of 2.5 H. What current must flow to set up a flux of 20 mWb? [4.8 A]
21. When a current of 2 A flows in a coil, the flux linking with the coil is 80 μWb. If the coil inductance is 0.5 H, calculate the number of turns of the coil. [12 500]
22. A coil of 1200 turns has a flux of 15 mWb linking with it when carrying a current of 4 A. Calculate the coil inductance and the e.m.f. induced in the coil when the current collapses to zero in 25 ms. [4.5 H, 720 V]
23. A coil has 300 turns and an inductance of 4.5 mH. How many turns would be needed to produce a 0.72 mH coil assuming the same core is used? [120 turns]
24. A steady current of 5 A when flowing in a coil of 1000 turns produces a magnetic flux of 500 μWb. Calculate the inductance of the coil. The current of 5 A is then reversed in 12.5 ms. Calculate the e.m.f. induced in the coil. [0.1 H, 80 V]

Mutual inductance

25. The mutual inductance between two coils is 150 mH. Find the e.m.f. induced in one coil when the current in the other is increasing at a rate of 30 A/s. [4.5 V]
26. Determine the mutual inductance between two coils when a current changing at 50 A/s in one coil induces an e.m.f. of 80 mV in the other. [1.6 mH]
27. Two coils have a mutual inductance of 0.75 H. Calculate the e.m.f. induced in one

coil when a current of 2.5 A in the other coil is reversed in 15 ms. [250 V]

28. The mutual inductance between two coils is 240 mH. If the current in one coil changes from 15 A to 6 A in 12 ms, calculate (a) the average e.m.f. induced in the other coil, (b) the change of flux linked with the other coil if it is wound with 400 turns. [(a) 180 V (b) 5.4 mWb]

29. A mutual inductance of 0.6 H exists between two coils. If a current of 6 A in one coil is reversed in 0.8 s calculate (a) the average e.m.f. induced in the other coil, (b) the number of turns on the other coil if the flux change linking with the other coil is 5 mWb. [(a) 0.9 V (b) 144]

10

Electrical measuring instruments and measurements

At the end of this chapter you should be able to:

- recognize the importance of testing and measurements in electric circuits
- appreciate the essential devices comprising an analogue instrument
- explain the operation of an attraction and a repulsion type of moving-iron instrument
- explain the operation of a moving-coil rectifier instrument
- compare moving-coil, moving-iron and moving-coil rectifier instruments
- calculate values of shunts for ammeters and multipliers for voltmeters
- understand the advantages of electronic instruments
- understand the operation of an ohmmeter/megger
- appreciate the operation of multimeters/Avometers
- understand the operation of a wattmeter
- appreciate instrument 'loading' effect
- understand the operation of a CRO for d.c. and a.c. measurements
- calculate periodic time, frequency, peak to peak values from waveforms on a CRO
- recognize harmonics present in complex waveforms
- determine ratios of powers, currents and voltages in decibels
- understand null methods of measurement for a Wheatstone bridge and d.c. potentiometer
- understand the operation of a.c. bridges
- understand the operation of a Q-meter
- appreciate the most likely source of errors in measurements
- appreciate calibration accuracy of instruments

10.1 Introduction

Tests and measurements are important in designing, evaluating, maintaining and servicing electrical circuits and equipment. In order to detect electrical quantities such as current, voltage, resistance or power, it is necessary to transform an electrical quantity or condition into a visible indication. This is done with the aid of instruments (or meters) that indicate the magnitude of quantities either by the position of a pointer moving over a graduated scale (called an analogue instrument) or in the form of a decimal number (called a digital instrument).

10.2 Analogue instruments

All analogue electrical indicating instruments require three essential devices:

(a) **A deflecting** or **operating device**. A mechanical force is produced by the current or voltage which causes the pointer to deflect from its zero position.

(b) **A controlling device**. The controlling force acts in opposition to the deflecting force and ensures that the deflection shown on the meter is always the same for a given measured quantity. It also prevents the pointer always going to the maximum deflection. There are two main types of controlling device – spring control and gravity control.

(c) **A damping device**. The damping force ensures that the pointer comes to rest in its final position quickly and without undue oscillation. There are three main types of damping used – eddy-current damping, air-friction damping and fluid-friction damping.

There are basically two types of scale – linear and non-linear. A **linear scale** is shown in Fig. 10.1(a), where the divisions or graduations are evenly spaced. The voltmeter shown has a range 0–100 V, i.e. a full-scale deflection (f.s.d.) of 100 V. A **non-linear scale** is shown in Fig. 10.1(b). The scale is cramped at the beginning and the graduations are uneven throughout the range. The ammeter shown has a f.s.d. of 10 A.

10.3 Moving-iron instrument

(a) An **attraction type** of moving-iron instrument is shown diagrammatically in Fig. 10.2(a).

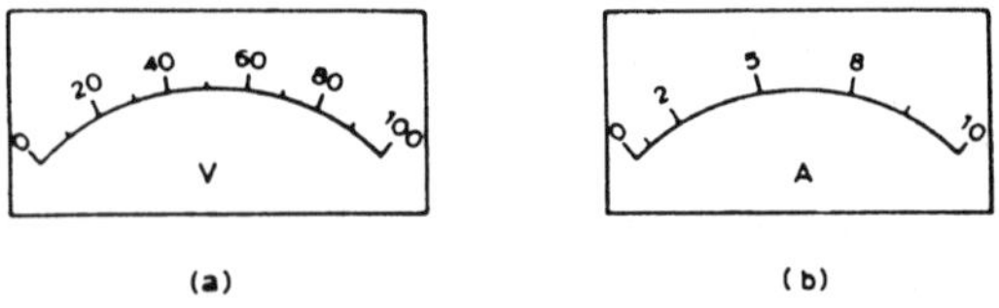

Figure 10.1

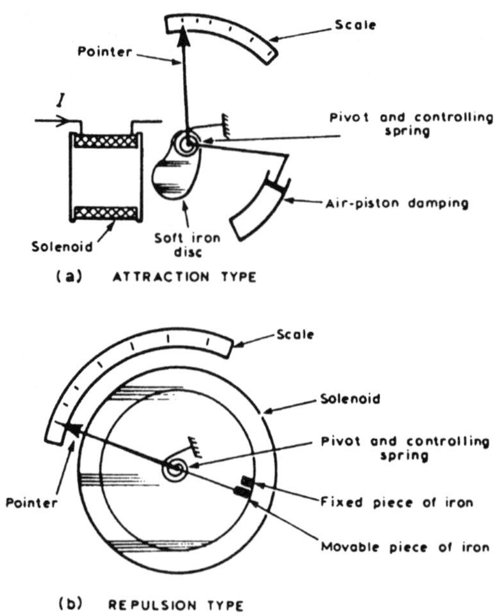

Figure 10.2

When current flows in the solenoid, a pivoted soft-iron disc is attracted towards the solenoid and the movement causes a pointer to move across a scale.

(b) In the **repulsion type** moving-iron instrument shown diagrammatically in Fig. 10.2(b), two pieces of iron are placed inside the solenoid, one being fixed, and the other attached to the spindle carrying the pointer. When current passes through the solenoid, the two pieces of iron are magnetized in the same direction and therefore repel each other. The pointer thus moves across the scale. The force

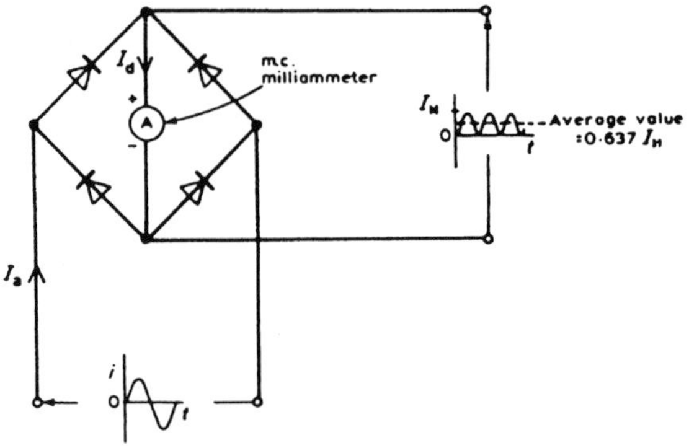

Figure 10.3

moving the pointer is, in each type, proportional to I^2. Because of this the direction of current does not matter and the moving-iron instrument can be used on d.c. or a.c. The scale, however, is non-linear.

10.4 The moving-coil rectifier instrument

A moving-coil instrument, which measures only d.c., may be used in conjunction with a bridge rectifier circuit as shown in Fig. 10.3 to provide an indication of alternating currents and voltages (see Chapter 12). The average value of the full-wave rectified current is $0.637\,I_m$. However, a meter being used to measure a.c. is usually calibrated in r.m.s. values. For sinusoidal quantities the indication is $(0.707\,I_m)/(0.637\,I_m)$, i.e. 1.11 times the mean value. Rectifier instruments have scales calibrated in r.m.s. quantities and it is assumed by the manufacturer that the a.c. is sinusoidal.

10.5 Comparison of moving-coil, moving-iron and moving-coil rectifier instruments

See Table 10.1 below.

Table 10.1

Type of instrument	*Moving coil*	*Moving iron*	*Moving-coil rectifier*
Suitable for measuring	Direct current and voltage	Direct and alternating currents and voltage (reading in r.m.s. value)	Alternating current and voltage (reads average value but scale is adjusted to give r.m.s. value for sinusoidal waveforms)
Scale	Linear	Non-linear	Linear
Method of control	Hairsprings	Hairsprings	Hairsprings
Method of damping	Eddy current	Air	Eddy current
Frequency limits	—	20–200 Hz	20–100 kHz
Advantages	1 Linear scale 2 High sensitivity 3 Well shielded from stray magnetic fields 4 Lower power consumption	1 Robust construction 2 Relatively cheap 3 Measures d.c. and a.c. 4 In frequency range 20–100 Hz reads r.m.s. correctly regardless of supply waveform	1 Linear scale 2 High sensitivity 3 Well shielded from stray magnetic fields 4 Low power consumption 5 Good frequency range
Disadvantages	1 Only suitable for d.c. 2 More expensive than moving-iron type 3 Easily damaged	1 Non-linear scale 2 Affected by stray magnetic fields 3 Hysteresis errors in d.c. circuits 4 Liable to temperature errors 5 Due to the inductance of the solenoid, readings can be affected by variation of frequency	1 More expensive than moving-iron type 2 Errors caused when supply is non-sinusoidal

(For the principle of operation of a moving-coil instrument, see Chapter 8, page 79.)

10.6 Shunts and multipliers

An **ammeter**, which measures current, has a low resistance (ideally zero) and must be connected in series with the circuit.

A **voltmeter**, which measures p.d., has a high resistance (ideally infinite) and must be connected in parallel with the part of the circuit whose p.d. is required.

There is no difference between the basic instrument used to measure current and voltage since both use a milliammeter as their basic part. This is a sensitive instrument which gives f.s.d. for currents of only a few milliamperes. When an ammeter is required to measure currents of larger magnitude, a proportion of the current is diverted through a low-value resistance connected in parallel with the meter. Such a diverting resistor is called a **shunt**.

From Fig. 10.4(a),

$V_{PQ} = V_{RS}$. Hence $I_a r_a = I_S R_S$

Thus the value of the shunt,

$$R_S = \frac{I_a r_a}{I_s} \text{ ohms}$$

The milliammeter is converted into a voltmeter by connecting a high resistance (called a **multiplier**) in series with it as shown in Fig. 10.4(b). From Fig. 10.4(b),

$V = V_a + V_M = Ir_a + IR_M$

Thus the value of the multiplier,

$$R_M = \frac{V - Ir_a}{I} \text{ ohms}$$

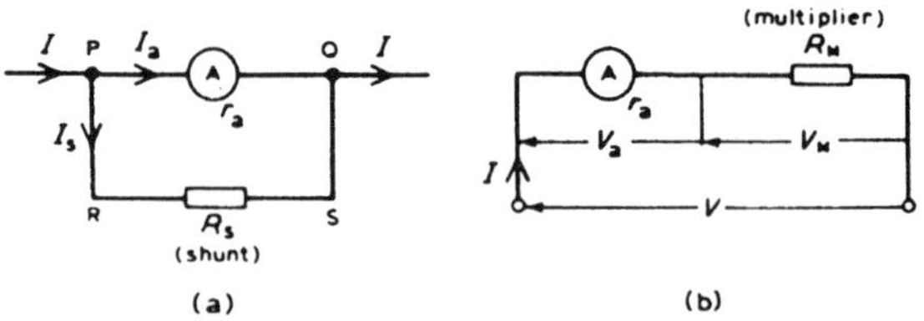

Figure 10.4

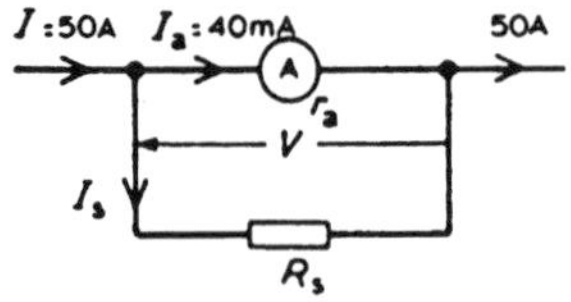

Figure 10.5

Problem 1. A moving-coil instrument gives a f.s.d. when the current is 40 mA and its resistance is 25 Ω. Calculate the value of the shunt to be connected in parallel with the meter to enable it to be used as an ammeter for measuring currents up to 50 A.

The circuit diagram is shown in Fig. 10.5, where r_a = resistance of instrument = 25 Ω. R_s = resistance of shunt; I_a = maximum permissible current flowing in instrument = 40 mA = 0.04 A. I_s = current flowing in shunt; I = total circuit current required to give f.s.d. = 50 A.

Since $I = I_a + I_s$ then $I_s = I - I_a = 50 - 0.04 = 49.96$ A. $V = I_a r_a = I_s R_s$, hence

$$R_s = \frac{I_a r_a}{I_s} = \frac{(0.04)(25)}{49.96} = 0.020\,02\ \Omega$$

$$= \mathbf{20.02\ m\Omega}$$

Thus for the moving-coil instrument to be used as an ammeter with a range 0–50 A, a resistance of value 20.02 mΩ needs to be connected in parallel with the instrument.

Problem 2. A moving-coil instrument having a resistance of 10 Ω, gives a f.s.d. when the current is 8 mA. Calculate the value of the multiplier to be connected in series with the instrument so that it can be used as a voltmeter for measuring p.d.s. up to 100 V.

The circuit diagram is shown in Fig. 10.6, where r_a = resistance of instrument = 10 Ω, R_M = resistance of multiplier, I = total permissible instrument current = 8 mA = 0.008 A, V = total p.d. required to give f.s.d. = 100 V.

$V = V_a + V_M = Ir_a + IR_M$, i.e. $100 = (0.008)(10) + (0.008)R_M$, or $100 - 0.08 = 0.008\,R_M$, thus

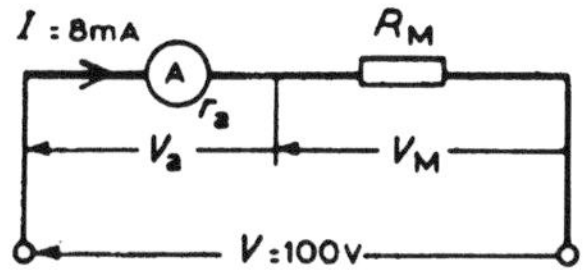

Figure 10.6

$$R_M = \frac{99.92}{0.008} = 12\,490\ \Omega = \mathbf{12.49\ k\Omega}$$

Hence for the moving-coil instrument to be used as a voltmeter with a range 0–100 V, a resistance of value 12.49 kΩ needs to be connected in series with the instrument.

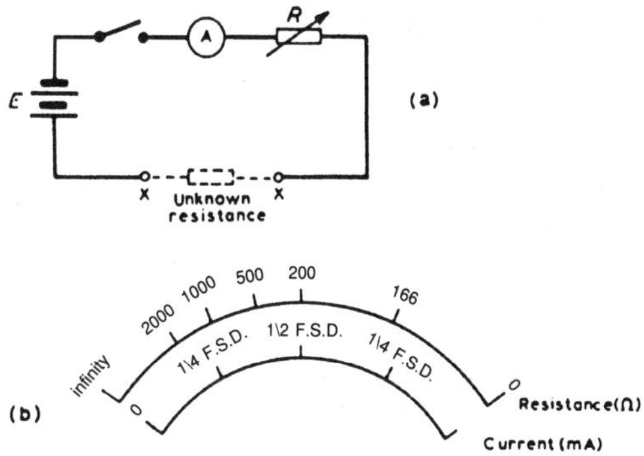

Figure 10.7

10.7 Electronic instruments

Electronic measuring instruments have advantages over instruments, such as the moving-iron or moving-coil meters, in that they have a much higher input resistance (some as high as 1000 MΩ) and can handle a much wider range of frequency (from d.c. up to megahertz).

The digital voltmeter (DVM) is one which provides a digital display of the voltage being measured. Advantages of a DVM over analogue instruments include higher accuracy and resolution, no observational or parallex errors (see section 10.20) and a very high input resistance, constant on all ranges.

A digital multimeter is a DVM with additional circuitry which makes it capable of measuring a.c. voltage, d.c. and a.c. current and resistance.

Instruments for a.c. measurements are generally calibrated with a sinusoidal alternating waveform to indicate r.m.s. values when a sinusoidal signal is applied to the instrument. Some instruments, such as the moving-iron and electro-dynamic instruments, give a true r.m.s. indication. With other instruments the indication is either scaled up from the mean value (such as with the rectifier moving-coil instrument) or scaled down from the peak value.

Sometimes quantities to be measured have complex waveforms (see section 10.13), and whenever a quantity is non-sinusoidal, errors in instrument readings can occur if the instrument has been calibrated for sine waves only. Such waveform errors can be largely eliminated by using electronic instruments.

10.8 The ohmmeter

An **ohmmeter** is an instrument for measuring electrical resistance. A simple ohmmeter circuit is shown in Fig. 10.7(a). Unlike the ammeter or voltmeter, the ohmmeter circuit does not receive the energy necessary for its operation from the circuit under test. In the ohmmeter this energy is supplied by a self-contained source of voltage, such as a battery. Initially, terminals XX are short-circuited and R adjusted to give f.s.d. on the milliammeter. If current I is at a maximum value and voltage E is constant, then resistance $R = E/I$ is at a minimum value. Thus f.s.d. on the milliammeter is made zero on the resistance scale. When terminals XX are open-circuited no current flows and R (= E/O) is infinity, ∞.

The milliammeter can thus be calibrated directly in ohms. A cramped (non-linear) scale results and is 'back to front', as shown in Fig. 10.7(b). When calibrated, an unknown resistance is placed between terminals XX and its value determined from the position of the pointer on the scale. An ohmmeter designed for measuring low values of resistance is called a **continuity tester**. An ohmmeter designed for measuring high values of resistance (i.e. megohms) is called an **insulation resistance tester** (e.g. **'Megger'**).

10.9 Multimeters

Instruments are manufactured that combine a moving-coil meter with a number of shunts and series multipliers, to provide a range of readings

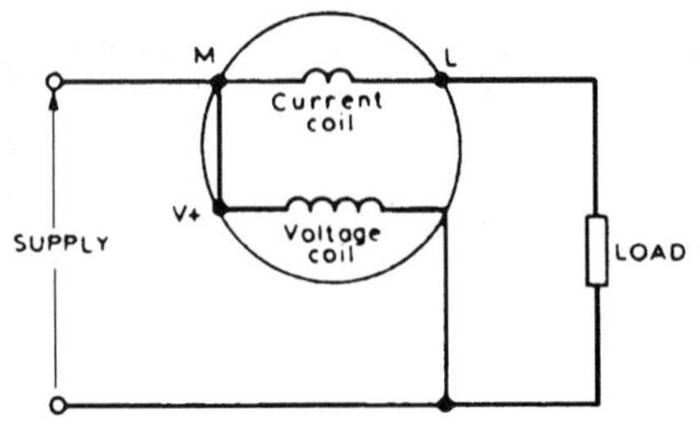

Figure 10.8

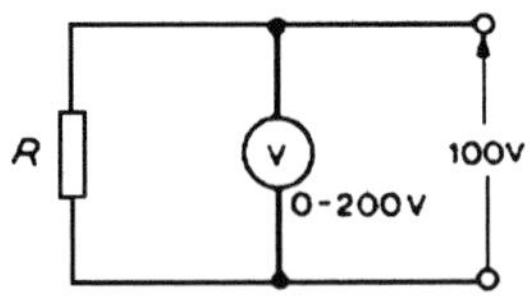

Figure 10.9

on a single scale graduated to read current and voltage. If a battery is incorporated then resistance can also be measured. Such instruments are called **multimeters** or **universal instruments** or **multirange instruments**. An 'Avometer' is a typical example. A particular range may be selected either by the use of separate terminals or by a selector switch. Only one measurement can be performed at one time. Often such instruments can be used in a.c. as well as d.c. circuits when a rectifier is incorporated in the instrument.

10.10 Wattmeters

A **wattmeter** is an instrument for measuring electrical power in a circuit. Figure 10.8 shows typical connections of a wattmeter used for measuring power supplied to a load. The instrument has two coils:

(i) a current coil, which is connected in series with the load, like an ammeter; and
(ii) a voltage coil, which is connected in parallel with the load, like a voltmeter.

10.11 Instrument 'loading' effect

Some measuring instruments depend for their operation on power taken from the circuit in which measurements are being made. Depending on the 'loading' effect of the instrument (i.e. the current taken to enable it to operate), the prevailing circuit conditions may change.

The resistance of voltmeters may be calculated since each has a stated sensitivity (or 'figure of merit'), often stated in 'kΩ per volt' of f.s.d. A voltmeter should have as high a resistance as possible (ideally infinite). In a.c. circuits the impedance of the instrument varies with frequency and thus the loading effect of the instrument can change.

Problem 3. Calculate the power dissipated by the voltmeter and by resistor R in Fig. 10.9 when (a) $R = 250\ \Omega$, (b) $R = 2\ \text{M}\Omega$.

Assume that the voltmeter sensitivity (sometimes called figure of merit) is 10 kΩ/V.

(a) Resistance of voltmeter, R_v = sensitivity × f.s.d. Hence, $R_v = 10\ \text{k}\Omega/\text{V} \times 200\ \text{V} = 2000\ \text{k}\Omega = 2\ \text{m}\Omega$. Current flowing in voltmeter,

$$I_v = \frac{V}{R_v} = \frac{100}{2 \times 10^6} = \mathbf{50 \times 10^{-6}\ A}$$

Power dissipated by voltmeter $= VI_v = (100)(50 \times 10^{-6}) = \mathbf{5\ mW}$.

When $R = 250\ \Omega$, current in resistor,

$$I_R = \frac{V}{R} = \frac{100}{250} = \mathbf{0.4\ A}$$

Power dissipated in load resistor $R = VI_R = (100)(0.4) = \mathbf{40\ W}$.

Thus the power dissipated in the voltmeter is insignificant in comparison with the power dissipated in the load.

(b) When $R = 2\ \text{M}\Omega$, current in resistor,

$$I_R = \frac{V}{R} = \frac{100}{2 \times 10^6} = 50 \times 10^{-6}\ \text{A}$$

Power dissipated in load resistor $R = VI_R = 100 \times 50 \times 10^{-6} = \mathbf{5\ mW}$. In this case the higher load resistance reduced the power dissipated such that the voltmeter is using as much power as the load.

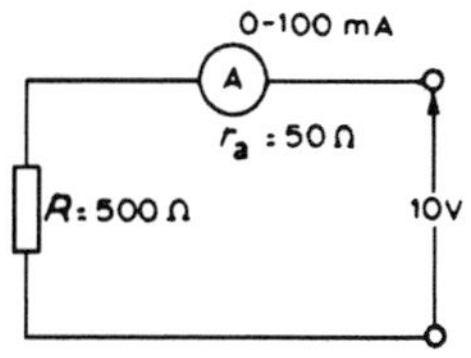

Figure 10.10

> Problem 4. An ammeter has a f.s.d. of 100 mA and a resistance of 50 Ω. The ammeter is used to measure the current in a load of resistance 500 Ω when the supply voltage is 10 V. Calculate (a) the ammeter reading expected (neglecting its resistance), (b) the actual current in the circuit, (c) the power dissipated in the ammeter, (d) the power dissipated in the load.

From Fig. 10.10,

(a) expected ammeter reading = V/R = 10/500 = **20 mA**.
(b) Actual ammeter reading = $V/(R + r_a)$ = 10/(500 + 50) = **18.18 mA**. Thus the ammeter itself has caused the circuit conditions to change from 20 mA to 18.18 mA.
(c) Power dissipated in the ammeter = I^2R = $(18.18 \times 10^{-3})^2(50)$ = **16.53 mW**.
(d) Power dissipated in the load resistor = I^2R = $(18.18 \times 10^{-3})^2(500)$ = **165.3 mW**.

> Problem 5. A voltmeter having a f.s.d. of 100 V and a sensitivity of 1.6 kΩ/V is used to measure voltage V_1 in the circuit of Fig. 10.11. Determine (a) the value of voltage V_1 with the voltmeter not connected, and (b) the voltage indicated by the voltmeter when connected between A and B.

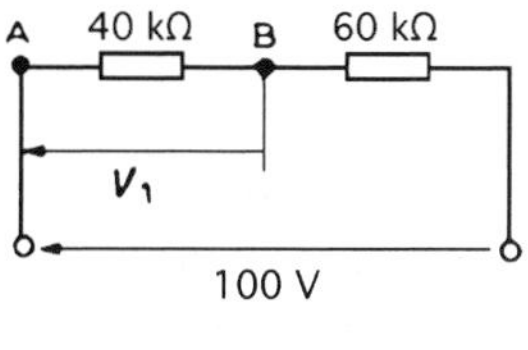

Figure 10.11

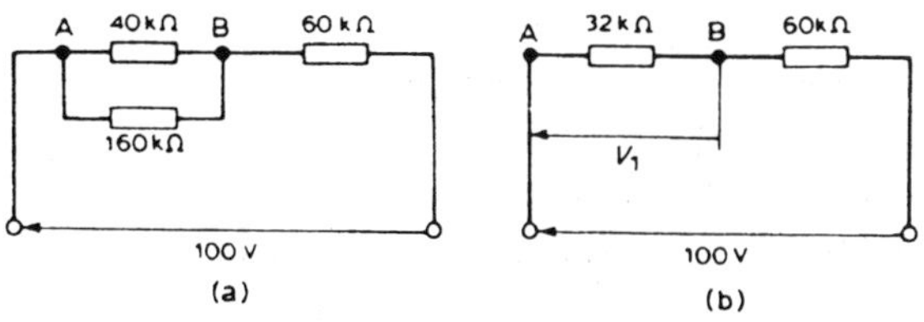

Figure 10.12

(a) By voltage division,

$$V_1 = \left(\frac{40}{40 + 60}\right) 100 = \mathbf{40\ V}$$

(b) The resistance of a voltmeter having a 100 V f.s.d. and sensitivity 1.6 kΩ/V is 100 V × 1.6 kΩ/V = 160 kΩ. When the voltmeter is connected across the 40 kΩ resistor the circuit is as shown in Fig. 10.12(a) and the equivalent resistance of the parallel network is given by

$$\left(\frac{40 \times 160}{40 + 160}\right) \text{k}\Omega, \text{ i.e. } \left(\frac{40 \times 160}{200}\right) \text{k}\Omega = 32\ \text{k}\Omega$$

The circuit is now effectively as shown in Fig. 10.12(b). Thus the voltage indicated on the voltmeter is

$$\left(\frac{32}{32 + 60}\right) 100\ \text{V} = \mathbf{34.78\ V}$$

A considerable error is thus caused by the loading effect of the voltmeter on the circuit. The error is reduced by using a voltmeter with a higher sensitivity.

> Problem 6. (a) A current of 20 A flows through a load having a resistance of 2 Ω. Determine the power dissipated in the load. (b) A wattmeter, whose current coil has a resistance of 0.01 Ω, is connected as shown in Fig. 10.13. Determine the wattmeter reading.

(a) Power dissipated in the load, $P = I^2R$ = $(20)^2(2)$ = **800 W**.
(b) With the wattmeter connected in the circuit the total resistance R_T is 2 + 0.01 = 2.01 Ω. The wattmeter reading is thus I^2R_T = $(20)^2(2.01)$ = **804 W**.

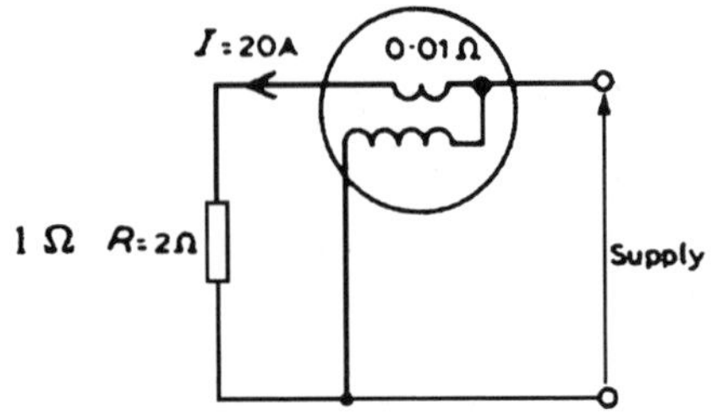

Figure 10.13

10.12 The cathode ray oscilloscope

The **cathode ray oscilloscope** (c.r.o.) may be used in the observation of waveforms and for the measurement of voltage, current, frequency, phase and periodic time. For examining periodic waveforms the electron beam is deflected horizontally (i.e. in the X direction) by a sawtooth generator acting as a timebase. The signal to be examined is applied to the vertical deflection system (Y direction) usually after amplification.

Oscilloscopes normally have a transparent grid of 10 mm by 10 mm squares in front of the screen, called a graticule. Among the timebase controls is a 'variable' switch which gives the sweep speed as time per centimetre. This may be in s/cm, ms/cm or μs/cm, a large number of switch positions being available. Also on the front panel of a c.r.o. is a Y amplifier switch marked in volts per centimetre, with a large number of available switch positions.

(i) With **direct voltage measurements**, only the Y amplifier 'volts/cm' switch on the c.r.o. is used. With no voltage applied to the Y plates the position of the spot trace on the screen is noted. When a direct voltage is applied to the Y plates the new position of the spot trace is an indication of the magnitude of the voltage. For example, in Fig. 10.14(a), with no voltage applied to the Y plates, the spot trace is in the centre of the screen (initial position) and then the spot trace moves 2.5 cm to the final position shown, on application of a d.c. voltage. With the 'volts/cm' switch on 10 volts/cm the magnitude of the direct voltage is 2.5 cm × 10 volts/cm, i.e. 25 volts.

(ii) With **alternating voltage measurements**, let a sinusoidal waveform be displayed on a c.r.o. screen as shown in Fig. 10.14(b). If the s/cm switch is on, say, 5 ms/cm then the **periodic time *T*** of the sinewave is 5 ms/cm × 4 cm, i.e. 20 ms or 0.02 s. Since frequency,

$$f = \frac{1}{T}, \textbf{ frequency} = \frac{\mathbf{1}}{\mathbf{0.02}} = \mathbf{50\ Hz}$$

If the 'volts/cm' switch is on, say, 20 volts/cm then the **amplitude** or **peak value** of the sinewave shown is 20 volts/cm × 2 cm, i.e. 40 V. Since

$$\text{r.m.s. voltage} = \frac{\text{peak voltage}}{\sqrt{2}}$$

(see Chapter 12)

r.m.s. voltage = $40/\sqrt{2}$ = **28.28 volts**

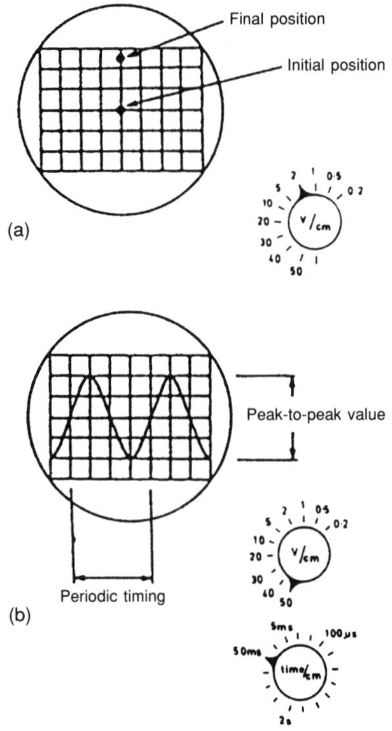

Figure 10.14

Double beam oscilloscopes are useful whenever two signals are to be compared simultaneously. The c.r.o. demands reasonable skill in adjustment and use. However, its greatest advantage is in observing the shape of a waveform – a feature not possessed by other measuring instruments.

Problem 7. Describe how a simple c.r.o. is adjusted to give (a) a spot trace, (b) a continuous horizontal trace on the screen, explaining the functions of the various controls.

(a) To obtain a spot trace on a typical c.r.o. screen:
   (i) Switch on the c.r.o.
   (ii) Switch the timebase control to off. This control is calibrated in time per centimetres – for example, 5 ms/cm or 100 μs/cm. Turning it to zero ensures no signal is applied to the X plates. The Y-plate input is left open-circuited.
   (iii) Set the intensity, X-shift and Y-shift controls to about the mid-range positions.
   (iv) A spot trace should now be observed on the screen. If not, adjust either or both of the X- and Y-shift controls. The X-shift control varies the position of the spot trace in a horizontal direction while the Y-shift control varies its vertical position.
   (v) Use the X- and Y-shift controls to bring the spot to the centre of the screen and use the focus control to focus the electron beam into a small circular spot.

(b) To obtain a continuous horizontal trace on the screen the same procedure as in (a) is initially adopted. Then the timebase control is switched to a suitable position, initially the millisecond timebase range, to ensure that the repetition rate of the sawtooth is sufficient for the persistence of the vision time of the screen phosphor to hold a given trace.

Problem 8. For the c.r.o. square voltage waveform shown in Fig. 10.15 determine (a) the periodic time, (b) the frequency and (c) the peak-to-peak voltage. The 'time/cm' (or timebase control) switch is on 100 μs/cm and the 'volts/cm' (or signal amplitude control) switch is on 20 V/cm.

(In Figs. 10.15 to 10.18 assume that the squares shown are 1 cm by 1 cm.)

(a) The width of one complete cycle is 5.2 cm. Hence the periodic time, $T = 5.2\text{ cm} \times 100 \times 10^{-6}\text{ s/cm}$ = **0.52 ms**.
(b) Frequency, $f = 1/T = 1/(0.52 \times 10^{-3})$ = **1.92 kHz**.

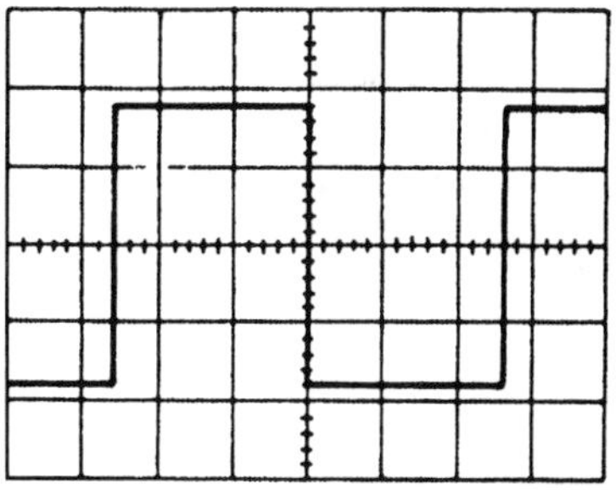

Figure 10.15

(c) The peak-to-peak height of the display is 3.6 cm, hence the peak-to-peak voltage = 3.6 cm × 20 V/cm = **72 V**.

Problem 9. For the c.r.o. display of a pulse waveform shown in Fig. 10.16 the 'time/cm' switch is on 50 ms/cm and the 'volts/cm' switch is on 0.2 V/cm. Determine (a) the periodic time, (b) the frequency, (c) the magnitude of the pulse voltage.

(a) The width of one complete cycle is 3.5 cm. Hence the periodic time, T = 3.5 cm × 50 ms/cm = **175 ms**.
(b) Frequency, $f = 1/T = 1/0.175$ = **5.71 Hz**.
(c) The height of a pulse is 3.4 cm. Hence the magnitude of the pulse voltage = 3.4 cm × 0.2 V/cm = **0.68 V**.

Problem 10. A sinusoidal voltage trace displayed by a c.r.o. is shown in Fig. 10.17. If the 'time/cm' switch is on 500 μs/cm and the 'volts/cm' switch is on 5 V/cm, find, for the waveform, (a) the frequency, (b) the peak-to-peak voltage, (c) the amplitude, (d) the r.m.s. value.

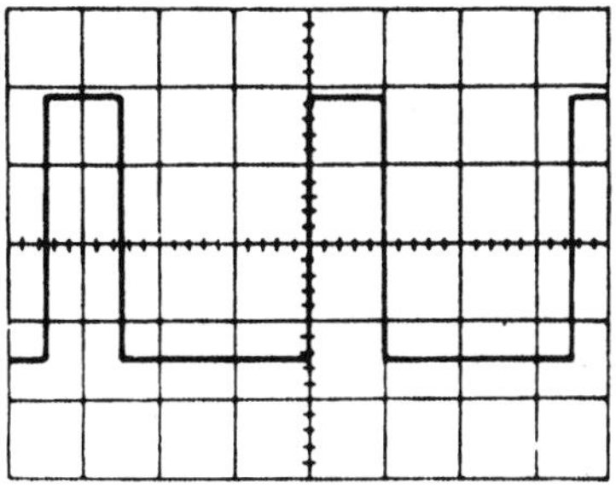

Figure 10.16

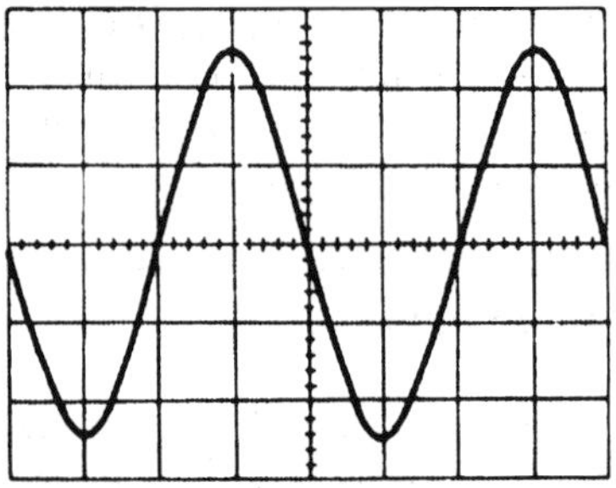

Figure 10.17

(a) The width of one complete cycle is 4 cm. Hence the periodic time, T is 4 cm × 500 μs/cm, i.e. 2 ms. Frequency, $f = 1/T = 1/(2 \times 10^{-3})$ = **500 Hz**.
(b) The peak-to-peak height of the waveform is 5 cm. Hence the peak-to-peak voltage = 5 cm × 5 V/cm = **25 V**.
(c) Amplitude = $\frac{1}{2}$ × 25 V = **12.5 V**.
(d) The peak value of voltage is the amplitude, i.e. 12.5 V, and r.m.s. voltage = peak voltage/($\sqrt{2}$) = 12.5/($\sqrt{2}$) = **8.84 V**.

Problem 11. For the double-beam oscilloscope displays shown in Fig. 10.18, determine (a) their frequency, (b) their r.m.s. values, (c) their phase difference. The 'time/cm' switch is on 100 μs/cm and the 'volts/cm' switch on 2 V/cm.

(a) The width of each complete cycle is 5 cm for both waveforms. Hence the periodic time, T, of each waveform is 5 cm × 100 μs/cm, i.e. 0.5 ms.
Frequency of each waveform, $f = 1/T = 1/(0.5 \times 10^{-3})$ = **2 kHz**.

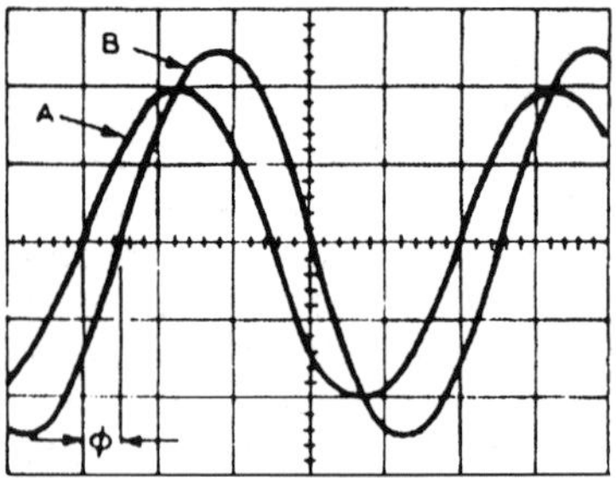

Figure 10.18

(b) The peak value of waveform A is 2 cm × 2 V/cm = **4 V**.
Hence the r.m.s. value of waveform A = 4/($\sqrt{2}$) = **2.83 V**.
The peak value of waveform B is 2.5 cm × 2 V/cm = 5 V, hence the r.m.s. value of waveform B = 5/($\sqrt{2}$) = **3.54 V**.
(c) Since 5 cm represents 1 cycle, then 5 cm represents 360°, i.e. 1 cm represents 360/5 = 72°.
The phase angle ϕ = 0.5 cm × 72°/cm = 36°.
Hence waveform A leads waveform B by 36°.

10.13 Waveform harmonics

(i) Let an instantaneous voltage v be represented by $v = V_m \sin 2\pi ft$ volts. This is a waveform which varies sinusoidally with time t, has a frequency f, and a maximum value V_m. Alternating voltages are usually assumed to have wave-shapes which are sinusoidal where only one frequency is present. If the waveform is not sinusoidal it is called a **complex wave**, and, whatever its shape, it may be split up mathematically into components called the **fundamental** and a number of **harmonics**. This process is called harmonic analysis. The fundamental (or first harmonic) is sinusoidal and has the supply frequency, f; the other harmonics are also sine waves having frequencies which are integer multiples of f. Thus, if the supply frequency is 50 Hz, then the third harmonic frequency is 150 Hz, the fifth 250 Hz, and so on.
(ii) A complex waveform comprising the sum of the fundamental and a third harmonic of about half the amplitude of the fundamental is shown in Fig. 10.19(a), both waveforms being initially in phase with each other. If further odd harmonic waveforms of the appropriate amplitudes are added, a good approximation to a square wave results. In Fig. 10.19(b), the third harmonic is shown having an initial phase displacement from the fundamental. The positive and negative half-cycles of each of the complex waveforms shown in Figs. 10.19(a) and (b) are identical in shape, and this is a feature of waveforms containing the fundamental and only odd harmonics.
(iii) A complex waveform comprising the sum of the fundamental and a second harmonic of about half the amplitude of the fundamental

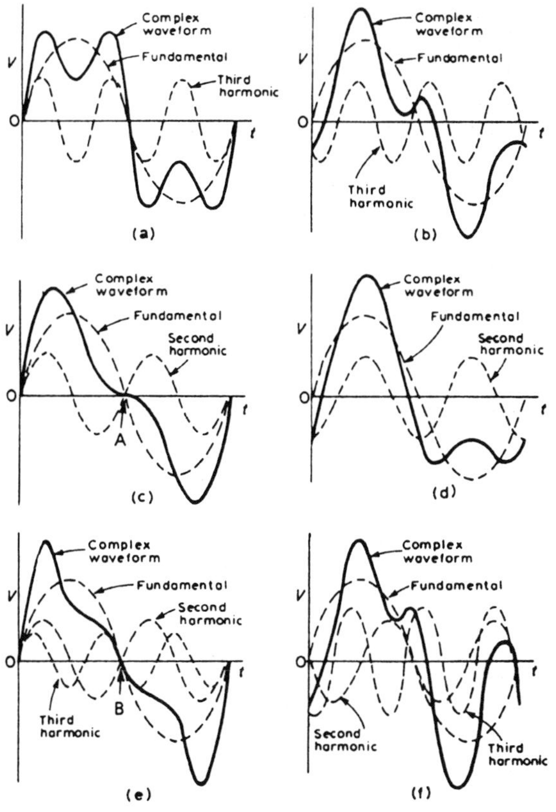

Figure 10.19

is shown in Fig. 10.19(c), each waveform being initially in phase with each other. If further even harmonics of appropriate amplitudes are added a good approximation to a triangular wave results. In Fig. 10.19(c) the negative cycle appears as a mirror image of the positive cycle about point A. In Fig. 10.19(d) the second harmonic is shown with an initial phase displacement from the fundamental and the positive and negative half-cycles are dissimilar.

(iv) A complex waveform comprising the sum of the fundamental, a second harmonic and a third harmonic is shown in Fig. 10.19(e), each waveform being initially 'in-phase'. The negative half-cycle appears on a mirror image of the positive cycle about point B. In Fig. 10.19(f) a complex waveform comprising the sum of the fundamental, a second harmonic and a third harmonic is shown with initial phase displacement. The positive and negative half-cycles are seen to be dissimilar.

The features mentioned relative to Figs. 10.19(a) to (f) make it possible to recognize the harmonics present in a complex waveform displayed on a c.r.o.

10.14 Logarithmic ratios

In electronic systems, the ratio of two similar quantities measured at different points in the system are often expressed in logarithmic units. By definition, if the ratio of two powers P_1 and P_2 is to be expressed in **decibel (dB) units** then the number of decibels, X, is given by

$$\mathbf{X = 10 \log \left(\frac{P_2}{P_1}\right) dB}$$

Thus, when the power ratio, $P_2/P_1 = 1$ then the decibel power ratio $= 10 \log 1 = 0$; when the power ratio, $P_2/P_1 = 100$ then the decibel power ratio $= 10 \log 100 = +20$ (i.e. a power gain); and when the power ratio, $P_2/P_1 = 1/10$ then the decibel power ratio $= 10 \log (1/10) = -10$ (i.e. a power loss or attenuation).

Logarithmic units may also be used for voltage and current ratios. Power, P is given by $P = I^2R$ or $P = V^2/R$. Substituting in equation (10.1) gives:

$$X = 10 \log \left(\frac{I_2^2 R_2}{I_1^2 R_1}\right) \text{dB or}$$

$$X = 10 \log \left(\frac{V_2^2 R_2}{V_1^2 R_1}\right) \text{dB}$$

If $R_1 = R_2$ then $X = 10 \log (I_2^2/I_1^2)$ dB or $X = 10 \log (V_2^2/V_1^2)$ dB

i.e. $\mathbf{X = 20 \log \left(\frac{I_2}{I_1}\right) dB}$ or $\mathbf{X = 20 \log \left(\frac{V_2}{V_1}\right) dB}$

(from the laws of logarithms).

From equation (10.1), X decibels is a logarithmic ratio of two similar quantities and is not an absolute unit of measurement. It is therefore necessary to state a **reference level** to measure a number of decibels above or below that reference. The most widely used reference level for power is 1 mW, and when power levels are expressed in decibels, above or below the 1 mW reference level, the unit given to the new power level is dBm.

A voltmeter can be re-scaled to indicate the power level directly in decibels. The scale is generally calibrated by taking a reference level of 0 dB when a power of 1 mW is dissipated in a 600 Ω resistor (this being the natural impedance of a simple transmission line). The reference voltage V is then obtained from $P = V^2/R$, i.e. $1 \times 10^{-3} = V^2/600$ from which, $V = 0.775$ volts. In general, the number of dBm,

$$X = 20 \log \left(\frac{V}{0.775}\right)$$

Thus $V = 0.20$ V corresponds to 20 log (0.20/0.775) = −11.77 dBm and $V = 0.90$ V corresponds to 20 log (0.90/0.775) = +1.3 dBm, and so on.

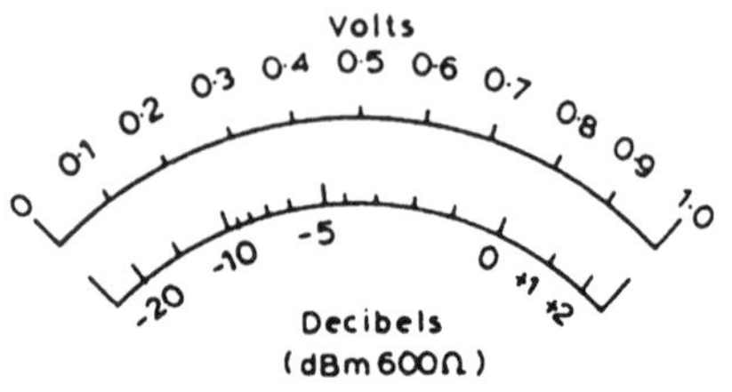

Figure 10.20

A typical **decibelmeter**, or **dB meter**, scale is shown in Fig. 10.20. Errors are introduced with dB meters when the circuit impedance is not 600 Ω.

> Problem 12. The ratio of two powers is (a) 2, (b) 30, (c) 1000, (d) 1/20. Determine the decibel power ratio in each case.

From above, the power ratio in decibels, X, is given by: $X = 10 \log (P_2/P_1)$

(a) When $P_2/P_1 = 2$, $X = 10 \log (2) = 10(0.30)$ = **3 dB**.
(b) When $P_2/P_1 = 30$, $X = 10 \log (30) = 10(1.48)$ = **14.8 dB**.
(c) When $P_2/P_1 = 1000$, $X = 10 \log (1000)$ = 10(3.0) = **30 dB**.
(d) When $P_2/P_1 = 1/20 = 0.05$, $X = 10 \log (0.05)$ = 10(−1.30) = **−13 dB**.

(a), (b) and (c) represent power gains and (d) represents a power loss or attenuation.

> Problem 13. The current input to a system is 5 mA and the current output is 20 mA. Find the decibel current ratio assuming the input and load resistances of the system are equal.

From above, the decibel current ratio is

$$20 \log \left(\frac{I_2}{I_1}\right) = 20 \log \left(\frac{20}{5}\right) = 20 \log 4$$

$$= 20(0.60) = \mathbf{12\ dB\ gain}$$

> Problem 14. 6% of the power supplied to a cable appears at the output terminals. Determine the power loss in decibels.

If P_1 = input power and P_2 = output power then $P_2/P_1 = 6/100 = 0.06$. Decibel power ratio = $10 \log (P_2/P_1) = 10 \log (0.06) = 10(-0.1222)$ = **−12.22 dB. Hence the decibel power loss, or attenuation, is 12.22 dB.**

> Problem 15. An amplifier has a gain of 14 dB. Its input power is 8 mW. Find its output power.

Decibel power ratio = $10 \log (P_2/P_1)$ where P_1 = input power = 8 mW, and P_2 = output power. Hence

$$14 = 10 \log (P_2/P_1)$$

$$1.4 = \log (P_2/P_1)$$

and

$$10^{1.4} = \frac{P_2}{P_1} \text{ (from the definition of a logarithm)}$$

i.e.

$$25.12 = \frac{P_2}{P_1}$$

Output power, $P_2 = 25.12P_1 = (25.12)(8)$ = **201 mW** or **0.201 W.**

> Problem 16. Determine, in decibels, the ratio of output power to input power of a three-stage communications system, the stages having gains of 12 dB, 15 dB and −8 dB. Find also the overall power gain.

The decibel ratio may be used to find the overall power ratio of a chain simply by adding the decibel power ratios together. Hence the overall decibel power ratio = 12 + 15 − 8 = **19 dB gain**.

$$19 = 10 \log (P_2/P_1)$$

Hence

$$1.9 = \log (P_2/P_1) \text{ and } 10^{1.9} = P_2/P_1 = 79.4$$

Thus the overall power gain, P_2/P_1 = 79.4.
For the first stage, $12 = 10 \log (P_2/P_1)$, from which $P_2/P_1 = 10^{1.2} = 15.85$. Similarly for the second stage, $P_2/P_1 = 31.62$ and for the third stage, $P_2/P_1 = 0.1585$. The overall power ratio is thus $15.85 \times 31.62 \times 0.1585 = 79.4$.

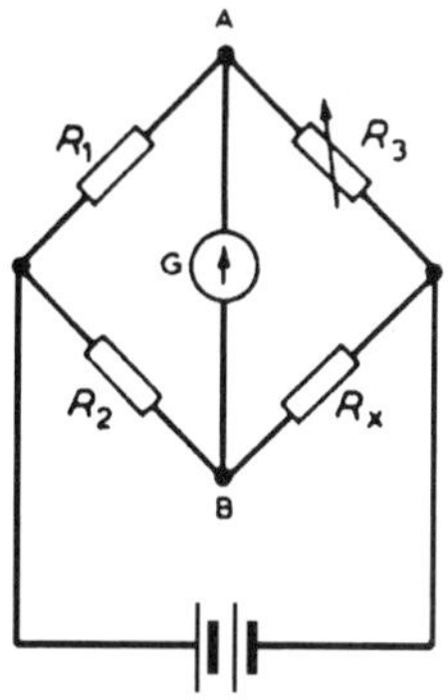

Figure 10.21

Problem 17. The output voltage from an amplifier is 4 V. If the voltage gain is 27 dB, calculate the value of the input voltage assuming that the amplifier input resistance and load resistance are equal.

Voltage gain in decibels = 27 = 20 log (V_2/V_1) = 20 log $(4/V_1)$, hence

$$\frac{27}{20} = \log\left(\frac{4}{V_1}\right)$$

$$1.35 = \log\left(\frac{4}{V_1}\right)$$

$$10^{1.35} = \frac{4}{V_1}$$

from which

$$V_1 = \frac{4}{10^{1.35}} = \frac{4}{22.39} = 0.179 \text{ V}$$

Hence the input voltage V_1 is 0.179 V.

10.15 Null method of measurement

A null method of measurement is a simple, accurate and widely used method which depends on an instrument reading being adjusted to read zero current only. The method assumes:

(i) if there is any deflection at all, then some current is flowing;
(ii) if there is no deflection, then no current flows (i.e. a null condition).

Hence it is unnecessary for a meter sensing current flow to be calibrated when used in this way. A sensitive milliammeter or microammeter with centre-zero position setting is called a **galvanometer**. Examples where the method is used are in the Wheatstone bridge (see section 10.16), in the d.c. potentiometer (see section 10.17) and with a.c. bridges (see section 10.18).

10.16 Wheatstone bridge

Figure 10.21 shows a **Wheatstone bridge** circuit which compares an unknown resistance R_x with others of known values, i.e. R_1 and R_2, which have fixed values, and R_3, which is variable. R_3 is varied until zero deflection is obtained on the galvanometer G. No current then flows through the meter, $V_A = V_B$, and the bridge is said to be 'balanced'. At balance,

$$R_1R_x = R_2R_3, \text{ i.e. } \boxed{\mathbf{R_x = \frac{R_2R_3}{R_1} \text{ ohms}}}$$

Problem 18. In a Wheatstone bridge ABCD, a galvanometer is connected between A and C, and a battery between B and D. A resistor of unknown value is connected between A and B. When the bridge is balanced, the resistance between B and C is 100 Ω, that between C and D is 10 Ω and that between D and A is 400 Ω. Calculate the value of the unknown resistance.

The Wheatstone bridge is shown in Fig. 10.22 where R_x is the unknown resistance. At balance, equating the products of opposite ratio arms gives:

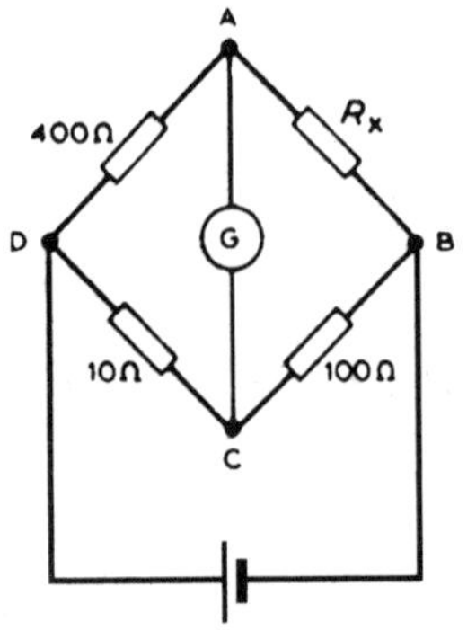

Figure 10.22

$$(R_x)(10) = (100)(400)$$

and

$$R_x = \frac{(100)(400)}{10} = 4000\ \Omega$$

Hence the unknown resistance, R_x = 4 kΩ.

10.17 D.c. potentiometer

The **d.c. potentiometer** is a null-balance instrument used for determining values of e.m.f.s and p.d.s by comparison with a known e.m.f. or p.d. In Fig. 10.23(a), using a standard cell of known e.m.f. E_1, the slider S is moved along the slide wire until balance is obtained (i.e. the galvanometer deflection is zero), shown as length l_1.

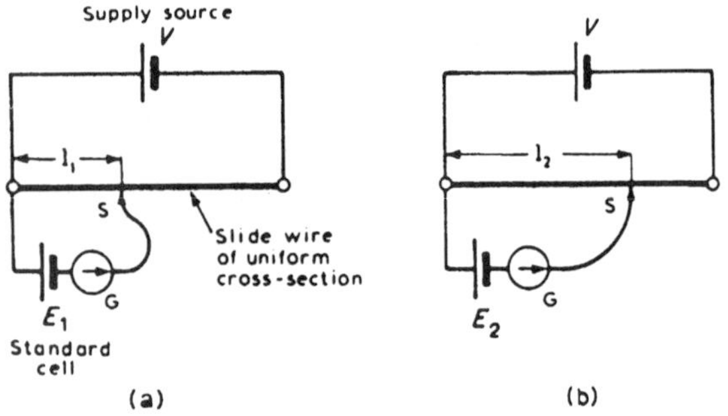

Figure 10.23

The standard cell is now replaced by a cell of unknown e.m.f. E_2 (see Fig. 10.23(b)) and again balance is obtained (shown as l_2). Since $E_1 \propto l_1$ and $E_2 \propto l_2$ then $E_1/E_2 = l_1/l_2$ and

$$\boldsymbol{E_2 = E_1\left(\frac{l_2}{l_1}\right)} \textbf{ volts}$$

A potentiometer may be arranged as a resistive two-element potential divider in which the division ratio is adjustable to give a simple variable d.c. supply. Such devices may be constructed in the form of a resistive element carrying a sliding contact which is adjusted by a rotary or linear movement of the control knob.

Problem 19. In a d.c. potentiometer balance is obtained at a length of 400 mm when using a standard cell of 1.0186 volts. Determine the e.m.f. of a dry cell if balance is obtained with a length of 650 mm.

E_1 = 1.0186 V, l_1 = 400 mm, l_2 = 650 mm. With reference to Fig. 10.23, $E_1/E_2 = l_1/l_2$, from which

$$E_2 = E_1\left(\frac{l_2}{l_1}\right) = (1.0186)\left(\frac{650}{400}\right) = \mathbf{1.655\ volts}$$

10.18 A.c. bridges

A Wheatstone bridge type circuit, shown in Fig. 10.24, may be used in a.c. circuits to determine unknown values of inductance and capacitance, as well as resistance.

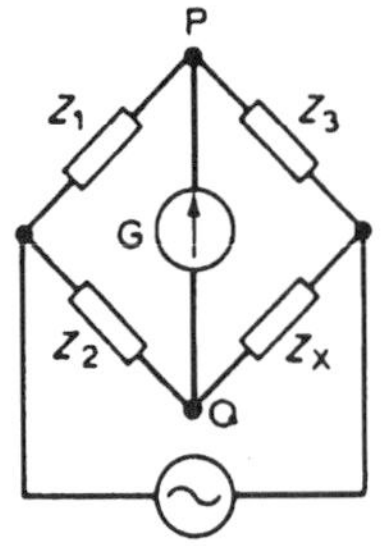

Figure 10.24

When the potential differences across Z_3 and Z_x (or across Z_1 and Z_2) are equal in magnitude and phase, then the current flowing through the galvanometer, G, is zero. At balance, $Z_1Z_x = Z_2Z_3$, from which,

$$\mathbf{Z_x = \frac{Z_2 Z_3}{Z_1}\ \Omega}$$

There are many forms of a.c. bridge, and these include: the Maxwell, Hay, Owen and Heaviside bridges for measuring inductance, and the De Sauty, Schering and Wien bridges for measuring capacitance. A **commercial or universal bridge** is one which can be used to measure resistance, inductance or capacitance.

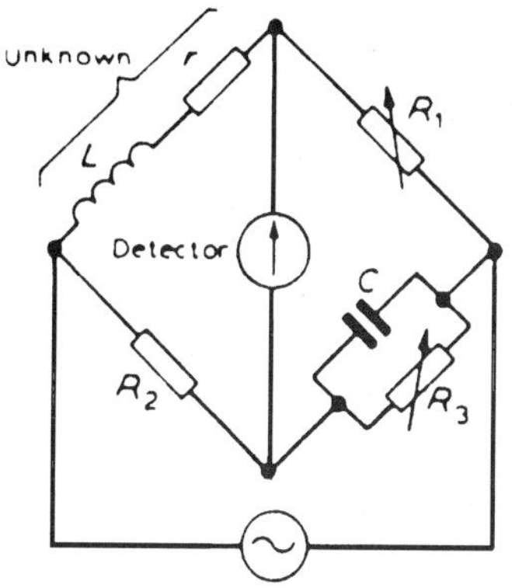

Figure 10.25

Maxwell's bridge for measuring the inductance L and resistance r of an inductor is shown in Fig. 10.25. At balance the products of diagonally opposite impedances are equal, thus $Z_1Z_2 = Z_3Z_4$. Using complex quantities, $Z_1 = R_1$, $Z_2 = R_2$,

$$Z_3 = \frac{R_3(-jX_c)}{R_3 - jX_c}\left(\text{i.e. } \frac{\text{product}}{\text{sum}}\right)$$

and $Z_4 = r + jX_L$. Hence

$$R_1R_2 = \frac{R_3(-jX_C)}{(R_3 - jX_C)}(r + jX_L)$$

i.e.

$$R_1R_2(R_3 - jX_C) = (-jR_3X_C)(r + jX_L)$$

$$R_1R_2R_3 - jR_1R_2X_C = -jrR_3X_C - j^2R_3X_CX_L$$

i.e.

$$R_1R_2R_3 - jR_1R_2X_C = -jrR_3X_C + R_3X_CX_L$$

(since $j^2 = -1$)

Equating the real parts gives: $R_1R_2R_3 = R_3X_CX_L$, from which,

$$X_L = \frac{R_1R_2}{X_C}$$

i.e.

$$2\pi fL = \frac{R_1R_2}{\frac{1}{2\pi fC}} = R_1R_2(2\pi fC)$$

Hence inductance, $L = R_1R_2C$ henry. (10.2)

Equating the imaginary parts gives: $-R_1R_2X_C = -rR_3X_C$ from which,

$$\textbf{resistance } r = \frac{R_1R_2}{R_3} \textbf{ ohms} \qquad (10.3)$$

Problem 20. For the a.c. bridge shown in Fig. 10.25 determine the values of the inductance and resistance of the coil when $R_1 = R_2 = 400\ \Omega$, $R_3 = 5\ \text{k}\Omega$ and $C = 7.5\ \mu\text{F}$.

From equation (10.2) above,

$$\text{inductance } L = R_1R_2C$$
$$= (400)(400)(7.5 \times 10^{-6})$$
$$= \mathbf{1.2\ H}$$

From equation (10.3) above,

$$\text{resistance } r = \frac{R_1R_2}{R_3} = \frac{(400)(400)}{(5000)} = \mathbf{32\ \Omega}$$

From equation (10.2), $R_2 = L/R_1C$ and from equation (10.3), $R_3 = (R_1/r)R_2$, hence

$$R_3 = \frac{R_1}{r}\left(\frac{L}{R_1C}\right) = \frac{L}{Cr}$$

If the frequency is constant then $R_3 \propto (L/r) \propto (\omega L/r) \propto$ Q-factor (see Chapters 13 and 14). Thus the bridge can be adjusted to give a direct indication of Q-factor.

10.19 Q-meter

The **Q-factor** for a series L–C–R circuit is the voltage magnification at resonance, i.e.

$$\text{Q-factor} = \frac{\text{voltage across capacitor}}{\text{supply voltage}}$$

$$= \frac{V_c}{V} \text{ (see Chapter 13)}$$

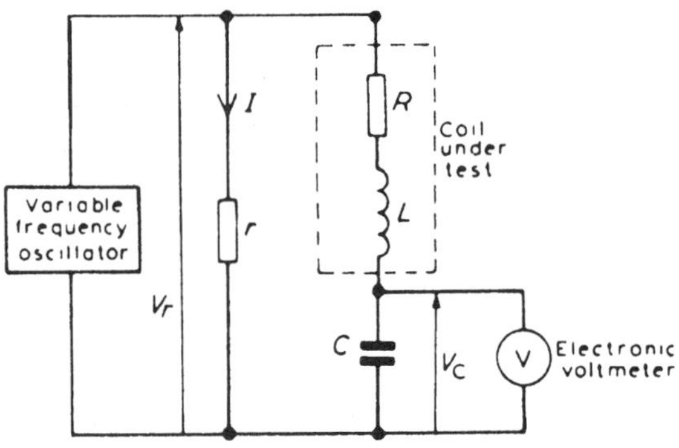

Figure 10.26

The simplified circuit of a **Q-meter**, used for measuring Q-factor, is shown in Fig. 10.26. Current from a variable frequency oscillator flowing through a very low resistance r develops a variable frequency voltage, V_r, which is applied to a series L–R–C circuit. The frequency is then varied until resonance causes voltage V_c to reach a maximum value. At resonance V_r and V_c are noted. Then

$$\text{Q-factor} = \frac{V_c}{V_r} = \frac{V_c}{Ir}$$

In a practical Q-meter, V_r is maintained constant and the electronic voltmeter can be calibrated to indicate the Q-factor directly. If a variable capacitor C is used and the oscillator is set to a given frequency, then C can be adjusted to give resonance. In this way inductance L may be calculated using

$$f_r = \frac{1}{2\pi\sqrt{(LC)}}$$

Since $Q = 2\pi fL/R$, then R may be calculated.

Q-meters operate at various frequencies and instruments exist with frequency ranges from 1 kHz to 50 MHz. Errors in measurement can exist with Q-meters since the coil has an effective parallel self-capacitance due to capacitance between turns. The accuracy of a Q-meter is approximately ±5%.

Problem 21. When connected to a Q-meter an inductor is made to resonate at 400 kHz. The Q-factor of the circuit is found to be 100 and the capacitance of the Q-meter capacitor is set to 400 pF. Determine (a) the inductance, and (b) the resistance of the inductor.

Resonant frequency, $f_r = 400\text{ kHz} = 400 \times 10^3$ Hz, Q-factor = 100, capacitance, $C = 400\text{ pF} = 400 \times 10^{-12}$ F. The circuit diagram of a Q-meter is shown in Fig. 10.26.

(a) At resonance,

$$f_r = \frac{1}{2\pi\sqrt{(LC)}}$$

for a series L–C–R circuit,

hence

$$2\pi f_r = \frac{1}{\sqrt{(LC)}}$$

from which

$$(2\pi f_r)^2 = \frac{1}{LC}$$

and

$$\text{inductance, } L = \frac{1}{(2\pi f_r)^2 C}$$

$$= \frac{1}{(2\pi 400 \times 10^3)^2(400 \times 10^{-12})}\ \text{H}$$

$= \mathbf{396\ \mu H}$ or $\mathbf{0.396\ mH}$

(b) $$\text{Q-factor at resonance} = \frac{2\pi f_r L}{R}$$

$$\text{from which resistance } R = \frac{2\pi f_r L}{Q}$$

$$= \frac{2\pi(400 \times 10^3)(0.396 \times 10^{-3})}{100}$$

$= \mathbf{9.95\ \Omega}$

10.20 Measurement errors

Errors are always introduced when using instruments to measure electrical quantities. The errors most likely to occur in measurements are those due to:

(i) the limitations of the instrument;
(ii) the operator;
(iii) the instrument disturbing the circuit.

Errors in the limitations of the instrument

The **calibration accuracy** of an instrument depends on the precision with which it is

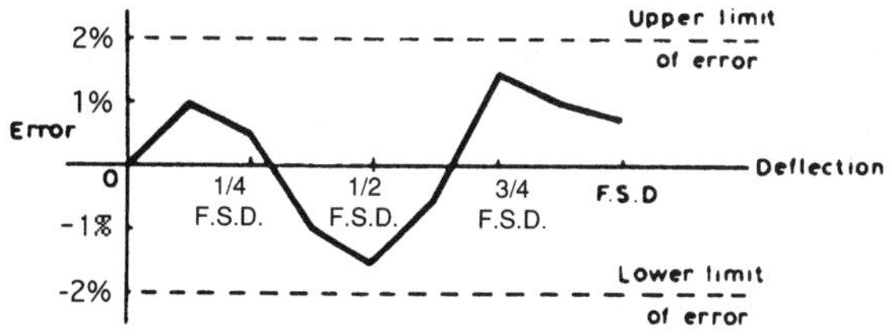

Figure 10.27

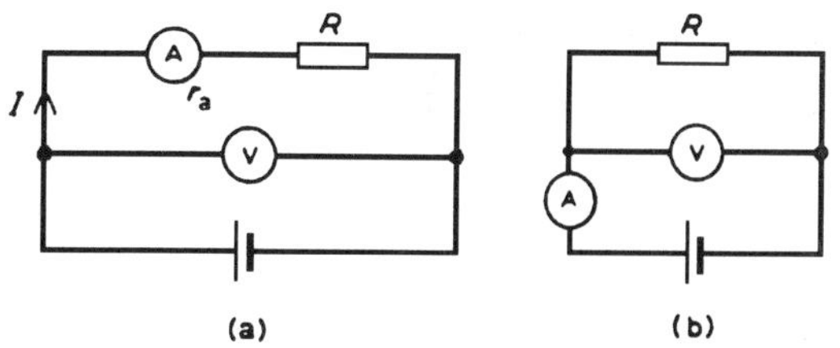

Figure 10.28

constructed. Every instrument has a margin of error which is expressed as a percentage of the instruments full-scale deflection. For example, industrial grade instruments have an accuracy of ±2% of f.s.d. Thus if a voltmeter has a f.s.d. of 100 V and it indicates 40 V say, then the actual voltage may be anywhere between 40 ± (2% of 100), or 40 ± 2, i.e. between 38 V and 42 V.

When an instrument is calibrated, it is compared against a standard instrument and a graph is drawn of 'error' against 'meter deflection'. A typical graph is shown in Fig. 10.27 where it is seen that the accuracy varies over the scale length. Thus a meter with a ±2% f.s.d. accuracy would tend to have an accuracy which is much better than ±2% f.s.d. over much of the range.

Errors by the operator

It is easy for an operator to misread an instrument reading. With linear scales the values of the sub-divisions are reasonably easy to determine; non-linear scale graduations are more difficult to estimate. Also, scales differ from instrument to instrument and some meters have more than one scale (as with multimeters) and mistakes in reading indications are easily made. When reading a meter scale it should be viewed from an angle perpendicular to the surface of the scale at the location of the pointer; a meter scale should not be viewed 'at an angle'.

Errors due to the instrument disturbing the circuit

Any instrument connected into a circuit will affect that circuit to some extent. Meters require some power to operate, but provided this power is small compared with the power in the measured circuit, then little error will result. Incorrect positioning of instruments in a circuit can be a source of errors. For example, let a resistance be measured by the voltmeter–ammeter method as shown in Fig. 10.28. Assuming 'perfect' instruments, the resistance should be given by the voltmeter reading divided by the ammeter reading (i.e. $R = V/I$). However, in Fig. 10.28(a), $V/I = R + r_a$ and in Fig. 10.28(b) the current through the ammeter is that through the resistor plus that through the voltmeter. Hence the voltmeter reading divided by the ammeter reading will not give the true value of the resistance R for either methods of connection.

Problem 22. The current flowing through a resistor of 5 kΩ ± 0.4% is measured as 2.5 mA with an accuracy of measurement of ±0.5%. Determine the nominal value of the voltage across the resistor and its accuracy.

Voltage, $V = IR = (2.5 \times 10^{-3})(5 \times 10^{3}) = 12.5$ V. The maximum possible error is 0.4% + 0.5% = 0.9%. Hence the voltage, V = 12.5 V ± 0.9% of 12.5 V = 0.9/100 × 12.5 = 0.1125 V = 0.11 V correct to 2 significant figures. Hence the voltage V may also be expressed as **12.5 ± 0.11 volts** (i.e. a voltage lying between 12.39 V and 12.61 V).

Problem 23. The current I flowing in a resistor R is measured by a 0–10 A ammeter which gives an indication of 6.25 A. The voltage V across the resistor is measured by a 0–50 V voltmeter, which gives an indication of 36.5 V. Determine the resistance of the resistor, and its accuracy of measurement if both instruments have a limit of error of 2% of f.s.d. Neglect any loading effects of the instruments.

Resistance, $R = V/I$ = 36.5/6.25 = 5.84 Ω. Voltage error is ±2% of 50 V = ±1.0 V and expressed as a percentage of the voltmeter reading gives (±1/36.5) × 100% = ±2.74%. Current error is

±2% of 10 A = ±0.2 A and expressed as a percentage of the ammeter reading gives (±0.2/6.25) × 100% = ±3.2%.
Maximum relative error = sum of errors = 2.74% + 3.2% = ±5.94%; 5.94% of 5.84 Ω = 0.347 Ω, hence the resistance of the resistor may be expressed as:

5.84 Ω ± 5.94%, or **5.84 ± 0.35 Ω** (rounding off)

> Problem 24. The arms of a Wheatstone bridge ABCD have the following resistances: AB: R_1 = 1000 Ω ± 1.0%; BC: R_2 = 100 Ω ± 0.5%; CD: unknown resistance R_x; DA: R_3 = 432.5 Ω ± 0.2%. Determine the value of the unknown resistance and its accuracy of measurement.

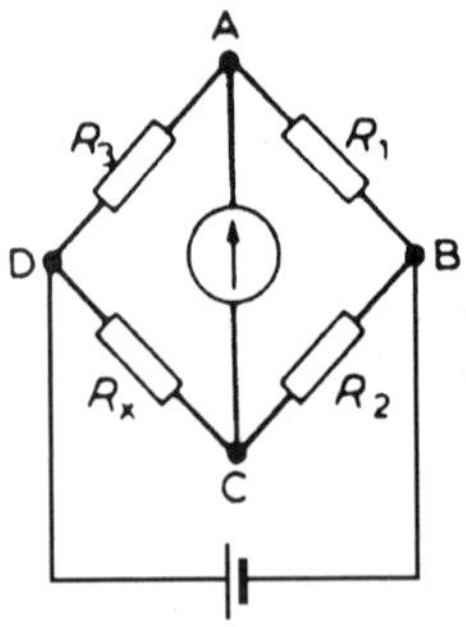

Figure 10.29

The Wheatstone bridge network is shown in Fig. 10.29 and at balance:

$$R_1R_x = R_2R_3$$

i.e.

$$R_x = \frac{R_2R_3}{R_1} = \frac{(100)(432.5)}{1000} = 43.25\ \Omega$$

The maximum relative error of R_x is given by the sum of the three individual errors, i.e. 1.0% + 0.5% + 0.2% = 1.7%.

Hence

$$\boldsymbol{R_x = 43.25\ \Omega \pm 1.7\%}$$

1.7% of 43.25 Ω = 0.74 Ω (rounding off), thus R_x may also be expressed as

$$\boldsymbol{R_x = 43.25 \pm 0.74\ \Omega}$$

10.21 Multi-choice questions on electrical measuring instruments and measurements

(Answers on page 283.)

1. Which of the following would apply to a moving-coil instrument?
   (a) An uneven scale, measuring d.c.
   (b) An even scale, measuring a.c.
   (c) An uneven scale, measuring a.c.
   (d) An even scale, measuring d.c.
2. In question 1, which would refer to a moving-iron instrument?
3. In question 1, which would refer to a moving-coil rectifier instrument?
4. Which of the following is needed to extend the range of a milliammeter to read voltages of the order of 100 V?
   (a) a parallel high-value resistance;
   (b) a series high-value resistance;
   (c) a parallel low-value resistance;
   (d) a series low-value resistance.
5. Figure 10.30 shows a scale of a multi-range ammeter. What is the current indicated when switched to a 25 A scale?
   (a) 84 A; (b) 5.6 A; (c) 14 A; (d) 8.4 A.

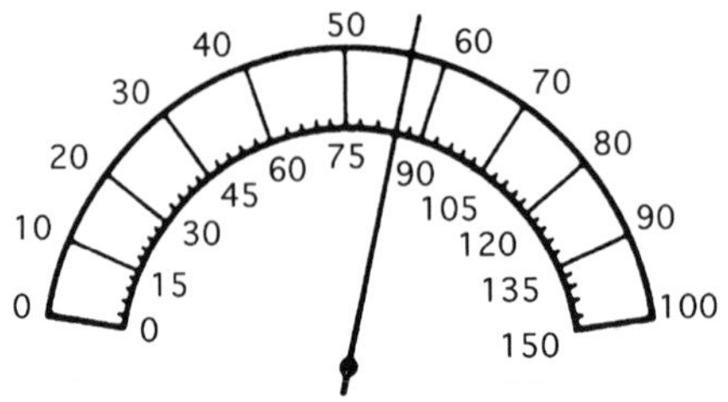

Figure 10.30

A sinusoidal waveform is displayed on a c.r.o. screen. The peak-to-peak distance is 5 cm and the distance between cycles is 4 cm. The 'variable' switch is on 100 μs/cm and the 'volts/cm' switch is on 10 V/cm. In questions 6 to 10, select the correct answer from the following:
(a) 25 V (b) 5 V (c) 0.4 ms (d) 35.4 V
(e) 4 ms (f) 50 V (g) 250 Hz (h) 2.5 V
(i) 2.5 kHz (j) 17.7 V

6. Determine the peak-to-peak voltage.
7. Determine the periodic time of the waveform.

8. Determine the maximum value of the voltage.
9. Determine the frequency of the waveform.
10. Determine the r.m.s. value of the waveform.

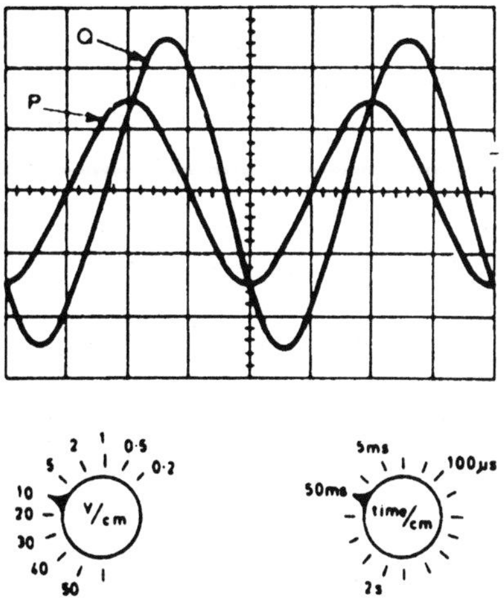

Figure 10.31

Figure 10.31 shows double-beam c.r.o. waveform traces. For the quantities stated in questions 11 to 17, select the correct answer from the following:

(a) 30 V (b) 0.2 s (c) 50 V (d) $\frac{15}{\sqrt{2}}$ V
(e) 54° leading (f) $\frac{250}{\sqrt{2}}$ V (g) 15 V (h) 100 μs
(i) $\frac{50}{\sqrt{2}}$ V (j) 250 V (k) 10 kHz (l) 75 V
(m) 40 μs (n) $\frac{3\pi}{10}$ rads lagging (o) $\frac{25}{\sqrt{2}}$ V
(p) 5 kHz (q) $\frac{30}{\sqrt{2}}$ V (r) 25 kHz (s) $\frac{75}{\sqrt{2}}$ V
(t) $\frac{3\pi}{10}$ rads leading

11. Amplitude of waveform P.
12. Peak-to-peak value of waveform Q.
13. Periodic time of both waveforms.
14. Frequency of both waveforms.
15. R.m.s. value of waveform P.
16. R.m.s. value of waveform Q.
17. Phase displacement of waveform Q relative to waveform P.
18. The input and output powers of a system are 2 mW and 18 mW, respectively. The decibel power ratio of output power to input power is (a) 9; (b) 9.54; (c) 1.9; (d) 19.08.
19. The input and output voltages of a system are 500 μV and 500 mV, respectively. The decibel voltage ratio of output to input voltage (assuming input resistance equals load resistance) is
(a) 1000; (b) 30; (c) 0; (d) 60.
20. The input and output currents of a system are 3 mA and 18 mA, respectively. The decibel current ratio of output to input current (assuming the input and load resistances are equal) is
(a) 15.56; (b) 6; (c) 1.6; (d) 7.78.
21. Which of the following statements is false?
(a) The Schering bridge is normally used for measuring unknown capacitances.
(b) A.c. electronic measuring instruments can handle a much wider range of frequency than the moving-coil instrument.
(c) A complex waveform is one which is non-sinusoidal.
(d) A square-wave normally contains the fundamental and even harmonics.
22. A voltmeter has a f.s.d. of 100 V, a sensitivity of 1 kΩ/V and an accuracy of ±2% of f.s.d. When the voltmeter is connected into a circuit it indicates 50 V. Which of the following statements is false?
(a) Voltage reading is 50 ± 2 V.
(b) Voltmeter resistance is 100 kΩ.
(c) Voltage reading is 50 V ± 2%.
(d) Voltage reading is 50 V ± 4%.

10.22 Short answer questions on electrical measuring instruments and measurements

1. What is the main difference between an analogue and a digital type measuring instrument?
2. Name the three essential devices for all analogue electrical indicating instruments.
3. Complete the following statements:
(a) An ammeter has a .... resistance and is connected .... with the circuit.
(b) A voltmeter has a .... resistance and is connected .... with the circuit.
4. State two advantages and two disadvantages of a moving-coil instrument.
5. What effect does the connection of (a) a shunt (b) a multiplier have on a milliammeter?
6. State two advantages and two disadvantages of a moving-iron instrument.
7. Name two advantages of electronic measuring instruments compared with moving-coil or moving-iron instruments.

8. Briefly explain the principle of operation of an ohmmeter.
9. Name a type of ohmmeter used for measuring (a) low resistance values (b) high resistance values.
10. What is a multimeter?
11. When may a rectifier instrument be used in preference to either the moving-coil or moving-iron instrument?
12. Name five quantities that a c.r.o. is capable of measuring?
13. What is harmonic analysis?
14. What is a feature of waveforms containing the fundamental and odd harmonics?
15. Express the ratio of two powers P_1 and P_2 in decibel units.
16. What does a power level unit of dBm indicate?
17. What is meant by a null method of measurement?
18. Sketch a Wheatstone bridge circuit used for measuring an unknown resistance in a d.c. circuit and state the balance condition.
19. How may a d.c. potentiometer be used to measure p.d.s?
20. Name five types of a.c. bridge used for measuring unknown inductance, capacitance or resistance.
21. What is a universal bridge?
22. State the name of an a.c. bridge used for measuring inductance.
23. Briefly describe how the measurement of Q-factor may be achieved.
24. Why do instrument errors occur when measuring complex waveforms?
25. Define 'calibration accuracy' as applied to a measuring instrument.
26. State three main areas where errors are most likely to occur in measurements.

10.23 Further questions on electrical measuring instruments and measurements

1. A moving-coil instrument gives f.s.d. for a current of 10 mA. Neglecting the resistance of the instrument, calculate the approximate value of series resistance needed to enable the instrument to measure up to (a) 20 V; (b) 100 V; (c) 250 V.
   [(a) 2 kΩ (b) 10 kΩ (c) 25 kΩ]
2. A meter of resistance 50 Ω has a f.s.d. of 4 mA. Determine the value of shunt resistance required in order that f.s.d. should be (a) 15 mA; (b) 20 A; (c) 100 A.
   [(a) 18.18 Ω (b) 10.00 mΩ (c) 2.00 mΩ]
3. A moving-coil instrument having a resistance of 20 Ω, gives a f.s.d. when the current is 5 mA. Calculate the value of the multiplier to be connected in series with the instrument so that it can be used as a voltmeter for measuring p.d.s up to 200 V. [39.98 kΩ]
4. A moving-coil instrument has a f.s.d. current of 20 mA and a resistance of 25 Ω. Calculate the values of resistance required to enable the instrument to be used (a) as a 0–10 A ammeter, and (b) as a 0–100 V voltmeter. State the mode of resistance connection in each case.
   [(a) 50.10 mΩ in parallel (b) 4.975 kΩ in series]
5. A meter has a resistance of 40 Ω and registers a maximum deflection when a current of 15 mA flows. Calculate the value of resistance that converts the movement into (a) an ammeter with a maximum deflection of 50 A; (b) a voltmeter with a range 0–250 V.
   [(a) 12.00 mΩ in parallel (b) 16.63 kΩ in series]
6. (a) Describe, with the aid of diagrams, the principle of operation of a moving-iron instrument.
   (b) Draw a circuit diagram showing how a moving-coil instrument may be used to measure alternating current.
   (c) Discuss the advantages and disadvantages of moving-coil rectifier instruments when compared with moving-iron instruments.
7. A 0–1 A ammeter having a resistance of 50 Ω is used to measure the current flowing in a 1 kΩ resistor when the supply voltage is 250 V. Calculate: (a) the approximate value of current (neglecting the ammeter resistance); (b) the actual current in the circuit; (c) the power dissipated in the ammeter; (d) the power dissipated in the 1 kΩ resistor.
   [(a) 0.250 A (b) 0.238 A (c) 2.832 W (d) 56.64 W]
8. For the square voltage waveform displayed on a c.r.o. shown in Fig. 10.32, find (a) its frequency; (b) its peak-to-peak voltage.
   [(a) 41.7 Hz (b) 176 V]
9. For the pulse waveform shown in Fig. 10.33, find (a) its frequency, (b) the magnitude of the pulse voltage. [(a) 0.56 Hz (b) 8.4 V]

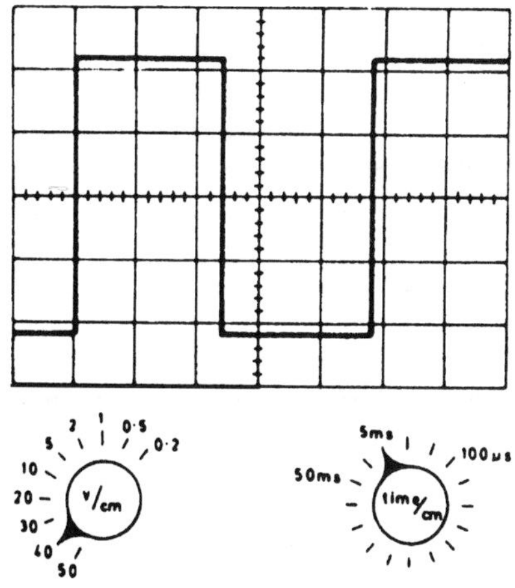

Figure 10.32

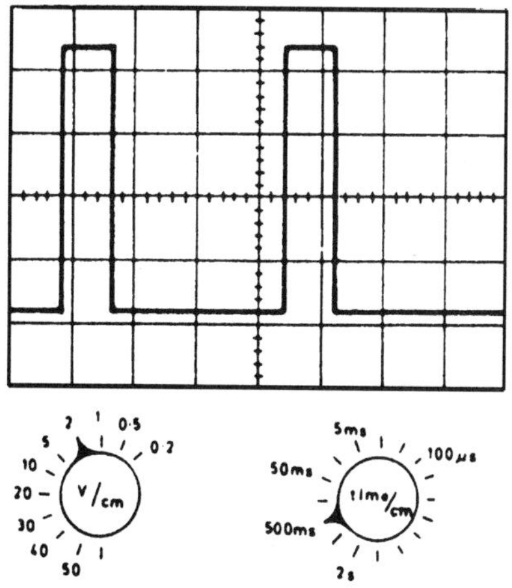

Figure 10.33

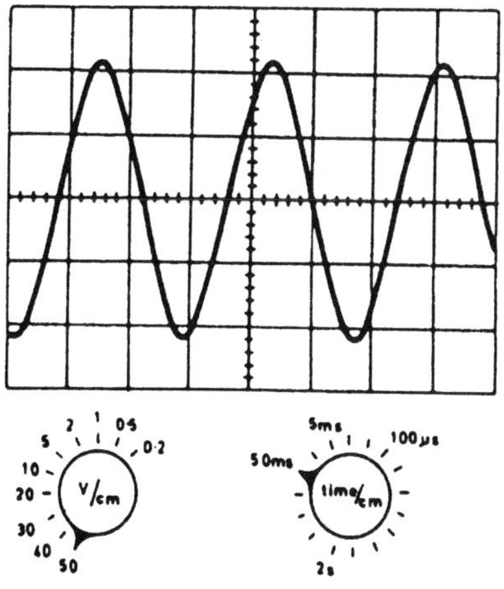

Figure 10.34

10. For the sinusoidal waveform shown in Fig. 10.34, determine (a) its frequency; (b) the peak-to-peak voltage, (c) the r.m.s. voltage. [(a) 7.14 Hz (b) 220 V (c) 77.78 V]
11. The ratio of two powers is (a) 3; (b) 10; (c) 20; (d) 10 000. Determine the decibel power ratio for each. [(a) 4.77 dB (b) 10 dB (c) 13 dB (d) 40 dB]
12. The ratio of two powers is (a) $\frac{1}{10}$ (b) $\frac{1}{3}$ (c) $\frac{1}{40}$ (d) $\frac{1}{100}$. Determine the decibel power ratio for each. [(a) –10 dB (b) –4.77 dB (c) –16.02 dB (d) –20 dB]
13. The input and output currents of a system are 2 mA and 10 mA, respectively. Determine the decibel current ratio of output to input current assuming input and output resistances of the system are equal. [13.98 dB]
14. 5% of the power supplied to a cable appears at the output terminals. Determine the power loss in decibels. [13 dB]
15. An amplifier has a gain of 24 dB. Its input power is 10 mW. Find its output power. [2.51 W]
16. Determine, in decibels, the ratio of the output power to input power of a four-stage system, the stages having gains of 10 dB, 8 dB, –5 dB and 7 dB. Find also the overall power gain. [20 dB, 100]
17. The output voltage from an amplifier is 7 mV. If the voltage gain is 25 dB calculate the value of the input voltage assuming that the amplifier input resistance and load resistance are equal. [0.39 mV]
18. The voltage gain of a number of cascaded amplifiers are 23 dB, –5.8 dB, –12.5 dB and 3.8 dB. Calculate the overall gain in decibels assuming that input and load resistances for each stage are equal. If a voltage of 15 mV is applied to the input of the system, determine the value of the output voltage. [8.5 dB, 39.91 mV]
19. The scale of a voltmeter has a decibel scale added to it, which is calibrated by taking a reference level of 0 dB when a power of 1 mW is dissipated in a 600 Ω resistor. Determine the voltage at (a) 0 dB; (b) 1.5 dB; and (c) –15 dB. (d) What decibel reading corresponds to 0.5 V? [(a) 0.775 V (b) 0.921 V (c) 0.138 V (d) –3.807 dB]
20. (a) Describe, with the aid of a diagram, the principle of the Wheatstone bridge and

hence deduce the balance condition giving the unknown resistance in terms of known values of resistance.

(b) In a Wheatstone bridge PQRS, a galvanometer is connected between Q and S and a voltage source between P and R. An unknown resistor R_x is connected between P and Q. When the bridge is balanced, the resistance between Q and R is 200 Ω, that between R and S is 10 Ω and that between S and P is 150 Ω. Calculate the value of R_x.

[3 kΩ]

21. (a) Describe, with the aid of a diagram, how a d.c. potentiometer can be used to measure the e.m.f. of a cell.

(b) Balance is obtained in a d.c. potentiometer at a length of 31.2 cm when using a standard cell of 1.0186 volts. Calculate the e.m.f. of a dry cell if balance is obtained with a length of 46.7 cm.

[1.525 V]

22. A Wheatstone bridge PQRS has the following arm resistances: PQ, 1 kΩ ± 2%; QR, 100 Ω ± 0.5%; RS, unknown resistance; SP, 273.6 Ω ± 0.1%. Determine the value of the unknown resistance, and its accuracy of measurement.

[27.36 Ω ± 2.6% or 27.36 ± 0.71 Ω]

23. A Maxwell bridge circuit ABCD has the following arm impedances: AB, 250 Ω resistance; BC, 15 μF capacitor in parallel with a 10 kΩ resistor; CD, 400 Ω resistor; DA, unknown inductor having inductance L and resistance R. Determine the values of L and R assuming the bridge is balanced.

[1.5 H, 10 Ω]

24. A Q-meter measures the Q-factor of a series L–C–R circuit to be 200 at a resonant frequency of 250 kHz. If the capacitance of the Q-meter capacitor is set to 300 pF, determine (a) the inductance L and (b) the resistance R of the inductor.

[(a) 1.351 mH (b) 10.61 Ω]

25. (a) A current of 15 A flows through a load having a resistance of 4 Ω. Determine the power dissipated in the load. (b) A wattmeter, whose current coil has a resistance of 0.02 Ω is connected (as shown in Fig. 10.13) to measure the power in the load. Determine the wattmeter reading assuming the current in the load is still 15 A.

[(a) 900 W (b) 904.5 W]

26. A voltage of 240 V is applied to a circuit consisting of an 800 Ω resistor in series with a 1.6 kΩ resistor. What is the voltage across the 1.6 kΩ resistor? The p.d. across the 1.6 kΩ resistor is measured by a voltmeter of f.s.d. 250 V and sensitivity 100 Ω/V. Determine the voltage indicated.

[160 V; 156.7 V]

27. The p.d. across a resistor is measured as 37.5 V with an accuracy of ±0.5%. The value of the resistor is 6 kΩ ± 0.8%. Determine the current flowing in the resistor and its accuracy of measurement.

[6.25 mA ± 1.3% or 6.25 ± 0.08 mA]

28. The voltage across a resistor is measured by a 75 V f.s.d. voltmeter which gives an indication of 52 V. The current flowing in the resistor is measured by a 20 A f.s.d. ammeter which gives an indication of 12.5 A. Determine the resistance of the resistor and its accuracy if both instruments have an accuracy of ±2% of f.s.d.

[4.16 Ω ± 6.08% or 4.16 ± 0.25 Ω]

29. List the errors most likely to occur in the measurements of electrical quantities.

A 240 V supply is connected across a load resistance R. Also connected across R is a voltmeter having a f.s.d. of 300 V and a figure of merit (i.e. sensitivity) of 8 kΩ/V. Calculate the power dissipated by the voltmeter and by the load resistance if (a) R = 100 Ω; (b) R = 1 MΩ.

Comment on the results obtained.

[(a) 24 mW, 576 W (b) 24 mW, 57.6 mW]

Part 2

Electrical Principles

11

D.c. circuit theorems

At the end of this chapter you should be able to:

- state and use Kirchhoff's laws to determine unknown currents and voltages in d.c. circuits
- understand the superposition theorem and apply it to find currents in d.c. circuits
- understand general d.c. circuit theory
- understand Thévenin's theorem and apply a procedure to determine unknown currents in d.c. circuits
- recognize the circuit diagram symbols for ideal voltage and current sources
- understand Norton's theorem and apply a procedure to determine unknown currents in d.c. circuits
- appreciate and use the equivalence of the Thévenin and Norton equivalent networks
- state the maximum power transfer theorem and use it to determine maximum power in a d.c. circuit

11.1 Introduction

The laws which determine the currents and voltage drops in d.c. networks are: (a) Ohm's law (see Chapter 2); (b) the laws for resistors in series and in parallel (see Chapter 5); and (c) Kirchhoff's laws (see section 11.2). In addition, there are a number of circuit theorems which have been developed for solving problems in electrical networks. These include:

(i) the superposition theorem (see section 11.3);
(ii) Thévenin's theorem (see section 11.5);
(iii) Norton's theorem (see section 11.7); and
(iv) the maximum power transfer theorem (see section 11.8).

11. 2 Kirchhoff's laws

Kirchhoff's laws state:

(a) **Current law.** *At any junction in an electric circuit the total current flowing towards that junction is equal to the total current flowing away from the junction, i.e.* $\Sigma I = 0$.

Thus, referring to Fig. 11.1:

$$I_1 + I_2 = I_3 + I_4 + I_5 \text{ or } I_1 + I_2 - I_3 - I_4 - I_5 = 0$$

(b) **Voltage law.** *In any closed loop in a network, the algebraic sum of the voltage drops (i.e. products of current and resistance) taken around the loop is equal to the resultant e.m.f. acting in that loop.*

Thus, referring to Fig. 11.2:

$$E_1 - E_2 = IR_1 + IR_2 + IR_3$$

(Note that if current flows away from the positive terminal of a source, that source is considered by convention to be positive. Thus moving anticlockwise around the loop of Fig. 11.2, E_1 is positive and E_2 is negative.)

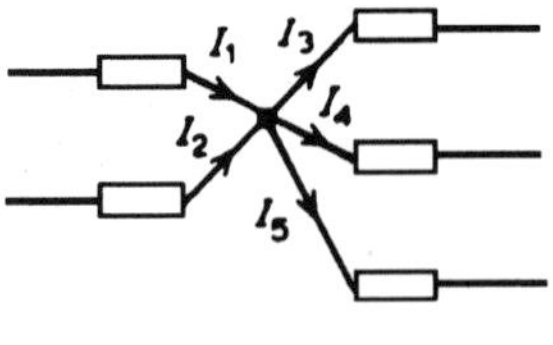

Figure 11.1

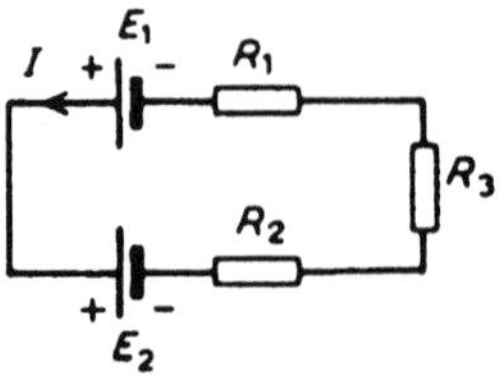

Figure 11.2

> Problem 1. (a) Find the unknown currents marked in Fig. 11.3(a). (b) Determine the value of e.m.f. E in Fig. 11.3(b).

(a) Applying Kirchhoff's current law:
For junction B: $50 = 20 + I_1$. Hence $\boldsymbol{I_1 = 30}$ **A**.
For junction C: $20 + 15 = I_2$. Hence $\boldsymbol{I_2 = 35}$ **A**.
For junction D:

$$I_1 = I_3 + 120$$

i.e. $30 = I_3 + 120$. Hence $\boldsymbol{I_3 = -90}$ **A**

(i.e. in the opposite direction to that shown in Fig. 11.3(a)).
For junction E:

$$I_4 = 15 - I_3$$

i.e. $I_4 = 15 - (-90)$. Hence $\boldsymbol{I_4 = 105}$ **A**.

For junction F: $120 = I_5 + 40$. Hence $\boldsymbol{I_5 = 80}$ **A**.

(b) Applying Kirchhoff's voltage law and moving clockwise around the loop of Fig. 11.3(b) starting at point A:

$$3 + 6 + E - 4 = (I)(2) + (I)(2.5) + (I)(1.5) + (I)(1)$$

$$= I(2 + 2.5 + 1.5 + 1)$$

i.e. $5 + E = 2(7)$

Hence $E = 14 - 5 = \mathbf{9\ V}$

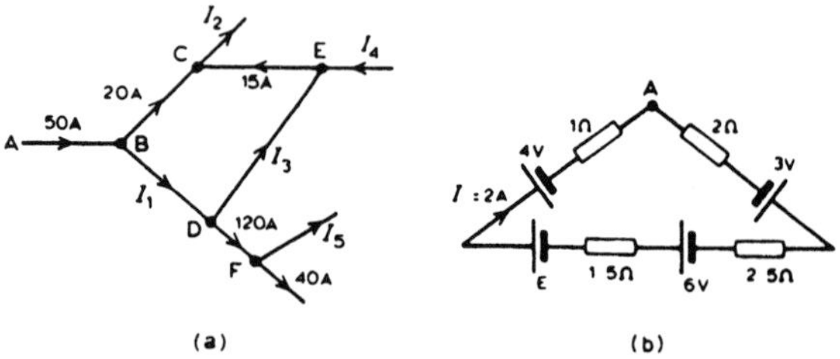

Figure 11.3

> Problem 2. Use Kirchhoff's laws to determine the currents flowing in each branch of the network shown in Fig. 11.4.

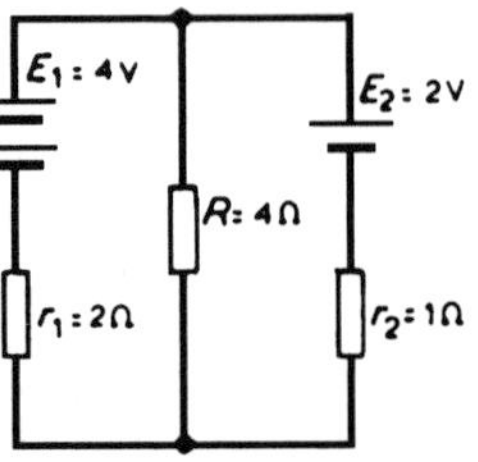

Figure 11.4

Procedure

1. Use Kirchhoff's current law and label current directions on the original circuit diagram. The directions chosen are arbitrary, but it is usual, as a starting point, to assume that current flows from the positive terminals of the batteries. This is shown in Fig. 11.5 where the three branch currents are expressed in terms of I_1 and I_2 only, since the current through R is $I_1 + I_2$.

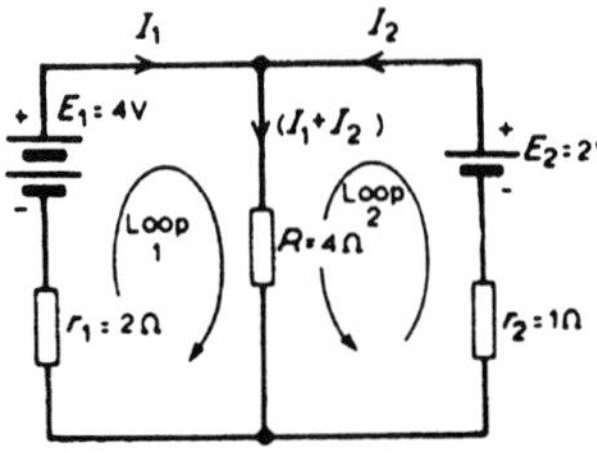

Figure 11.5

2. Divide the circuit into two loops and apply Kirchhoff's voltage law to each. From loop 1 of Fig. 11.5, and moving in a clockwise direction as indicated (the direction chosen does not matter), gives

$E_1 = I_1r_1 + (I_1 + I_2)R$

i.e. $4 = 2I_1 + 4(I_1 + I_2)$

i.e. $6I_1 + 4I_2 = 4$ (1)

From loop 2 of Fig. 11.5, and moving in an anticlockwise direction as indicated (once again, the choice of direction does not matter; it does not have to be in the same direction as that chosen for the first loop), gives:

$E_2 = I_2r_2 + (I_1 + I_2)R$

i.e. $2 = I_2 + 4(I_1 + I_2)$

i.e. $4I_1 + 5I_2 = 2$ (2)

3. Solve equations (1) and (2) for I_1 and I_2.

$2 \times (1)$ gives: $12I_1 + 8I_2 = 8$ (3)

$3 \times (2)$ gives: $12I_1 + 15I_2 = 6$ (4)

(3) – (4) gives: $-7I_2 = 2$ hence $I_2 = -2/7 =$ **–0.286 A**

(i.e. I_2 is flowing in the opposite direction to that shown in Fig. 11.5). From (1)

$6I_1 + 4(-0.286) = 4$

$6I_1 = 4 + 1.144$

Hence

$I_1 = \dfrac{5.144}{6} = \mathbf{0.857\ A}$

Current flowing through resistance R is

$I_1 + I_2 = 0.857 + (-0.286)$

$= \mathbf{0.571\ A}$

Note that a third loop is possible, as shown in Fig. 11.6, giving a third equation which can be used as a check:

$E_1 - E_2 = I_1r_1 - I_2r_2$

$4 - 2 = 2I_1 - I_2$

$2 = 2I_1 - I_2$

(Check: $2I_1 - I_2 = 2(0.857) - (-0.286) = 2$.)

Problem 3. Determine, using Kirchhoff's laws, each branch current for the network shown in Fig. 11.7.

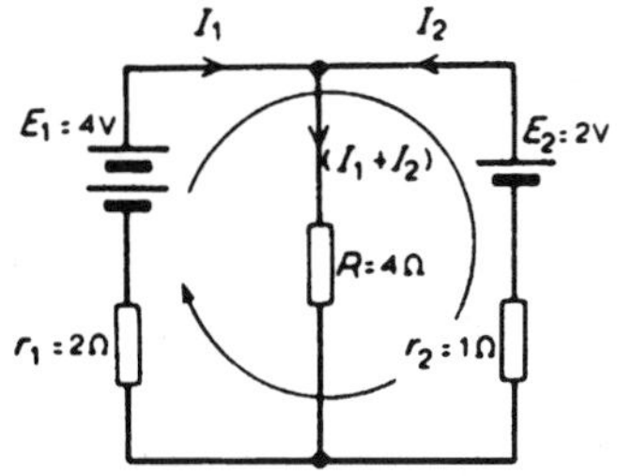

Figure 11.6

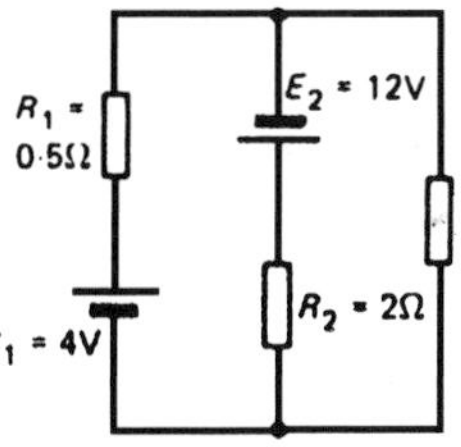

Figure 11.7

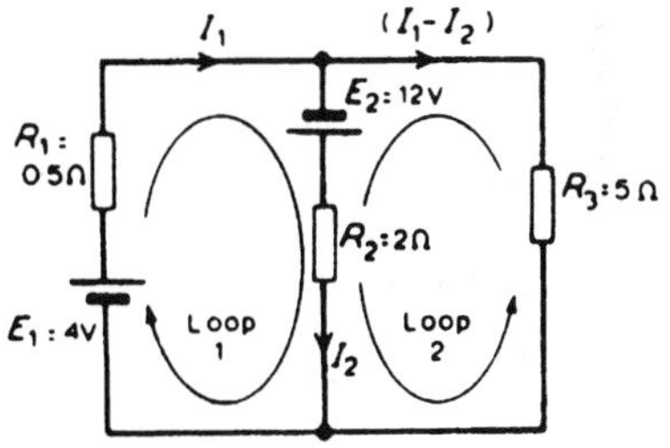

Figure 11.8

1. Currents and their directions are shown labelled in Fig. 11.8 following Kirchhoff's current law. It is usual, although not essential, to follow conventional current flow with current flowing from the positive terminal of the source.
2. The network is divided into two loops as shown in Fig. 11.8. Applying Kirchhoff's voltage law gives:

For loop 1:

$E_1 + E_2 = I_1R_1 + I_2R_2$

i.e. $16 = 0.5I_1 + 2I_2$ (1)

For loop 2:

$E_2 = I_2R_2 - (I_1 - I_2)R_3$

Note that since loop 2 is in the opposite direction to current $(I_1 - I_2)$ the volt drop across R_3 (i.e. $(I_1 - I_2)(R_3)$) is by convention negative. Thus

$$12 = 2I_2 - 5(I_1 - I_2)$$

i.e. $12 = -5I_1 + 7I_2$ (2)

3. Solving equations (1) and (2) to find I_1 and I_2:

$10 \times (1)$ gives $160 = 5I_1 + 20I_2$ (3)

$(2) + (3)$ gives $172 = 27I_2$, hence

$$I_2 = \frac{172}{27} = 6.37\text{ A}$$

From (1):

$$16 = 0.5I_1 + 2(6.37)$$

$$I_1 = \frac{16 - 2(6.37)}{0.5} = \mathbf{6.52\ A}$$

Current flowing in R_3 $= I_1 - I_2 = 6.52 - 6.37 = \mathbf{0.15\ A}$.

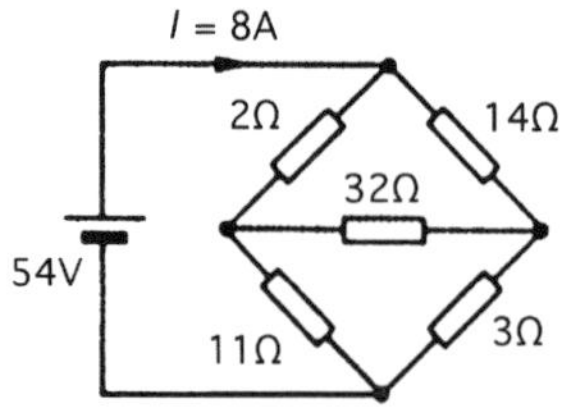

Figure 11.9

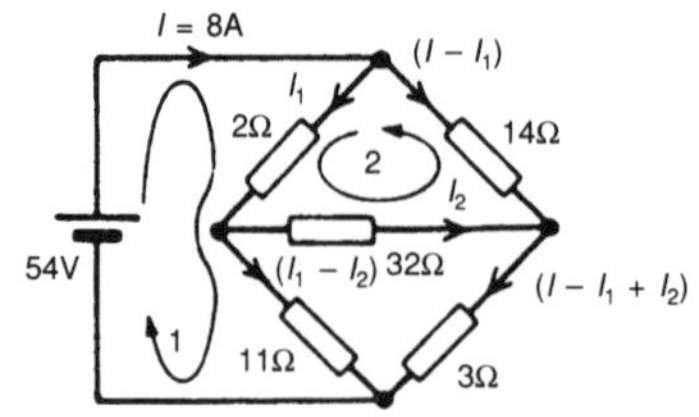

Figure 11.10

Problem 4. For the bridge network shown in Fig. 11.9 determine the currents in each of the resistors.

Let the current in the 2 Ω resistor be I_1, then by Kirchhoff's current law, the current in the 14 Ω resistor is $(I - I_1)$. Let the current in the 32 Ω resistor be I_2 as shown in Fig. 11.10. Then the current in the 11 Ω resistor is $(I_1 - I_2)$ and that in the 3 Ω resistor is $(I - I_1 + I_2)$. Applying Kirchhoff's voltage law to loop 1 and moving in a clockwise direction as shown in Fig. 11.10 gives:

$$54 = 2I_1 + 11(I_1 - I_2)$$

i.e. $13I_1 - 11I_2 = 54$ (1)

Applying Kirchhoff's voltage law to loop 2 and moving in an anticlockwise direction as shown in Fig. 11.10 gives:

$$0 = 2I_1 + 32I_2 - 14(I - I_1)$$

However $I = 8$ A; hence
$0 = 2I_1 + 32I_2 - 14(8 - I_1)$, i.e.

$16I_1 + 32I_2 = 112$ (2)

Equations (1) and (2) are simultaneous equations with two unknowns, I_1 and I_2.

$16 \times (1)$ gives: $208I_1 - 176I_2 = 864$ (3)

$13 \times (2)$ gives: $208I_1 + 416I_2 = 1456$ (4)

$(4) - (3)$ gives: $592I_2 = 592$

$I_2 = 1$ A

Substituting for I_2 in (1) gives:

$$13I_1 - 11 = 54$$

$$I_1 = \frac{65}{13} = 5\text{ A}$$

Hence,
the current flowing in the 2 Ω resistor $= I_1 = \mathbf{5\ A}$;
the current flowing in the 14 Ω resistor $= I - I_1 = 8 - 5 = \mathbf{3\ A}$;
the current flowing in the 32 Ω resistor $= I_2 = \mathbf{1\ A}$;
the current flowing in the 11 Ω resistor $= I_1 - I_2 = 5 - 1 = \mathbf{4\ A}$; and
the current flowing in the 3 Ω resistor $= I - I_1 + I_2 = 8 - 5 + 1 = \mathbf{4\ A}$.

11.3 The superposition theorem

The **superposition theorem** states:

In any network made up of linear resistances and containing more than one source of e.m.f., the resultant current flowing in any branch is the algebraic sum of the currents that would flow in that branch if each source was considered

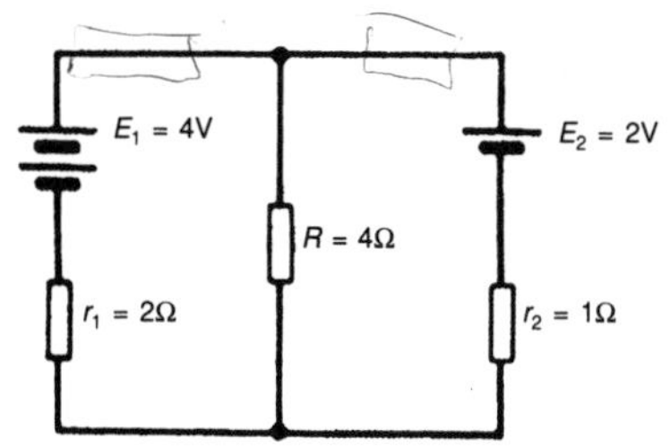

Figure 11.11

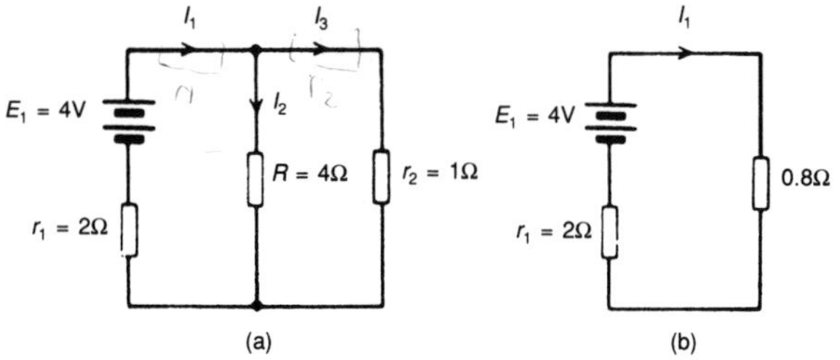

Figure 11.12

separately, all other sources being replaced at that time by their respective internal resistances.

> Problem 5. Figure 11.11 shows a circuit containing two sources of e.m.f., each with their internal resistance. Determine the current in each branch of the network by using the superposition theorem.

Procedure:

1. Redraw the original circuit with source E_2 removed, being replaced by r_2 only, as shown in Fig. 11.12(a).
2. Label the currents in each branch and their directions as shown in Fig. 11.12(a) and determine their values. (Note that the choice of current directions depends on the battery polarity, which by convention is taken as flowing from the positive battery terminal as shown.) R in parallel with r_2 gives an equivalent resistance of $(4 \times 1)/(4 + 1) = 0.8\ \Omega$. From the equivalent circuit of Fig. 11.12(b)

$$I_1 = \frac{E_1}{r_1 + 0.8} = \frac{4}{2 + 0.8} = 1.429\text{ A}$$

From Fig. 11.12(a)

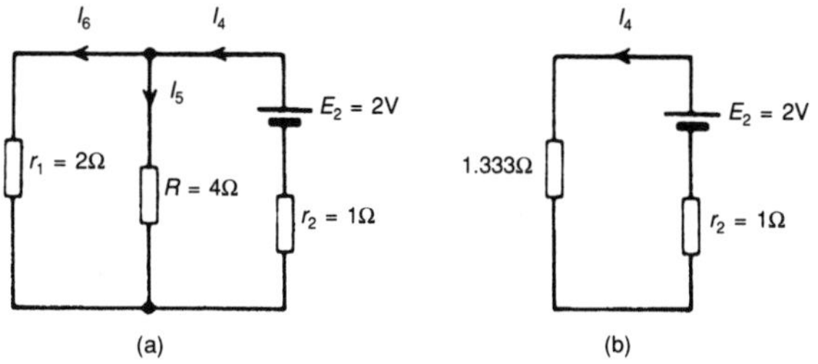

Figure 11.13

$$I_2 = \left(\frac{1}{4 + 1}\right)(I_1) = \frac{1}{5}(1.429) = 0.286\text{ A}$$

and

$$I_3 = \left(\frac{4}{4 + 1}\right)(I_1) = \frac{4}{5}(1.429) = 1.143\text{ A}$$

3. Redraw the original circuit with source E_1 removed, being replaced by r_1 only, as shown in Fig. 11.13(a).
4. Label the currents in each branch and their directions as shown in Fig. 11.13(a) and determine their values. r_1 in parallel with R gives an equivalent resistance of

$$\frac{2 \times 4}{2 + 4} = \frac{8}{6} = 1.333\ \Omega$$

From the equivalent circuit of Fig. 11.13(b)

$$I_4 = \frac{E_2}{1.333 + r_2} = \frac{2}{1.333 + 1} = 0.857\text{ A}$$

From Fig. 11.13(a)

$$I_5 = \left(\frac{2}{2 + 4}\right)(I_4) = \frac{2}{6}(0.857) = 0.286\text{ A}$$

$$I_6 = \left(\frac{4}{2 + 4}\right)(I_4) = \frac{4}{6}(0.857) = 0.571\text{ A}$$

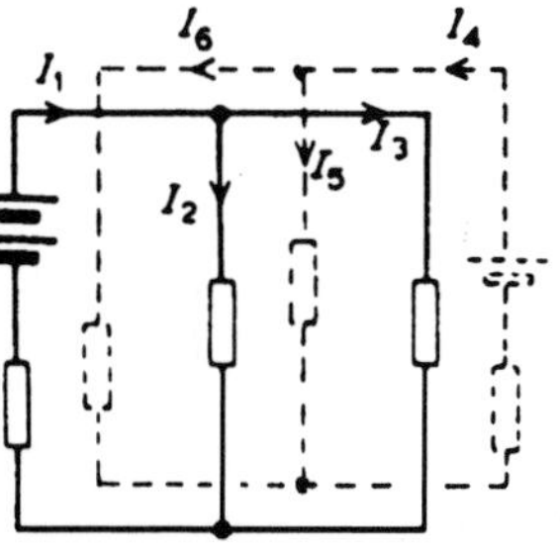

Figure 11.14

5. Superimpose Fig. 11.13(a) on to Fig. 11.12(a) as shown in Fig. 11.14.
6. Determine the algebraic sum of the currents flowing in each branch. Resultant current flowing through source 1, i.e.

$$I_1 - I_6 = 1.429 - 0.571$$
$$= \mathbf{0.858\ A\ (discharging)}$$

Resultant current flowing through source 2, i.e.

$$I_4 - I_3 = 0.857 - 1.143$$
$$= \mathbf{-0.286\ A\ (charging)}$$

Resultant current flowing through resistor R, i.e.

$$I_2 + I_5 = 0.286 + 0.286$$
$$= \mathbf{0.572\ A}$$

The resultant currents with their directions are shown in Fig. 11.15.

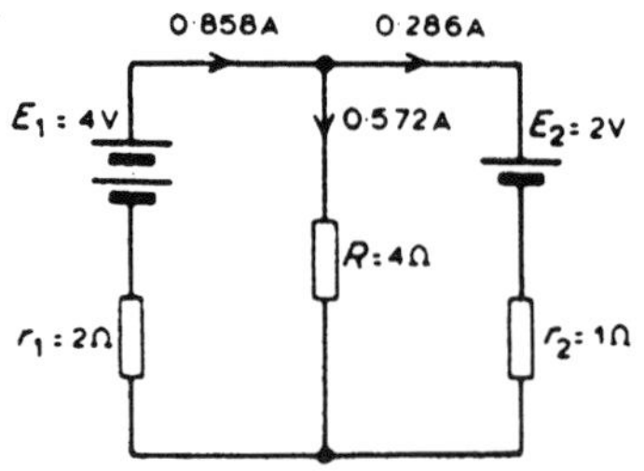

Figure 11.15

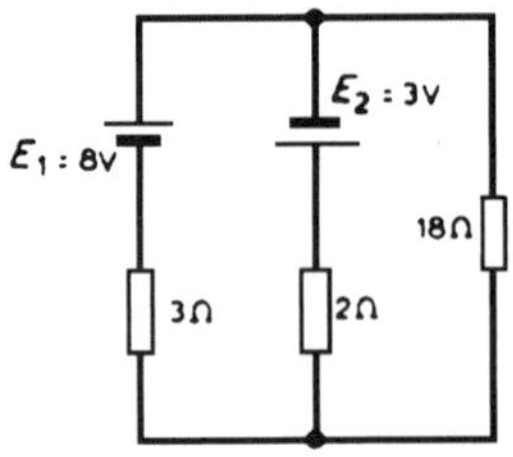

Figure 11.16

> Problem 6. For the circuit shown in Fig. 11.16, find, using the superposition theorem, (a) the current flowing in and the p.d. across the 18 Ω resistor, (b) the current flowing from the 8 V battery and (c) the current flowing from the 3 V battery.

1. Removing source E_2 gives the circuit of Fig. 11.17(a).
2. The current directions are labelled as shown in Fig. 11.17(a), I_1 flowing from the positive terminal of E_1. From Fig. 11.17(b),

$$I_1 = \frac{E_1}{3 + 1.8} = \frac{8}{4.8} = 1.667\ \text{A}$$

From Fig. 11.17(a),

$$I_2 = \left(\frac{18}{2 + 18}\right) I_1 = \frac{18}{20}(1.667) = 1.500\ \text{A}$$

and

$$I_3 = \left(\frac{2}{2 + 18}\right) I_1 = \frac{2}{20}(1.667) = 0.167\ \text{A}$$

3. Removing source E_1 gives the circuit of Fig. 11.18(a) (which is the same as Fig. 11.18(b)).
4. The current directions are labelled as shown in Figs. 11.18(a) and 11.18(b), I_4 flowing from the positive terminal of E_2.
From Fig. 11.18(c),

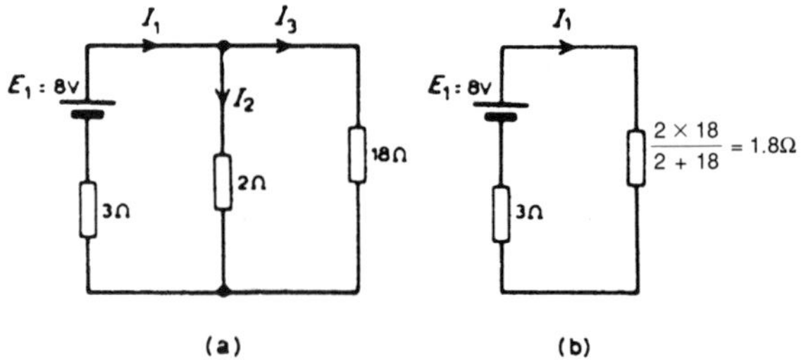

Figure 11.17

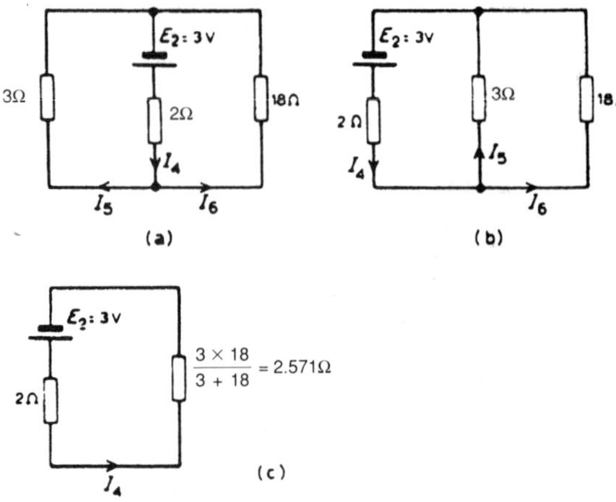

Figure 11.18

$$I_4 = \frac{E_2}{2 + 2.571} = \frac{3}{4.571} = 0.656 \text{ A}$$

From Fig. 11.18(b),

$$I_5 = \left(\frac{18}{3 + 18}\right)(I_4) = \frac{18}{21}(0.656) = 0.562 \text{ A}$$

$$I_6 = \left(\frac{3}{3 + 18}\right)(I_4) = \frac{3}{21}(0.656) = 0.094 \text{ A}$$

5. Superimposing Fig. 11.18(a) on to Fig. 11.17(a) gives the circuit in Fig. 11.19.
6 (a) Resultant current in the 18 Ω resistor

$= I_3 - I_6$

$= 0.167 - 0.094 =$ **0.073 A**

p.d. across the 18 Ω resistor = 0.073 × 18 = **1.314 V**.

(b) Resultant current in the 8 V battery

$= I_1 + I_5 = 1.667 + 0.562$

= **2.229 A (discharging)**

(c) Resultant current in the 3 V battery

$= I_2 + I_4 = 1.500 + 0.656$

= **2.156 A (discharging)**

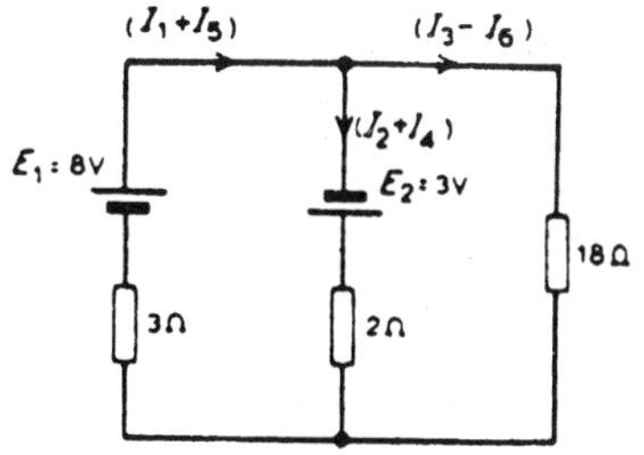

Figure 11.19

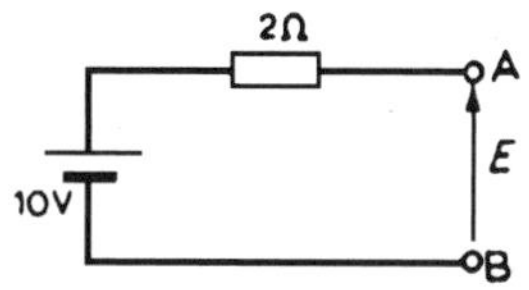

Figure 11.20

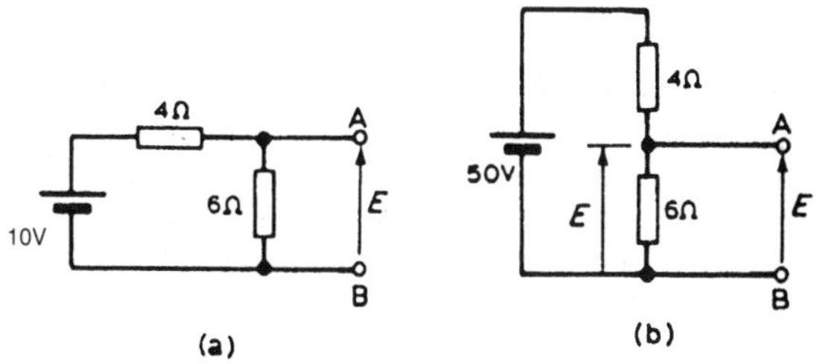

Figure 11.21

11.4 General d.c. circuit theory

The following points involving d.c. circuit analysis need to be appreciated before proceeding with problems using Thévenin's and Norton's theorems:

(i) The open-circuit voltage, E, across terminals AB in Fig. 11.20 is equal to 10 V, since no current flows through the 2 Ω resistor and hence no voltage drop occurs.

(ii) The open-circuit voltage, E, across terminals AB in Fig. 11.21(a) is the same as the voltage across the 6 Ω resistor. The circuit may be redrawn as shown in Fig. 11.21(b)

$$E = \left(\frac{6}{6 + 4}\right)(50)$$

by voltage division in a series circuit, i.e. ***E* = 30 V**.

(iii) For the circuit shown in Fig. 11.22(a) representing a practical source supplying energy, $V = E - Ir$, where E is the battery e.m.f., V is the battery terminal voltage and r is the internal resistance of the battery (as shown in section 4.6). For the circuit shown in Fig. 11.22(b), $V = E - (-I)r$, i.e. $V = E + Ir$.

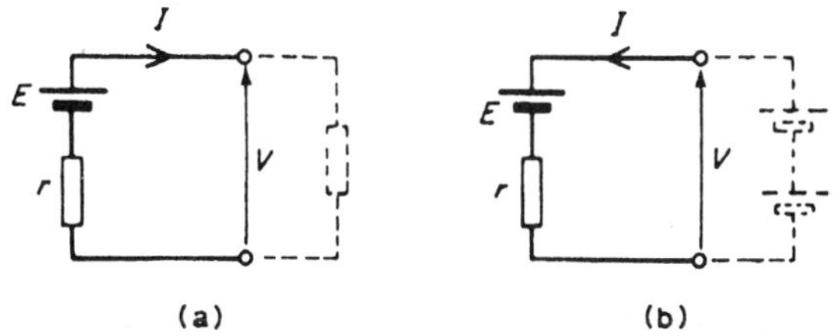

Figure 11.22

(iv) The resistance 'looking-in' at terminals AB in Fig. 11.23(a) is obtained by reducing the circuit in stages as shown in Figs. 11.23(b) to (d). Hence the equivalent resistance across AB is 7 Ω.

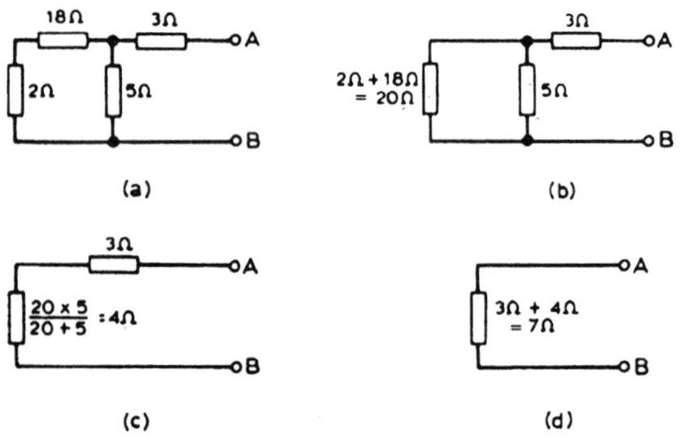

Figure 11.23

(v) For the circuit shown in Fig. 11.24(a), the 3 Ω resistor carries no current and the p.d. across the 20 Ω resistor is 10 V. Redrawing the circuit gives Fig. 11.24(b), from which

$$E = \left(\frac{4}{4+6}\right) \times 10 = \mathbf{4\ V}$$

(vi) If the 10 V battery in Fig. 11.24(a) is removed and replaced by a short-circuit, as shown in Fig. 11.24(c), then the 20 Ω resistor may be removed. The reason for this is that a short-circuit has zero resistance, and 20 Ω in parallel with zero ohms gives an equivalent resistance of (20 × 0)/(20 + 0), i.e. 0 Ω. The circuit is then as shown in Fig. 11.24(d), which is redrawn in Fig. 11.24(e).

From Fig. 11.24(e), the equivalent resistance across AB is given by

$$r = \frac{6 \times 4}{6 + 4} + 3 = 2.4 + 3 = \mathbf{5.4\ \Omega}$$

(vii) To find the voltage across AB in Fig. 11.25: Since the 20 V supply is across the 5 Ω and 15 Ω resistors in series then, by voltage division, the voltage drop across AC,

$$V_{AC} = \left(\frac{5}{5+15}\right)(20) = 5\ V$$

Similarly, $V_{CB} = (12/(12 + 3))(20) = 16$ V. V_C is at a potential of +20 V. $V_A = V_C - V_{AC} = +20 - 5 = 15$ V and $V_B = V_C - V_{BC} = +20 - 16 = 4$ V. Hence the voltage between AB is $V_A - V_B = 15 - 4 = 11$ V and current would flow from A to B since A has a higher potential than B.

(viii) In Fig. 11.26(a), to find the equivalent resistance across AB the circuit may be redrawn as in Figs. 11.26(b) and (c). From Fig. 11.26(c), the equivalent resistance across AB is given by

$$\frac{5 \times 15}{5 + 15} + \frac{12 \times 3}{12 + 3}$$

$$= 3.75 + 2.4 = \mathbf{6.15\ \Omega}$$

(ix) In the worked problems in sections 11.5 and 11.7 following, it may be considered that Thévenin's and Norton's theorems have no obvious advantages compared with, say,

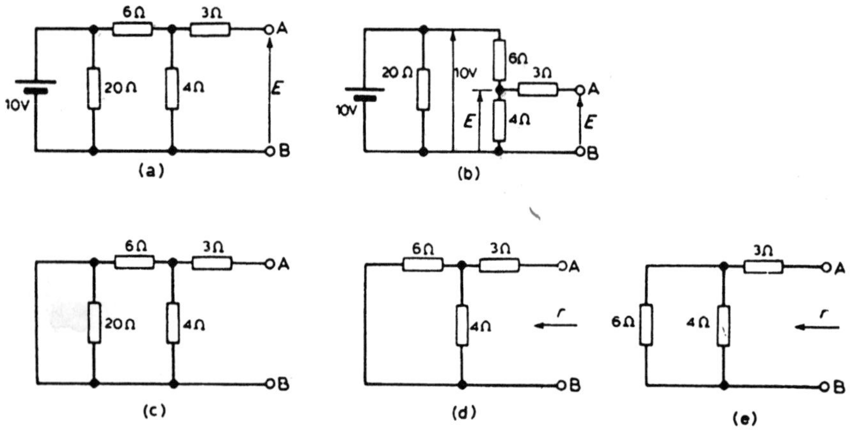

Figure 11.24

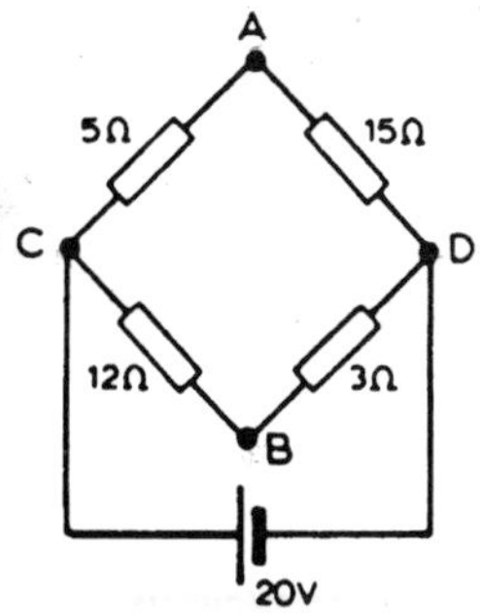

Figure 11.25

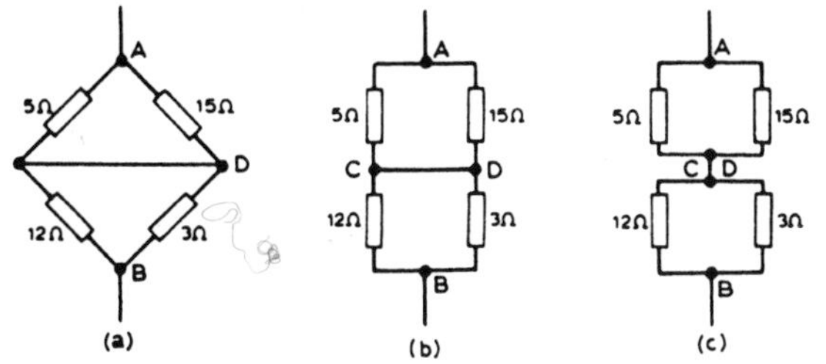

Figure 11.26

Kirchhoff's laws. However, these theorems can be used to analyse part of a circuit and in much more complicated networks the principle of replacing the supply by a constant voltage source in series with a resistance (or impedance) is very useful.

11.5 Thévenin's theorem

Thévenin's theorem states:

The current in any branch of a network is that which would result if an e.m.f. equal to the p.d. across a break made in the branch, were introduced into the branch, all other e.m.f.s being removed and represented by the internal resistances of the sources.

The procedure adopted when using Thévenin's theorem is summarized below. To determine the current in any branch of an active network (i.e. one containing a source of e.m.f.):

(i) remove the resistance R from that branch;
(ii) determine the open-circuit voltage, E, across the break;
(iii) remove each source of e.m.f. and replace them by their internal resistances and then determine the resistance, r, 'looking-in' at the break;
(iv) determine the value of the current from the equivalent circuit shown in Fig. 11.27, i.e.

$$\boldsymbol{I = \frac{E}{R + r}}$$

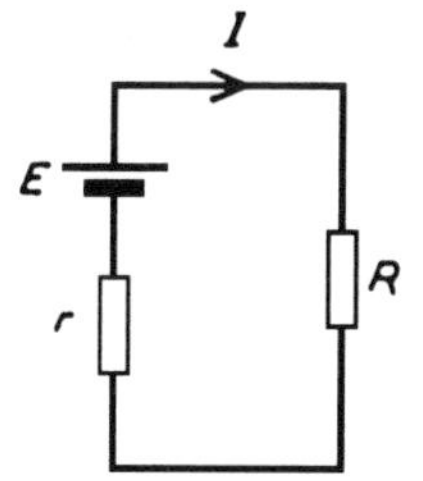

Figure 11.27

Problem 7. Use Thévenin's theorem to find the current flowing in the 10 Ω resistor for the circuit shown in Fig. 11.28(a).

Following the above procedure:

(i) The 10 Ω resistance is removed from the circuit as shown in Fig. 11.28(b).
(ii) There is no current flowing in the 5 Ω resistor and current I_1 is given by

$$I_1 = \frac{10}{R_1 + R_2} = \frac{10}{2 + 8} = 1\text{ A}$$

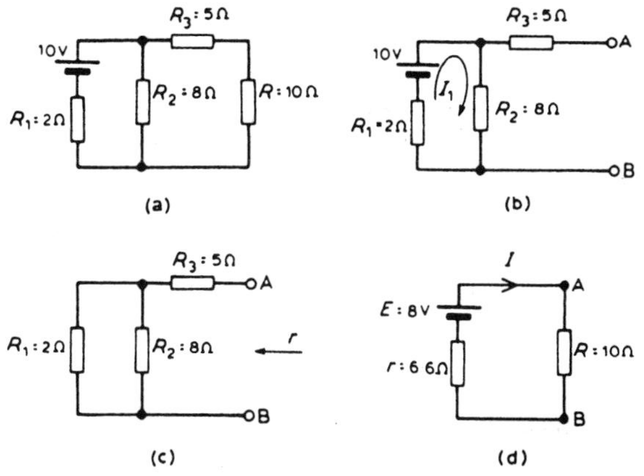

Figure 11.28

P.d. across $R_2 = I_1R_2 = 1 \times 8 = 8$ V. Hence p.d. across AB, i.e. the open-circuit voltage across the break, $E = 8$ V.

(iii) Removing the source of e.m.f. gives the circuit of Fig. 11.28(c).

$$\text{Resistance } r = R_3 + \frac{R_1R_2}{R_1 + R_2}$$

$$= 5 + \frac{2 \times 8}{2 + 8} = 5 + 1.6 = 6.6\ \Omega$$

(iv) The equivalent Thévenin's circuit is shown in Fig. 11.28(d).

$$\text{Current } I = \frac{E}{R + r} = \frac{8}{10 + 6.6} = \frac{8}{16.6}$$

$$= 0.482 \text{ A}$$

Hence the current flowing in the 10 Ω resistor of Fig. 11.28(a) is **0.482 A**.

> Problem 8. For the network shown in Fig. 11.29(a) determine the current in the 0.8 Ω resistor using Thévenin's theorem.

Following the procedure:

(i) The 0.8 Ω resistor is removed from the circuit as shown in Fig. 11.29(b).

(ii) $$\text{Current } I_1 = \frac{12}{1 + 5 + 4} = \frac{12}{10} = 1.2 \text{ A}$$

P.d. across 4 Ω resistor $= 4I_1 = (4)(1.2) = 4.8$ V. Hence p.d. across AB, i.e. the open-circuit voltage across AB, $E = 4.8$ V.

(iii) Removing the source of e.m.f. gives the circuit shown in Fig. 11.29(c). The equivalent circuit of Fig. 11.29(c) is shown in Fig. 11.29(d), from which,

$$\text{resistance } r = \frac{4 \times 6}{4 + 6} = \frac{24}{10} = 2.4\ \Omega$$

(iv) The equivalent Thévenin's circuit is shown in Fig. 11.29(e), from which,

$$\text{current } I = \frac{E}{r + R} = \frac{4.8}{2.4 + 0.8} = \frac{4.8}{3.2}$$

= 1.5 A = current in the 0.8 Ω resistor

> Problem 9. Use Thévenin's theorem to determine the current I flowing in the 4 Ω resistor shown in Fig. 11.30(a). Find also the power dissipated in the 4 Ω resistor.

Following the procedure:

(i) The 4 Ω resistor is removed from the circuit as shown in Fig. 11.30(b).

(ii) $$\text{Current } I_1 = \frac{E_1 - E_2}{r_1 + r_2} = \frac{4 - 2}{2 + 1} = \frac{2}{3} \text{ A}$$

p.d. across AB, $E = E_1 - I_1r_1 = 4 - \frac{2}{3}(2) = 2\frac{2}{3}$ V (see section 11.4(iii)). (Alternatively, p.d. across AB, $E = E_2 - I_1r_2 = 2 - (-2/3)(1) = 2\frac{2}{3}$ V.)

(iii) Removing the sources of e.m.f. gives the circuit shown in Fig. 11.30(c), from which,

$$\text{resistance } r = \frac{2 \times 1}{2 + 1} = \frac{2}{3}\ \Omega$$

(iv) The equivalent Thévenin's circuit is shown in Fig. 11.30(d), from which,

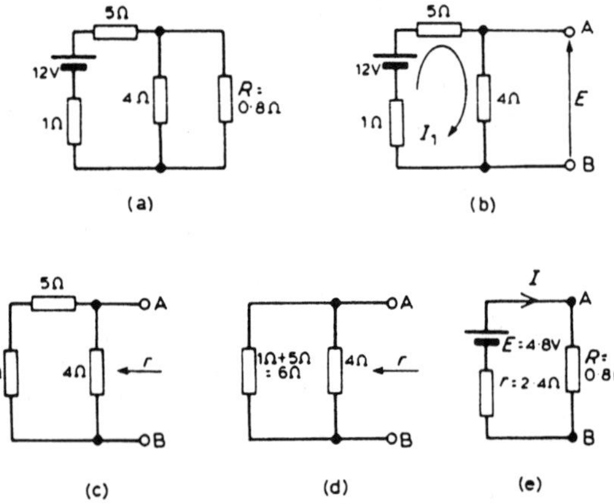

Figure 11.29

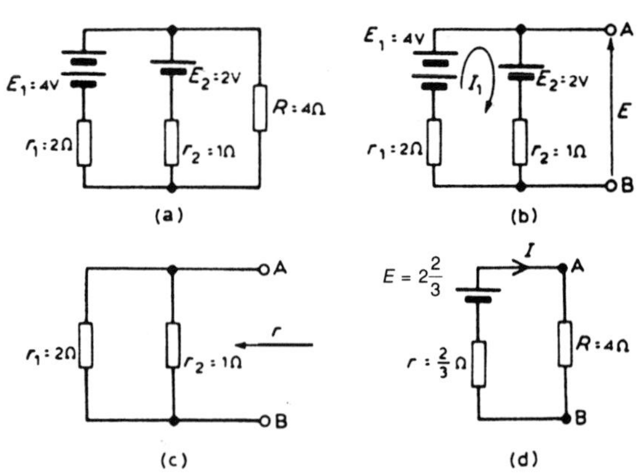

Figure 11.30

$$\text{current } I = \frac{E}{r+R} = \frac{2\frac{2}{3}}{\frac{2}{3}+4} = \frac{8/3}{14/3} = \frac{8}{14}$$

$= \mathbf{0.571\ A}$

$=$ **current in the 4 Ω resistor**

Power dissipated in the 4 Ω resistor, $P = I^2R = (0.571)^2(4) = \mathbf{1.304\ W}$.

> Problem 10. Determine the current in the 5 Ω resistance of the network shown in Fig. 11.31(a) using Thévenin's theorem. Hence find the currents flowing in the other two branches.

Following the procedure:

(i) The 5 Ω resistance is removed from the circuit as shown in Fig. 11.31(b).

(ii) $$\text{Current } I_1 = \frac{12+4}{0.5+2} = \frac{16}{2.5} = 6.4\text{ A}$$

p.d. across AB,

$$E = E_1 - I_1r_1 = 4 - (6.4)(0.5)$$
$$= 0.8\text{ V (see section 11.4(iii))}$$

(Alternatively, $E = E_2 - I_1r_1 = -12 - (-6.4)(2) = 0.8$ V.)

(iii) Removing the sources of e.m.f. gives the circuit shown in Fig. 11.31(c), from which,

$$\text{resistance } r = \frac{0.5 \times 2}{0.5+2} = \frac{1}{2.5} = 0.4\ \Omega$$

(vi) The equivalent Thévenin's circuit is shown in Fig. 11.31(d), from which,

$$\text{current } I = \frac{E}{r+R} = \frac{0.8}{0.4+5} = \frac{0.8}{5.4} = \mathbf{0.148\ A}$$

$=$ **current in the 5 Ω resistor**

From Fig. 11.31(e), voltage $V = IR_3 = (0.148)(5) = 0.74$ V. From section 11.4(iii), $V = E_1 - I_Ar_1$, i.e. $0.74 = 4 - (I_A)(0.5)$. Hence

$$\text{current } I_A = \frac{4-0.74}{0.5} = \frac{3.26}{0.5} = \mathbf{6.52\ A}$$

Also from Fig. 11.31(e), $V = -E_2 + I_Br_2$, i.e. $0.74 = -12 + (I_B)(2)$. Hence

$$\text{current } I_B = \frac{12+0.74}{2} = \frac{12.74}{2} = 6.37\text{ A}$$

(Check, from Fig. 11.31(e), $I_A = I_B + I$, correct to 2 significant figures (Kirchhoff's current law).)

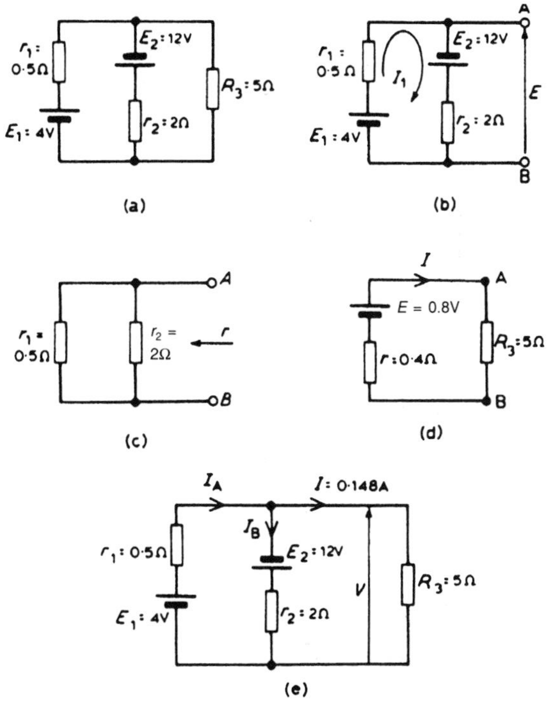

Figure 11.31

> Problem 11. Use Thévenin's theorem to determine the current flowing in the 3 Ω resistance of the network shown in Fig. 11.32(a). The voltage source has negligible internal resistance.

(Note the symbol for an ideal voltage source in Fig. 11.32(a) which may be used as an alternative to the battery symbol.)
Following the procedure:

(i) The 3 Ω resistance is removed from the circuit as shown in Fig. 11.32(b).
(ii) The $1\frac{2}{3}$ Ω resistance now carries no current. P.d. across 10 Ω resistor $= (10/(10+5))(24) = 16$ V (see section 11.4(v)). Hence p.d. across AB, $\boldsymbol{E} = \mathbf{16\ V}$.
(iii) Removing the source of e.m.f. and replacing it by its internal resistance means that the 20 Ω resistance is short-circuited as shown in Fig. 11.32(c) since its internal resistance is zero. The 20 Ω resistance may thus be removed as shown in Fig. 11.32(d) (see section 11.4(vi)). From Fig. 11.32(d),

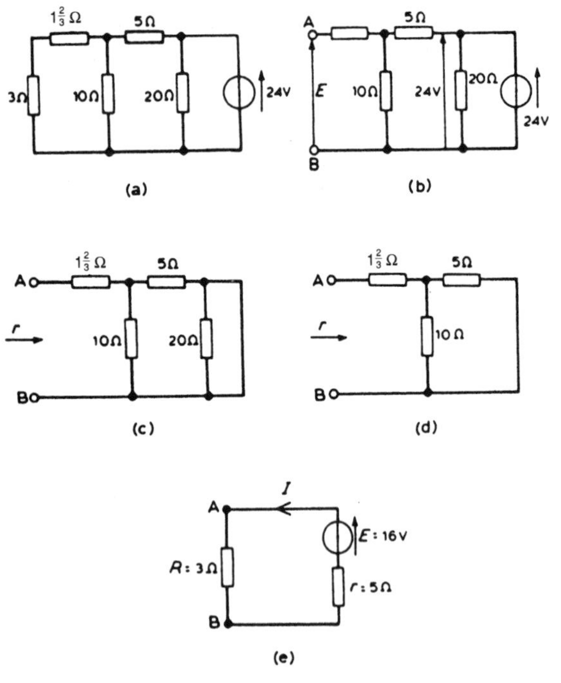

Figure 11.32

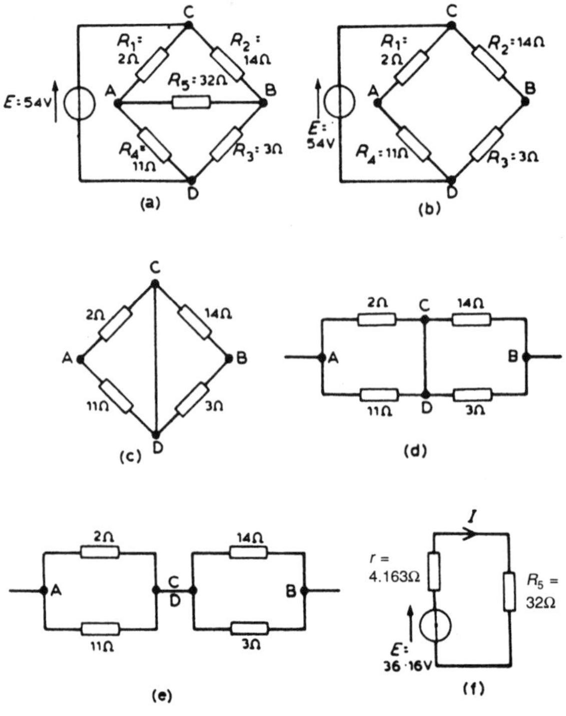

Figure 11.33

$$\text{resistance } r = 1\frac{2}{3} + \frac{10 \times 5}{10 + 5} = 1\frac{2}{3} + \frac{50}{15} = 5\ \Omega$$

(iv) The equivalent Thévenin's circuit is shown in Fig. 11.32(e), from which,

$$\text{current } I = \frac{E}{r + R} = \frac{16}{3 + 5} = \frac{16}{8} = \mathbf{2\ A}$$

$$= \textbf{current in the 3}\ \mathbf{\Omega}\ \textbf{resistance}$$

> Problem 12. A Wheatstone bridge network is shown in Fig. 11.33(a). Calculate the current flowing in the 32 Ω resistor, and its direction, using Thévenin's theorem. Assume the source of e.m.f. to have negligible resistance.

Following the procedure:

(i) The 32 Ω resistor is removed from the circuit as shown in Fig. 11.33(b).

(ii) The p.d. between A and C is given by

$$V_{AC} = \left(\frac{R_1}{R_1 + R_4}\right)(E) = \left(\frac{2}{2 + 11}\right)(54)$$

$$= 8.31\ \text{V}$$

The p.d. between B and C is given by

$$V_{BC} = \left(\frac{R_2}{R_2 + R_3}\right)(E) = \left(\frac{14}{14 + 3}\right)(54)$$

$$= 44.47\ \text{V}$$

Hence the p.d. between A and B = 44.47 – 8.31 = 36.16 V. Point C is at a potential of +54 V. Between C and A is a voltage drop of 8.31 V. Hence the voltage at point A is 54 – 8.31 = 45.69 V. Between C and B is a voltage drop of 44.47 V. Hence the voltage at point B is 54 – 44.47 = 9.53 V. Since the voltage at A is greater than at B, current must flow in the direction A to B (see section 11.4(vii)).

(iii) Replacing the source of e.m.f. with a short-circuit (i.e. zero internal resistance) gives the circuit shown in Fig. 11.33(c). The circuit is redrawn and simplified as shown in Figs. 11.33(d) and (e), from which the resistance between terminals A and B,

$$r = \frac{2 \times 11}{2 + 11} + \frac{14 \times 3}{14 + 3} = \frac{22}{13} + \frac{42}{17}$$

$$= 1.692 + 2.471 = 4.163\ \Omega$$

(iv) The equivalent Thévenin's circuit is shown in Fig. 11.33(f), from which,

$$\text{current } I = \frac{E}{r + R_5} = \frac{36.16}{4.163 + 32} = 1\ \text{A}$$

Hence the current in the 32 Ω resistor of Fig. 11.33(a) is 1 A, flowing from A to B.

11.6 Constant-current source

A source of electrical energy can be represented by a source of e.m.f. in series with a resistance. In section 11.5, the Thévenin constant-voltage source consisted of a constant e.m.f. E in series with an internal resistance r. However this is not the only form of representation. A source of electrical energy can also be represented by a constant-current source in parallel with a resistance. It may be shown that the two forms are equivalent. An **ideal constant-voltage generator** is one with zero internal resistance so that it supplies the same voltage to all loads. An **ideal constant-current generator** is one with infinite internal resistance so that it supplies the same current to all loads.

Note the symbol for an ideal current source (BS 3939, 1985), shown in Fig. 11.34.

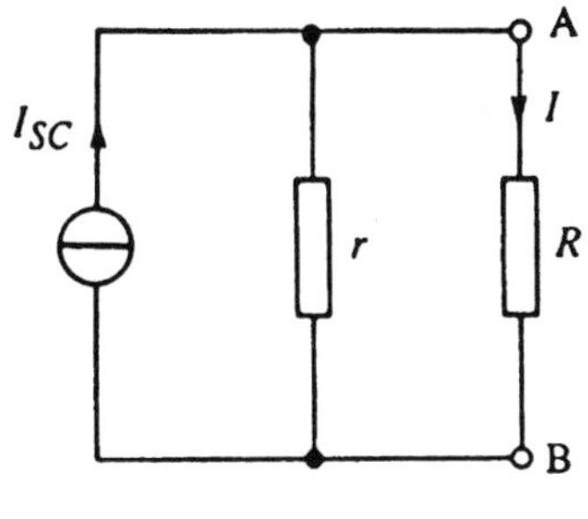

Figure 11.34

11.7 Norton's theorem

Norton's theorem states

The current that flows in any branch of a network is the same as that which would flow in the branch if it were connected across a source of electrical energy, the short-circuit current of which is equal to the current that would flow in a short-circuit across the branch, and the internal resistance of which is equal to the resistance which appears across the open-circuited branch terminals.

The procedure adopted when using Norton's theorem is summarized below. To determine the current flowing in a resistance R of a branch AB of an active network:

(i) short-circuit branch AB;
(ii) determine the short-circuit current I_{SC} flowing in the branch;
(iii) remove all sources of e.m.f. and replace them by their internal resistance (or, if a current source exists, replace with an open-circuit), then determine the resistance r, 'looking-in' at a break made between A and B;
(iv) determine the current I flowing in resistance R from the Norton equivalent network shown in Fig. 11.34, i.e.

$$\boldsymbol{I = \left(\frac{r}{r + R}\right) I_{SC}}$$

Problem 13. Use Norton's theorem to determine the current flowing in the 10 Ω resistance for the circuit shown in Fig. 11.35(a).

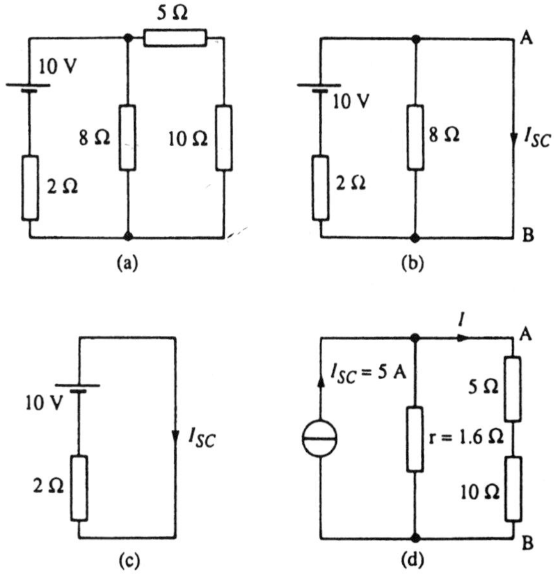

Figure 11.35

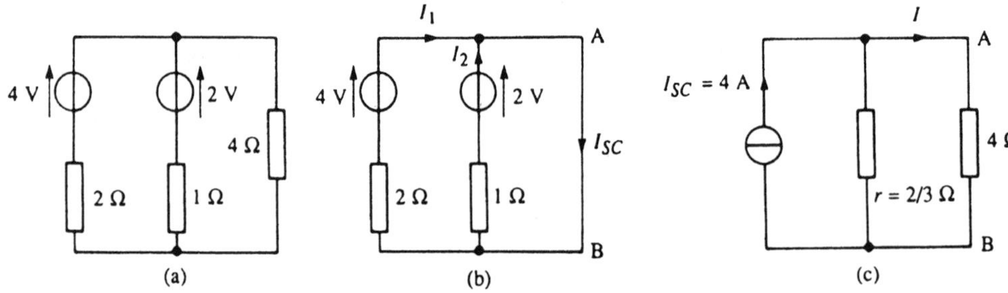

Figure 11.36

Following the above procedure:

(i) The branch containing the 10 Ω resistance is short-circuited as shown in Fig. 11.35(b).

(ii) Fig. 11.35(c) is equivalent to Fig. 11.35(b). Hence $I_{SC} = 10/2 = 5$ A.

(iii) If the 10 V source of e.m.f. is removed from Fig. 11.35(b) the resistance 'looking-in' at a break made between A and B is given by:

$$r = \frac{2 \times 8}{2 + 8} = 1.6\ \Omega$$

(iv) From the Norton equivalent network shown in Fig. 11.35(d) the current in the 10 Ω resistance, by current division, is given by

$$I = \left(\frac{1.6}{1.6 + 5 + 10}\right)(5) = \mathbf{0.482\ A}$$

as obtained previously in Problem 7 using Thévenin's theorem.

Problem 14. Use Norton's theorem to determine the current I flowing in the 4 Ω resistance shown in Fig. 11.36(a).

Following the procedure:

(i) The 4 Ω branch is short-circuited as shown in Fig. 11.36(b).

(ii) From Fig. 11.36(b),

$$I_{SC} = I_1 + I_2 = \frac{4}{2} + \frac{2}{1} = 4\ \text{A}$$

(iii) If the sources of e.m.f. are removed the resistance 'looking-in' at a break made between A and B is given by

$$r = \frac{2 \times 1}{2 + 1} = \frac{2}{3}\ \Omega$$

(iv) From the Norton equivalent network shown in Fig. 11.36(c) the current in the 4 Ω resistance is given by

$$I = \left(\frac{\frac{2}{3}}{\frac{2}{3} + 4}\right)(4) = \mathbf{0.571\ A}$$

as obtained previously in Problems 2, 5 and 9 using Kirchhoff's laws and the theorems of superposition and Thévenin.

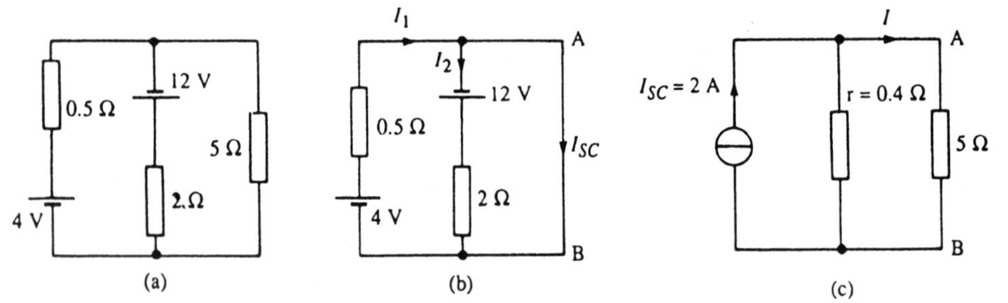

Figure 11.37

> Problem 15. Determine the current in the 5 Ω resistance of the network shown in Fig. 11.37(a) using Norton's theorem. Hence find the currents flowing in the other two branches.

Following the procedure:

(i) The 5 Ω branch is short-circuited as shown in Fig. 11.37(b).

(ii) From Fig. 11.37(b),

$$I_{SC} = I_1 - I_2 = \frac{4}{0.5} - \frac{12}{2} = 8 - 6 = \mathbf{2\ A}$$

(iii) If each source of e.m.f. is removed the resistance 'looking-in' at a break made between A and B is given by

$$r = \frac{0.5 \times 2}{0.5 + 2} = 0.4\ \Omega$$

(iv) From the Norton equivalent network shown in Fig. 11.37(c), the current in the 5 Ω resistance is given by

$$I = \left(\frac{0.4}{0.4 + 5}\right)(2) = \mathbf{0.148\ A}$$

as obtained previously in Problem 10 using Thévenin's theorem.
The currents flowing in the other two branches are obtained in the same way as in Problem 10. Hence the current flowing from the 4 V source is **6.52 A** and the current flowing from the 12 V source is **6.37 A**.

> Problem 16. Use Norton's theorem to determine the current flowing in the 3 Ω resistance of the network shown in Fig. 11.38(a). The voltage source has negligible internal resistance.

Following the procedure:

(i) The branch containing the 3 Ω resistance is short-circuited as shown in Fig. 11.38(b).

(ii) From the equivalent circuit shown in Fig. 11.38(c), $I_{SC} = 24/5 = \mathbf{4.8\ A}$.

(iii) If the 24 V source of e.m.f. is removed the resistance 'looking-in' at a break made between A and B is obtained from Fig. 11.38(d) and its equivalent circuit shown in Fig. 11.38(e) and is given by

$$r = \frac{10 \times 5}{10 + 5} = \frac{50}{15} = 3\frac{1}{3}\ \Omega$$

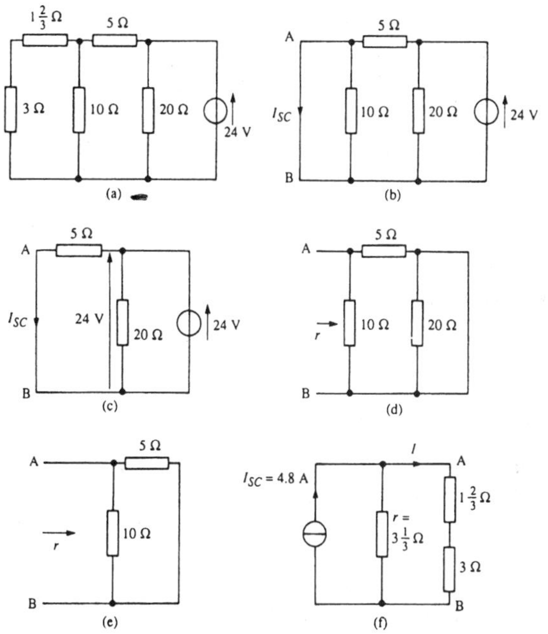

Figure 11.38

(iv) From the Norton equivalent network shown in Fig. 11.38(f) the current in the 3 Ω resistance is given by

$$I = \left(\frac{3\frac{1}{3}}{3\frac{1}{3} + 1\frac{2}{3} + 3}\right)(4.8) = \mathbf{2\ A}$$

as obtained previously in Problem 11 using Thévenin's theorem.

> Problem 17. Determine the current flowing in the 2 Ω resistance in the network shown in Fig. 11.39(a).

Following the procedure:

(i) The 2 Ω resistance branch is short-circuited as shown in Fig. 11.39(b).

(ii) Fig. 11.39(c) is equivalent to Fig. 11.39(b). Hence $I_{SC} = (6/(6 + 4))(15) = \mathbf{9\ A}$ by current division.

(iii) If the 15 A current source is replaced by an open-circuit then from Fig. 11.39(d) the resistance 'looking-in' at a break made between A and B is given by (6 + 4) Ω in parallel with (8 + 7) Ω, i.e.

$$r = \frac{(10)(15)}{10 + 15} = \frac{150}{25} = 6\ \Omega$$

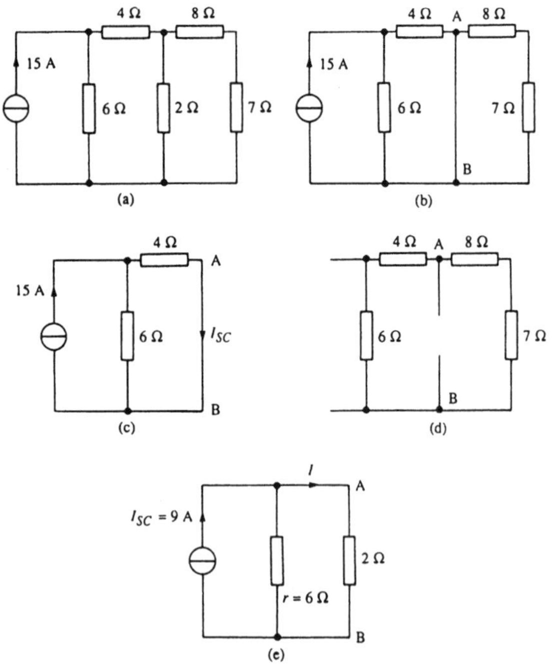

Figure 11.39

(iv) From the Norton equivalent network shown in Fig. 11.39(e) the current in the 2 Ω resistance is given by:

$$I = \left(\frac{6}{6+2}\right)(9) = \mathbf{6.75\ A}$$

11.8 Thévenin and Norton equivalent networks

The Thévenin and Norton networks shown in Fig. 11.40 are equivalent. The resistance 'looking-in' at terminals AB is the same in each of the networks, i.e. r.

If terminals AB in Fig. 11.40(a) are short-circuited, the short-circuit current is given by E/r. If terminals AB in Fig. 11.40(b) are short-circuited, the short-circuit current is I_{SC}. For the circuit shown in Fig. 11.40(a) to be equivalent to the circuit in Fig. 11.40(b) the same short-circuit current must flow. Thus $I_{SC} = E/r$. Figure 11.41 shows a source of e.m.f. E in series with a resistance r feeding a load resistance R. From Fig. 11.41,

$$I = \frac{E}{r+R} = \frac{E/r}{(r+R)/r} = \left(\frac{r}{r+R}\right)\frac{E}{r}$$

$$\text{i.e. } I = \left(\frac{r}{r+R}\right) I_{SC}$$

From Fig. 11.42 it can be seen that, when viewed from the load, the source appears as a source of current I_{SC} which is divided between r and R connected in parallel. Thus the two representations shown in Fig. 11.40 are equivalent.

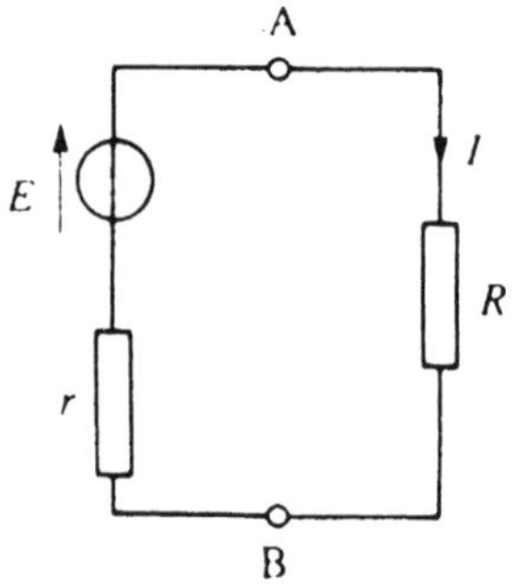

Figure 11.41

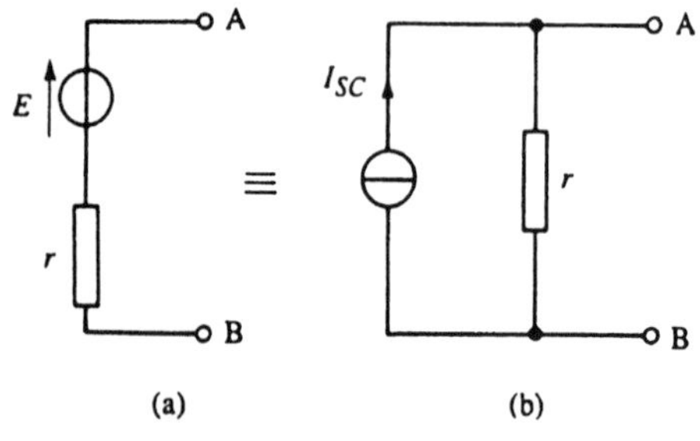

Figure 11.40

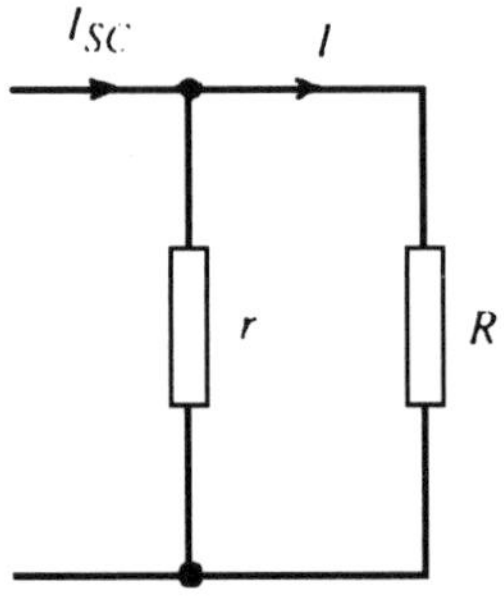

Figure 11.42

Problem 18. Convert the circuit shown in Fig. 11.43 to an equivalent Norton network.

If terminals AB in Fig. 11.43 are short-circuited, the short-circuit current $I_{SC} = 10/2 = 5$ A. The resistance 'looking-in' at terminals AB is 2 Ω. Hence the equivalent Norton network is as shown in Fig. 11.44.

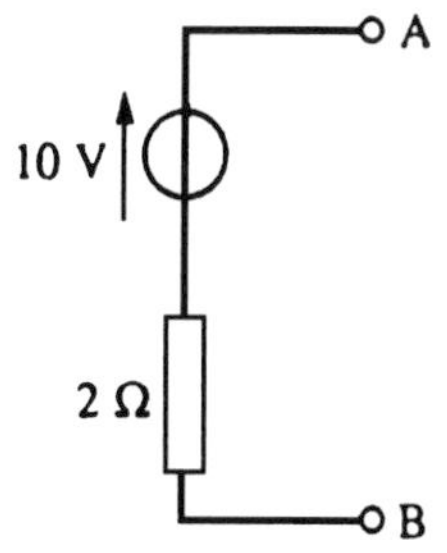

Figure 11.43

Problem 19. Convert the network shown in Fig. 11.45 to an equivalent Thévenin circuit.

The open-circuit voltage E across terminals AB in Fig. 11.45 is given by $E = (I_{SC})(r) = (4)(3) = 12$ V. The resistance 'looking-in' at terminals AB is 3 Ω. Hence the equivalent Thévenin circuit is as shown in Fig. 11.46.

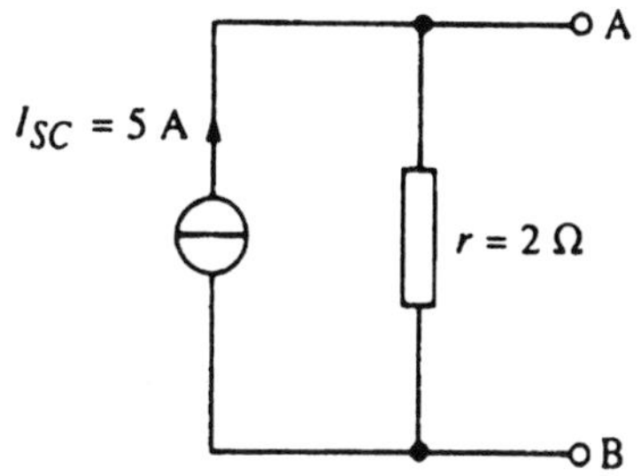

Figure 11.44

Problem 20. (a) Convert the circuit to the left of terminals AB in Fig. 11.47(a) to an equivalent Thévenin circuit by initially converting to a Norton equivalent circuit. (b) Determine the current flowing in the 1.8 Ω resistor.

(a) For the branch containing the 12 V source, converting to a Norton equivalent circuit gives $I_{SC_1} = 12/2 = 4$ A and $r_1 = 3$ Ω. For the branch containing the 24 V source, converting to a Norton equivalent circuit gives $I_{SC_2} = 24/2 = 12$ A and $r_2 = 2$ Ω. Thus Fig. 11.47(b) shows a network equivalent to Fig. 11.47(a).
From Fig. 11.47(b) the total short-circuit current is 4 + 12 = 16 A and the total resistance is given by: (3 × 2)/(3 + 2) = 1.2 Ω. Thus Fig. 11.47(b) simplifies to Fig. 11.47(c).
The open-circuit voltage across AB of Fig. 11.47(c), $E = (16)(1.2) = 19.2$ V, and the resistance 'looking-in' at AB is 1.2 Ω. Hence the Thévenin equivalent circuit is as shown in Fig. 11.47(d).

(b) When the 1.8 Ω resistance is connected between terminals A and B of Fig. 11.47(d) the current I flowing is given by

$$I = \frac{19.2}{1.2 + 1.8} = \mathbf{6.4\ A}$$

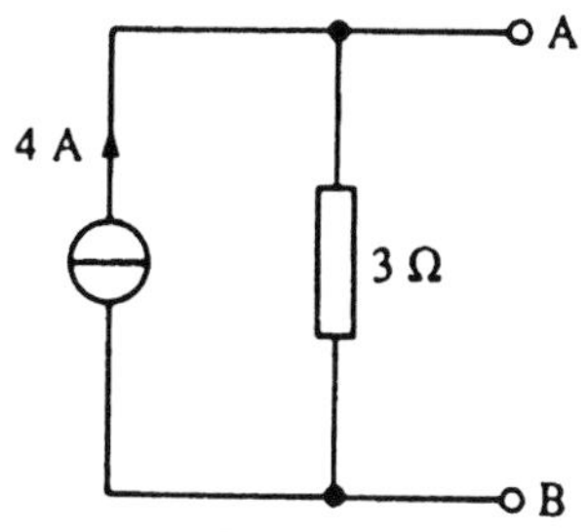

Figure 11.45

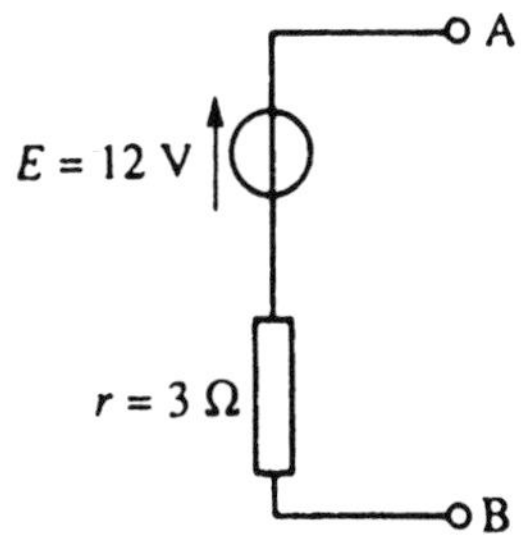

Figure 11.46

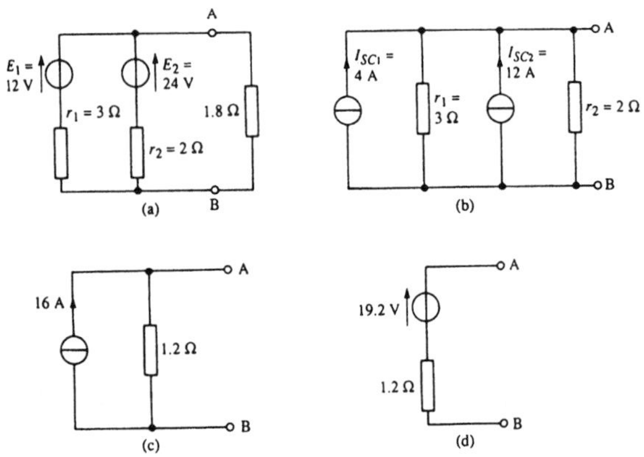

Figure 11.47

> Problem 21. Determine by successive conversions between Thévenin and Norton equivalent networks a Thévenin equivalent circuit for terminals AB of Fig. 11.48(a). Hence determine the current flowing in the 200 Ω resistance.

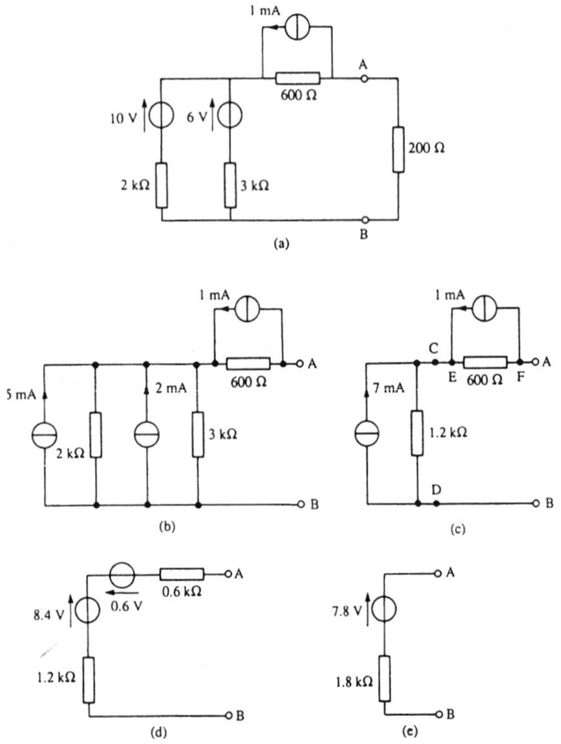

Figure 11.48

For the branch containing the 10 V source, converting to a Norton equivalent network gives $I_{SC1} = 10/2000 = 5$ mA and $r_1 = 2$ kΩ.

For the branch containing the 6 V source, converting to a Norton equivalent network gives $I_{SC2} = 6/3000 = 2$ mA and $r_2 = 3$ kΩ. Thus the network of Fig. 11.48(a) converts to Fig. 11.48(b). Combining the 5 mA and 2 mA current sources gives the equivalent network of Fig. 11.48(c) where the short-circuit current for the original two branches considered is 7 mA and the resistance is $(2 \times 3)/(2 + 3) = 1.2$ kΩ.

Both of the Norton equivalent networks shown in Fig. 11.48(c) may be converted to Thévenin equivalent circuits. The open-circuit voltage across CD is $(7 \times 10^{-3})(1.2 \times 10^3) = 8.4$ V and the resistance 'looking-in' at CD is 1.2 kΩ. The open-circuit voltage across EF is $(1 \times 10^{-3})(600) = 0.6$ V and the resistance 'looking-in' at EF is 0.6 kΩ. Thus Fig. 11.48(c) converts to Fig. 11.48(d). Combining the two Thévenin circuits gives $E = 8.4 - 0.6 =$ **7.8 V** and the resistance $r = (1.2 + 0.6)$ kΩ = **1.8 kΩ**. Thus the Thévenin equivalent circuit for terminals AB of Fig. 11.48(a) is as shown in Fig. 11.48(e). Hence the current I flowing in a 200 Ω resistance connected between A and B is given by

$$I = \frac{7.8}{1800 + 200} = \frac{7.8}{2000} = \mathbf{3.9\ mA}$$

11.9 Maximum power transfer theorem

The **maximum power transfer theorem** states:

> *The power transferred from a supply source to a load is at its maximum when the resistance of the load is equal to the internal resistance of the source.*

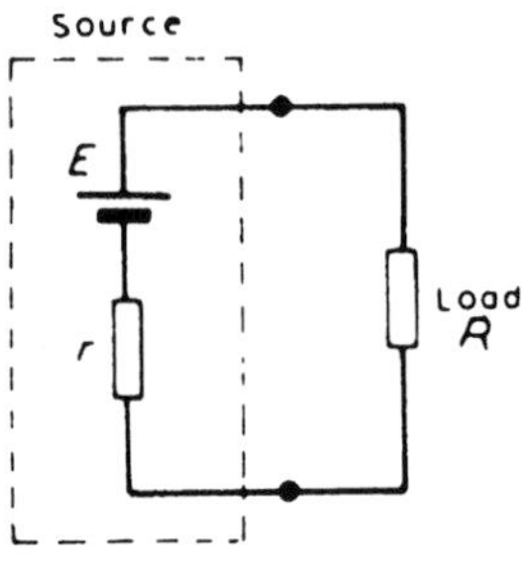

Figure 11.49

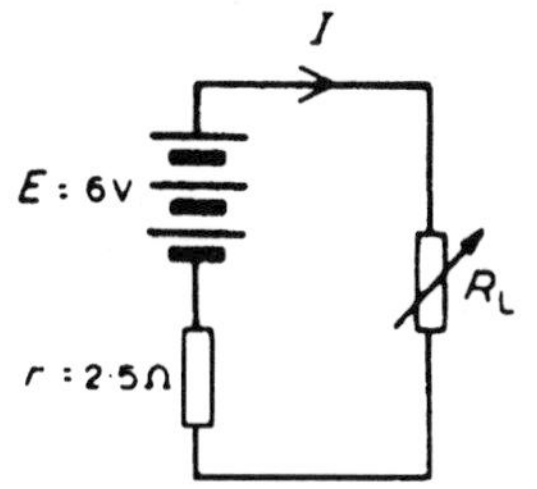

Figure 11.50

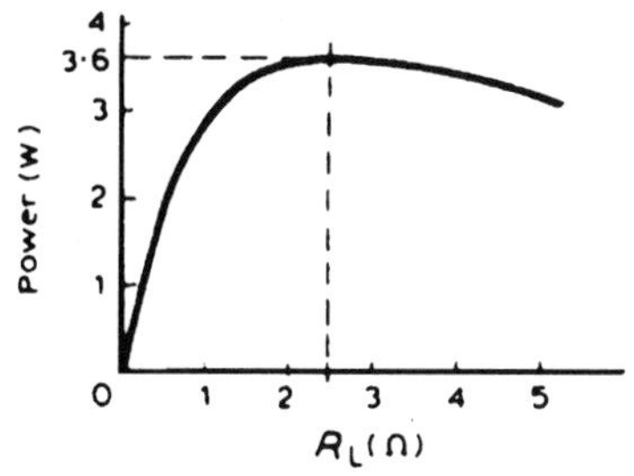

Figure 11.51

Hence, in Fig. 11.49, when $R = r$ the power transferred from the source to the load is a maximum.

> Problem 22. The circuit diagram of Fig. 11.50 shows dry cells of source e.m.f. 6 V, and internal resistance 2.5 Ω. If the load resistance R_L is varied from 0 to 5 Ω in 0.5 Ω steps, calculate the power dissipated by the load in each case. Plot a graph of R_L (horizontally) against power (vertically) and determine the maximum power dissipated.

When $R_L = 0$,

$$\text{current } I = \frac{E}{r + R_L} = \frac{6}{2.5} = 2.4 \text{ A}$$

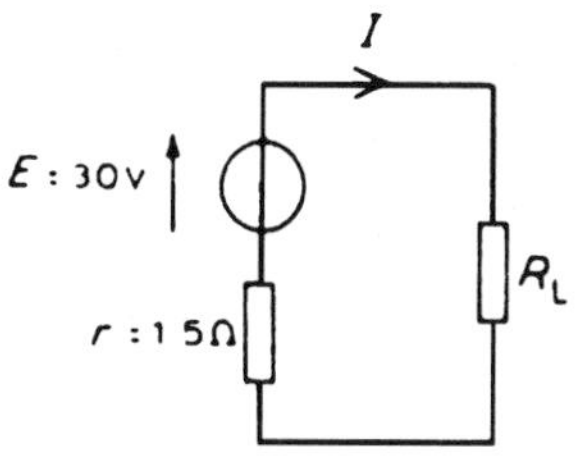

Figure 11.52

and power dissipated in R_L, $P = I^2R_L$, i.e. $P = (2.4)^2(0) = 0$ W.
When $R_L = 0.5\ \Omega$,

$$\text{current } I = \frac{E}{r + R_L} = \frac{6}{2.5 + 0.5} = 2 \text{ A}$$

and $P = I^2R_L = (2)^2(0.5) = 2$ W.
When $R_L = 1.0\ \Omega$,

$$\text{current } I = \frac{6}{2.5 + 1.0} = 1.714 \text{ A}$$

and $P = (1.714)^2(1.0) = 2.94$ W.
With similar calculations Table 11.1 is produced.

A graph of R_L against P is shown in Fig. 11.51. The **maximum value of power is 3.60 W** which occurs when R_L is 2.5 Ω, i.e. maximum power occurs when $R_L = r$, which is what the maximum power transfer theorem states.

> Problem 23. A d.c. source has an open-circuit voltage of 30 V and an internal resistance of 1.5 Ω. State the value of load resistance that gives maximum power dissipation and determine the value of this power.

The circuit diagram is shown in Fig. 11.52. From the maximum power transfer theorem: for maximum power dissipation, $\boldsymbol{R_L = r = 1.5\ \Omega}$.
From Fig. 11.52,

$$\text{current } I = \frac{E}{r + R_L} = \frac{30}{1.5 + 1.5} = 10 \text{ A}$$

Power $P = I^2R_L = (10)^2(1.5) =$ **150 W** = maximum power dissipated.

Table 11.1

R_L (Ω)	0	0.5	1.0	1.5	2.0	2.5	3.0	3.5	4.0	4.5	5.0
$I = \dfrac{E}{r + R_L}$ (A)	2.4	2.0	1.714	1.5	1.333	1.2	1.091	1.0	0.923	0.857	0.8
$P = I^2R_L$ (W)	0	2.00	2.94	3.38	3.56	3.60	3.57	3.50	3.41	3.31	3.20

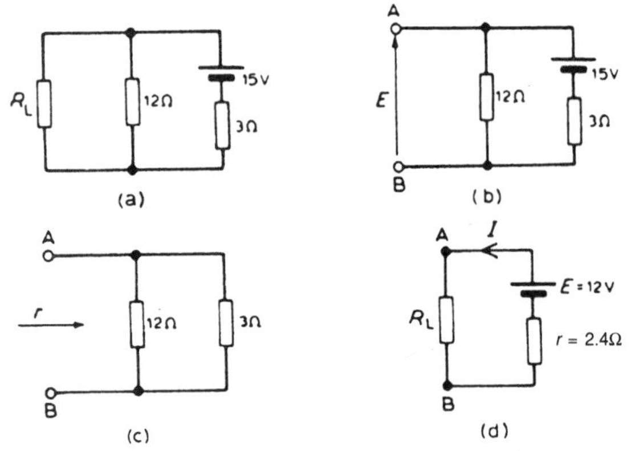

Figure 11.53

> Problem 24. Find the value of the load resistor R_L shown in Fig. 11.53(a) that gives maximum power dissipation and determine the value of this power.

Using the procedure for Thévenin's theorem:

(i) Resistance R_L is removed from the circuit as shown in Fig. 11.53(b).

(ii) The p.d. across AB is the same as the p.d. across the 12 Ω resistor. Hence

$$E = \left(\frac{12}{12+3}\right)(15) = 12 \text{ V}.$$

(iii) Removing the source of e.m.f. gives the circuit of Fig. 11.53(c), from which,

$$\text{resistance } r = \frac{12 \times 3}{12 + 3} = \frac{36}{15} = 2.4\ \Omega$$

(iv) The equivalent Thévenin's circuit supplying terminals AB is shown in Fig. 11.53(d), from which,

$$\text{current } I = \frac{E}{r + R_L}$$

For maximum power, $R_L = r =$ **2.4 Ω**. Thus current, $I = 12/(2.4 + 2.4) = 2.5$ A. Power, P, dissipated in load R_L, $P = I^2R_L = (2.5)^2(2.4)$ = **15 W**.

11.10 Multi-choice questions on d.c. circuit theorems

(Answers on page 283.)

1. Which of the following statements is true: For the junction in the network shown in Fig. 11.54:

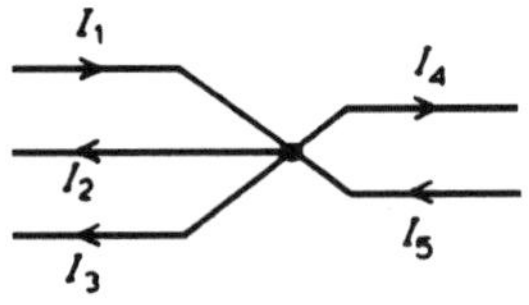

Figure 11.54

(a) $I_5 - I_4 = I_3 - I_2 + I_1$;
(b) $I_1 + I_2 + I_3 = I_4 + I_5$;
(c) $I_2 + I_3 + I_5 = I_1 + I_4$;
(d) $I_1 - I_2 - I_3 - I_4 + I_5 = 0$.

2. Which of the following statements is true? For the circuit shown in Fig. 11.55:

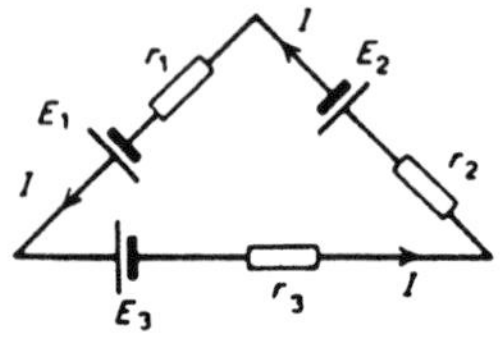

Figure 11.55

(a) $E_1 + E_2 + E_3 = Ir_1 + Ir_2 + Ir_3$;
(b) $E_2 + E_3 - E_1 - I(r_1 + r_2 + r_3) = 0$;
(c) $I(r_1 + r_2 + r_3) = E_1 - E_2 - E_3$;
(d) $E_2 + E_3 - E_1 = Ir_1 + Ir_2 + Ir_3$.

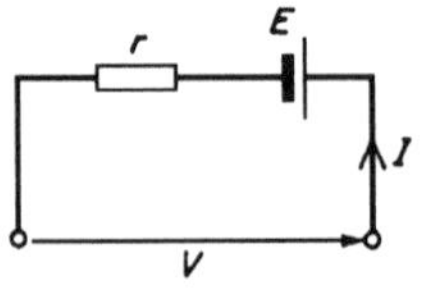

Figure 11.56

3. For the circuit shown in Fig. 11.56, the internal resistance, r, is given by:

(a) $\dfrac{I}{V - E}$; (b) $\dfrac{V - E}{I}$; (c) $\dfrac{I}{E - V}$;

(d) $\dfrac{E - V}{I}$.

4. For the circuit shown in Fig. 11.57, voltage V is:
(a) 12 V; (b) 2 V; (c) 10 V; (d) 0 V.
5. For the circuit shown in Fig. 11.57, current I_1 is:
(a) 2 A; (b) 14.4 A; (c) 0.5 A; (d) 0 A.

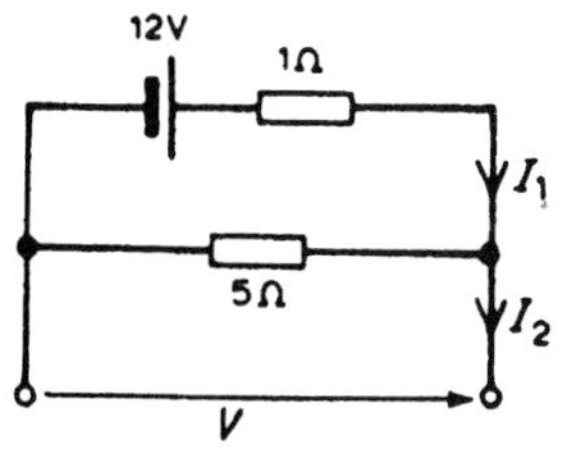

Figure 11.57

6. For the circuit shown in Fig. 11.57, current I_2 is:
(a) 2 A; (b) 14.4 A; (c) 0.5 A; (d) 0 A.
7. The equivalent resistance across terminals AB of Fig. 11.58 is:
(a) 9.31 Ω; (b) 7.24 Ω; (c) 10.0 Ω; (d) 6.75 Ω.

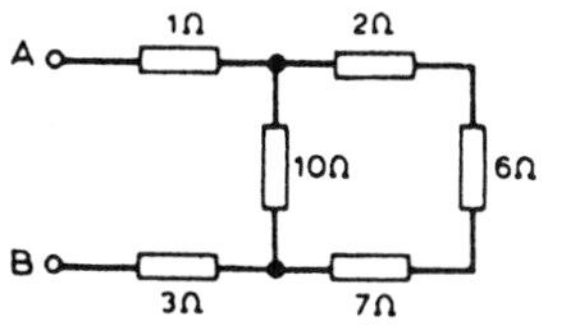

Figure 11.58

8. With reference to Fig. 11.59, which of the following statements is correct?
(a) $V_{PQ} = 2$ V;
(b) $V_{PQ} = 15$ V;
(c) When a load is connected between P and Q, current would flow from Q to P;
(d) $V_{PQ} = 20$ V.

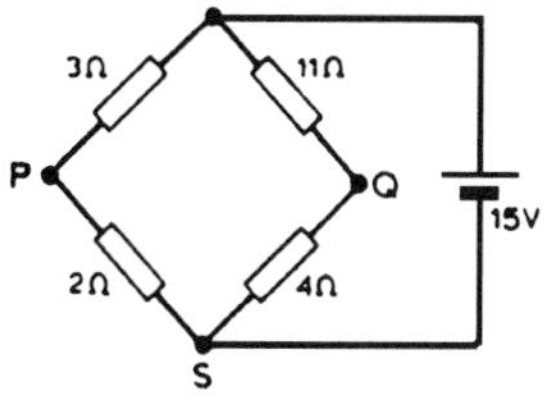

Figure 11.59

9. In Fig. 11.59, if the 15 V battery is replaced by a short-circuit, the equivalent resistance across terminal PQ is:
(a) 20 Ω; (b) 4.20 Ω; (c) 4.13 Ω; (d) 4.29 Ω.
10. For the circuit shown in Fig. 11.60, maximum power transfer from the source is required.

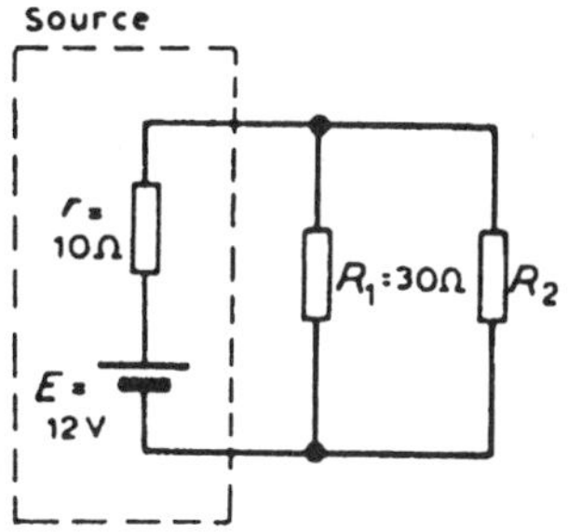

Figure 11.60

For this to be so, which of the following statements is true?
(a) $R_2 = 10$ Ω; (b) $R_2 = 30$ Ω; (c) $R_2 = 7.5$ Ω; (d) $R_2 = 15$ Ω.

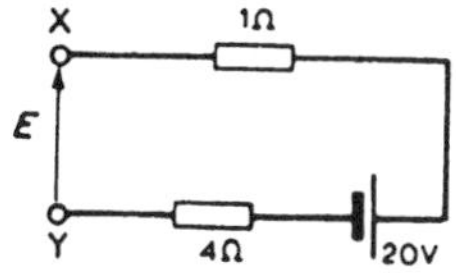

Figure 11.61

11. The open-circuit voltage E across terminals XY of Fig. 11.61 is:
(a) 0 V; (b) 20 V; (c) 4 V; (d) 16 V.

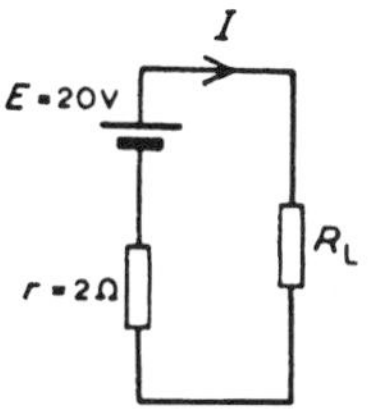

Figure 11.62

12. The maximum power transferred by the source in Fig. 11.62 is:
(a) 5 W; (b) 200 W; (c) 40 W; (d) 50 W.

11.11 Short answer questions on d.c. circuit theorems

1. Name two laws and three theorems which may be used to find currents and p.d.s in electrical circuits.

2. State Kirchhoff's current law.
3. State Kirchhoff's voltage law.
4. State, in your own words, the superposition theorem.
5. State, in your own words, Thévenin's theorem.
6. State, in your own words, Norton's theorem.
7. State the maximum power transfer theorem.

11.12 Further questions on d.c. circuit theorems

Kirchhoff's laws

1. Find currents I_3, I_4 and I_6 in Fig. 11.63.
[$I_3 = 2$ A , $I_4 = -1$ A, $I_6 = 3$ A]

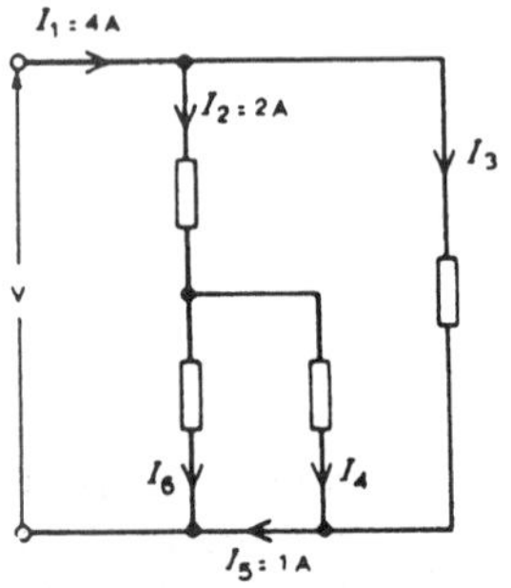

Figure 11.63

2. For the networks shown in Fig. 11.64, find the values of the currents marked.
[(a) $I_1 = 4$ A, $I_2 = -1$ A, $I_3 = 13$ A
(b) $I_1 = 40$ A, $I_2 = 60$ A, $I_3 = 120$ A
$I_4 = 100$ A, $I_5 = -80$ A]

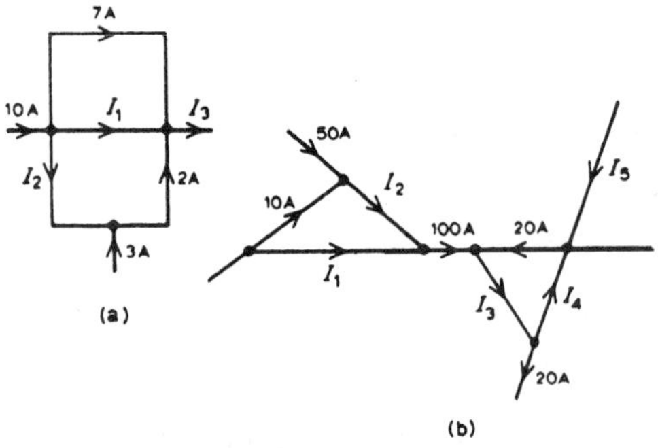

Figure 11.64

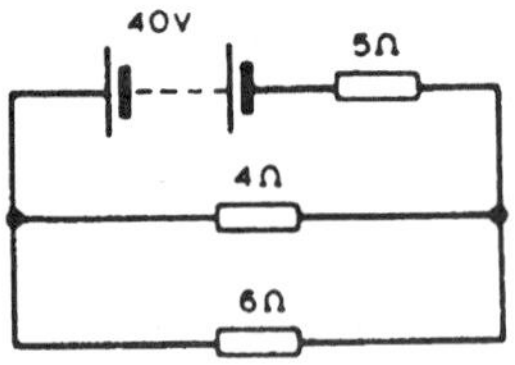

Figure 11.65

3. Use Kirchhoff's laws to find the current flowing in the 6 Ω resistor of Fig. 11.65 and the power dissipated in the 4 Ω resistor.
[2.162 A, 42.07 W]
4. Find the current flowing in the 3 Ω resistor for the network shown in Fig. 11.66(a). Find also the p.d. across the 10 Ω and 2 Ω resistors. [2.715 A, 7.410 V, 3.948 V]

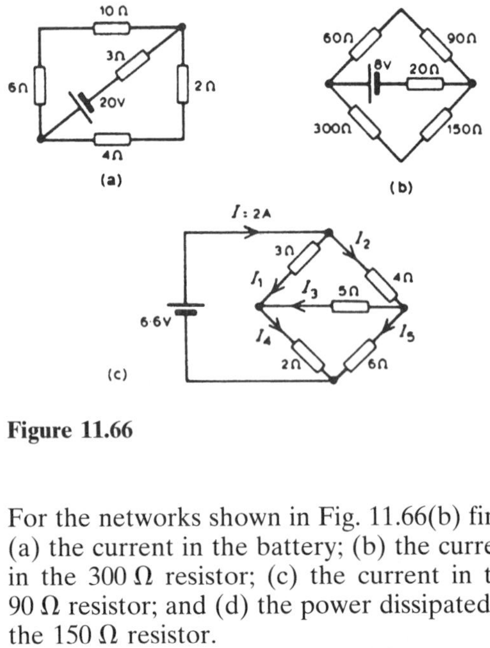

Figure 11.66

5. For the networks shown in Fig. 11.66(b) find: (a) the current in the battery; (b) the current in the 300 Ω resistor; (c) the current in the 90 Ω resistor; and (d) the power dissipated in the 150 Ω resistor.
[(a) 60.38 mA, (b) 15.10 mA, (c) 45.28 mA, (d) 34.20 mW]
6. For the bridge network shown in Fig. 11.66(c), find the currents I_1 to I_5.
[$I_1 = 1.25$ A, $I_2 = 0.75$ A, $I_3 = 0.15$ A, $I_4 = 1.40$ A, $I_5 = 0.60$ A]

Superposition theorem

7. Use the superposition theorem to find currents I_1, I_2 and I_3 of Fig. 11.67(a).
[$I_1 = 2$ A, $I_2 = 3$ A, $I_3 = 5$ A]

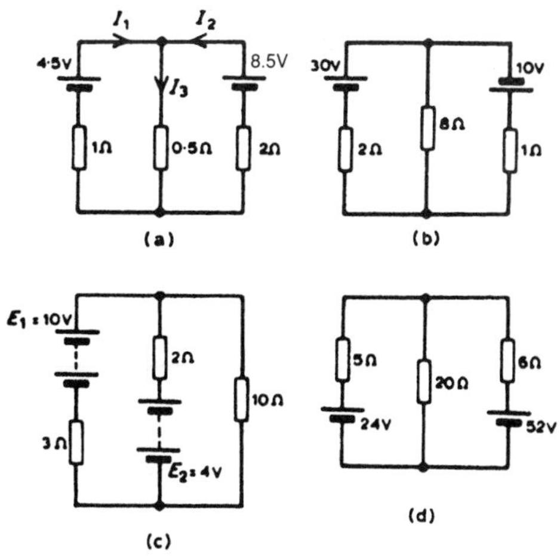

Figure 11.67

8. Use the superposition theorem to find the current in the 8 Ω resistor of Fig. 11.67(b). [0.385 A]
9. Use the superposition theorem to find the current in each branch of the network shown in Fig. 11.67(c).
[10 V battery discharges at 1.429 A
4 V battery charges at 0.857 A]
Current through 10 Ω resistor is 0.572 A
10. Use the superposition theorem to determine the current in each branch of the arrangement shown in Fig. 11.67(d).
[24 V battery charges at 1.664 A
52 V battery discharges at 3.280 A
Current in 20 Ω resistor is 1.616 A]

Thévenin's theorem

11. Use Thévenin's theorem to find the current flowing in the 14 Ω resistor of the network shown in Fig. 11.68. Find also the power dissipated in the 14 Ω resistor. [0.434 A, 2.64 W]

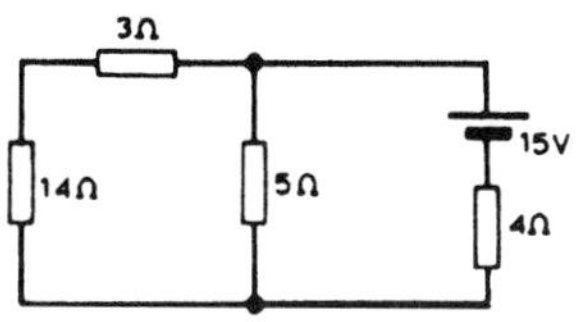

Figure 11.68

12. Use Thévenin's theorem to find the current flowing in the 6 Ω resistor shown in Fig. 11.69 and the power dissipated in the 4 Ω resistor. [2.162 A, 42.07 W]

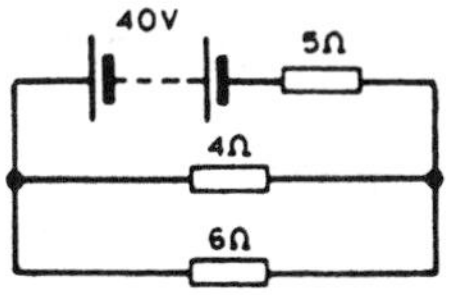

Figure 11.69

13. Repeat Problems 7–10 using Thévenin's theorem.
14. In the network shown in Fig. 11.70, the battery has negligible internal resistance. Find, using Thévenin's theorem, the current flowing in the 4 Ω resistor. [0.918 A]

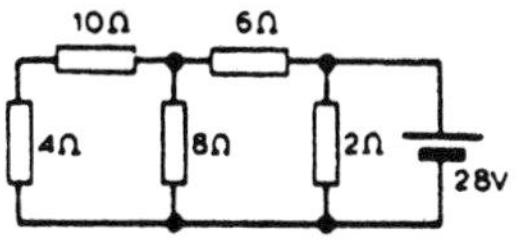

Figure 11.70

15. For the bridge network shown in Fig. 11.71, find the current in the 5 Ω resistor, and its direction, by using Thévenin's theorem. [0.153 A from B to A]

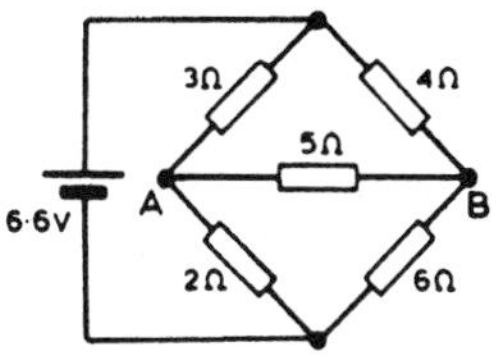

Figure 11.71

Norton's theorem

16. Repeat problems 7–12, 14 and 15 using Norton's theorem.
17. Determine the current flowing in the 6 Ω resistance of the network shown in Fig. 11.72 by using Norton's theorem. [2.5 mA]
18. Convert the circuits shown in Fig. 11.73 to Norton equivalent networks.
[(a) I_{SC} = 25 A, r = 2 Ω
(b) I_{SC} = 2 mA, r = 5 Ω]

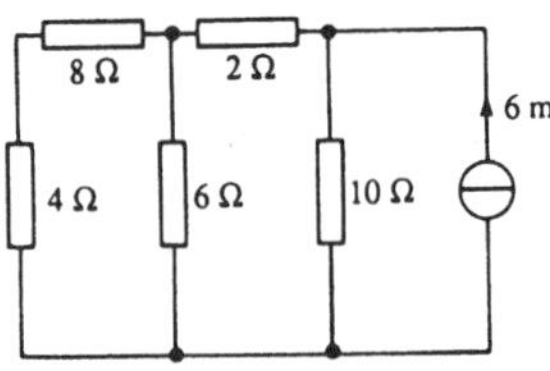

Figure 11.72

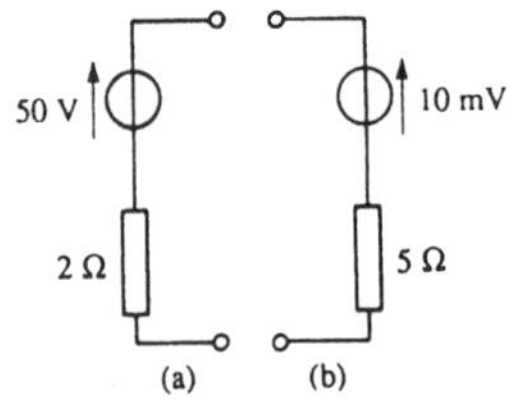

Figure 11.73

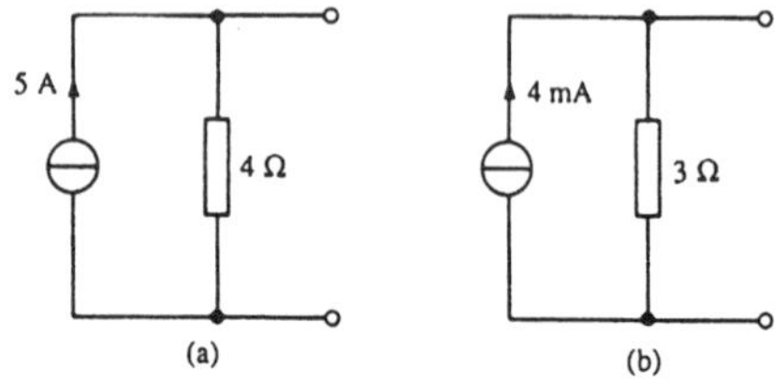

Figure 11.74

19. Convert the networks shown in Fig. 11.74 to Thévenin equivalent circuits.
[(a) $E = 20$ V, $r = 4\ \Omega$
(b) $E = 12$ mV, $r = 3\ \Omega$]

20. (a) Convert the network to the left of terminals AB in Fig. 11.75 to an equivalent Thévenin circuit by initially converting to a Norton equivalent network.
(b) Determine the current flowing in the 1.8 Ω resistance connected between A and B in Fig. 11.75.
[(a) $E = 18$ V, $r = 1.2\ \Omega$ (b) 6 A]

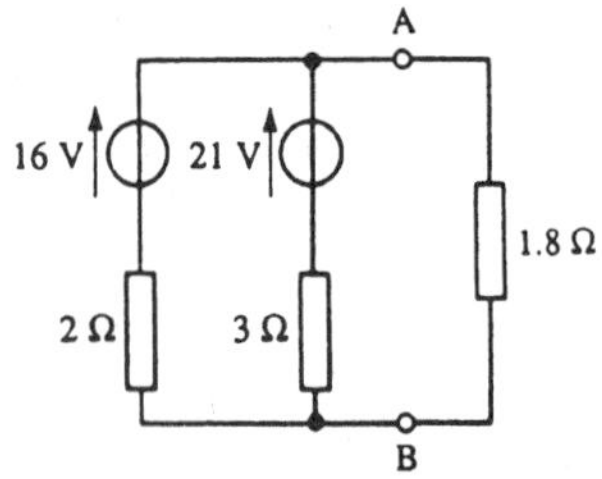

Figure 11.75

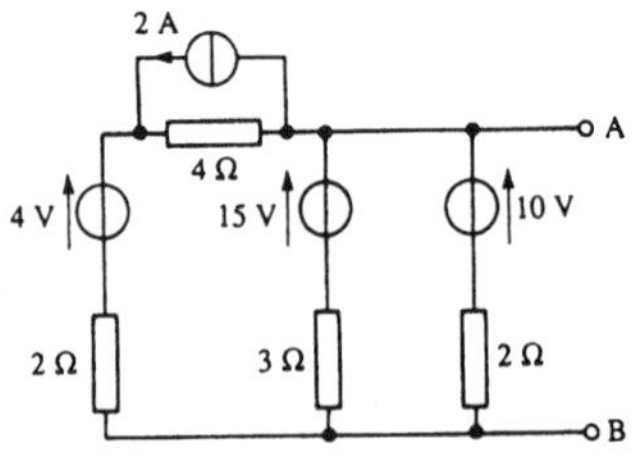

Figure 11.76

21. Determine, by successive conversions between Thévenin and Norton equivalent networks, a Thévenin equivalent circuit for terminals AB of Fig. 11.76. Hence determine the current flowing in a 6 Ω resistor connected between A and B.
[$E = 9\frac{1}{3}$ V, $r = 1\ \Omega$, $1\frac{1}{3}$ A]

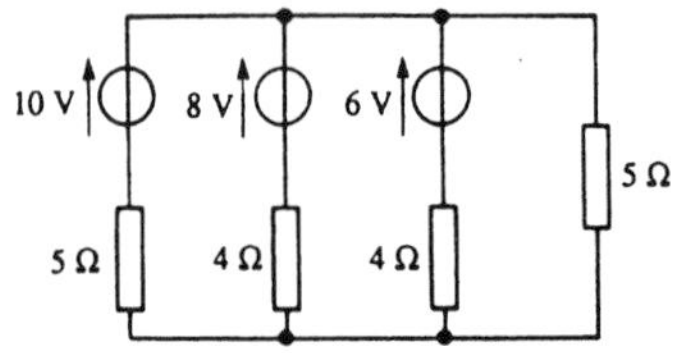

Figure 11.77

22. For the network shown in Fig. 11.77, convert each branch containing a voltage source to its Norton equivalent and hence determine the current flowing in the 5 Ω resistance.
[1.22 A]

Maximum power transfer theorem

23. A d.c. source has an open-circuit voltage of 20 V and an internal resistance of 2 Ω. Determine the value of the load resistance that gives maximum power dissipation. Find the value of this power. [2 Ω, 50 W]

24. Determine the value of the load resistance R_L shown in Fig. 11.78 that gives maximum power dissipation and find the value of the power. [$R_L = 1.6\ \Omega$, $P = 57.6$ W]

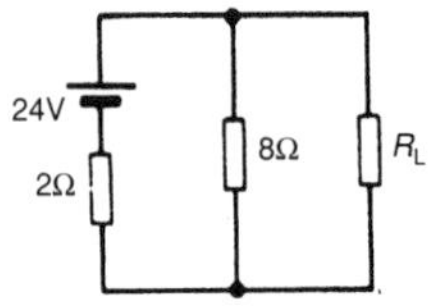

Figure 11.78

12

Alternating voltages and currents

At the end of this chapter you should be able to:

- appreciate why a.c. is used in preference to d.c.
- describe the principle of operation of an a.c. generator
- distinguish between unidirectional and alternating waveforms
- define cycle, period or periodic time T and frequency f of a waveform
- perform calculations involving $T = 1/f$
- define instantaneous, peak, mean and r.m.s. values, and form and peak factors for a sine wave
- calculate mean and r.m.s. values, form and peak factors for given waveforms
- understand and perform calculations on the general sinusoidal equation $v = V_m \sin(\omega t \pm \phi)$
- understand lagging and leading angles
- combine two sinusoidal waveforms (a) by plotting graphically, (b) by drawing phasors to scale and (c) by calculation
- perform calculations for unknown currents or voltages and power in a purely resistive a.c. circuit
- understand rectification, and describe methods of obtaining half-wave and full-wave rectification

12.1 Introduction

Electricity is produced by generators at power stations and then distributed by a vast network of transmission lines (called the National Grid system) to industry and for domestic use. It is easier and cheaper to generate alternating current (a.c.) than direct current (d.c.) and a.c. is more conveniently distributed than d.c. since its voltage can be readily altered using transformers. Whenever d.c. is needed in preference to a.c., devices called rectifiers are used for conversion (see section 12.7).

12.2 The a.c. generator

Let a single-turn coil be free to rotate at constant angular velocity ω symmetrically between the poles of a magnet system as shown in Fig. 12.1. An e.m.f. is generated in the coil (from Faraday's law) which varies in magnitude and reverses its direction at regular intervals. The reason for this is shown in Fig. 12.2. In positions (a), (e) and (i) the conductors of the loop are effectively moving along the magnetic field, no flux is cut and hence no e.m.f. is induced. In position (c) maximum flux is cut and hence maximum e.m.f. is induced. In

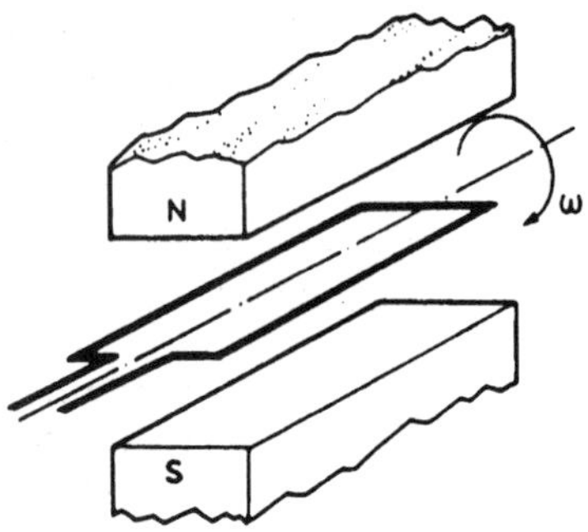

Figure 12.1

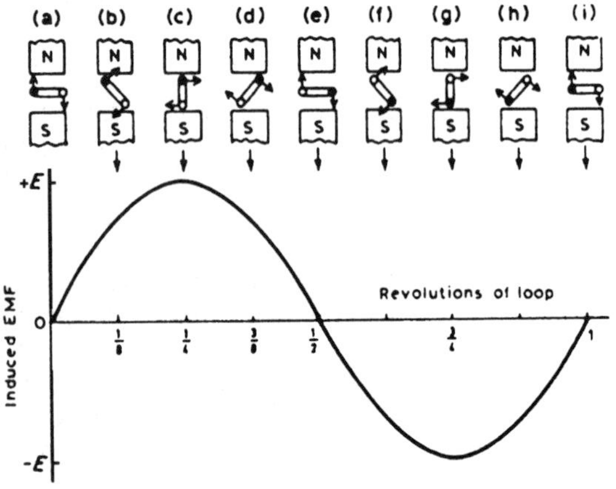

Figure 12.2

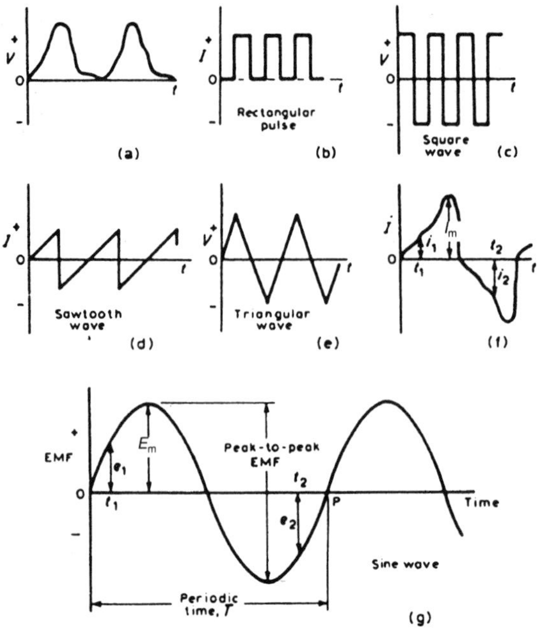

Figure 12.3

position (g), maximum flux is cut and hence maximum e.m.f. is again induced. However, using Fleming's right-hand rule, the induced e.m.f. is in the opposite direction to that in position (c) and is thus shown as $-E$. In positions (b), (d), (f) and (h) some flux is cut and hence some e.m.f. is induced. If all such positions of the coil are considered, in one revolution of the coil, one cycle of alternating e.m.f. is produced as shown. This is the principle of operation of the a.c. generator (i.e. the alternator).

12.3 Waveforms

If values of quantities which vary with time t are plotted to a base of time, the resulting graph is called a **waveform**. Some typical waveforms are shown in Fig. 12.3. Waveforms (a) and (b) are **unidirectional waveforms**, for, although they vary considerably with time, they flow in one direction only (i.e. they do not cross the time axis and become negative). Waveforms (c) to (g) are called **alternating waveforms** since their quantities are continually changing in direction (i.e. alternately positive and negative).

A waveform of the type shown in Fig. 12.3(g) is called a **sine wave**. It is the shape of the waveform of e.m.f. produced by an alternator and thus the mains electricity supply is of 'sinusoidal' form.

One complete series of values is called a **cycle** (i.e. from O to P in Fig. 12.3(g)). The time taken for an alternating quantity to complete one cycle is called the **period** or the **periodic time**, ***T***, of the waveform.

The number of cycles completed in one second is called the **frequency**, ***f***, of the supply and is measured in **hertz, Hz**. The standard frequency of the electricity supply in Great Britain is 50 Hz.

$$\mathbf{T = \frac{1}{f} \text{ or } f = \frac{1}{T}}$$

Problem 1. Determine the periodic time for frequencies of (a) 50 Hz and (b) 20 kHz.

(a) Periodic time $T = \frac{1}{f} = \frac{1}{50} =$ **0.02 s** or **20 ms**

(b) Periodic time $T = \frac{1}{f} = \frac{1}{20\,000} =$ **0.000 05 s** or **50 μs**

Problem 2. Determine the frequencies for periodic times of (a) 4 ms, (b) 4 μs.

(a) Frequency $f = \frac{1}{T} = \frac{1}{4 \times 10^{-3}}$

$= \frac{1000}{4} =$ **250 Hz**

(b) Frequency $f = \frac{1}{T} = \frac{1}{4 \times 10^{-6}}$

$= \frac{1\,000\,000}{4} =$ **250 000 Hz** or **250 kHz**

or **0.25 MHz**

Problem 3. An alternating current completes 5 cycles in 8 ms. What is its frequency?

Time for 1 cycle = (8/5) ms = 1.6 ms = periodic time T.

Frequency $f = \frac{1}{T} = \frac{1}{1.6 \times 10^{-3}}$

$= \frac{1000}{1.6} = \frac{10\,000}{16} =$ **625 Hz**

12.4 A.c. values

Instantaneous values are the values of the alternating quantities at any instant of time. They are represented by small letters, i, v, e, etc. (see Figs. 12.3(f) and (g)).

The largest value reached in a half-cycle is called the **peak value** or the **maximum value** or the **crest value** or the **amplitude** of the waveform. Such values are represented by V_m, I_m, E_m etc. (see Figs. 12.3(f) and (g)). A peak-to-peak value of e.m.f. is shown in Fig. 12.3(g) and is the difference between the maximum and minimum values in a cycle.

The **average** or **mean value** of a symmetrical alternating quantity (such as a sine wave) is the average value measured over a half-cycle (since over a complete cycle the average value is zero).

Average or mean value = $\dfrac{\textbf{area under the curve}}{\textbf{length of base}}$

The area under the curve is found by approximate methods such as the trapezoidal rule, the mid-ordinate rule or Simpson's rule. Average values are represented by V_{AV}, I_{AV}, etc.

For a sine wave,
average value = 0.637 × maximum value
(i.e. 2/π × maximum value)

The **effective value** of an alternating current is that current which will produce the same heating effect as an equivalent direct current. The effective value is called the **root mean square (r.m.s.) value** and whenever an alternating quantity is given, it is assumed to be the r.m.s. value. For example, the domestic mains supply in Great Britain is 240 V and is assumed to mean 240 V r.m.s. The symbols used for r.m.s. values are I, V, E, etc. For a non-sinusoidal waveform as shown in Fig. 12.4 the r.m.s. value is given by:

$$I = \sqrt{\left(\frac{i_1^2 + i_2^2 + \ldots + i_n^2}{n}\right)}$$

where n is the number of intervals used.

For a sine wave,
r.m.s. value = 0.707 × maximum value
(i.e. 1/√2 × maximum value)

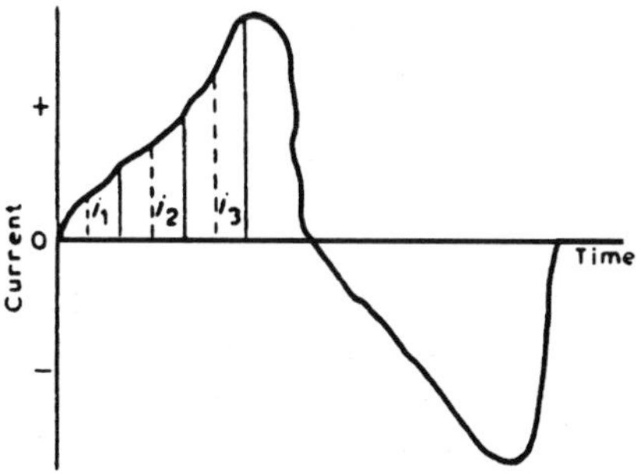

Figure 12.4

$$\text{Form factor} = \frac{\text{r.m.s. value}}{\text{average value}}$$

For a sine wave, form factor = 1.11.

$$\text{Peak factor} = \frac{\text{maximum value}}{\text{r.m.s. value}}$$

For a sine wave, peak factor = 1.41.

The values of form and peak factors give an indication of the shape of waveforms.

Problem 4. For the periodic waveforms shown in Fig. 12.5 determine for each: (i) frequency, (ii) average value over half a cycle, (iii) r.m.s. value, (iv) form factor and (v) peak factor.

(a) **Triangular waveform (Fig. 12.5(a))**

(i) Time for 1 complete cycle = 20 ms = periodic time, T. Hence

$$\text{frequency } f = \frac{1}{T} = \frac{1}{20 \times 10^{-3}} = \frac{1000}{20} = \mathbf{50\ Hz}$$

(ii) Area under the triangular waveform for a half-cycle = $\frac{1}{2}$ × base × height = $\frac{1}{2}$ × $(10 \times 10^{-3}) \times 200$ = 1 volt second.

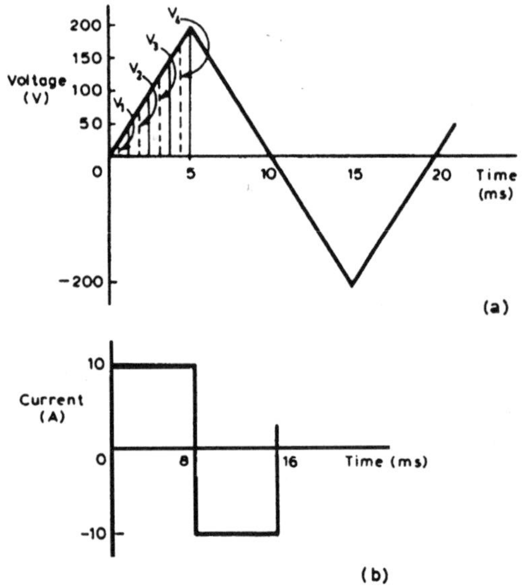

Figure 12.5

Average value of waveform

$$= \frac{\text{area under curve}}{\text{length of base}}$$

$$= \frac{1 \text{ volt second}}{10 \times 10^{-3} \text{ second}} = \frac{1000}{10} = 100 \text{ V}$$

(iii) In Fig. 12.5(a), the first 1/4 cycle is divided into four intervals. Thus

$$\text{r.m.s. value} = \sqrt{\left(\frac{i_1^2 + i_2^2 + i_3^2 + i_4^2}{4}\right)}$$

$$= \sqrt{\left(\frac{25^2 + 75^2 + 125^2 + 175^2}{4}\right)}$$

$$= \mathbf{114.6\ V}$$

(Note that the greater the number of intervals chosen, the greater the accuracy of the result. For example, if twice the number of ordinates as that chosen above are used, the r.m.s. value is found to be 115.6 V.)

(iv) Form factor $= \frac{\text{r.m.s. value}}{\text{average value}}$

$$= \frac{114.6}{100} = \mathbf{1.15}$$

(v) Peak factor $= \frac{\text{maximum value}}{\text{r.m.s. value}}$

$$= \frac{200}{114.6} = \mathbf{1.75}$$

(b) **Rectangular waveform (Fig. 12.5(b))**

(i) Time for 1 complete cycle = 16 ms = periodic time, T. Hence

$$\text{frequency } f = \frac{1}{T} = \frac{1}{16 \times 10^{-3}} = \frac{1000}{16}$$

$$= \mathbf{62.5\ Hz}$$

(ii) Average value over half a cycle

$$= \frac{\text{area under curve}}{\text{length of base}}$$

$$= \frac{10 \times (8 \times 10^{-3})}{8 \times 10^{-3}} = \mathbf{10\ A}$$

(iii) The r.m.s. value = $\sqrt{((i_1^2 + i_2^2 + \ldots + i_n^2)/n)}$ = **10 A** however many intervals are chosen, since the waveform is rectangular.

(iv) Form factor $= \frac{\text{r.m.s. value}}{\text{average value}} = \frac{10}{10} = \mathbf{1}$

(v) Peak factor $= \frac{\text{maximum value}}{\text{r.m.s. value}} = \frac{10}{10} = \mathbf{1}$

Problem 5. The following table gives the corresponding values of current and time for a half-cycle of alternating current.

time, t (ms)	0	0.5	1.0	1.5	2.0	
current, i (A)	0	7	14	23	40	
time, t (ms)	2.5	3.0	3.5	4.0	4.5	5.0
current, i (A)	56	68	76	60	5	0

Assuming the negative half-cycle is identical in shape to the positive half-cycle, plot the waveform and find (a) the frequency of the supply, (b) the instantaneous values of current after 1.25 ms and 3.8 ms, (c) the peak or maximum value, (d) the mean or average value, and (e) the r.m.s. value of the waveform.

The half-cycle of alternating current is shown plotted in Fig. 12.6.

(a) Time for a half-cycle = 5 ms. Hence the time for 1 cycle, i.e. the periodic time, T = 10 ms or 0.01 s.

$$\text{Frequency } f = \frac{1}{T} = \frac{1}{0.01} = \mathbf{100\ Hz}$$

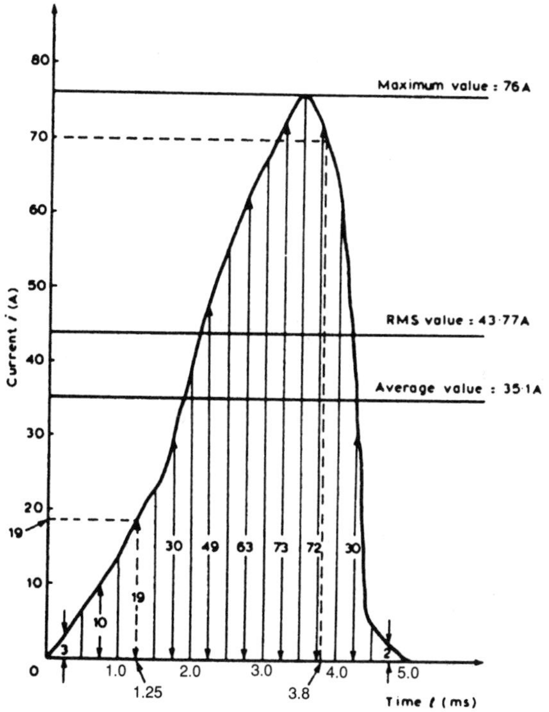

Figure 12.6

(b) Instantaneous value of current after 1.25 ms is **19 A**, from Fig. 12.6. Instantaneous value of current after 3.8 ms is **70 A**, from Fig. 12.6.

(c) Peak or maximum value = **76 A**.

(d) Mean or average value = (area under curve)/(length of base). Using the mid-ordinate rule with 10 intervals, each of width 0.5 ms, gives:

Area under curve

$= (0.5 \times 10^{-3})[3 + 10 + 19 + 30 + 49 + 63 + 73 + 72 + 30 + 2]$ (see Fig. 12.6)

$= (0.5 \times 10^{-3})(351)$

Hence

mean or average value

$$= \frac{(0.5 \times 10^{-3})(351)}{5 \times 10^{-3}} = \mathbf{35.1\ A}$$

(e) r.m.s. value

$$= \sqrt{\left(\frac{3^2+10^2+19^2+30^2+49^2+63^2+73^2+72^2+30^2+2^2}{10}\right)}$$

$$= \sqrt{\left(\frac{19\,157}{10}\right)} = \mathbf{43.8\ A}$$

Problem 6. Calculate the r.m.s. value of a sinusoidal current of maximum value 20 A.

For a sine wave,

r.m.s. value = 0.707 × maximum value

= 0.707 × 20 = **14.14 A**

Problem 7. Determine the peak and mean values for a 240 V mains supply.

For a sine wave, r.m.s. value of voltage $V = 0.707 \times V_m$. A 240 V mains supply means that 240 V is the r.m.s. value, hence

$$V_m = \frac{V}{0.707} = \frac{240}{0.707} = \mathbf{339.5\ V = peak\ value}$$

Mean value $V_{AV} = 0.637\ V_m = 0.637 \times 339.5$
$= \mathbf{216.3\ V}$

Problem 8. A supply voltage has a mean value of 150 V. Determine its maximum value and its r.m.s. value.

For a sine wave, mean value = 0.637 × maximum value. Hence

$$\text{maximum value} = \frac{\text{mean value}}{0.637} = \frac{150}{0.637}$$
$$= \mathbf{235.5\ V}$$

r.m.s. value = 0.707 × maximum value = 0.707 × 235.5 = **166.5 V**.

12.5 The equation of a sinusoidal waveform

In Fig. 12.7, OA represents a vector that is free to rotate anticlockwise about 0 at an angular velocity of ω rad/s. A rotating vector is known as a **phasor**. After time t seconds the vector OA has turned through an angle ωt. If the line BC is constructed perpendicular to OA as shown, then $\sin \omega t = BC/OB$, i.e. $BC = OB \sin \omega t$.

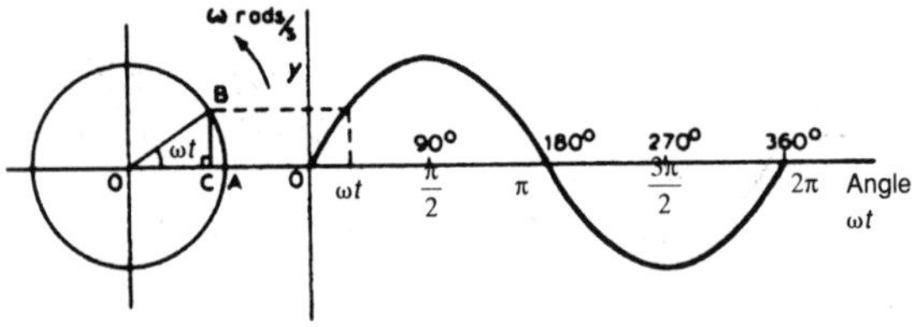

Figure 12.7

If all such vertical components are projected on to a graph of y against angle ωt (in radians), a sine curve results, of maximum value OA. Any quantity which varies sinusoidally can thus be represented as a phasor.

A sine curve may not always start at 0°. To show this a periodic function is represented by $y = \sin(\omega t \pm \phi)$, where ϕ is the phase (or angle) difference compared with $y = \sin \omega t$. In Fig. 12.8(a), $y_2 = \sin(\omega t + \phi)$ starts ϕ radians earlier than $y_1 = \sin \omega t$ and is thus said to **lead** y_1 by ϕ radians. Phasors y_1 and y_2 are shown in Fig. 12.8(b) at the time when $t = 0$. In Fig. 12.8(c), $y_4 = \sin(\omega t - \phi)$ starts ϕ radians later than $y_3 = \sin \omega t$ and is thus said to **lag** y_3 by ϕ radians. Phasors y_3 and y_4 are shown in Fig. 12.8(d) at the time when $t = 0$.

Given the general sinusoidal voltage, $\boldsymbol{v = V_m \sin(\omega t + \phi)}$, then

(i) Amplitude or maximum value = V_m.
(ii) Peak-to-peak value = $2\,V_m$.
(iii) Angular velocity = ω rad/s.

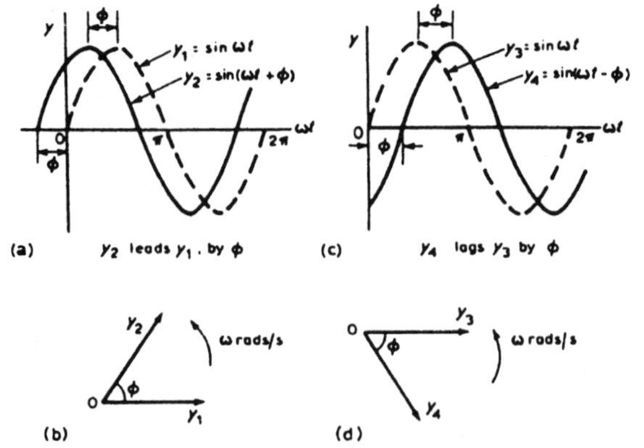

Figure 12.8

(iv) Periodic time, $T = 2\pi/\omega$ seconds.
(v) Frequency, $f = \omega/2\pi$ Hz (since $\omega = 2\pi f$).
(vi) ϕ = angle of lag or lead (compared with $v = V_m \sin \omega t$).

> Problem 9. An alternating voltage is given by $v = 282.8 \sin 314t$ volts. Find (a) the r.m.s. voltage, (b) the frequency and (c) the instantaneous value of voltage when $t = 4$ ms.

(a) The general expression for an alternating voltage is $v = V_m \sin(\omega t \pm \phi)$. Comparing $v = 282.8 \sin 314t$ with this general expression gives the peak voltage as 282.8 V. Hence the r.m.s. voltage = 0.707 × maximum value = 0.707 × 282.8 = **200 V**.

(b) Angular velocity, $\omega = 314$ rad/s, i.e. $2\pi f = 314$. Hence

$$\text{frequency } f = \frac{314}{2\pi} = \mathbf{50\ Hz}$$

(c) When $t = 4$ ms,

$$v = 282.8 \sin(314 \times 4 \times 10^{-3})$$
$$= 282.8 \sin(1.256)$$

(Note that 1.256 radians $= \left(1.256 \times \frac{180}{\pi}\right)^\circ$

$= 71.96° = 71°\,58'$

Hence

$$v = 282.8 \sin 71°58'$$
$$= \mathbf{268.9\ V})$$

Problem 10. An alternating voltage is given by $v = 75 \sin(200\pi t - 0.25)$ volts. Find (a) the amplitude, (b) the peak-to-peak value, (c) the r.m.s. value, (d) the periodic time, (e) the frequency, and (f) the phase angle (in degrees and minutes) relative to $75 \sin 200\pi t$.

Comparing $v = 75 \sin(200\pi t - 0.25)$ with the general expression $v = V_m \sin(\omega t \pm \phi)$ gives:

(a) Amplitude, or peak value = 75 V.
(b) Peak-to-peak value = 2 × 75 = 150 V.
(c) The r.m.s. value = 0.707 × maximum value = 0.707 × 75 = 53 V.
(d) Angular velocity, $\omega = 200\pi$ rad/s, hence

$$\text{periodic time } T = \frac{2\pi}{\omega} = \frac{2\pi}{200\pi} = \frac{1}{100}$$

$$= \mathbf{0.01\ s} \text{ or } \mathbf{10\ ms}$$

(e) Frequency $f = 1/T = 1/0.01 =$ **100 Hz.**
(f) Phase angle, ϕ = 0.25 radians lagging $75 \sin 200\pi t$

$$0.25 \text{ rads} = \left(0.25 \times \frac{180}{\pi}\right)^\circ = 14.32^\circ = 14^\circ 19'$$

Hence phase angle = **14°19′ lagging**.

Problem 11. An alternating voltage, v, has a periodic time of 0.01 s and a peak value of 40 V. When time t is zero, v = –20 V. Express the instantaneous voltage in the form $v = V_m \sin(\omega t \pm \phi)$.

Amplitude, V_m = 40 V. Periodic time $T = 2\pi/\omega$. Hence angular velocity, $\omega = (2\pi/T) = (2\pi/0.01) = 200\pi$ rad/s.
$v = V_m \sin(\omega t + \phi)$ thus becomes $v = 40 \sin(200\pi t + \phi)$ V. When time $t = 0$, $v = -20$ V, i.e. $-20 = 40 \sin\phi$ so that $\sin\phi = -20/40 = -0.5$. Hence

$$\phi = \arcsin(-0.5) = -30^\circ = \left(-30 \times \frac{\pi}{180}\right) \text{ rads}$$

$$= -\frac{\pi}{6} \text{ rads}$$

Thus

$$\boldsymbol{v = 40 \sin(200\pi t - \pi/6)\ \text{V}}$$

Problem 12. The current in an a.c. circuit at any time t seconds is given by: $i = 120 \sin(100\pi t + 0.36)$ amperes. Find:

(a) the peak value, the periodic time, the frequency and phase angle relative to $120 \sin 100\pi t$;
(b) the value of the current when $t = 0$;
(c) the value of the current when $t = 8$ ms;
(d) the time when the current first reaches 60 A; and
(e) the time when the current is first a maximum.

(a) Peak value = **120 A**.

$$\text{Periodic time } T = \frac{2\pi}{\omega} = \frac{2\pi}{100\pi}$$

(since $\omega = 100\pi$)

$$= \frac{1}{50} = \mathbf{0.02\ s} \text{ or } \mathbf{20\ ms}$$

Frequency, $f = (1/T) = (1/0.02) =$ **50 Hz.**
Phase angle = 0.36 rads = $(0.36 \times 180/\pi)^\circ$ = **20°38′ leading**.

(b) When $t = 0$, $i = 120 \sin(0 + 0.36) = 120 \sin 20^\circ 38' =$ **49.29 A**.

(c) When $t = 8$ ms,

$$i = 120 \sin\left(100\pi\left(\frac{8}{10^3}\right) + 0.36\right)$$

$$= 120 \sin 2.8733 \ (= 120 \sin 164^\circ 38')$$

$$= \mathbf{31.80\ A}$$

(d) When $i = 60$ A, $60 = 120 \sin(100\pi t + 0.36)$, thus $(60/120) = \sin(100\pi t + 0.36)$, so that $(100\pi t + 0.36) = \arcsin 0.5 = 30^\circ = (\pi/6)$ rads = 0.5236 rads. Hence

$$\text{time } t = \frac{0.5236 - 0.36}{100\pi} = \mathbf{0.5208\ ms}$$

(e) When the current is a maximum, $i = 120$ A. Thus

$$120 = 120 \sin(100\pi t + 0.36)$$

$$1 = \sin(100\pi t + 0.36)$$

$(100\pi t + 0.36) = \arcsin 1 = 90^\circ = (\pi/2)$ rads = 1.5708 rads. Hence

$$\text{time } t = \frac{1.5708 - 0.36}{100\pi} = \mathbf{3.854\ ms}$$

12.6 Combination of waveforms

The resultant of the addition (or subtraction) of two sinusoidal quantities may be determined either:

(a) by plotting the periodic functions graphically (see worked Problems 13 and 16); or
(b) by resolution of phasors by drawing or calculation (see worked Problems 14 and 15).

> Problem 13. The instantaneous values of two alternating currents are given by $i_1 = 20 \sin \omega t$ amperes and $i_2 = 10 \sin (\omega t + \pi/3)$ amperes. By plotting i_1 and i_2 on the same axes, using the same scale, over one cycle, and adding ordinates at intervals, obtain a sinusoidal expression for $i_1 + i_2$.

$i_1 = 20 \sin \omega t$ and $i_2 = 10 \sin (\omega t + \pi/3)$ are shown plotted in Fig. 12.9. Ordinates of i_1 and i_2 are added at, say, 15° intervals (a pair of dividers are useful for this). For example,

at 30°, $i_1 + i_2 = 10 + 10 = 20$ A
at 60°, $i_1 + i_2 = 8.7 + 17.3 = 26$ A
at 150°, $i_1 + i_2 = 10 + (-5) = 5$ A, and so on.

The resultant waveform for $i_1 + i_2$ is shown by the broken line in Fig. 12.9. It has the same period, and hence frequency, as i_1 and i_2. The amplitude or peak value is 26.5 A.

ωt (degrees)	0	30	60	90
$\sin\omega t$	0	0.866	1	
$i_1 = 20 \sin \omega t$	0	10	17.32	20
$(\omega t + 60)$	60	90	120	150
$\sin(\omega t + \frac{\pi}{2})$	0.866	1	0.866	0.5
$i_2 = 10 \sin(\omega t + \frac{\pi}{2})$	8.66	10	8.66	5

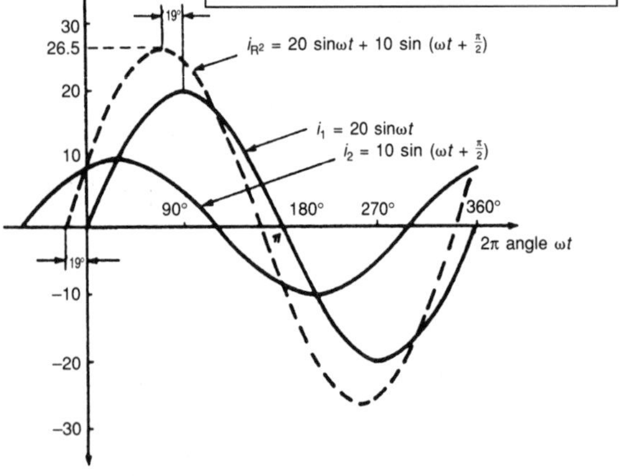

Figure 12.9

The resultant waveform leads the curve $i_1 = 20 \sin \omega t$ by 19°, i.e. $(19 \times \pi/180)$ rads $= 0.3323$ rads. Hence the sinusoidal expression for the resultant $i_1 + i_2$ is given by

$$\mathbf{i_R = i_1 + i_2 = 26.5 \sin (\omega t + 0.332)\ A}$$

> Problem 14. Two alternating voltages are represented by $v_1 = 50 \sin \omega t$ volts and $v_2 = 100 \sin (\omega t - \pi/6)$ V. Draw the phasor diagram and find, by calculation, a sinusoidal expression to represent $v_1 + v_2$.

Phasors are usually drawn at the instant when time $t = 0$. Thus v_1 is drawn horizontally 50 units long and v_2 is drawn 100 units long lagging v_1 by $\pi/6$ rads, i.e. 30°. This is shown in Fig. 12.10(a) where 0 is the point of rotation of the phasors.

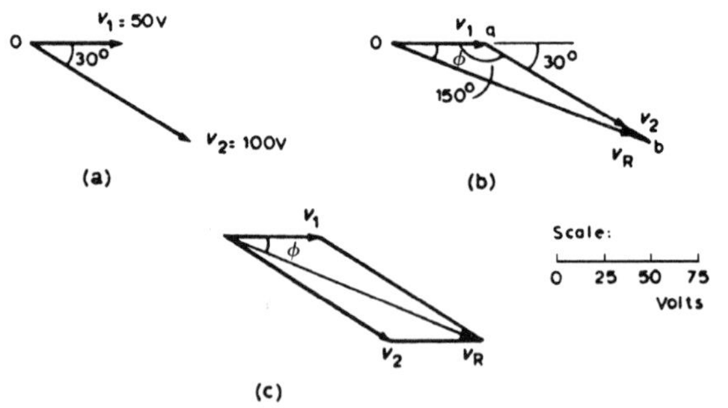

Figure 12.10

Procedure to draw phasor diagram to represent $v_1 + v_2$:

(i) Draw v_1 horizontal 50 units long, i.e. oa of Fig. 12.10(b).
(ii) Join v_2 to the end of v_1 at the appropriate angle, i.e. ab of Fig. 12.10(b).
(iii) The resultant $v_R = v_1 + v_2$ is given by the length ob and its phase angle may be measured with respect to v_1.

Alternatively, when two phasors are being added the resultant is always the diagonal of the parallelogram, as shown in Fig. 12.10(c).

From the drawing, by measurement, $v_R = 145$ V and angle $\phi = 20°$ lagging v_1. A more accurate solution is obtained by calculation, using the cosine and sine rules. Using the cosine rule on triangle oab of Fig. 12.10(b) gives

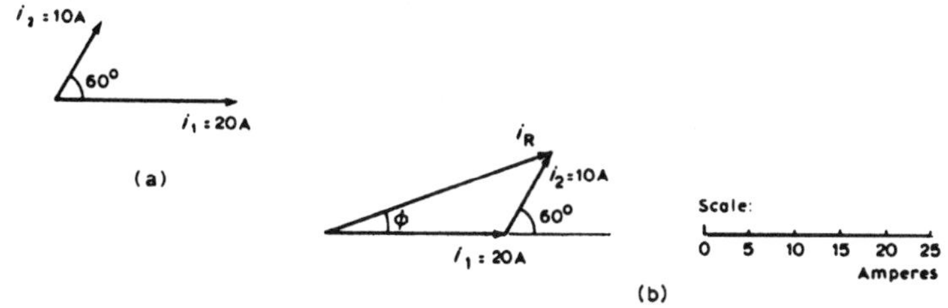

Figure 12.11

$$v_R{}^2 = v_1{}^2 + v_2{}^2 - 2v_1v_2 \cos 150°$$
$$= 50^2 + 100^2 - 2(50)(100) \cos 150°$$
$$= 2500 + 10\,000 - (-8660)$$
$$v_R = \sqrt{(21\,160)} = 145.5 \text{ V}$$

Using the sine rule,

$$\frac{100}{\sin \phi} = \frac{145.5}{\sin 150°}$$

$$\sin \phi = \frac{100 \sin 150°}{145.5} \; 0.3436$$

$\phi = \arcsin 0.3436 = 20°6' = 0.35$ radians, and lags v_1. Hence

$$v_R = v_1 + v_2 = \mathbf{145.5 \sin (\omega t - 0.35) \text{ V}}$$

> Problem 15. Find a sinusoidal expression for $(i_1 + i_2)$ of Problem 13, (a) by drawing phasors, (b) by calculation.

(a) The relative positions of i_1 and i_2 at time $t = 0$ are shown as phasors in Fig. 12.11(a). The phasor diagram in Fig. 12.11(b) shows the resultant i_R, and i_R is measured as 26 A and angle ϕ as 19° (i.e. 0.33 rads) leading i_1.

Hence, by drawing, $i_R = 26 \sin (\omega t + 0.33)$ A.

(b) From Fig. 12.11(b), by the cosine rule:

$$i_R{}^2 = 20^2 + 10^2 - 2(20)(10)(\cos 120°)$$

from which $i_R =$ **26.46 A.**
By the sine rule:

$$\frac{10}{\sin \phi} = \frac{26.46}{\sin 120°}$$

from which

$$\phi = 19°10' \text{ (i.e. 0.333 rads)}$$

Hence, by calculation
$i_R = 26.46 \sin (\omega t + 0.333)$ A.

> Problem 16. Two alternating voltages are given by $v_1 = 120 \sin \omega t$ volts and $v_2 = 200 \sin (\omega t - \pi/4)$ volts. Obtain sinusoidal expressions for $v_1 - v_2$ (a) by plotting waveforms, and (b) by resolution of phasors.

(a) $v_1 = 120 \sin \omega t$ and $v_2 = 200 \sin (\omega t - \pi/4)$ are shown plotted in Fig. 12.12. Care must be taken when subtracting values of ordinates especially when at least one of the ordinates is negative. For example

at 30°, $v_1 - v_2 = 60 - (-52) = 112$ V
at 60°, $v_1 - v_2 = 104 - 52 = 52$ V
at 150°, $v_1 - v_2 = 60 - 193 = -133$ V
and so on.

The resultant waveform, $v_R = v_1 - v_2$, is shown by the broken line in Fig. 12.12. The maximum value of v_R is 143 V and the

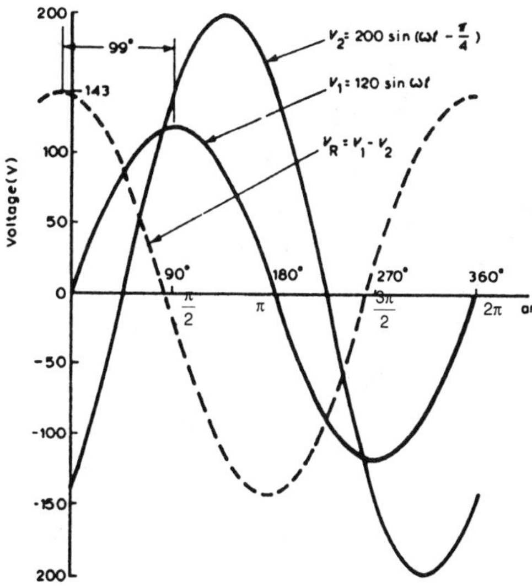

Figure 12.12

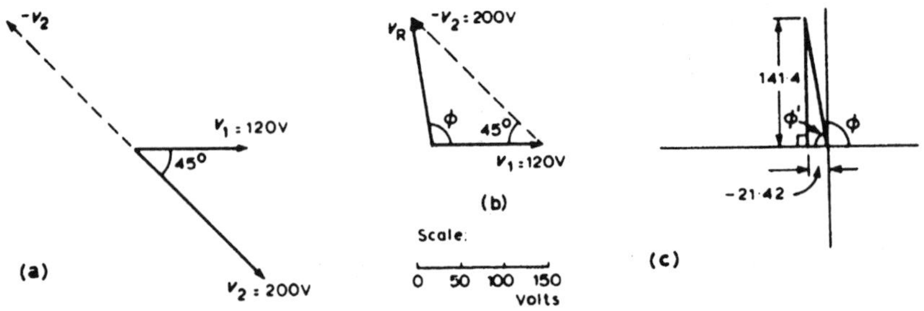

Figure 12.13

waveform is seen to lead v_1 by 99° (i.e. 1.73 radians).

Hence, by drawing,

$\boldsymbol{v_R = v_1 - v_2 = 143 \sin(\omega t + 1.73)}$ **volts.**

(b) The relative positions of v_1 and v_2 are shown at time $t = 0$ as phasors in Fig. 12.13(a). Since the resultant of $v_1 - v_2$ is required, $-v_2$ is drawn in the opposite direction to $+v_2$ (shown by the broken line in Fig. 12.13(a)). The phasor diagram with the resultant is shown in Fig. 12.13(b) where $-v_2$ is added phasorially to v_1.

By resolution:

Sum of horizontal components of v_1 and v_2 $= 120 \cos 0° - 200 \cos 45° = -21.42$.

Sum of vertical components of v_1 and v_2 $= 120 \sin 0° + 200 \sin 45° = 141.4$.

From Fig. 12.13(c), resultant $v_R = \sqrt{[(-21.42)^2 + (141.4)^2]} = 143.0$, and

$$\tan \phi' = \frac{141.4}{21.42} = \tan 6.6013$$

from which

$$\phi' = \arctan 6.6013 = 81°23'$$

and

$$\phi = 98°37' \text{ or } 1.721 \text{ radians}$$

Hence, by resolution of phasors,

$\boldsymbol{v_R = v_1 - v_2 = 143.0 \sin(\omega t + 1.721)}$ **volts**

12.7 Rectification

The process of obtaining unidirectional currents and voltages from alternating currents and voltages is called **rectification**. Automatic switching in circuits is carried out by devices called diodes.

Using a single diode, as shown in Fig. 12.14, half-wave rectification is obtained. When P is sufficiently positive with respect to Q, diode D is switched on and current i flows. When P is negative with respect to Q, diode D is switched off. Transformer T isolates the equipment from direct connection with the mains supply and enables the mains voltage to be changed.

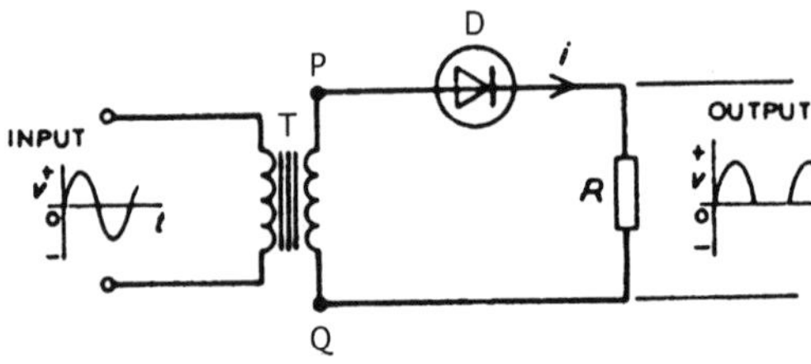

Figure 12.14

Two diodes may be used as shown in Fig. 12.15 to obtain **full-wave rectification**. A centre-tapped transformer T is used. When P is sufficiently positive with respect to Q, diode D_1 conducts and current flows (shown by the broken line in Fig. 12.15). When S is positive with respect to Q, diode D_2 conducts and current flows (shown by the continuous line in Fig. 12.15). The current flowing in R is in the same direction for both half-cycles of the input. The output waveform is thus as shown in Fig. 12.15.

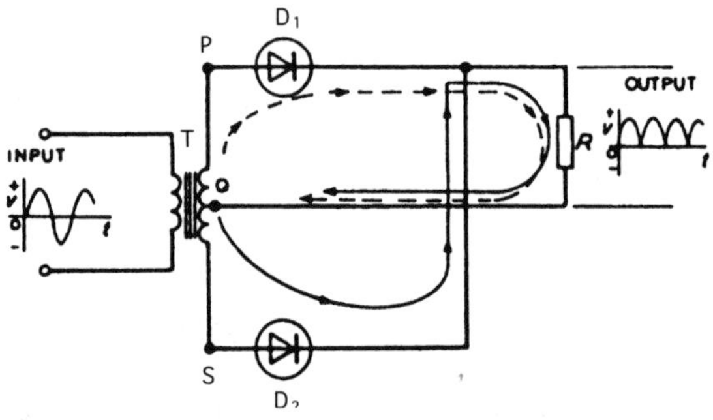

Figure 12.15

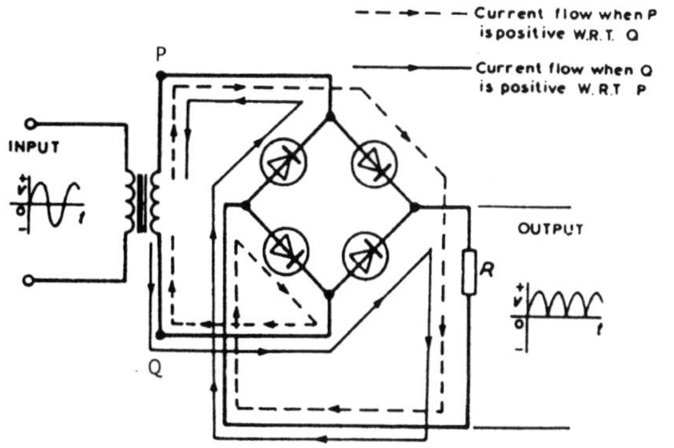

Figure 12.16

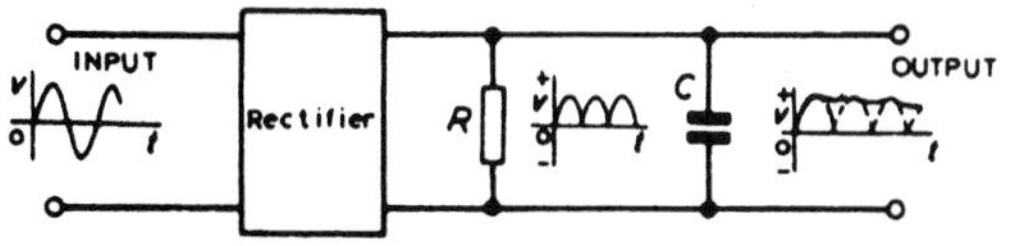

Figure 12.17

Four diodes may be used in a **bridge rectifier** circuit, as shown in Fig. 12.16, to obtain **full-wave rectification**. As for the rectifier shown in Fig. 12.15, the current flowing in R is in the same direction for both half-cycles of the input giving the output waveform shown.

To smooth the output of the rectifiers described above, capacitors having a large capacitance may be connected across the load resistors R. The effect of this is shown on the output in Fig. 12.17.

12.8 Multi-choice questions on alternating voltages and currents

(Answers on page 284.)

1. The value of an alternating current at any given instant is (a) a maximum value; (b) a peak value; (c) an instantaneous value; (d) an r.m.s. value.
2. An alternating current completes 100 cycles in 0.1 s. Its frequency is
   (a) 20 Hz; (b) 100 Hz; (c) 0.002 Hz;
   (d) 1 kHz.
3. In Fig. 12.18, at the instant shown, the generated e.m.f. will be

Figure 12.18

   (a) zero; (b) an r.m.s. value; (c) an average value; (d) a maximum value.
4. The supply of electrical energy for a consumer is usually by a.c. because
   (a) transmission and distribution are more easily effected;
   (b) it is most suitable for variable speed motors;
   (c) the volt drop in cables is minimal;
   (d) cable power losses are negligible.
5. Which of the following statements is false?
   (a) It is cheaper to use a.c. than d.c.
   (b) Distribution of a.c. is more convenient than with d.c. since voltages may be readily altered using transformers.
   (c) An alternator is an a.c. generator.
   (d) A rectifier changes d.c. into a.c.
6. An alternating voltage of maximum value 100 V is applied to a lamp. Which of the following direct voltages, if applied to the lamp, would cause the lamp to light with the same brilliance?
   (a) 100 V; (b) 63.7 V; (c) 70.7 V;
   (d) 141.4 V.
7. The value normally stated when referring to alternating currents and voltages is the
   (a) instantaneous value; (b) r.m.s. value; (c) average value; (d) peak value.
8. State which of the following is false. For a sine wave:
   (a) the peak factor is 1.414;
   (b) the r.m.s. value is 0.707 × peak value;
   (c) the average value is 0.637 × r.m.s. value;
   (d) the form factor is 1.11.
9. An a.c. supply is 70.7 V, 50 Hz. Which of the following statements is false?
   (a) The periodic time is 20 ms.
   (b) The peak value of the voltage is 70.7 V.
   (c) The r.m.s. value of the voltage is 70.7 V.
   (d) The peak value of the voltage is 100 V.
10. An alternating voltage is given by $v = 100 \sin(50\pi t - 0.30)$ V. Which of the following statements is true?

(a) The r.m.s. voltage is 100 V.
(b) The periodic time is 20 ms.
(c) The frequency is 25 Hz.
(d) The voltage is leading $v = 100 \sin 50\pi t$ by 0.30 radians.

12.9 Short answer questions on alternating voltages and currents

1. Briefly explain the principle of operation of the simple alternator.
2. What is the difference between an alternating and a unidirectional waveform?
3. What is meant by (a) waveform, (b) cycle.
4. The time to complete one cycle of a waveform is called the ....
5. What is frequency? Name its unit.
6. The mains supply voltage has a special shape of waveform called a ....
7. Define peak value.
8. What is meant by the r.m.s. value?
9. The domestic mains electricity supply voltage in Great Britain is ....
10. What is the mean value of a sinusoidal alternating e.m.f. which has a maximum value of 100 V?
11. The effective value of a sinusoidal waveform is .... × maximum value.
12. What is a phasor quantity?
13. Complete the statement: Form factor = ...÷........, and for a sine wave, form factor = ...
14. Complete the statement: Peak factor = ...÷........, and for a sine wave, peak factor = ...
15. A sinusoidal current is given by $i = I_m \sin(\omega t \pm \alpha)$. What do the symbols I_m, ω and α represent?
16. How is switching obtained when converting a.c. to d.c.?
17. Draw an appropriate circuit diagram suitable for half-wave rectification.
18. How may full-wave rectification be achieved?
19. What is a simple method of smoothing the output of a rectifier?

12.10 Further questions on alternating voltages and currents

Frequency and periodic time

1. Determine the periodic time for the following frequencies:
(a) 2.5 Hz; (b) 100 Hz; (c) 40 kHz.
[(a) 0.4 s (b) 10 ms (c) 25 μs]
2. Calculate the frequency for the following periodic times:
(a) 5 ms; (b) 50 μs; (c) 0.2 s.
[(a) 0.2 kHz (b) 20 kHz (c) 5 Hz]
3. An alternating current completes 4 cycles in 5 ms. What is its frequency? [800 Hz]

A.c. values of non-sinusoidal waveforms

4. An alternating current varies with time over half a cycle as follows:

Current (A)	0	0.7	2.0	4.2	8.4	8.2	2.5	1.0	0.4	0.2	0
time (ms)	0	1	2	3	4	5	6	7	8	9	10

The negative half-cycle is similar. Plot the curve and determine:
(a) the frequency; (b) the instantaneous values at 3.4 ms and 5.8 ms; (c) its mean value; and (d) its r.m.s. value.
[(a) 50 Hz (b) 5.5 A, 3.4 A (c) 2.8 A (d) 4.0 A]

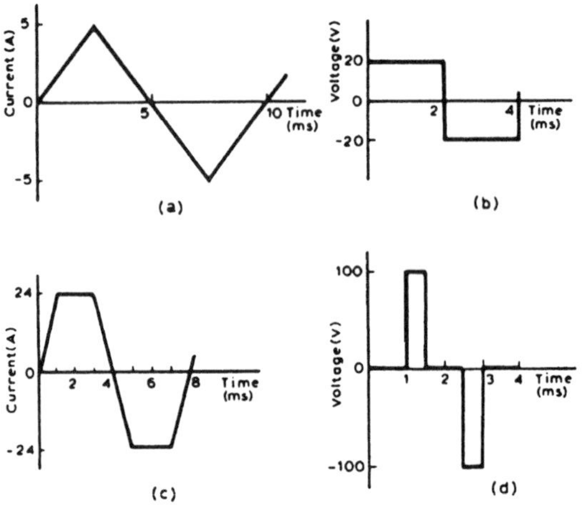

Figure 12.19

5. For the waveforms shown in Fig. 12.19 determine for each (i) the frequency, (ii) the average value over half a cycle, (iii) the r.m.s. value, (iv) the form factor, (v) the peak factor.
[(a)(i) 100 Hz (ii) 2.50 A (iii) 2.88 A (iv) 1.15 (v) 1.74
(b)(i) 250 Hz (ii) 20 V (iii) 20 V (iv) 1.0 (v) 1.0]

(c)(i) 125 Hz (ii) 18 A (iii) 19.56 A (iv) 1.09 (v) 1.23
(d)(i) 250 Hz (ii) 25 V (iii) 50 V (iv) 2.0 (v) 2.0]

6. An alternating voltage is triangular in shape, rising at a constant rate to a maximum of 300 V in 8 ms and then falling to zero at a constant rate in 4 ms. The negative half cycle is identical in shape to the positive half-cycle. Calculate (a) the mean voltage over half a cycle, and (b) the r.m.s. voltage. [(a) 150 V (b) 170 V]
7. An alternating e.m.f. varies with time over half a cycle as follows:

e.m.f. (V)	0	45	80	155	215	320	210	95	0
time (ms)	0	1.5	3.0	4.5	6.0	7.5	9.0	10.5	12.0

The negative half-cycle is identical in shape to the positive half-cycle. Plot the waveform and determine (a) the periodic time and frequency, (b) the instantaneous value of voltage at 3.75 ms, (c) the times when the voltage is 125 V, (d) the mean value, and (e) the r.m.s. value.
[(a) 24 ms, 41.67 Hz (b) 115 V (c) 4 ms and 10.1 ms (d) 142 V (e) 171 V]

A.c. values of sinusoidal waveforms

8. Calculate the r.m.s. value of a sinusoidal curve of maximum value 300 V. [212.1 V]
9. Find the peak and mean values for a 200 V mains supply. [282.9 V, 180.2 V]
10. Plot a sine wave of peak value 10.0 A. Show that the average value of the waveform is 6.37 A over half a cycle, and that the r.m.s. value is 7.07 A.
11. A sinusoidal voltage has a maximum value of 120 V. Calculate its r.m.s. and average values. [84.8 V, 76.4 V]
12. A sinusoidal current has a mean value of 15.0 A. Determine its maximum and r.m.s. values. [23.55 A, 16.65 A]

$v = V_m \sin(\omega t \pm \phi)$

13. An alternating voltage is represented by $v = 20 \sin 157.1t$ volts. Find (a) the maximum value, (b) the frequency, (c) the periodic time. (d) What is the angular velocity of the phasor representing this waveform?
[(a) 20 V (b) 25 Hz (c) 0.04 s (d) 157.1 rad/s]
14. Find the peak value, the r.m.s. value, the periodic time, the frequency and the phase angle (in degrees and minutes) of the following alternating quantities:
(a) $v = 90 \sin 400\pi t$ volts
[90 V, 63.63 V, 5 ms, 200 Hz, 0°]
(b) $i = 50 \sin(100\pi t + 0.30)$ amperes
[50 A, 35.35 A, 0.02 s, 50 Hz, 17°11′ lead]
(c) $e = 200 \sin(628.4t - 0.41)$ volts
[200 V, 141.4 V, 0.01 s, 100 Hz, 23°29′ lag]
15. A sinusoidal current has a peak value of 30 A and a frequency of 60 Hz. At time $t = 0$, the current is zero. Express the instantaneous current i in the form $i = I_m \sin \omega t$.
[$i = 30 \sin 120\pi t$]
16. An alternating voltage v has a periodic time of 20 ms and a maximum value of 200 V. When time $t = 0$, $v = -75$ volts. Deduce a sinusoidal expression for v and sketch one cycle of the voltage showing important points. [$v = 200 \sin(100\pi t - 0.384)$]
17. The voltage in an alternating current circuit at any time t seconds is given by $v = 60 \sin 40t$ volts. Find the first time when the voltage is (a) 20 V, (b) –30 V.
[(a) 8.496 ms (b) 91.63 ms]
18. The instantaneous value of voltage in an a.c. circuit at any time t seconds is given by $v = 100 \sin(50\pi t - 0.523)$ V. Find:
(a) the peak-to-peak voltage, the periodic time, the frequency and the phase angle;
(b) the voltage when $t = 0$;
(c) the voltage when $t = 8$ ms;
(d) the times in the first cycle when the voltage is 60 V;
(e) the times in the first cycle when the voltage is –40 V, and
(f) the first time when the voltage is a maximum.
Sketch the curve for one cycle showing relevant points.
[(a) 200 V, 0.04 s, 25 Hz, 29°58′ lagging (b) –49.95 V (c) 66.96 V (d) 7.426 ms, 19.23 ms (e) 25.95 ms, 40.71 ms (f) 13.33 ms]

Combination of periodic functions

19. The instantaneous values of two alternating voltages are given by $v_1 = 5 \sin \omega t$ and $v_2 = 8 \sin(\omega t - (\pi/6))$. By plotting v_1 and v_2 on the same axes, using the same scale, over one cycle, obtain expressions for (a) $v_1 + v_2$, and (b) $v_1 - v_2$.
[(a) $v_1 + v_2 = 12.58 \sin(\omega t - 0.325)$ V
(b) $v_1 - v_2 = 4.44 \sin(\omega t + 2.02)$ V]

20. Repeat Problem 19 by resolution of phasors.
21. Construct a phasor diagram to represent $i_1 + i_2$ where $i_1 = 12 \sin \omega t$ and $i_2 = 15 \sin (\omega t + \pi/3)$. By measurement, or by calculation, find a sinusoidal expression to represent $i_1 + i_2$.
[$23.43 \sin (\omega t + 0.588)$]
22. Determine, either by plotting graphs and adding ordinates at intervals, or by calculation, the following periodic functions in the form $v = V_m \sin (\omega t \pm \phi)$:
(a) $10 \sin \omega t + 4 \sin (\omega t + \pi/4)$
[$13.14 \sin (\omega t + 0.217)$]
(b) $80 \sin (\omega t + \pi/3) + 50 \sin (\omega t - \pi/6)$
[$94.34 \sin (\omega t + 0.489)$]
(c) $100 \sin \omega t - 70 \sin (\omega t - \pi/3)$
[$88.88 \sin (\omega t + 0.751)$]

Rectifiers

23. Describe, with appropriate circuit diagrams, the principle of operation of (a) a half-wave, and (b) a full-wave rectifier. Explain a simple method of smoothing the output waveform.

13

Single-phase series a.c. circuits

At the end of this chapter you should be able to:

- draw phasor diagrams and current and voltage waveforms for (a) purely resistive, (b) purely inductive, and (c) purely capacitive a.c. circuits
- perform calculations involving $X_L = 2\pi fC$ and $X_C = 1/(2\pi fC)$.
- draw circuit diagrams, phasor diagrams and voltage and impedance triangles for R–L, R–V and R–L–C series a.c. circuits and perform calculations using Pythagoras' theorem, trigonometric ratios and $Z = V/I$
- understand resonance
- derive the formula for resonant frequency and use it in calculations
- understand Q-factor and perform calculations using
  $\frac{V_L(\text{or } V_C)}{V}$ or $\frac{\omega_r L}{R}$ or $\frac{1}{\omega_r CR}$ or $\frac{1}{R}\sqrt{\left(\frac{L}{C}\right)}$
- understand bandwidth and half-power points
- perform calculations involving $(f_2 - f_1) = f_r/Q$
- understand selectivity and typical values of Q-factor
- appreciate that power P in an a.c. circuit is given by $P = VI\cos\phi$ or $I_R{}^2R$ and perform calculations using these formulae
- understand true, apparent and reactive power and power factor and perform calculations involving these quantities

13.1 Purely resistive a.c. circuit

In a purely resistive a.c. circuit, the current I_R and applied voltage V_R are in phase. See Fig. 13.1.

13.2 Purely inductive a.c. circuit

In a purely inductive a.c. circuit, the current I_L **lags** the applied voltage V_L by 90° (i.e. $\pi/2$ rad). See Fig. 13.2.

In a purely inductive circuit the opposition to the flow of alternating current is called the **inductive reactance**, X_L.

$$X_L = \frac{V_L}{I_L} = 2\pi fL\ \Omega$$

where f is the supply frequency, in hertz, and L is the inductance, in henrys. X_L is proportional to f as shown in Fig. 13.3.

Problem 1. (a) Calculate the reactance of a coil of inductance 0.32 H when it is connected to a 50 Hz supply. (b) A coil has a reactance of 124 Ω in a circuit with a supply of frequency 5 kHz. Determine the inductance of the coil.

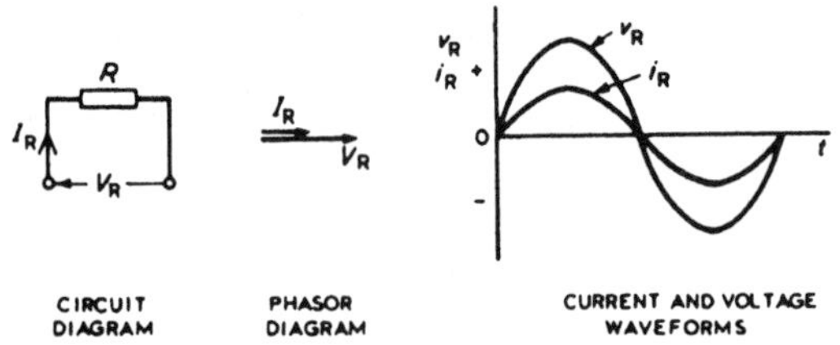

Figure 13.1

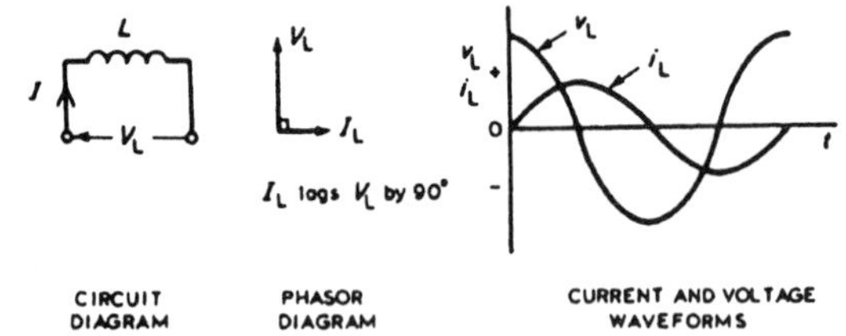

Figure 13.2

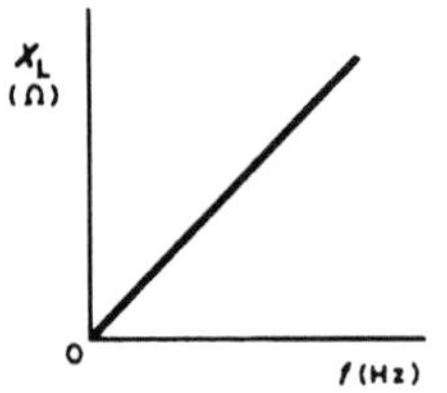

Figure 13.3

(a) Inductive reactance,
$X_L = 2\pi fL = 2\pi(50)(0.32) = \mathbf{100.5\ \Omega}$.
(b) Since $X_L = 2\pi fL$,

$$\text{inductance } L = \frac{X_L}{2\pi f} = \frac{124}{2\pi(5000)} \text{ H} = \mathbf{3.95\ mH}$$

Problem 2. A coil has an inductance of 40 mH and negligible resistance. Calculate its inductive reactance and the resulting current if connected to (a) a 240 V, 50 Hz supply, and (b) a 100 V, 1 kHz supply.

(a) Inductive reactance, $X_L = 2\pi fL$
$= 2\pi(50)(40 \times 10^{-3}) = \mathbf{12.57\ \Omega}$.
Current, $I = (V/X_L) = (240/12.57) = \mathbf{19.09\ A}$.
(b) Inductive reactance,
$X_L = 2\pi(1000)(40 \times 10^{-3}) = 251.3\ \Omega$.
Current, $I = (V/X_L) = (100/251.3) = \mathbf{0.398\ A}$.

13.3 Purely capacitive a.c. circuit

In a purely capacitive a.c. circuit, the current I_C **leads** the applied voltage V_C by 90° (i.e. $\pi/2$ rad). See Fig. 13.4.

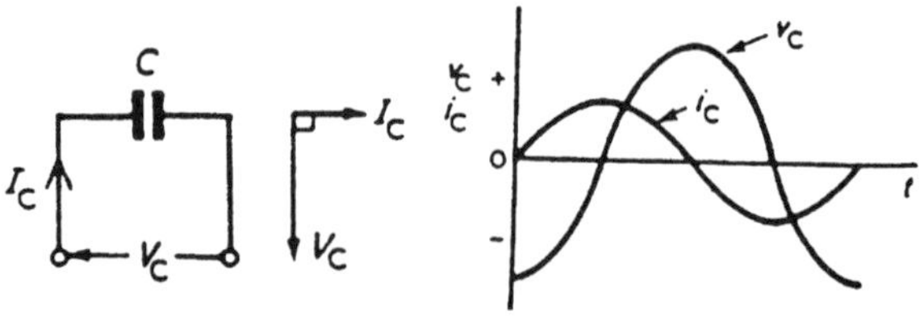

Figure 13.4

In a purely capacitive circuit the opposition to the flow of alternating current is called the **capacitive reactance**, X_C.

$$X_C = \frac{V_C}{I_C} = \frac{1}{2\pi fC}\ \Omega$$

where C is the capacitance in farads. X_C varies with frequency f as shown in Fig. 13.5.

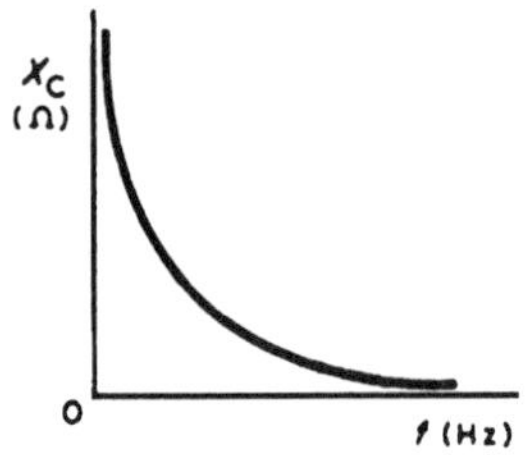

Figure 13.5

Problem 3. Determine the capacitive reactance of a capacitor of 10 μF when connected to a circuit of frequency (a) 50 Hz, (b) 20 kHz.

(a) Capacitive reactance X_C

$$= \frac{1}{2\pi fC} = \frac{1}{2\pi(50)(10 \times 10^{-6})}$$

$$= \frac{10^6}{2\pi(50)(10)} = \mathbf{318.3\ \Omega}$$

(b) $X_C = \dfrac{1}{2\pi fC} = \dfrac{1}{2\pi(20 \times 10^3)(10 \times 10^{-6})}$

$$= \frac{10^6}{2\pi(20 \times 10^3)(10)} = \mathbf{0.796\ \Omega}$$

Hence as the frequency is increased from 50 Hz to 20 kHz, X_C decreases from 318.3 Ω to 0.796 Ω (see Fig. 13.5).

> Problem 4. A capacitor has a reactance of 40 Ω when operated on a 50 Hz supply. Determine the value of its capacitance.

Since $X_C = 1/2\pi fC$

$$\text{capacitance } C = \frac{1}{2\pi fX_C} = \frac{1}{2\pi(50)(40)}\ \text{F}$$

$$= \frac{10^6}{2\pi(50)(40)}\ \mu\text{F} = \mathbf{79.58\ \mu F}$$

> Problem 5. Calculate the current taken by a 23 μF capacitor when connected to a 240 V, 50 Hz supply.

$$\text{Current } I = \frac{V}{X_C} = \frac{V}{\dfrac{1}{2\pi fC}} = 2\pi fCV$$

$$= 2\pi(50)(23 \times 10^{-6})(240)$$

$$= \mathbf{1.73\ A}$$

13.4 *R–L* series a.c. circuit

In an a.c. circuit containing inductance L and resistance R, the applied voltage V is the phasor sum of V_R and V_L (see Fig. 13.6), and thus the current I lags the applied voltage V by an angle lying between 0° and 90° (depending on the values of V_R and V_L), shown as angle ϕ. In any a.c. series circuit the current is common to each component and is thus taken as the **reference phasor**.

From the phasor diagram of Fig. 13.6, the **voltage triangle** is derived. For the R–L circuit:

$$V = \sqrt{(V_R^2 + V_L^2)} \quad \text{(by Pythagoras' theorem)}$$

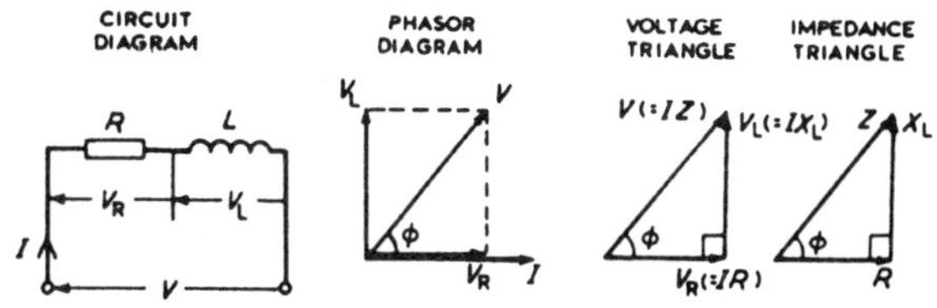

Figure 13.6

and

$$\tan\phi = \frac{V_L}{V_R} \quad \text{(by trigonometric ratios)}$$

In an a.c. circuit, the ratio of applied voltage V to current I is called the **impedance *Z***, i.e.

$$\boldsymbol{Z = \frac{V}{I}\ \Omega}$$

If each side of the voltage triangle in Fig. 13.6 is divided by current I then the **impedance triangle** is derived. For the R–L circuit:

$$Z = \sqrt{(R^2 + X_L^2)} \qquad \tan\phi = \frac{X_L}{R}$$

$$\sin\phi = \frac{X_L}{Z} \text{ and } \cos\phi = \frac{R}{Z}$$

> Problem 6. In a series R–L circuit the p.d. across the resistance R is 12 V and the p.d. across the inductance L is 5 V. Find the supply voltage and the phase angle between current and voltage.

From the voltage triangle of Fig. 13.6, supply voltage $V = \sqrt{(12^2 + 5^2)}$, i.e. $\boldsymbol{V = 13\ V}$.
(Note that in a.c. circuits, the supply voltage is not the arithmetic sum of the p.d.s across components. It is, in fact, the **phasor sum**.)
$\tan\phi = V_L/V_R = 5/12$, from which $\phi = \arctan(5/12) =$ **22°37′ lagging**.
('Lagging' infers that the current is 'behind' the voltage, since phasors revolve anticlockwise.)

> Problem 7. A coil has a resistance of 4 Ω and an inductance of 9.55 mH. Calculate (a) the reactance, (b) the impedance, and (c) the current taken from a 240 V, 50 Hz supply. Determine also the phase angle between the supply voltage and current.

$R = 4\ \Omega$; $L = 9.55$ mH $= 9.55 \times 10^{-3}$ H; $f = 50$ Hz; $V = 240$ V.

(a) Inductive reactance, $X_L = 2\pi fL$ $= 2\pi(50)(9.55 \times 10^{-3}) =$ **3 Ω**.

(b) Impedance, $Z = \sqrt{(R^2 + X_L^2)} = \sqrt{(4^2 + 3^2)} =$ **5 Ω**.

(c) Current, $I = V/Z = 240/5 =$ **48 A**.

The circuit and phasor diagrams and the voltage and impedance triangles areas shown in Fig. 13.6.

Since $\tan\phi = X_L/R$, $\phi = \arctan(X_L/R) = \arctan(3/4) = 36.87° =$ **36°52′ lagging**.

> Problem 8. A coil takes a current of 2 A from a 12 V d.c. supply. When connected to a 240 V, 50 Hz supply the current is 20 A. Calculate the resistance, impedance, inductive reactance and inductance of the coil.

$$\text{Resistance } R = \frac{\text{d.c. voltage}}{\text{d.c. current}} = \frac{12}{2} = 6\ \Omega$$

$$\text{Impedance } Z = \frac{\text{a.c. voltage}}{\text{a.c. current}} = \frac{240}{20} = 12\ \Omega$$

Since $Z = \sqrt{(R^2 + X_L^2)}$,

inductive reactance, $X_L = \sqrt{(Z^2 - R^2)}$

$= \sqrt{(12^2 - 6^2)} = 10.39\ \Omega$

Since $X_L = 2\pi fL$, inductance $L = X_L/2\pi f = 10.39/(2\pi(50)) =$ **33.1 mH**.

This problem indicates a simple method for finding the inductance of a coil, i.e. first to measure the current when the coil is connected to a d.c. supply of known voltage, and then to repeat the process with an a.c. supply.

> Problem 9. A coil of inductance 318.3 mH and negligible resistance is connected in series with a 200 Ω resistor to a 240 V, 50 Hz supply. Calculate (a) the inductive reactance of the coil, (b) the impedance of the circuit, (c) the current in the circuit, (d) the p.d. across each component, and (e) the circuit phase angle.

$L = 318.3$ mH $= 0.3183$ H; $R = 200\ \Omega$; $V = 240$ V; $f = 50$ Hz. The circuit diagram is as shown in Fig. 13.6.

(a) Inductive reactance $X_L = 2\pi fL = 2\pi(50)(0.3183) =$ **100 Ω**.

(b) Impedance $Z = \sqrt{(R^2 + X_L^2)} = \sqrt{[(200)^2 + (100)^2]} =$ **223.6 Ω**.

(c) Current $I = V/Z = 240/223.6 =$ **1.073 A**.

(d) The p.d. across the coil, $V_L = IX_L = 1.073 \times 100 =$ **107.3 V**.
The p.d. across the resistor, $V_R = IR = 1.073 \times 200 =$ **214.6 V**.
(Check: $\sqrt{(V_R^2 + V_L^2)} = \sqrt{[(214.6)^2 + (107.3)^2]} = 240$ V, the supply voltage.)

(e) From the impedance triangle, angle $\phi = \arctan(X_L/R) = \arctan(100/200)$, **hence the phase angle ϕ = 26°34′ lagging.**

> Problem 10. A coil consists of a resistance of 100 Ω and an inductance of 200 mH. If an alternating voltage, v, given by $v = 200 \sin 500t$ volts is applied across the coil, calculate (a) the circuit impedance, (b) the current flowing, (c) the p.d. across the resistance, (d) the p.d. across the inductance and (e) the phase angle between voltage and current.

Since $v = 200 \sin 500t$ volts then $V_m = 200$ V and $\omega = 2\pi f = 500$ rad/s. Hence r.m.s. voltage $V = 0.707 \times 200 = 141.4$ V.

Inductive reactance, $X_L = 2\pi fL = \omega L = 500 \times 200 \times 10^{-3} = 100\ \Omega$.

(a) Impedance $Z = \sqrt{(R^2 + X_L^2)} = \sqrt{(100^2 + 100^2)} =$ **141.4 Ω**.

(b) Current $I = V/Z = 141.4/141.4 =$ **1 A**.

(c) p.d. across the resistance $V_R = IR = 1 \times 100 =$ **100 V**.
p.d. across the inductance $V_L = IX_L = 1 \times 100 =$ **100 V**.

(d) Phase angle between voltage and current, $\phi = X_L/R = \arctan(100/100)$, **hence ϕ = 45° or (π/4) rad**.

> Problem 11. A pure inductance of 1.273 mH is connected in series with a pure resistance of 30 Ω. If the frequency of the sinusoidal supply is 5 kHz and the p.d. across the 30 Ω resistor is 6 V, determine the value of the supply voltage and the voltage across the 1.273 mH inductance. Draw the phasor diagram.

The circuit is shown in Fig. 13.7(a).

Supply voltage, $V = IZ$. Current $I = (V_R/R) = (6/30) = 0.20$ A.

Inductive reactance $X_L = 2\pi fL$ $= 2\pi(5 \times 10^3)(1.273 \times 10^{-3}) = 40\ \Omega$.

(c) For a purely capacitive a.c. circuit, the average power is zero. See Fig. 13.21(c).

Figure 13.22 shows current and voltage waveforms for an R–L circuit where the current lags the voltage by angle ϕ. The waveform for power (where $p = vi$) is shown by the broken line, and its shape, and hence average power, depends on the value of angle ϕ.

For an R–L, R–C or R–L–C series a.c. circuit, the average power P is given by:

$$\boldsymbol{P = VI \cos\phi \text{ watts}}$$

or

$$\boldsymbol{P = I^2R \text{ watts}}$$

(V and I being r.m.s. values).

> Problem 24. An instantaneous current, $i = 250 \sin \omega t$ mA flows through a pure resistance of 5 kΩ. Find the power dissipated in the resistor.

Power dissipated, $P = I^2R$ where I is the r.m.s. value of current. If $i = 250 \sin \omega t$ mA, then $I_m = 0.250$ A and r.m.s. current, $I = (0.707 \times 0.250)$ A. Hence power $P = (0.707 \times 0.250)^2(5000) =$ **156.2 watts.**

> Problem 25. A series circuit of resistance 60 Ω and inductance 75 mH is connected to a 110 V, 60 Hz supply. Calculate the power dissipated.

Inductive reactance, $X_L = 2\pi fL = 2\pi(60)(75 \times 10^{-3}) = 28.27\ \Omega$.
Impedance, $Z = \sqrt{(R^2 + X_L^2)} = \sqrt{[(60)^2 + (28.27)^2]} = 66.33\ \Omega$.
Current, $I = V/Z = 110/66.33 =$ **1.658 A.**
To calculate power dissipation in an a.c. circuit two formulae may be used:

(i) $P = I^2R = (1.658)^2(60) =$ **165 W.**

or

(ii) $P = VI \cos\phi$ where $\cos\phi = R/Z = 60/66.33 = 0.9046$; hence

$P = (110)(1.658)(0.9046) =$ **165 W.**

13.12 Power triangle and power factor

Figure 13.23(a) shows a phasor diagram in which the current I lags the applied voltage V by angle ϕ. The horizontal component of V is $V \cos\phi$ and the vertical component of V is $V \sin\phi$. If each of the voltage phasors is multiplied by I, Fig. 13.23(b) is obtained and is known as the **power triangle**.

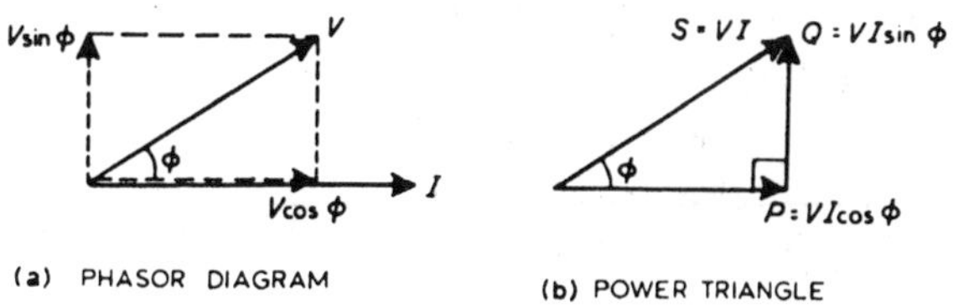

Figure 13.23

Apparent power, $\boldsymbol{S = VI}$ **voltamperes (VA)**
True or active power, $\boldsymbol{P = VI \cos\phi}$ **watts (W)**
Reactive power, $\boldsymbol{Q = VI \sin\phi}$ **reactive voltamperes (var)**

$$\text{Power factor} = \frac{\text{True power } P}{\text{Apparent power } S.}$$

For sinusoidal voltages and currents,

$$\text{power factor} = \frac{P}{S} = \frac{VI \cos\phi}{VI}$$

i.e.

$$\text{p.f.} = \cos\phi = \frac{R}{Z} \text{ (from Fig. 13.6)}$$

(The relationships stated above are also true when current I leads voltage V.)

> Problem 26. A pure inductance is connected to a 150 V, 50 Hz supply, and the apparent power of the circuit is 300 VA. Find the value of the inductance.

Apparent power $S = VI$. Hence current $I = S/V = 300/150 = 2$ A.
Inductive reactance $X_L = V/I = 150/2 = 75\ \Omega$.
Since $X_L = 2\pi fL$,

$$\text{inductance } L = \frac{X_L}{2\pi f} = \frac{75}{2\pi(50)} = \mathbf{0.239\ H}$$

> Problem 27. A transformer has a rated output of 200 kVA at a power factor of 0.8. Determine the rated power output and the corresponding reactive power.

$VI = 200$ kVA $= 200 \times 10^3$; p.f. $= 0.8 = \cos\phi$. Power output, $P = VI\cos\phi = (200 \times 10^3)(0.8)$ $=$ **160 kW**.

Reactive power, $Q = VI\sin\phi$. If $\cos\phi = 0.8$, then $\phi = \arccos 0.8 = 36°52'$. Hence $\sin\phi = \sin 36°52'$ $= 0.6$. Hence reactive power, $Q = (200 \times 10^3)(0.6)$ $=$ **120 kvar**.

Problem 28. A load takes 90 kW at a power factor of 0.5 lagging. Calculate the apparent power and the reactive power.

True power $P = 90$ kW $= VI\cos\phi$. Power factor $= 0.5 = \cos\phi$.

Apparent power $S = VI = 90$ kW/0.5
$=$ **180 kVA**

Angle $\phi = \arccos 0.5 = 60°$. Hence $\sin\phi = \sin 60°$ $= 0.866$. Hence reactive power, $Q = VI\sin\phi = 180 \times 10^3 \times 0.866 =$ **156 kvar**.

Problem 29. The power taken by an inductive circuit when connected to a 120 V, 50 Hz supply is 400 W and the current is 8 A. Calculate (a) the resistance, (b) the impedance, (c) the reactance, (d) the power factor, and (e) the phase angle between voltage and current.

(a) Power $P = I^2R$. Hence $R = P/I^2 = 400/(8)^2 =$ **6.25 Ω**.
(b) Impedance $Z = V/I = 120/8 =$ **15 Ω**.
(c) Since $Z = \sqrt{(R^2 + X_L^2)}$, then $X_L = \sqrt{(Z^2 - R^2)}$ $= \sqrt{[(15)^2 - (6.25)^2]} =$ **13.64 Ω**.
(d) Power factor $= \dfrac{\text{true power}}{\text{apparent power}}$

$$= \frac{VI\cos\phi}{VI} = \frac{400}{(120)(8)} = \mathbf{0.4167}$$

(e) p.f. $= \cos\phi = 0.4167$. Hence

phase angle $\phi = \arccos 0.4167$
$=$ **65°22′** lagging.

Problem 30. A circuit consisting of a resistor in series with a capacitor takes 100 watts at a power factor of 0.5 from a 100 V, 60 Hz supply. Find (a) the current flowing, (b) the phase angle, (c) the resistance, (d) the impedance, and (e) the capacitance.

(a) Power factor $=$ (true power)/(apparent power), i.e. $0.5 = 100/100I$. Hence

$$I = \frac{100}{(0.5)(100)} = \mathbf{2\ A}$$

(b) Power factor $= 0.5 = \cos\phi$. Hence phase angle $\phi = \arccos 0.5 =$ **60° leading**.
(c) Power $P = I^2R$. Hence resistance $R = P/I^2 = 100/(2)^2 =$ **25 Ω**.
(d) Impedance $Z = V/I = 100/2 =$ **50 Ω**.
(e) Capacitive reactance, $X_C = \sqrt{(Z^2 - R^2)} = \sqrt{(50^2 - 25^2)} = 43.30$ Ω. $X_C = 1/2\pi fC$. Hence

$$\text{capacitance } C = \frac{1}{2\pi f X_C} = \frac{1}{2\pi(60)(43.30)}\ \text{F}$$

$$= \mathbf{61.26\ \mu F}$$

13.13 Multi-choice questions on single-phase a.c. circuits

(Answers on page 284.)

1. An inductance of 10 mH connected across a 100 V, 50 Hz supply has an inductive reactance of (a) 10π Ω, (b) 1000π Ω, (c) π Ω,(d) π HΩ.
2. When the frequency of an a.c. circuit containing resistance and inductance is increased, the current (a) decreases, (b) increases, (c) stays the same.
3. In question 2, the phase angle of the circuit (a) decreases, (b) increases, (c) stays the same.
4. When the frequency of an a.c. circuit containing resistance and capacitance is decreased, the current (a) increases, (b) decreases, (c) stays the same.
5. In question 4, the phase angle of the circuit (a) increases, (b) decreases, (c) stays the same.
6. A capacitor of 1 μF is connected to a 50 Hz supply. The capacitive reactance is (a) 50 MΩ, (b) (10/π) kΩ, (c) ($\pi/10^4$) Ω, (d) (10/π) Ω.
7. In a series a.c. circuit the voltage across a pure inductance is 12 V and the voltage across a pure resistance is 5 V. The supply voltage is (a) 13 V, (b) 17 V, (c) 7 V, (d) 2.4 V.
8. Inductive reactance results in a current that
(a) leads the voltage by 90°;
(b) is in phase with the voltage;
(c) leads the voltage by π rad;
(d) lags the voltage by π/2 rad.

9. State which of the following statements is false.
   (a) Impedance is at a minimum at resonance in an a.c. circuit.
   (b) The product of r.m.s. current and voltage gives the apparent power in an a.c. circuit.
   (c) Current is at a maximum at resonance in an a.c. circuit.
   (d) (Apparent power)/(True power) gives power factor.
10. The impedance of a coil, which has a resistance of X ohms and an inductance of Y henrys, connected across a supply of frequency K Hz, is
   (a) $2\pi KY$, (b) $X + Y$, (c) $\sqrt{(X^2 + Y^2)}$,
   (d) $\sqrt{\{X^2 + (2\pi KY)^2\}}$.
11. In question 10, the phase angle between the current and the applied voltage is given by:
   (a) arctan (Y/X); (b) arctan $(2\pi KY/X)$;
   (c) arctan $(X/2\pi KY)$; (d) tan $(2\pi KY/X)$.
12. When a capacitor is connected to an a.c. supply the current
   (a) leads the voltage by 180°;
   (b) is in phase with the voltage;
   (c) leads the voltage by $\pi/2$ rad;
   (d) lags the voltage by 90°.
13. When the frequency of an a.c. circuit containing resistance and capacitance is decreased the current (a) increases, (b) decreases, (c) stays the same.
14. In an R–L–C series a.c. circuit a current of 5 A flows when the supply voltage is 100 V. The phase angle between current and voltage is 60° lagging. Which of the following statements is false?
   (a) The circuit is effectively inductive.
   (b) The apparent power is 500 VA.
   (c) The equivalent circuit reactance is 20 Ω.
   (d) The true power is 250 W.
15. A series a.c. circuit comprising a coil of inductance 100 mH and resistance 1 Ω and a 10 μF capacitor is connected across a 10 V supply. At resonance the p.d. across the capacitor is
   (a) 10 kV, (b) 1 kV, (c) 100 V, (d) 10 V.

13.14 Short answer questions on single-phase a.c. circuits

1. Complete the following statements:
   (a) In a purely resistance a.c. circuit the current is .... with the voltage.
   (b) In a purely inductive a.c. circuit the current .... the voltage by .... degrees.
   (c) In a purely capacitive a.c. circuit the current .... the voltage by .... degrees.
2. Draw phasor diagrams to represent (a) a purely resistive a.c. circuit, (b) a purely inductive a.c. circuit, and (c) a purely capacitive a.c. circuit.
3. What is inductive reactance? State the symbol and formula for determining inductive reactance.
4. What is capacitive reactance? State the symbol and formula for determining capacitive reactance.
5. Draw phasor diagrams to represent (a) a coil (having both inductance and resistance), and (b) a series capacitive circuit containing resistance.
6. What does 'impedance' mean when referring to an a.c. circuit?
7. Draw an impedance triangle for an R–L circuit. Derive from the triangle an expression for (a) impedance, and (b) phase angle.
8. Draw an impedance triangle for an R–C circuit. From the triangle derive an expression for (a) impedance, and (b) phase angle.
9. What is series resonance?
10. Derive a formula for resonant frequency f_r in terms of L and C.
11. What does the Q-factor of a series circuit mean?
12. State three formulae used to calculate the Q-factor of a series circuit at resonance.
13. State an advantage of a high Q-factor in a series high-frequency circuit.
14. State a disadvantage of a high Q-factor in a series power circuit.
15. State two formulae which may be used to calculate power in an a.c. circuit.
16. Show graphically that for a purely inductive or purely capacitive a.c. circuit the average power is zero.
17. Define 'power factor'.
18. Define (a) apparent power, (b) reactive power.
19. Define (a) bandwidth, (b) selectivity.

13.15 Further questions on single-phase a.c. circuits

A.c. circuits containing pure inductance and pure capacitance

1. Calculate the reactance of a coil of inductance 0.2 H when it is connected to

(a) a 50 Hz, (b) a 600 Hz, and (c) a 40 kHz supply.
[(a) 62.83 Ω (b) 754 Ω (c) 50.27 kΩ]

2. A coil has a reactance of 120 Ω in a circuit with a supply frequency of 4 kHz. Calculate the inductance of the coil. [4.77 mH]
3. A supply of 240 V, 50 Hz is connected across a pure inductance and the resulting current is 1.2 A. Calculate the inductance of the coil. [0.637 H]
4. An e.m.f. of 200 V at a frequency of 2 kHz is applied to a coil of pure inductance 50 mH. Determine (a) the reactance of the coil, and (b) the current flowing in the coil. [(a) 628 Ω (b) 0.318 A]
5. A 120 mH inductor has a 50 mA, 1 kHz alternating current flowing through it. Find the p.d. across the inductor. [37.7 V]
6. Calculate the capacitive reactance of a capacitor of 20 μF when connected to an a.c. circuit of frequency (a) 20 Hz, (b) 500 Hz, (c) 4 kHz. [(a) 397.9 Ω (b) 15.92 Ω (c) 1.989 Ω]
7. A capacitor has a reactance of 80 Ω when connected to a 50 Hz supply. Calculate the value of its capacitance. [39.79 μF]
8. Calculate the current taken by a 10 μF capacitor when connected to a 200 V, 100 Hz supply. [1.257 A]
9. A capacitor has a capacitive reactance of 400 Ω when connected to a 100 V, 25 Hz supply. Determine its capacitance and the current taken from the supply. [15.92 μF, 0.25 A]
10. Two similar capacitors are connected in parallel to a 200 V, 1 kHz supply. Find the value of each capacitor if the circuit current is 0.628 A. [0.25 μF]

R–L a.c. circuits

11. Determine the impedance of a coil which has a resistance of 12 Ω and a reactance of 16 Ω. [20 Ω]
12. A coil of inductance 80 mH and resistance 60 Ω is connected to a 200 V, 100 Hz supply. Calculate the circuit impedance and the current taken from the supply. Find also the phase angle between the current and the supply voltage. [78.27 Ω, 2.555 A, 39° 57′ lagging]
13. An alternating voltage given by $v = 100 \sin 240t$ volts is applied across a coil of resistance 32 Ω and inductance 100 mH. Determine (a) the circuit impedance, (b) the current flowing, (c) the p.d. across the resistance, and (d) the p.d. across the inductance. [(a) 40 Ω (b) 1.77 A (c) 56.64 V (d) 42.48 V]
14. A coil takes a current of 5 A from a 20 V d.c. supply. When connected to a 200 V, 50 Hz a.c. supply the current is 25 A. Calculate the (a) resistance, (b) impedance, and (c) inductance of the coil. [(a) 4 Ω (b) 8 Ω (c) 22.06 mH]
15. A resistor and an inductor of negligible resistance are connected in series to an a.c. supply. The p.d. across the resistor is 18 V and the p.d. across the inductor is 24 V. Calculate the supply voltage and the phase angle between voltage and current. [30 V, 53°8′ lagging]
16. A coil of inductance 636.6 mH and negligible resistance is connected in series with a 100 Ω resistor to a 240 V, 50 Hz supply. Calculate (a) the inductive reactance of the coil, (b) the impedance of the circuit, (c) the current in the circuit, (d) the p.d. across each component, and (e) the circuit phase angle. [(a) 200 Ω (b) 223.6 Ω (c) 1.118 A (d) 223.6 V, 111.8 V (c) 63°26′ lagging]

R–C a.c. circuits

17. A voltage of 35 V is applied across a *C–R* series circuit. If the voltage across the resistor is 21 V, find the voltage across the capacitor. [28 V]
18. A resistance of 50 Ω is connected in series with a capacitance of 20 μF. If a supply of 200 V, 100 Hz is connected across the arrangement find (a) the circuit impedance, (b) the current flowing, and (c) the phase angle between voltage and current. [(a) 93.98 Ω (b) 2.128 A (c) 57° 51′ leading]
19. A 24.87 μF capacitor and a 30 Ω resistor are connected in series across a 150 V supply. If the current flowing is 3 A find (a) the frequency of the supply, (b) the p.d. across the resistor, and (c) the p.d. across the capacitor. [(a) 160 Hz (b) 90 V (c) 120 V]
20. An alternating voltage $v = 250 \sin 800t$ volts is applied across a series circuit containing a 30 Ω resistor and 50 μF capacitor. Calculate (a) the circuit impedance, (b) the current flowing, (c) the p.d. across the resistor, (d) the p.d. across the capacitor, and (e) the phase angle between voltage and current. [(a) 39.05 Ω (b) 4.526 A (c) 135.8 V (d) 113.2 V (e) 39°48′]

21. A 400 Ω resistor is connected in series with a 2358 pF capacitor across a 12 V a.c. supply. Determine the supply frequency if the current flowing in the circuit is 24 mA. [225 kHz]

R–L–C a.c. circuits

22. A 40 μF capacitor in series with a coil of resistance 8 Ω and inductance 80 mH is connected to a 200 V, 100 Hz supply. Calculate (a) the circuit impedance, (b) the current flowing, (c) the phase angle between voltage and current, (d) the voltage across the coil, and (e) the voltage across the capacitor.
[(a) 13.18 Ω (b) 15.17 A (c) 52°38′ (d) 772.1 V (e) 603.6 V]

23. Three impedances are connected in series across a 100 V, 2 kHz supply. The impedances comprise:
(i) an inductance of 0.45 mH and 2 Ω resistance,
(ii) an inductance of 570 μH and 5 Ω resistance, and
(iii) a capacitor of 10 μF and 3 Ω resistance.
Assuming no mutual inductive effects between the two inductances calculate (a) the circuit impedance, (b) the circuit current, (c) the circuit phase angle, and (d) the voltage across each impedance. Draw the phasor diagram.
[(a) 11.12 Ω (b) 8.99 A (c) 25°56′ lagging (d) 53.92 V, 78.53 V, 76.46 V]

24. For the circuit shown in Fig. 13.24 determine the voltages V_1 and V_2 if the supply frequency is 1 kHz. Draw the phasor diagram and hence determine the supply voltage V and the circuit phase angle.
[V_1 = 26.0 V, V_2 = 67.05 V, V = 50 V, 53°8′ leading]

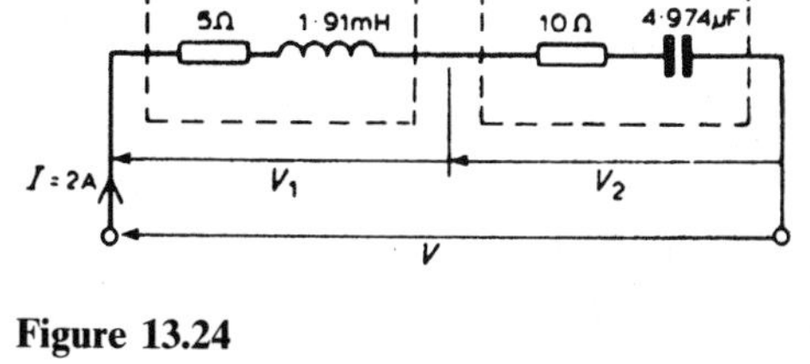

Figure 13.24

Series resonance and Q-factor

25. Find the resonant frequency of a series a.c. circuit consisting of a coil of resistance 10 Ω and inductance 50 mH and capacitance 0.5 μF. Find also the current flowing at resonance if the supply voltage is 100 V.
[3.183 kHz, 10 A]

26. The current at resonance in a series *L–C–R* circuit is 0.2 mA. If the applied voltage is 250 mV at a frequency of 100 kHz and the circuit capacitance is 0.04 μF, find the circuit resistance and inductance.
[1.25 kΩ, 63.3 μH]

27. A coil of resistance 25 Ω and inductance 100 mH is connected in series with a capacitance of 0.12 μF across a 200 V, variable frequency supply. Calculate (a) the resonant frequency, (b) the current at resonance, and (c) the factor by which the voltage across the reactance is greater than the supply voltage.
[(a) 1.453 kHz (b) 8 A (c) 36.52]

28. A coil of 0.5 H inductance and 8 Ω resistance is connected in series with a capacitor across a 200 V, 50 Hz supply. If the current is in phase with the supply voltage, determine the value of the capacitor and the p.d. across its terminals. [20.26 μF, 3.928 kV]

29. Calculate the inductance which must be connected in series with a 1000 pF capacitor to give a resonant frequency of 400 kHz.
[0.158 mH]

30. A series circuit comprises a coil of resistance 20 Ω and inductance 2 mH and a 500 pF capacitor. Determine the Q-factor of the circuit at resonance. If the supply voltage is 1.5 V, what is the voltage across the capacitor? [100, 150 V]

Power in a.c. circuits

31. A voltage $v = 200 \sin \omega t$ volts is applied across a pure resistance of 1.5 Ω. Find the power dissipated in the resistor. [13.33 W]

32. A 50 μF capacitor is connected to a 100 V, 200 Hz supply. Determine the true power and the apparent power.
[0, 628.3 VA]

33. A motor takes a current of 10 A when supplied from a 250 V a.c. supply. Assuming a power factor of 0.75 lagging find the power consumed. Find also the cost of running the motor for 1 week continuously if 1 kWh of electricity costs 7.20p. [1875 W, £22.68]

34. A motor takes a current of 12 A when supplied from a 240 V a.c. supply. Assuming a power factor of 0.75 lagging, find the power consumed. [2.16 kW]
35. A transformer has a rated output of 100 kVA at a power factor of 0.6. Determine the rated power output and the corresponding reactive power. [60 kW, 80 kvar]
36. A substation is supplying 200 kVA and 150 kvar. Calculate the corresponding power and power factor. [132 kW, 0.66]
37. A load takes 50 kW at a power factor of 0.8 lagging. Calculate the apparent power and the reactive power. [62.5 kVA, 37.5 kvar]
38. A coil of resistance 400 Ω and inductance H is connected to a 75 V, 400 Hz supply. ulate the power dissipated in the coil. [5.452 W]

Ω resistor and a 6 μF capacitor are ed in series across a 150 V, 200 Hz Calculate (a) the circuit impedance, urrent flowing, and (c) the power in the circuit. [(a) 154.9 Ω (b) 0.968 A (c) 75 W]

40. taken by a series circuit contain- in e and inductance is 240 W when connected to a 200 V, 50 Hz supply. If the current flowing is 2 A find the values of the resistance and inductance. [60 Ω, 255 mH]

41. The power taken by a *C–R* series circuit, when connected to a 105 V, 2.5 kHz supply, is 0.9 kW and the current is 15 A. Calculate (a) the resistance, (b) the impedance, (c) the reactance, (d) the capacitance, (e) the power factor, and (f) the phase angle between voltage and current. [(a) 4 Ω (b) 7 Ω (c) 5.745 Ω (d) 11.08 μF (e) 0.571 (f) 55°9′ leading]
42. A circuit consisting of a resistor in series with an inductance takes 210 W at a power factor of 0.6 from a 50 V, 100 Hz supply. Find (a) the current flowing, (b) the circuit phase angle, (c) the resistance, (d) the impedance, and (e) the inductance. [(a) 7 A (b) 53°8′ lagging (c) 4.286 Ω (d) 7.143 Ω (e) 9.095 mH]
43. A 200 V, 60 Hz supply is applied to a capacitive circuit. The current flowing is 2 A and the power dissipated is 150 W. Calculate the values of the resistance and capacitance. [37.5 Ω, 28.61 μF]

14

Single-phase parallel a.c. circuits

At the end of this chapter you should be able to:

- calculate unknown currents, impedances and circuit phase angle from phasor diagrams for (a) *R–L* parallel a.c. circuit (b) *R–C* parallel a.c. circuit (c) *L–C* parallel a.c. circuit (d) *LR–C* parallel a.c. circuit
- state the condition for parallel resonance in an *LR–C* circuit
- derive the resonant frequency equation for an *LR–C* parallel a.c. circuit
- determine the current and dynamic resistance at resonance in an *LR–C* parallel circuit
- understand and calculate Q-factor in an *LR–C* parallel circuit
- understand how power factor may be improved

14.1 Introduction

In parallel circuits, such as those shown in Figs. 14.1 and 14.2, the voltage is common to each branch of the network and is thus taken as the reference phasor when drawing phasor diagrams.

For any parallel a.c. circuit:

True or active power, $P = VI\cos\phi$ watts (W)

or $P = I_R^2 R$ watts

Apparent power, $S = VI$ voltamperes (VA)

Reactive power, $Q = VI\sin\phi$ reactive voltamperes (var)

Power factor =

$$\frac{\text{true power}}{\text{apparent power}} = \frac{P}{S} = \cos\phi$$

(These formulae are the same as for series a.c. circuits as shown in Chapter 13.)

14.2 *R–L* parallel a.c. circuit

In the two-branch parallel circuit containing resistance R and inductance L shown in Fig. 14.1, the current flowing in the resistance, I_R, is in phase with the supply voltage V and the current flowing in the inductance, I_L, lags the supply voltage by 90°. The supply current I is the phasor sum of I_R and I_L and thus the current I lags the applied voltage V by an angle lying between 0° and 90° (depending on the values of I_R and I_L), shown as angle ϕ in the phasor diagram. From the phasor diagram: $I = \sqrt{(I_R^2 + I_L^2)}$ (by Pythagoras' theorem) where

$$I_R = \frac{V}{R} \text{ and } I_L = \frac{V}{X_L}$$

$$\tan\phi = \frac{I_L}{I_R},\ \sin\phi = \frac{I_L}{I} \text{ and } \cos\phi = \frac{I_R}{I}$$

(by trigonometric ratios)

$$\text{Circuit impedance, } Z = \frac{V}{I}$$

Problem 1. A 20 Ω resistor is connected in parallel with an inductance of2.387 mH across a 60 V, 1 kHz supply. Calculate (a) the current in each branch, (b) the supply current, (c) the circuit phase angle, (d) the circuit impedance, and (e) the power consumed.

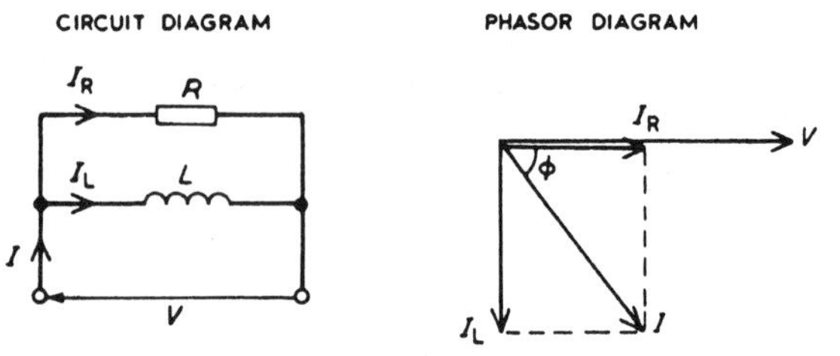

Figure 14.1

The circuit and phasor diagrams are as shown in Fig. 14.1.

(a) Current flowing in the resistor

$$I_R = \frac{V}{R} = \frac{60}{20} = \mathbf{3\ A}$$

Current flowing in the inductance,

$$I_L = \frac{V}{X_L} = \frac{V}{2\pi f L}$$
$$= \frac{60}{2\pi(1000)(2.387 \times 10^{-3})} = \mathbf{4\ A}$$

(b) From the phasor diagram,

supply current, $I = \sqrt{(I_R^2 + I_L^2)} = \sqrt{(3^2 + 4^2)}$
$= \mathbf{5\ A}$

(c) Circuit phase angle $\phi = \arctan(I_L/I_R) = \arctan(4/3) =$ **53°8′ lagging**.

(d) Circuit impedance, $Z = V/I = 60/5 =$ **12 .V.**

(e) Power consumed $P = VI \cos\phi = (60)(5)(\cos 53°8') =$ **180 W**. (Alternatively, power consumed $P = I_R^2 R = (3)^2(20) =$ **180 W**.)

14.3 *R–C* parallel a.c. circuit

In the two branch parallel circuit containing resistance R and capacitance C shown in Fig. 14.2, I_R is in-phase with the supply voltage V and the current flowing in the capacitor, I_C, leads V by 90°. The supply current I is the phasor sum of I_R and I_C and thus the current I leads the applied voltage V by an angle lying between 0° and 90° (depending on the values of I_R and I_C), shown as angle α in the phasor diagram.

From the phasor diagram:

$$I = \sqrt{(I_R^2 + I_C^2)} \quad \text{(by Pythagoras' theorem)}$$

where

$$I_R = \frac{V}{R} \text{ and } I_C = \frac{V}{X_C}$$

$$\tan\alpha = \frac{I_C}{I_R}, \ \sin\alpha = \frac{I_C}{I} \text{ and } \cos\alpha = \frac{I_R}{I}$$

(by trigonometric ratios)

$$\text{Circuit impedance } Z = \frac{V}{I}$$

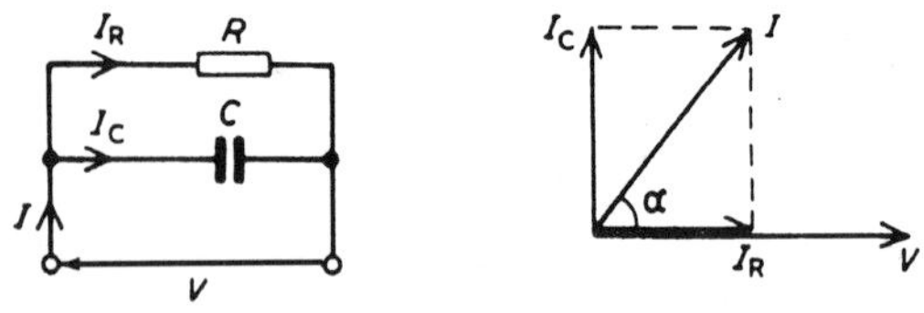

Figure 14.2

> **Problem 2.** A 30 μF capacitor is connected in parallel with an 80 Ω resistor across a 240 V, 50 Hz supply. Calculate (a) the current in each branch, (b) the supply current, (c) the circuit phase angle, (d) the circuit impedance, (e) the power dissipated, and (f) the apparent power.

The circuit and phasor diagrams are as shown in Fig. 14.2.

(a) Current in resistor,

$$I_R = \frac{V}{R} = \frac{240}{80} = \mathbf{3\ A}$$

Current in capacitor,

$$I_C = \frac{V}{X_C} = \frac{V}{\frac{1}{2\pi f C}} = 2\pi f C V$$
$$= 2\pi(50)(30 \times 10^{6})(240)$$
$$= \mathbf{2.262\ A}$$

(b) Supply current, $I = \sqrt{(I_R^2 + I_C^2)} = \sqrt{(3^2 + 2.262^2)} =$ **3.757 A**.

(c) Circuit phase angle, $\alpha = \arctan(I_C/I_R) = \arctan(2.262/3) =$ **37°1′ leading**.

(d) Circuit impedance, $Z = V/I = 240/3.757 =$ **63.88 Ω.**.

(e) True or active power dissipated,

$$P = VI \cos\alpha = (240)(3.757) \cos 37°1'$$
$$= \mathbf{720\ W}$$

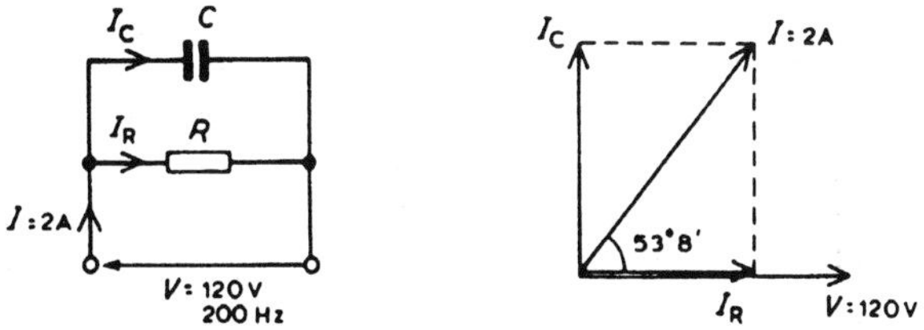

Figure 14.3

(Alternatively, true power $P = I_R^2R = (3)^2(80) = 720$ W.)

(f) Apparent power, $S = VI = (240)(3.757) =$ **901.7 VA**.

> Problem 3. A capacitor C is connected in parallel with a resistor R across a 120 V, 200 Hz supply. The supply current is 2 A at a power factor of 0.6 leading. Determine the values of C and R.

The circuit diagram is shown in Fig. 14.3(a).

Power factor = cos ϕ = 0.6 leading. Hence ϕ = arccos 0.6 = 53°8′ leading. From the phasor diagram shown in Fig. 14.3(b),

$$I_R = I \cos 53°8' = (2)(0.6) = \mathbf{1.2\ A}$$

and

$$I_C = I \sin 53°8' = (2)(0.8) = \mathbf{1.6\ A}$$

(Alternatively, I_R and I_C can be measured from the scaled phasor diagram.)

From the circuit diagram, $I_R = (V/R)$ from which, $R = (V/I_R) = 120/1.2 =$ **100 Ω**, and $I_C = (V/X_C) = 2\pi fCV$, from which

$$C = \frac{I_C}{2\pi fV} = \frac{1.6}{2\pi(200)(120)} = \mathbf{10.61\ \mu F}$$

14.4 *L–C* parallel a.c. circuit

In the two-branch parallel circuit containing inductance L and capacitance C shown in Fig. 14.4, I_L lags V by 90° and I_C leads V by 90°.

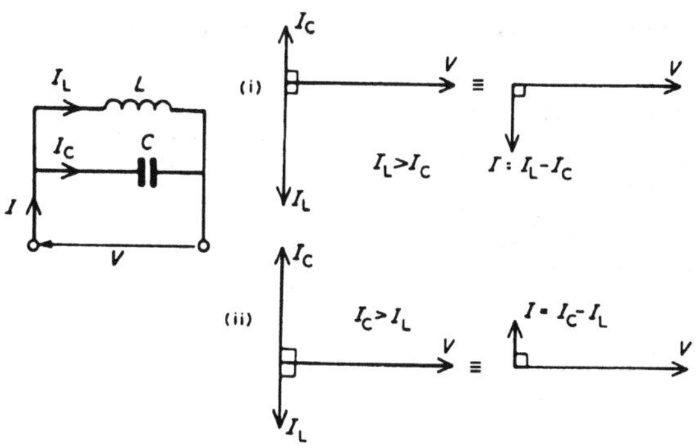

Figure 14.4

Theoretically there are three phasor diagrams possible – each depending on the relative values of I_L and I_C:

(i) $I_L > I_C$ (giving a supply current, $I = I_L - I_C$ lagging V by 90°).
(ii) $I_C > I_L$ (giving a supply current, $I = I_C - I_L$ leading V by 90°).
(iii) $I_L = I_C$ (giving a supply current, $I = 0$).

The latter condition is not possible in practice due to circuit resistance inevitably being present (as in the circuit described in section 14.5).

For the L–C parallel circuit,

$$I_L = \frac{V}{X_L},\ I_C = \frac{V}{X_C}$$

I = phasor difference between I_L and I_C, and $Z = V/I$.

> Problem 4. A pure inductance of 120 mH is connected in parallel with a 25 μF capacitor and the network is connected to a 100 V, 50 Hz supply. Determine (a) the branch currents, (b) the supply current and its phase angle, (c) the circuit impedance, and (d) the power consumed.

The circuit and phasor diagrams are as shown in Fig. 14.4.

(a) Inductive reactance, $X_L = 2\pi fL = 2\pi(50)(120 \times 10^{-3}) = 37.70\ \Omega$.
Capacitive reactance,

$$X_C = \frac{1}{2\pi fC} = \frac{1}{2\pi(50)(25 \times 10^{-6})} = 127.3\ \Omega$$

Current flowing in inductance,

$$I_L = \frac{V}{X_L} = \frac{100}{37.70} = \mathbf{2.653\ A}$$

Current flowing in capacitor,

$$I_C = \frac{V}{X_C} = \frac{100}{127.3} = \mathbf{0.786\ A}$$

(b) I_L and I_C are anti-phase. Hence supply current, $I = I_L - I_C = 2.653 - 0.786 =$ **1.867 A. The current *I* lags the supply voltage *V* by 90°** (see Fig. 14.4(i)).
(c) Circuit impedance, $Z = V/I = 100/1.867 =$ **53.56 Ω.**
(d) Power consumed, $P = VI \cos \phi = (100)(1.867)(\cos 90°) =$ **0 W.**

> Problem 5. Repeat Problem 4 for the condition when the frequency is changed to 150 Hz.

(a) Inductive reactance,

$$X_L = 2\pi(150)(120 \times 10^{-3}) = 113.1\ \Omega$$

Capacitive reactance,

$$X_C = \frac{1}{2\pi(150)(25 \times 10^{-6})} = 42.44\ \Omega$$

Current flowing in inductance,

$$I_L = \frac{V}{X_L} = \frac{100}{113.1} = \mathbf{0.884\ A}$$

Current flowing in capacitor,

$$I_C = \frac{V}{X_C} = \frac{100}{42.44} = \mathbf{2.356\ A}$$

(b) Supply current, $I = I_C - I_L = 2.356 - 0.884$ = **1.472 A leading *V* by 90°** (see Fig. 4(ii)).
(c) Circuit impedance, $Z = V/I = 100/1.472 =$ 67.93 Ω.
(d) Power consumed, $P = VI \cos \phi =$ **0 W** (since $\phi = 90°$).

From Problems 4 and 5:

(i) When $X_L < X_C$ then $I_L > I_C$ and I lags V by 90°.
(ii) When $X_L > X_C$ then $I_L < I_C$ and I leads V by 90°.
(iii) In a parallel circuit containing no resistance the power consumed is zero.

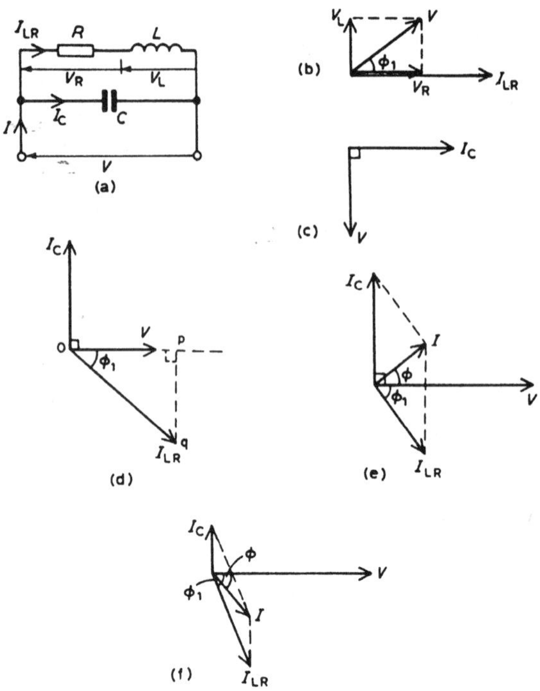

Figure 14.5

14.5 *LR–C* parallel a.c. circuit

In the two-branch circuit containing capacitance C in parallel with inductance L and resistance R in series (such as a coil) shown in Fig. 14.5(a), the phasor diagram for the LR branch alone is shown in Fig. 14.5(b) and the phasor diagram for the C branch alone is shown in Fig. 14.5(c). Rotating each and superimposing on one another gives the complete phasor diagram shown in Fig. 14.5(d).

The current I_{LR} of Fig. 14.5(d) may be resolved into horizontal and vertical components. The horizontal components, shown as op is $I_{LR} \cos \phi_1$ and the vertical component, shown as pq is $I_{LR} \sin \phi_1$. There are three possible conditions for this circuit:

(i) $I_C > I_{LR} \sin \phi_1$ (giving a supply current I leaving V by angle ϕ – as shown in Fig. 14.5(e)).
(ii) $I_{LR} \sin \phi > I_C$ (giving I lagging V by angle ϕ – as shown in Fig. 14.5(f)).
(iii) $I_C = I_{LR} \sin \phi_1$ (this is called **parallel resonance**, see section 14.6).

There are two methods of finding the phasor sum of currents I_{LR} and I_C in Figs 14.5(e) and (f).

These are: (i) by a scaled phasor diagram, or (ii) by resolving each current into their 'in-phase' (i.e. horizontal) and 'quadrature' (i.e. vertical) components, as demonstrated in Problems 6 and 7.

With reference to the phasor diagrams of Fig. 14.5: impedance of LR branch, $Z_{LR} = \sqrt{(R^2 + X_L^2)}$. Current, $I_{LR} = V/Z_{LR}$ and $I_C = V/X_C$.

Supply current I = phasor sum of I_{LR} and I_C (by drawing) and $\sqrt{\{(I_{LR}\cos\phi_1)^2 + (I_{LR}\sin\phi_1 \sim I_C)^2\}}$ (by calculation) where $\sim$ means 'the difference between'.

Circuit impedance $Z = V/I$; $\tan\phi_1 = (V_L/V_R) = (X_L/R)$; $\sin\phi_1 = (X_L/Z_{LR})$ and $\cos\phi_1 = R/Z_{LR}$.

$$\tan\phi = \frac{I_{LR}\sin\phi_1 \sim I_C}{I_{LR}\cos\phi_1}$$

$$\text{and } \cos\phi = \frac{I_{LR}\cos\phi_1}{I}$$

> Problem 6. A coil of inductance 159.2 mH and resistance 40 Ω is connected in parallel with a 30 μF capacitor across a 240 V, 50 Hz supply. Calculate (a) the current in the coil and its phase angle, (b) the current in the capacitor and its phase angle, (c) the supply current and its phase angle, (d) the circuit impedance, (e) the power consumed, (f) the apparent power, and (g) the reactive power. Draw the phasor diagram.

The circuit diagram is shown in Fig. 14.6(a).

(a) For the coil, inductive reactance $X_L = 2\pi fL = 2\pi(50)(159.2 \times 10^{-3}) = 50\ \Omega$. Impedance $Z_1 = \sqrt{(R^2 + X_L^2)} = \sqrt{(40^2 + 50^2)} = 64.03\ \Omega$. Current in coil, $I_{LR} = (V/Z_1) = 240/64.03 =$ **3.748 A**.
Branch phase angle $\phi_2 = \arctan(X_L/R) = \arctan(50/40) = \arctan 1.25 =$ **51°20′ lagging** (see phasor diagram in Fig. 14.6(b)).

(b) Capacitive reactance,

$$X_C = \frac{1}{2\pi fC} = \frac{1}{2\pi(50)(30 \times 10^{-6})} = 106.1\ \Omega$$

Current in capacitor,

$$I_C = \frac{V}{X_C} = \frac{240}{106.1}$$

= **2.262 A leading the supply voltage by 90°**

(see phasor diagram of Fig. 14.6(b)).

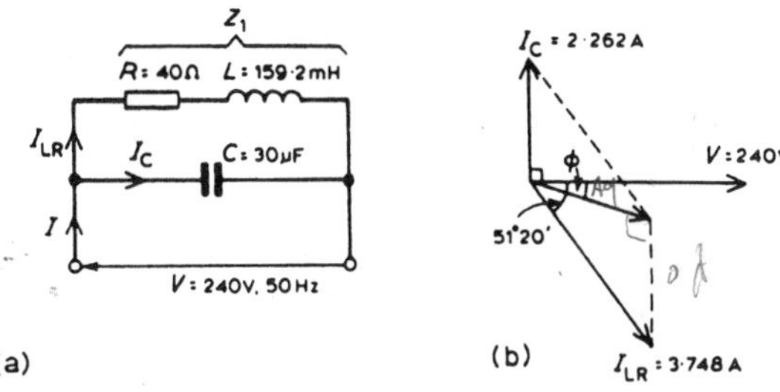

Figure 14.6

(c) The supply current I is the phasor sum of I_{LR} and I_C. This may be obtained by drawing the phasor diagram to scale and measuring the current I and its phase angle relative to V. (Current I will always be the diagonal of the parallelogram formed as in Fig. 14.6(b).) Alternatively the current I_{LR} and I_C may be resolved into their horizontal (or 'in-phase') and vertical (or 'quadranture') components. The horizontal component of I_{LR} is $I_{LR}\cos(51°20') = 3.748\cos 51°20' = 2.342$ A. The horizontal component of I_C is $I_C\cos 90° = 0$. Thus the total horizontal component, $I_H =$ **2.342 A**. The vertical component of $I_{LR} = -I_{LR}\sin(51°20') = -3.748\sin 51°20' = -2.926$ A. The vertical component of $I_C = I_C\sin 90° = 2.262\sin 90° = 2.262$ A. Thus the total vertical component, $I_V = -2.926 + 2.262 =$ **−0.664 A**. I_H and I_V are shown in Fig. 14.7, from which, $I = \sqrt{[(2.342)^2 + (-0.664)^2]} = 2.434$ A. Angle $\phi = \arctan(0.664/2.342) = 15°\ 50'$ lagging. **Hence the supply current I = 2.434 A lagging V by 15° 50′.**

(d) Circuit impedance, $Z = V/I = 240/2.434 =$ **98.60 Ω**.

(e) Power consumed, $P = VI\cos\phi = (240)(2.434)\cos 15°50' =$ **562 W**. (Alternatively, $P = I_R^2R = I_{LR}^2R$ (in this case) $= (3.748)^2(40) =$ **562 W**)

(f) Apparent power, $S = VI = (240)(2.434) =$ **584.2 VA**.

(g) Reactive power, $Q = VI\sin\phi = (240)(2.434)(\sin 15°50') =$ **159.4 var**.

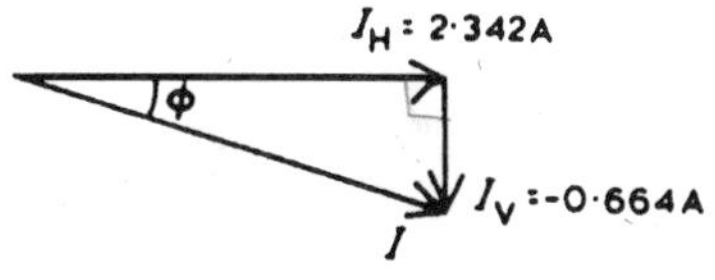

Figure 14.7

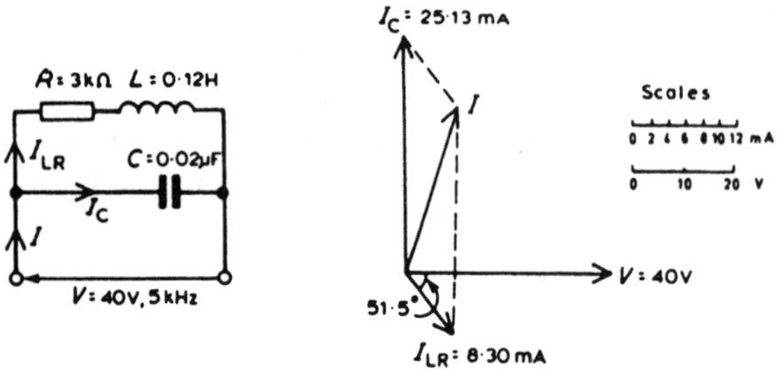

Figure 14.8

> Problem 7. A coil of inductance 0.12 H and resistance 3 kΩ is connected in parallel with a 0.02 μF capacitor and is supplied at 40 V at a frequency of 5 kHz. Determine (a) the current in the coil, and (b) the current in the capacitor. (c) Draw to scale the phasor diagram and measure the supply current and its phase angle. Calculate (d) the circuit impedance and (e) the power consumed.

The circuit diagram is shown in Fig. 14.8(a).

(a) Inductive reactance, $X_L = 2\pi fL = 2(5000)(0.12) = 3770\ \Omega$. Impedance of coil, $Z_1 = \sqrt{(R^2 + X_L^2)} = \sqrt{[(3000)^2 + (3770)^2]} = 4818\ \Omega$. Current in coil, $I_{LR} = V/Z_1 = 40/4818 =$ **8.30 mA**.
Branch phase angle $\phi = \arctan X_L/R = \arctan(3770/3000) =$ **51.5° lagging**.

(b) Capacitive reactance,

$$X_C = \frac{1}{2\pi fC} = \frac{1}{2\pi(5000)(0.02 \times 10^{-6})}$$

$$= 1592\ \Omega$$

Capacitor current,

$$I_C = \frac{V}{X_C} = \frac{40}{1592}$$

$$= \mathbf{25.13\ mA\ leading\ } \boldsymbol{V} \mathbf{\ by\ 90°}$$

(c) Currents I_{LR} and I_C are shown in the phasor diagram of Fig. 14.8(b). The parallelogram is completed as shown and the supply current is given by the diagonal of the parallelogram. The current I is measured as **19.3 mA** leading voltage V by **74.5°**.
(By calculation, $I = \sqrt{[(I_{LR}\cos 51.5°)^2 + (I_C - I_{LR}\sin 51.5°)^2]} = 19.34$ A and

$$\phi = \arctan\left(\frac{I_C - I_{LR}\sin 51.5°}{I_{LR}\cos 51.5°}\right) = 74.5°)$$

(d) Circuit impedance,

$$Z = \frac{V}{I} = \frac{40}{19.3 \times 10^{-3}} = \mathbf{2.073\ k\Omega}$$

(e) Power consumed, $P = VI\cos\phi = (40)(19.3 \times 10^{-3})(\cos 74.5°) =$ **206.3 mW**.
(Alternatively, $P = I_R^2 R = I_{LR}^2 R = (8.30 \times 10^{-3})^2(3000) = 206.7$ mW.)

14.6 Parallel resonance and Q-factor

Parallel resonance

Resonance occurs in the two-branch circuit containing capacitance C in parallel with inductance L and resistance R in series (see Fig. 14.5(a)) when the quadrature (i.e. vertical) component of current I_{LR} is equal to I_C. At this condition the supply current I is in phase with the supply voltage V.

Resonant frequency

When the quadrature component of I_{LR} is equal to I_C then: $I_C = I_{LR}\sin\phi_1$ (see Fig. 14.9). Hence

$$\frac{V}{X_C} = \left(\frac{V}{Z_{LR}}\right)\left(\frac{X_L}{Z_{LR}}\right) \text{ (from section 14.5)}$$

from which,

$$Z_{LR}^2 = X_C X_L = (2\pi f_r L)\left(\frac{1}{2\pi f_r C}\right) = \frac{L}{C} \quad (14.1)$$

Hence

$$[\sqrt{(R^2 + X_L^2)}]^2 = \frac{L}{C} \text{ and } R^2 + X_L^2 = \frac{L}{C}$$

Figure 14.9

$$(2\pi f_r L)^2 = \frac{L}{C} - R^2$$

$$f_r = \frac{1}{2\pi L}\sqrt{\left(\frac{L}{C} = R^2\right)} = \frac{1}{2\pi}\sqrt{\left(\frac{L}{L^2C} - \frac{R^2}{L^2}\right)}$$

i.e. parallel resonant frequency,

$$\boldsymbol{f_r = \frac{1}{2\pi}\sqrt{\left(\frac{1}{LC} - \frac{R^2}{L^2}\right)}\ \text{Hz}}$$

(When R is negligible, then

$$f_r = \frac{1}{2\pi\sqrt{(LC)}}$$

(which is the same as for series resonance).

Current at resonance

Current at resonance,

$$I_r = I_{LR}\cos\phi_1 \text{ (from Fig. 14.9)}$$

$$= \left(\frac{V}{Z_{LR}}\right)\left(\frac{R}{Z_{LR}}\right) \text{ (from section 14.5)}$$

$$= \frac{VR}{Z_{LR}^{\ 2}}$$

However, from equation (14.1), $Z_{LR}^2 = L/C$, hence

$$\boldsymbol{I_r = \frac{VR}{\frac{L}{C}} = \frac{VRC}{L}}$$

The current is at a **minimum** at resonance.

Dynamic resistance

Since the current at resonance is in phase with the voltage the impedance of the circuit acts as a resistance. This resistance is known as the **dynamic resistance**, $\boldsymbol{R_D}$ (or, sometimes, the dynamic impedance).

From equation (14.2),

impedance at resonance

$$= \frac{V}{I_r} = \left(\frac{V}{\frac{VRC}{L}}\right) = \frac{L}{RC}$$

i.e.

dynamic resistance, $\boldsymbol{R_D = \frac{L}{RC}}$ **ohms**

Rejector circuit

The parallel resonant circuit is often described as a **rejector** circuit since it presents its maximum impedance at the resonant frequency and the resultant current is a minimum.

Q-factor

Currents higher than the supply current can circulate within the parallel branches of a parallel resonant circuit, the current leaving the capacitor and establishing the magnetic field of the inductor, this then collapsing and recharging the capacitor, and so on. The **Q-factor** of a parallel resonant circuit is the ratio of the current circulating in the parallel branches of the circuit to the supply current, i.e. the current magnification.

Q-factor at resonance

= current magnification

$$= \frac{\text{circulating current}}{\text{supply current}}$$

$$= \frac{I_C}{I_r} = \frac{I_{LR}\sin\phi_1}{I_r}$$

$$= \frac{I_{LR}\sin\phi_1}{I_{LR}\cos\phi_1} = \frac{\sin\phi_1}{\cos\phi_1} = \tan\phi_1 = \frac{X_L}{R}$$

i.e.

Q-factor at resonance = $\boldsymbol{\frac{2\pi f_r L}{R}}$

(which is the same as for a series circuit).

Note that in a **parallel** circuit the Q-factor is a measure of **current magnification**, whereas in a **series** circuit it is a measure of **voltage magnification**.

At mains frequencies the Q-factor of a parallel circuit is usually low, typically less than 10, but in radio-frequency circuits the Q-factor can be very high.

Problem 8. A pure inductance of 150 mH is connected in parallel with a 40 μF capacitor across a 50 V, variable frequency supply. Determine (a) the resonant frequency of the circuit and (b) the current circulating in the capacitor and inductance at resonance.

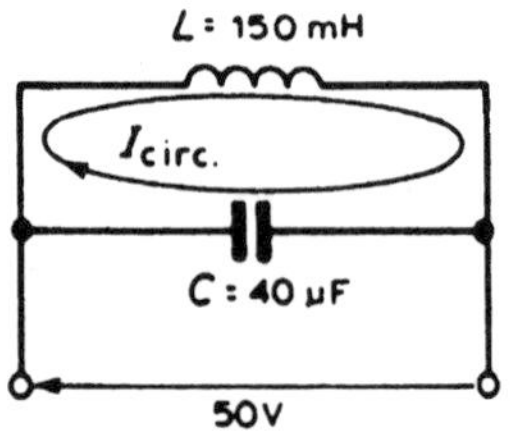

Figure 14.10

The circuit diagram is shown in Fig. 14.10.

(a) Parallel resonant frequency is given by

$$f_r = \frac{1}{2\pi}\sqrt{\left(\frac{1}{LC} - \frac{R^2}{L^2}\right)}$$

However, resistance $R = 0$. Hence,

$$f_r = \frac{1}{2\pi}\sqrt{\left(\frac{1}{LC}\right)}$$

$$= \frac{1}{2\pi}\sqrt{\left(\frac{1}{(150 \times 10^{-3})(40 \times 10^{-6})}\right)}$$

$$= \frac{1}{2\pi}\sqrt{\left(\frac{10^7}{(15)(4)}\right)} = \frac{10^3}{2\pi}\sqrt{\left(\frac{1}{6}\right)} = \mathbf{64.97\ Hz}$$

(b) Current circulating in L and C at resonance,

$$I_{CIRC} = \frac{V}{X_C} = \frac{V}{\frac{1}{2\pi f_r C}} = 2\pi f_r CV$$

Hence $I_{CIRC} = 2\pi(64.97)(40 \times 10^{-6})(50) =$ **0.816 A**.

(Alternatively,

$$I_{CIRC} = \frac{V}{X_L} = \frac{V}{2\pi f_r L} = \frac{50}{2\pi(64.97)(0.15)}$$

$$= \mathbf{0.817\ A})$$

Problem 9. A coil of inductance 0.20 H and resistance 60 Ω is connected in parallel with a 20 µF capacitor across a 20 V, variable frequency supply. Calculate (a) the resonant frequency, (b) the dynamic resistance, (c) the current at resonance, and (d) the circuit Q-factor at resonance.

(a) Parallel resonant frequency,

$$f_r = \frac{1}{2\pi}\sqrt{\left(\frac{1}{LC} - \frac{R^2}{L^2}\right)}$$

$$= \frac{1}{2\pi}\sqrt{\left(\frac{1}{(0.20)(20 \times 10^{-6})} - \frac{(60)^2}{(0.2)^2}\right)}$$

$$= \frac{1}{2\pi}\sqrt{(250\,000 - 90\,000)} = \frac{1}{2\pi}\sqrt{(160\,000)}$$

$$= \frac{1}{2\pi}(400)$$

$$= \mathbf{63.66\ Hz}$$

(b) Dynamic resistance,

$$R_D = \frac{L}{RC} = \frac{0.2}{(60)(20 \times 10^{-6})} = \mathbf{166.7\ \Omega}$$

(c) Current at resonance,

$$I_r = \frac{V}{R_D} = \frac{20}{166.7} = \mathbf{0.12\ A}$$

(d) Circuit Q-factor at resonance

$$= \frac{2\pi f_r L}{R} = \frac{2\pi(63.66)(0.2)}{60} = \mathbf{1.33}$$

Alternatively, Q-factor at resonance = current magnification (for a parallel circuit) $= I_c/I_r$.

$$I_c = \frac{V}{X_c} = \frac{V}{\left(\frac{1}{2\pi f_r C}\right)} = 2\pi f_r CV$$

$$= 2\pi(63.66)(20 \times 10^{-6})(20) = 0.16\ \text{A}$$

Hence Q-factor $= I_c/I_r = 0.16/0.12 =$ **1.33**, as obtained above.

Problem 10. A coil of inductance 100 mH and resistance 800 Ω is connected in parallel with a variable capacitor across a 12 V, 5 kHz supply. Determine for the condition when the supply current is a minimum: (a) the capacitance of the capacitor, (b) the dynamic resistance, (c) the supply current, and (d) the Q-factor.

(a) The supply current is a minimum when the parallel circuit is at resonance. Resonant frequency,

$$f_r = \frac{1}{2\pi}\sqrt{\left(\frac{1}{LC} - \frac{R^2}{L^2}\right)}$$

Transposing for C gives:

$$(2\pi f_r)^2 = \frac{1}{LC} - \frac{R^2}{L^2}$$

$$(2\pi f_r)^2 + \frac{R^2}{L^2} = \frac{1}{LC}$$

$$C = \frac{1}{L\left\{(2\pi f_r)^2 + \dfrac{R^2}{L^2}\right\}}$$

When $L = 100$ mH, $R = 800\ \Omega$ and $f_r = 5000$ Hz,

$$C = \frac{1}{100 \times 10^{-3}\left\{(2\pi(5000)^2 + \dfrac{800^2}{(100 \times 10^{-3})^2}\right\}}$$

$$= \frac{1}{0.1[\pi^2 10^8 + (0.64)10^8]}\ \text{F}$$

$$= \frac{10^6}{0.1(10.51 \times 10^8)}\ \mu\text{F}$$

$= \mathbf{0.009\ 515\ \mu F}$ or $\mathbf{9.515\ nF}$

(b) Dynamic resistance,

$$R_D = \frac{L}{CR} = \frac{100 \times 10^{-3}}{(9.515 \times 10^{-9})(800)} = \mathbf{13.14\ k\Omega}$$

(c) Supply current at resonance,

$$I_r = \frac{V}{R_D} = \frac{12}{13.14 \times 10^3} = \mathbf{0.913\ mA}$$

(d) Q-factor at resonance $= 2\pi f_r L/R = 2\pi(5000)(100 \times 10^{-3})/800 = \mathbf{3.93}$.
Alternatively, Q-factor at resonance is given by

$$\frac{I_c}{I_r} = \frac{V/X_c}{I_r} = \frac{2\pi f_r CV}{I_r}$$

$$= \frac{2\pi(5000)(9.515 \times 10^{-9})(12)}{0.913 \times 10^{-3}} = \mathbf{3.93}$$

14.7 Power factor improvement

For a particular power supplied, a high power factor reduces the current flowing in a supply system and therefore reduces the cost of cables, switchgear, transformers and generators. Supply authorities use tariffs which encourage electricity consumers to operate at a reasonably high power factor.

Industrial loads such as a.c. motors are essentially inductive (R–L) and may have a low power factor. One method of improving (or correcting) the power factor of an inductive load is to connect a static capacitor C in parallel with the load (see Fig. 14.11(a)). The supply current is reduced from I_{LR} to I, the phasor sum of I_{LR} and I_C, and the circuit power factor improves from $\cos\phi_1$ to $\cos\phi_2$ (see Fig. 14.11(b)).

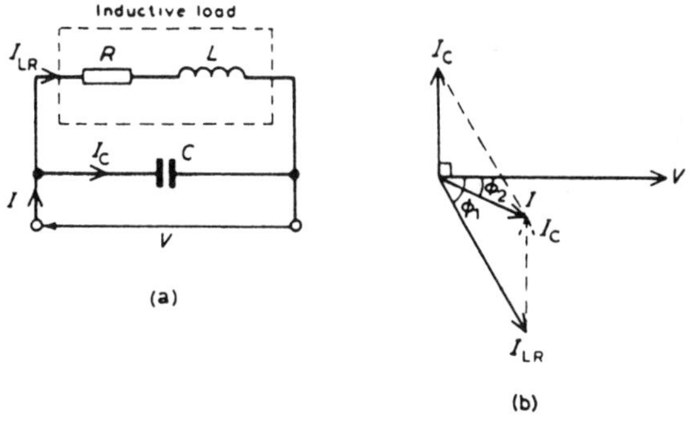

Figure 14.11

> Problem 11. A single-phase motor takes 50 A at a power factor of 0.6 lagging from a 240 V, 50 Hz supply. Determine (a) the current taken by a capacitor connected in parallel with the motor to correct the power factor to unity, and (b) the value of the supply current after power factor correction.

The circuit diagram is shown in Fig. 14.12(a).

(a) A power factor of 0.6 lagging means that $\cos\phi = 0.6$, i.e. $\phi = \arccos 0.6 = 53°\ 8'$. Hence I_M lags V by 53°8′ as shown in Fig. 14.12(b).
If the power factor is to be improved to unity then the phase difference between supply current I and voltage V is 0°, i.e. I is in phase with V as shown in Fig. 14.12(c). For this to be so, I_C must equal the length ab, such that the phasor sum of I_M and I_C is I.

$$ab = I_M \sin 53°\ 8' = 50(0.8) = 40\ \text{A}$$

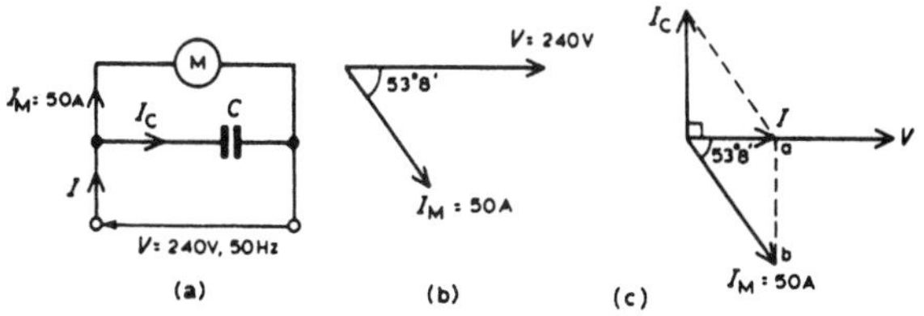

Figure 14.12

Hence the capacitor current I_c must be 40 A for the power factor to be unity.

(b) Supply current $I = I_M \cos 53°8' = 50(0.6) =$ **30 A**.

> Problem 12. A 400 V alternator is supplying a load of 42 kW at a power factor of 0.7 lagging. Calculate (a) the kVA loading and (b) the current taken from the alternator. (c) If the power factor is now raised to unity find the new kVA loading.

(a) Power $= VI\cos\phi = (VI)$(power factor). Hence

$$VI = \frac{\text{power}}{\text{p.f.}} = \frac{42 \times 10^3}{0.7} = \mathbf{60\ kVA}$$

(b) $VI = 60\,000$ VA. Hence $I = (60\,000/V) = 60\,000/400 =$ **150 A.**

(c) The kVA loading remains at **60 kVA** irrespective of changes in power factor.

> Problem 13. A motor has an output of 4.8 kW, an efficiency of 80% and a power factor of 0.625 lagging when operated from a 240 V, 50 Hz supply. It is required to improve the power factor to 0.95 lagging by connecting a capacitor in parallel with the motor. Determine (a) the current taken by the motor, (b) the supply current after power factor correction, (c) the current taken by the capacitor, (d) the capacitance of the capacitor, and (e) the kvar rating of the capacitor.

(a) Efficiency = (power output)/(power input). Hence

$$\frac{80}{100} = \frac{4800}{\text{power input}}$$

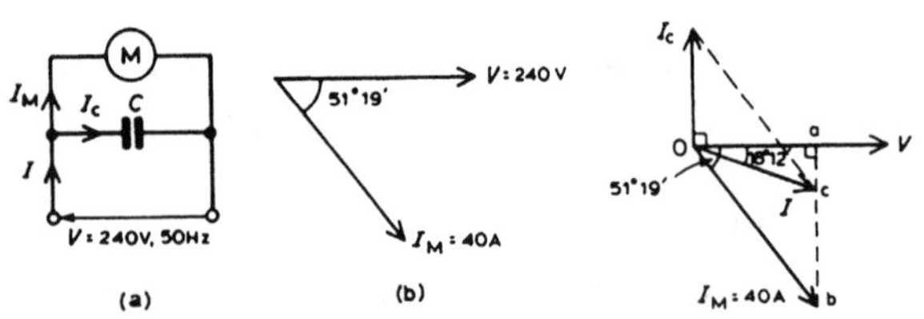

Figure 14.13

Power input $= 4800/0.8 = 6000$ W. Hence, $6000 = VI_M \cos\phi = (240)(I_M)(0.625)$, since $\cos\phi = \text{p.f.} = 0.625$.
Thus current taken by the motor,

$$I_M = \frac{6000}{(240)(0.625)} = \mathbf{40\ A}$$

The circuit diagram is shown in Fig. 14.13(a). The phase angle between I_M and V is given by $\phi = \arccos 0.625 = 51°19'$, hence the phasor diagram is as shown in Fig. 14.13(b).

(b) When a capacitor C is connected in parallel with the motor a current I_C flows which leads V by 90°. The phasor sum of I_M and I_C gives the supply current I, and has to be such as to change the circuit power factor to 0.95 lagging, i.e. a phase angle of arccos 0.95 or 18°12′ lagging, as shown in Fig. 14.13(c). The horizontal component of I_M (shown as oa)

$= I_M \cos 51°19'$

$= 40 \cos 51°19' = 25$ A

The horizontal component of I (also given by oa)

$= I \cos 18°12'$

$= 0.95\, I$

Equating the horizontal components gives: $25 = 0.95I$. Hence the supply current after p.f. correction,

$$I = \frac{25}{0.95} = \mathbf{26.32\ A}$$

(c) The vertical component of I_M (shown as ab)

$= I_M \sin 51°19'$

$= 40 \sin 51°19' = 31.22$ A

The vertical component of I (shown as ac)

$= I \sin 18°12'$

$= 26.32 \sin 18°12' = 8.22$ A

The magnitude of the capacitor current I_C (shown as bc) is given by ab − ac, i.e. $31.22 - 8.22 =$ **23 A.**

(d) Current $I_C = \dfrac{V}{X_c} = \dfrac{V}{\left(\dfrac{1}{2\pi fC}\right)} = 2\pi fCV$,

from which,

$$C = \frac{I_C}{2\pi fV} = \frac{23}{2\pi(50)(240)}\ \text{F} = \mathbf{305\ \mu F}$$

(e) kvar rating of the capacitor is given by

$$\frac{VI_c}{1000} = \frac{(240)(23)}{1000} = \mathbf{5.52\ kvar}$$

In this problem the supply current has been reduced from 40 A to 26.32 A without altering the current or power taken by the motor. This means that the size of generating plant and the cross-sectional area of conductors supplying both the factory and the motor can be less – with an obvious saving in cost.

Problem 14. A 250 V, 50 Hz single-phase supply feeds the following loads:

(i) incandescent lamps taking a current of 10 A at unity power factor;
(ii) fluorescent lamps taking 8 A at a power factor of 0.7 lagging;
(iii) a 3 kVA motor operating at full load and at a power factor of 0.8 lagging, and
(iv) a static capacitor.

Determine, for the lamps and motor, (a) the total current, (b) the overall power factor, and (c) the total power. (d) Find the value of the static capacitor to improve the overall power factor to 0.975 lagging.

A phasor diagram is constructed as shown in Fig. 14.14(a), where 8 A is lagging voltage V by arccos 0.7, i.e. 45.57°, and the motor current is 3000/250, i.e. 12 A lagging V by arccos 0.8, i.e. 36.87°.

(a) The horizontal component of the currents

$$= 10 \cos 0° + 12 \cos 36.87° + 8 \cos 45.57°$$

$$= 10 + 9.6 + 5.6 = 25.2 \text{ A}$$

The vertical component of the currents

$$= 10 \sin 0° + 12 \sin 36.87° + 8 \sin 45.57°$$

$$= 0 + 7.2 + 5.713 = 12.91 \text{ A}$$

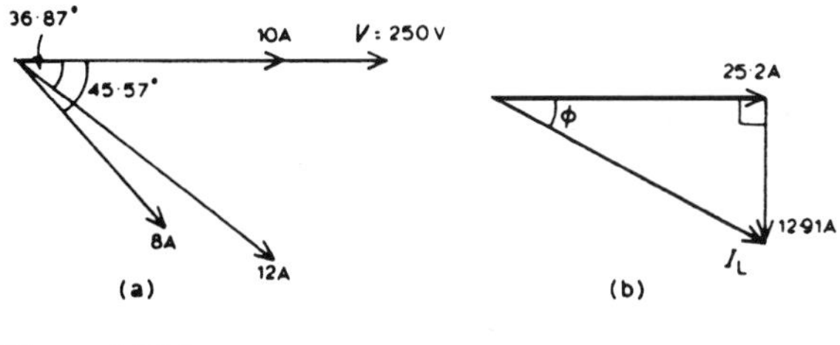

Figure 14.14

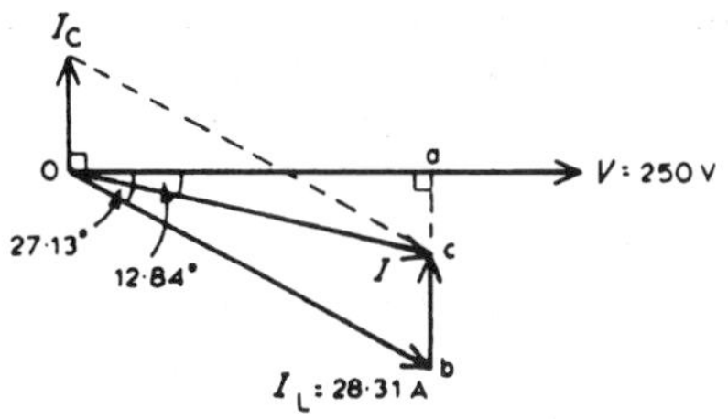

Figure 14.15

From Fig. 14.14(b), total current, $I_L = \sqrt{[(25.2)^2 + (12.91)^2]}$ = **28.31 A** at a phase angle of ϕ = arctan (12.91/25.2), i.e. 27.13° lagging.

(b) Power factor = cos ϕ = cos 27.13° = **0.890 lagging**.

(c) Total power, $P = VI_L \cos \phi$ = (250)(28.31) (0.890) = **6.3 kW**.

(d) To improve the power factor, a capacitor is connected in parallel with the loads. The capacitor takes a current I_C such that the supply current falls from 28.31 A to I, lagging V by arccos 0.975, i.e. 12.84°. The phasor diagram is shown in Fig. 14.15.

oa = 28.31 cos 27.13° = I cos 12.84°. Hence

$$I = \frac{28.31 \cos 27.13°}{\cos 12.84°} = 25.84 \text{ A}$$

Current I_C = bc = (ab – ac) = 28.31 sin 27.13° – 25.84 sin 12.84° = 12.91 – 5.742 = 7.168 A.

$$I_C = \frac{V}{X_C} = \frac{V}{\left(\frac{1}{2\pi fC}\right)} = 2\pi fCV$$

Hence

$$\text{capacitance } C = \frac{I_C}{2\pi fV} = \frac{7.168}{2\pi(50)(250)} \text{ F}$$

$$= \mathbf{91.27\ \mu F}$$

Thus to improve the power factor from 0.890 to 0.975 lagging, a 91.27 μF capacitor is connected in parallel with the loads.

14.8 Multi-choice questions on single-phase parallel a.c. circuits

(Answers on page 284.)

A two-branch parallel circuit containing a 10 Ω resistance in one branch and a 100 μF capacitor in the other, has a 120 V, 2/3π kHz supply

connected across it. Determine the quantities stated in questions 1 to 8, selecting the correct answer from the following list.
(a) 24 A (b) 6 Ω (c) 7.5 kΩ (d) 12 A
(e) arctan 3/4 leading (f) 0.8 leading (g) 7.5 Ω
(h) arctan 4/3 leading (i) 16 A
(j) arctan 5/3 lagging (k) 1.44 kW (l) 0.6 leading
(m) 22.5 Ω (n) 2.4 kW (o) arctan 4/3 lagging
(p) 0.6 lagging (q) 0.8 lagging (r) 1.92 kW
(s) 20 A

1. The current flowing in the resistance.
2. The capacitive reactance of the capacitor.
3. The current flowing in the capacitor.
4. The supply current.
5. The supply phase angle.
6. The circuit impedance.
7. The power consumed by the circuit.
8. The power factor of the circuit.
9. A two-branch parallel circuit consists of a 15 mH inductance in one branch and a 50 μF capacitor in the other across a 120 V, 1/π kHz supply. The supply current is:
   (a) 8 A leading by π/2 rad;
   (b) 16 A lagging by 90°;
   (c) 8 A lagging by 90°;
   (d) 16 A leading by π/2 rad.

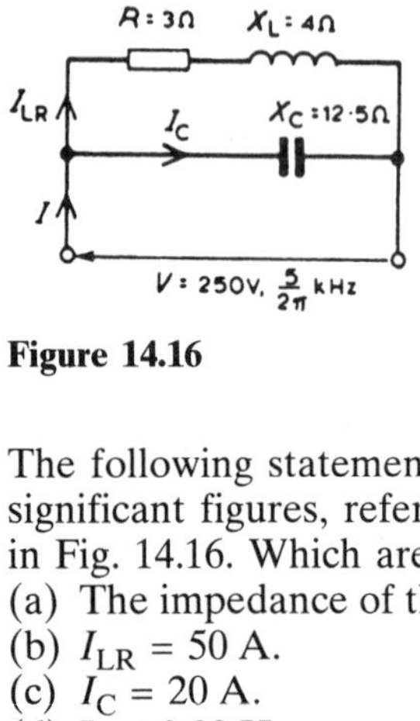

Figure 14.16

10. The following statements, taken correct to 2 significant figures, refer to the circuit shown in Fig. 14.16. Which are false?
    (a) The impedance of the R–L branch is 5 Ω.
    (b) I_{LR} = 50 A.
    (c) I_C = 20 A.
    (d) L = 0.80 H.
    (e) C = 16 μF.
    (f) The 'in-phase' component of the supply current is 30 A.
    (g) The 'quadrature' component of the supply current is 40 A.
    (h) I = 36 A.
    (i) Circuit phase angle = 33°41′ leading.
    (j) Circuit impedance = 6.9 Ω.
    (k) Circuit power factor = 0.83 lagging.
    (l) Power consumed = 9.0 kW.
11. Which of the following statements is false?
    (a) The supply current is a minimum at resonance in a parallel circuit.
    (b) The Q-factor at resonance in a parallel circuit is the voltage magnification.
    (c) Improving power factor reduces the current flowing through a system.
    (d) The circuit impedance is a maximum at resonance in a parallel circuit.
12. An LR–C parallel circuit has the following component values:
    R = 10 Ω, L = 10 mH, C = 10 μF, V = 100 V.
    Which of the following statements is false?
    (a) The resonant frequency f_r is 1.5/π kHz.
    (b) The current at resonance is 1 A.
    (c) The dynamic resistance is 100 Ω.
    (d) The circuit Q-factor at resonance is 30.

14.9 Short answer questions on single-phase parallel a.c. circuits

1. Draw a phasor diagram for a two-branch parallel circuit containing capacitance C in one branch and resistance R in the other, connected across a supply voltage V.
2. Draw the phasor diagram for a two-branch parallel circuit containing inductance L and resistance R in series in one branch, and capacitance C in the other, connected across a supply voltage V.
3. Draw the phasor diagram for a two-branch parallel circuit containing inductance L in one branch and capacitance C in the other for the condition in which inductive reactance is greater than capacitive reactance.
4. State two methods of determining the phasor sum of two currents.
5. State two formulae which may be used to calculate power in a parallel a.c. circuit.
6. State the condition for resonance for a two-branch circuit containing capacitance C in parallel with a coil of inductance L and resistance R.
7. Develop a formula for parallel resonance in terms of resistance R, inductance L and capacitance C.
8. What does the Q-factor of a parallel circuit mean?
9. Develop a formula for the current at resonance in an LR–C parallel circuit in terms of resistance R, inductance L, capacitance C and supply voltage V.

10. What is dynamic resistance? State a formula for dynamic resistance.
11. Explain a simple method of improving the power factor of an inductive circuit.
12. Why is it advantageous to improve power factor?

14.10 Further questions on single-phase parallel a.c. circuits

R–L parallel a.c. circuit

1. A 30 Ω resistor is connected in parallel with a pure inductance of 3 mH across a 110 V, 2 kHz supply. Calculate (a) the current in each branch, (b) the circuit current, (c) the circuit phase angle, (d) the circuit impedance, (e) the power consumed, and (f) the circuit power factor.
[(a) $I_R = 3.67$ A, $I_L = 2.92$ A (b) 4.69 A (c) 38°30′ lagging (d) 23.45 Ω (e) 404 W (f) 0.783 lagging]
2. A 40 Ω resistance is connected in parallel with a coil of inductance L and negligible resistance across a 200 V, 50 Hz supply and the supply current is found to be 8 A. Draw a phasor diagram to scale and determine the inductance of the coil. [102 mH]

R–C parallel a.c. circuit

3. A 1500 nF capacitor is connected in parallel with a 16 Ω resistor across a 10 V, 10 kHz supply. Calculate (a) the current in each branch, (b) the supply current, (c) the circuit phase angle, (d) the circuit impedance, (e) the power consumed, (f) the apparent power, and (g) the circuit power factor. Draw the phasor diagram.
[(a) $I_R = 0.625$ A, $I_C = 0.942$ A (b) 1.13 A (c) 56°28′ leading (d) 8.85 Ω (e) 6.25 W (f) 11.3 VA (g) 0.55 leading]
4. A capacitor C is connected in parallel with a resistance R across a 60 V, 100 Hz supply. The supply current is 0.6 A at a power factor of 0.8 leading. Calculate the value of R and C. [$R = 125$ Ω, $C = 9.55$ μF]

L–C parallel a.c. circuit

5. An inductance of 80 mH is connected in parallel with a capacitance of 10 μF across a 60 V, 100 Hz supply. Determine (a) the branch currents, (b) the supply current, (c) the circuit phase angle, (d) the circuit impedance, and (e) the power consumed.
[(a) $I_C = 0.377$ A, $I_L = 1.194$ A (b) 0.817 A (c) 90° lagging (d) 73.44 Ω (e) 0 W]
6. Repeat question 5 for a supply frequency of 200 Hz.
[(a) $I_C = 0.754$ A, $I_L = 0.597$ A (b) 0.157 A (c) 90° leading (d) 382.2 Ω (e) 0 W]

LR–C parallel a.c. circuit

7. A coil of resistance 60 Ω and inductance 318.4 mH is connected in parallel with a 15 μF capacitor across a 200 V, 50 Hz supply. Calculate (a) the current in the coil, (b) the current in the capacitor, (c) the supply current and its phase angle, (d) the circuit impedance, (e) the power consumed, (f) the apparent power and (g) the reactive power. Draw the phasor diagram.
[(a) 1.715 A (b) 0.943 A (c) 1.028 A at 30° 53′ lagging (d) 194.6 Ω (e) 176.4 W (f) 205.6 VA (g) 105.5 var]
8. A 25 nF capacitor is connected in parallel with a coil of resistance 2 kΩ and inductance 0.20 H across a 100 V, 4 kHz supply. Determine (a) the current in the coil, (b) the current in the capacitor, (c) the supply current and its phase angle (by drawing a phasor diagram to scale, and also by calculation), (d) the circuit impedance, and (e) the power consumed.
[(a) 18.48 mA (b) 62.83 mA (c) 46.17 mA at 81°29′ leading (d) 2.166 kΩ (e) 0.684 W]

Parallel resonance and Q-factor

9. A 0.15 μF capacitor and a pure inductance of 0.01 H are connected in parallel across a 10 V, variable frequency supply. Determine (a) the resonant frequency of the circuit, and (b) the current circulating in the capacitor and inductance. [(a) 4.11 kHz (b) 38.74 mA]
10. A 30 μF capacitor is connected in parallel with a coil of inductance 50 mH and unknown resistance R across a 120 V, 50 Hz supply. If the circuit has an overall power factor of 1 find (a) the value of R, (b) the current in the coil, and (c) the supply current.
[(a) 37.7 Ω (b) 2.94 A (c) 2.714 A]
11. A coil of resistance 25 Ω and inductance 150 mH is connected in parallel with a 10 μF capacitor across a 60 V, variable frequency

supply. Calculate (a) the resonant frequency, (b) the dynamic resistance, (c) the current at resonance, and (d) the Q-factor at resonance.
[(a) 127.2 Hz (b) 600 Ω (c) 0.10 A (d) 4.80]

12. A coil having resistance R and inductance 80 mH is connected in parallel with a 5 nF capacitor across a 25 V, 3 kHz supply. Determine for the condition when the current is a minimum, (a) the resistance R of the coil, (b) the dynamic resistance, (c) the supply current, and (d) the Q-factor.
[(a) 3.705 kΩ (b) 4.318 kΩ (c) 5.79 mA (d) 0.40]

Power factor improvement

13. A 415 V alternator is supplying a load of 55 kW at a power factor of 0.65 lagging. Calculate (a) the kVA loading and (b) the current taken from the alternator. (c) If the power factor is now raised to unity find the new kVA loading.
[(a) 84.6 kVA (b) 203.9 A (c) 84.6 kVA]
14. A coil of resistance 1.5 kΩ and 0.25 H inductance is connected in parallel with a variable capacitance across a 10 V, 8 kHz supply. Calculate (a) the capacitance of the capacitor when the supply current is a minimum, (b) the dynamic resistance, and (c) the supply current.
[(a) 1561 pF (b) 106.8 kΩ (c) 93.66 μA]
15. A single phase motor takes 30 A at a power factor of 0.65 lagging from a 240 V, 50 Hz supply. Determine (a) the current taken by the capacitor connected in parallel to correct the power factor to unity, and (b) the value of the supply current after power factor correction.
[(a) 22.80 A (b) 19.5 A]
16. A motor has an output of 6 kW, an efficiency of 75% and a power factor of 0.64 lagging when operated from a 250 V, 60 Hz supply. It is required to raise the power factor to 0.925 lagging by connecting a capacitor in parallel with the motor. Determine (a) the current taken by the motor, (b) the supply current after power factor correction, (c) the current taken by the capacitor, (d) the capacitance of the capacitor, and (e) the kvar rating of the capacitor.
[(a) 50 A (b) 34.59 A (c) 25.28 A (d) 268.2 μF (e) 6.32 kvar]
17. A supply of 250 V, 80 Hz is connected across an inductive load and the power consumed is 2 kW, when the supply current is 10 A. Determine the resistance and inductance of the circuit. What value of capacitance connected in parallel with the load is needed to improve the overall power factor to unity?
[R = 20 Ω, L = 29.84 mH, C = 47.75 μF]
18. A 200 V, 50 Hz single-phase supply feeds the following loads: (i) fluorescent lamps taking a current of 8 A at a power factor of 0.9 leading; (ii) incandescent lamps taking a current of 6 A at unity power factor; (iii) a motor taking a current of 12 A at a power factor of 0.65 lagging. Determine the total current taken from the supply and the overall power factor. Find also the value of a static capacitor connected in parallel with the loads to improve the overall power factor to 0.98 lagging.
[21.74 A, 0.966 lagging, 21.68 μF]

15

D.c. transients

At the end of this chapter you should be able to:

- understand the term 'transient'
- describe the transient response of capacitor and resistor voltage and current in a series *C–R* d.c. circuit
- define the term 'time constant'
- calculate time constant in a *C–R* circuit
- draw transient growth and decay curves for a *C–R* circuit
- use equations $v_C = V(1 - e^{-t/\tau})$, $v_R = Ve^{-t/\tau}$ and $i = Ie^{-t/\tau}$ for a *CR* circuit
- describe the transient response when discharging a capacitor
- describe the transient response of inductor and resistor voltage and current in a series *L–R* d.c. circuit
- calculate time constant in an *L–R* circuit
- draw transient growth and decay curves for an *L–R* circuit
- use equations $v_L = Ve^{-t/\tau}$, $v_R = V(1 - e^{-t/\tau})$ and $i = I(1 - e^{-t/\tau})$
- describe the transient response for current decay in an *L–R* circuit
- understand the switching of inductive circuits
- describe the effects of time constant on a rectangular waveform via integrator and differentiator circuits

15.1 Introduction

When a d.c. voltage is applied to a capacitor *C*, and resistor *R* connected in series, there is a short period of time immediately after the voltage is connected during which the current flowing in the circuit and voltages across *C* and *R* are changing.

Similarly, when a d.c. voltage is connected to a circuit having inductance *L* connected in series with resistance *R*, there is a short period of time immediately after the voltage is connected, during which the current flowing in the circuit and the voltages across *L* and *R* are changing.

These changing values are called **transients**.

15.2 Charging a capacitor

(a) The circuit diagram for a series connected *C–R* circuit is shown in Fig. 15.1. When switch S is closed then by Kirchhoff's voltage law:

$$V = v_C + v_R \qquad (15.1)$$

(b) The battery voltage *V* is constant. The capacitor voltage v_C is given by q/C, where *q* is the charge on the capacitor. The voltage drop across *R* is given by *iR*, where *i* is the current flowing in the circuit. Hence at all times:

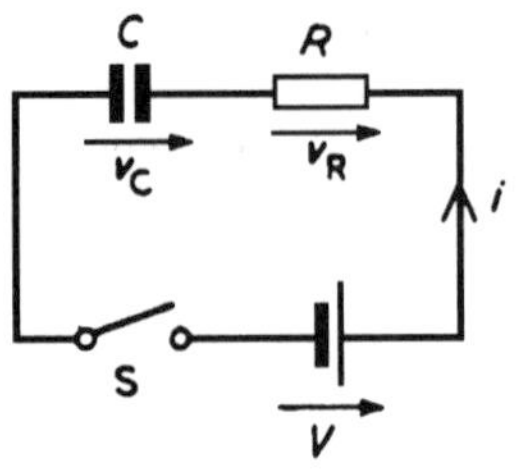

Figure 15.1

$$V = \frac{q}{C} + iR \qquad (15.2)$$

At the instant of closing S (initial circuit condition), assuming there is no initial charge on the capacitor, q_0 is zero, hence v_{Co} is zero. Thus from equation (15.1), $V = 0 + v_{Ro}$, i.e. $v_{Ro} = V$. This shows that the resistance to current is solely due to R, and the initial current flowing, $i_0 = I = V/R$.

(c) A short time later at time t_1 seconds after closing S, the capacitor is partly charged to, say, q_1 coulombs because current has been flowing. The voltage v_{C1} is now q_1/C volts. If the current flowing is i_1 amperes, then the voltage drop across R has fallen to i_1R volts. Thus, equation (15.2) is now $V = (q_1/C) + i_1R$.

(d) A short time later still, say at time t_2 seconds after closing the switch, the charge has increased to q_2 coulombs and v_C has increased q_2/C volts. Since $V = v_C + v_R$ and V is a constant, then v_R decreases to i_2R. Thus v_C is increasing and i and v_R are decreasing as time increases.

(e) Ultimately, a few seconds after closing S (final condition or **steady-state** condition), the capacitor is fully charged to, say, Q coulombs, current no longer flows, i.e. $i = 0$, and hence $v_R = iR = 0$. It follows from equation (15.1) that $v_C = V$.

(f) Curves showing the changes in v_C, v_R and i with time are shown in Fig. 15.2.

The curve showing the variation of v_C with time is called an **exponential growth curve** and the graph is called the 'capacitor voltage/time' characteristic. The curves showing the variation of v_R and i with time are called **exponential decay curves**, and the graphs are called 'resistor voltage/time' and 'current/time' characteristics, respectively. (The name 'exponential' shows that the shape can be expressed mathematically by an exponential mathematical equation, as shown in section 15.4).

15.3 Time constant for a *C–R* circuit

(a) If a constant d.c. voltage is applied to a series connected C–R circuit, a transient curve of capacitor voltage v_C is as shown in Fig. 15.2(a).

(b) With reference to Fig. 15.3, let the constant voltage supply be replaced by a variable voltage supply at time t_1 seconds. Let the voltage be varied so that the **current** flowing in the circuit is **constant**.

(c) Since the current flowing is a constant, the curve will follow a tangent, AB, drawn to the curve at point A.

(d) Let the capacitor voltage v_C reach its final value of V at time t_2 seconds.

(e) The time corresponding to $(t_2 - t_1)$ seconds is called the **time constant** of the circuit, denoted by the Greek letter tau, τ. The value of the time constant is CR seconds, i.e. for a series connected C–R circuit,

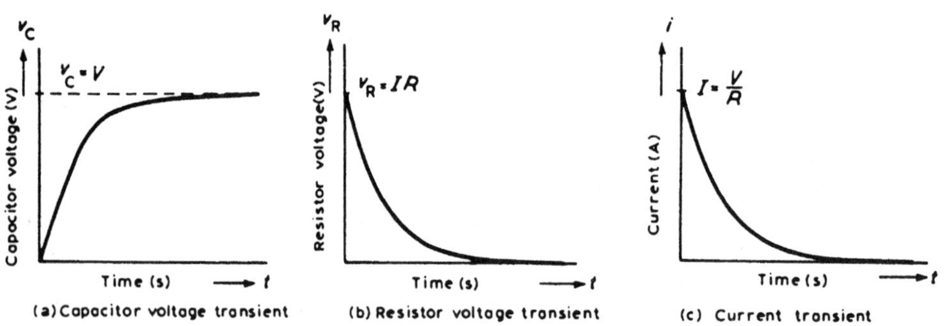

Figure 15.2

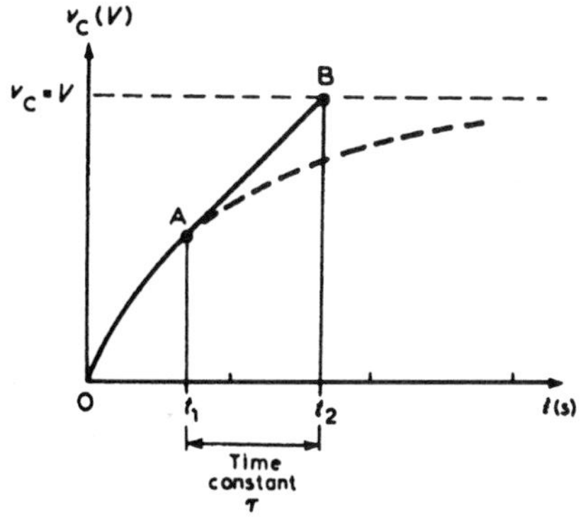

Figure 15.3

time constant τ = *CR* seconds

Since the variable voltage mentioned in para (b) above can be applied at any instant during the transient change, it may be applied at t = 0, i.e. at the instant of connecting the circuit to the supply. If this is done, then the time constant of the circuit may be defined as: '*the time taken for a transient to reach its final state if the initial rate of change is maintained*'.

15.4 Transient curves for a *C–R* circuit

There are two main methods of drawing transient curves graphically.

(a) The **tangent method** – this method is shown in Problem 1 below.

(b) The **initial slope and three-point method**, which is shown in Problem 2, and is based on the following properties of a transient exponential curve.

 (i) For a growth curve, the value of a transient at a time equal to one time constant is 0.632 of its steady-state value (usually taken as 63% of the steady-state value); at a time equal to two and a half time constants is 0.918 of its steady-state value (usually taken as 92% of its steady-state-value) and at a time equal to five time constants is equal to its steady-state value.

 (ii) For a decay curve, the value of a transient at a time equal to one time constant is 0.368 of its initial value (usually taken as 37% of its initial value), at a time equal to two and a half time constants is 0.082 of its initial value (usually taken as 8% of its initial value) and at a time equal to five time constants is equal to zero.

The transient curves shown in Fig. 15.2 have mathemtical equations, obtained by solving the differential equations representing the circuit. The equations of the curves are:

growth of capacitor voltage,
$v_C = V(1 - e^{-t/CR}) = V(1 - e^{-t/\tau})$

decay of resistor voltage,
$v_R = Ve^{-t/CR} = Ve^{-t/\tau}$

decay of current flowing,
$i = Ie^{-t/CR} = Ie^{-t/\tau}$

Problem 1. A 15 μF uncharged capacitor is connected in series with a 47 kΩ resistor across a 120 V, d.c. supply. Use the tangential graphical method to draw the capacitor voltage/time characteristic of the circuit. From the characteristic, determine the capacitor voltage for a time equal to one time constant after being connected to the supply and also two seconds after being connected to the supply. Also find the time for the capacitor voltage to reach one half of its steady-state value.

To construct an exponential curve, the time constant of the circuit and steady-state value need to be determined. Time constant = CR = 15 μF × 47 kΩ = 15 × 10^{-6} × 47 × 10^3 = 0.705 s. Steady-state value of $v_C = V$, i.e. v_C = 120 V.

With reference to Fig. 15.4, the scale of the horizontal axis is drawn so that it spans at least five time constants, i.e. 5 × 0.705 or about 3 1/2 seconds. The scale of the vertical axis spans the change in the capacitor voltage, that is, from 0 to 120 V. A broken line AB is drawn corresponding to the final value of v_C .

Point C is measured along AB so that AC is equal to 1τ, i.e. AC = 0.705 s. Straight line OC is drawn. Assuming that about five intermediate points are needed to draw the curve accurately, a point D is selected on OC corresponding to a v_C value of about 20 V. DE is drawn vertically. EF is made to correspond to 1τ, i.e. EF = 0.705 s. A straight line is drawn joining DF. This procedure of

(a) drawing a vertical line through point selected,
(b) at the steady-state value, drawing a horizontal line corresponding to 1τ, and
(c) joining the first and last points,

is repeated for v_C values of 40, 60, 80 and 100 V, giving points G, H, I and J.

The capacitor voltage effectively reaches its steady-state value of 120 V after a time equal to five time constants, shown as point K. Drawing a smooth curve through points O, D, G, H, I, J and K gives the exponential growth curve of capacitor voltage.

From the graph, the value of capacitor voltage at a time equal to the time constant is about **75 V**. It is a characteristic of all exponential growth curves, that after a time equal to one time constant, the value of the transient is 0.632 of its steady-state value. In this problem, $0.632 \times 120 = 75.84$ V. Also from the graph, when t is two seconds, v_C is about **115 volts**. (This value may be checked using the equation $v_C = V(1 - e^{-t/\tau})$, where $V = 120$ V, $\tau = 0.705$ s and $t = 2$ s. This calculation gives $v_C = 112.97$ V.)

The time for v_C to rise to one half of its final value (60 V) can be determined from the graph and is about **0.5 s**. (This value may be checked using $v_C = V(1 - e^{-t/\tau})$ where $V = 120$ V, $v_C = 60$ V and $\tau = 0.705$ s, giving $t = 0.489$ s.)

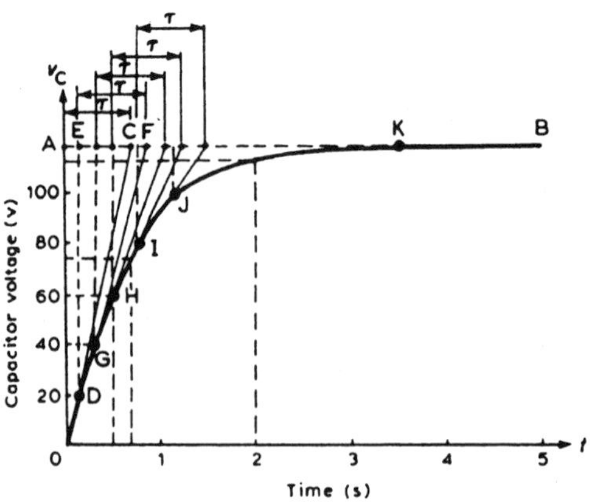

Figure 15.4

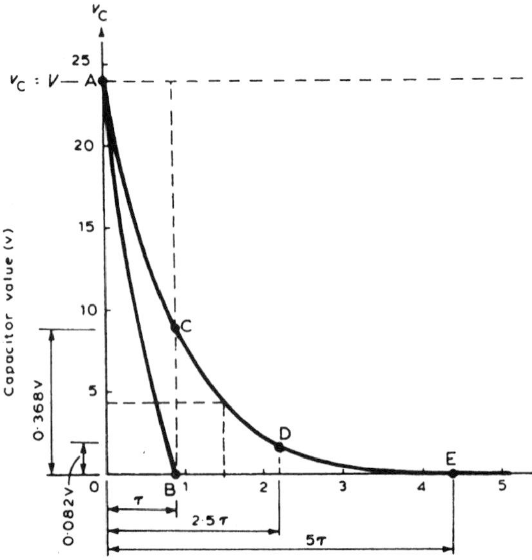

Figure 15.5

Problem 2. A 4 μF capacitor is charged to 24 V and then discharged through a 220 kΩ resistor. Use the 'initial slope and three-point' method to draw: (a) the capacitor voltage/time characteristic, (b) the resistor voltage/time characteristic, and (c) the current/time characteristic, for the transients which occur. From the characteristics determine the value of capacitor voltage, resistor voltage and current one and a half seconds after discharge has started.

To draw the transient curves, the time constant of the circuit and steady-state values are needed. Time constant, $\tau = CR = 4 \times 10^{-6} \times 220 \times 10^3 = 0.88$ s. Initially, capacitor voltage $v_C = v_R = 24$ V,

$$i = \frac{V}{R} = \frac{24}{220 \times 10^3} = 0.109 \text{ mA}$$

Finally, $v_C = v_R = i = 0$.

(a) The exponential decay of capacitor voltage is from 24 V to 0 V in a time equal to five time constants, i.e. $5 \times 0.88 = 4.4$ s. With reference to Fig. 15.5, to construct the decay curve:

(i) the horizontal scale is made so that it spans at least five time constants, i.e. 4.4 s;
(ii) the vertical scale is made to span the change in capacitor voltage, i.e. 0 to 24 V;
(iii) point A corresponds to the initial capacitor voltage, i.e. 24 V;
(iv) OB is made equal to one time constant and line AB is drawn. This gives the initial slope of the transient;
(v) the value of the transient after a time equal to one time constant is 0.368 of the initial value, i.e. $0.368 \times 24 = 8.83$ V. A vertical line is drawn through B and distance BC is made equal to 8.83 V;

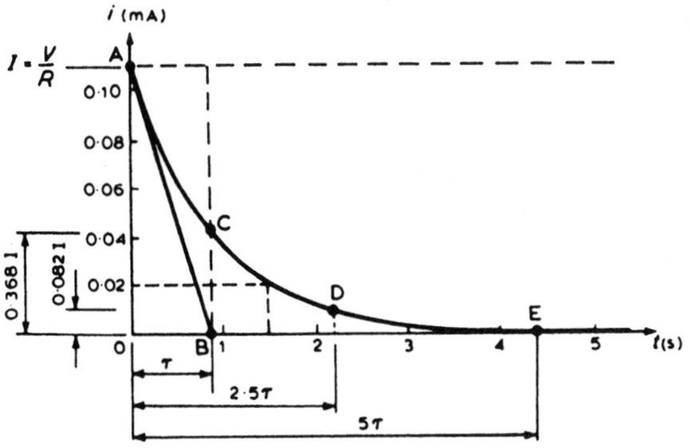

Figure 15.6

(vi) the value of the transient after a time equal to two and a half time constants is 0.082 of the initial value, i.e. 0.082 × 24 = 1.97 V, shown as point D in Fig. 15.5;

(vii) the transient effectively dies away to zero after a time equal to five time constants, i.e. 4.4 s, giving point E.

The smooth curve drawn through points A, C, D and E represents the decay transient. At $1\frac{1}{2}$ s after decay has started, $v_C \approx$ **4.4 V**.
(This may be checked using $v_C = Ve^{-t/\tau}$, where $V = 24$, $t = 1\frac{1}{2}$ and $\tau = 0.88$, giving $v_C = 4.36$ V.)

(b) The voltage drop across the resistor is equal to the capacitor voltage when a capacitor is discharging through a resistor, thus the resistor voltage/time characteristic is identical to that shown in Fig. 15.5. Since $v_R = v_C$, then at $1\frac{1}{2}$ seconds after decay has started, $v_R \approx$ **4.4 V** (see (a) above).

(c) The current/time characteristic is constructed in the same way as the capacitor voltage/time characteristic, shown in part (a) of this problem, and is as shown in Fig. 15.6. The values are:

point A: initial value of current = 0.109 mA
point C: at 1τ, i= 0.368 × 0.109 = 0.040 mA
point D: at 2.5τ, i = 0.082 × 0.109 = 0.009 mA
point E: at 5τ, $i = 0$

Hence the current transient is as shown. At a time of $1\frac{1}{2}$ seconds, the value of current, from the characteristic is **0.02 mA**. (This may be checked using $i = Ie^{(-t/\tau)}$ where $I = 0.109$, $t = 1\frac{1}{2}$ and $\tau = 0.88$, giving $i = 0.0198$ mA or 19.8 μA.)

Problem 3. A 20 μF capacitor is connected in series with a 50 kΩ resistor and the circuit is connected to a 20 V, d.c. supply. Determine

(a) the initial value of the current flowing,
(b) the time constant of the circuit,
(c) the value of the current one second after connection,
(d) the value of the capacitor voltage two seconds after connection, and
(e) the time after connection when the resistor voltage is 15 V.

Parts (c), (d) and (e) may be determined graphically, as shown in Problems 1 and 2 or by calculation as shown below.
$V = 20$ V, $C = 20\ \mu\text{F} = 20 \times 10^{-6}$ F, $R = 50\ \text{k}\Omega = 50 \times 10^3\ \Omega$.

(a) The initial value of the current flowing is given by

$$I = \frac{V}{R}, \text{ i.e. } \frac{20}{50 \times 10^3} = \mathbf{0.4\ mA}$$

(b) From section 15.3, the time constant, $\tau = CR = (20 \times 10^{-6}) \times (50 \times 10^3) = $ **1 s**.

(c) Current, $i = Ie^{-t/\tau}$. Working in mA units, $i = 0.4e^{-1/1} = 0.4 \times 0.368 =$ **0.147 mA**.

(d) Capacitor voltage,

$$v_C = V(1 - e^{-t/\tau}) = 20(1 - e^{-2/1})$$
$$= 20(1 - 0.135) = 20 \times 0.865$$
$$= \mathbf{17.3\ V}$$

(e) Resistor voltage, $v_R = Ve^{-t/\tau}$. Thus

$$15 = 20e^{-t/1}$$
$$\frac{15}{20} = e^{-t}, \text{ i.e. } e^t = \frac{20}{15} = \frac{4}{3}$$

Taking natural logarithms of each side of the equation gives

$$t = \ln\frac{4}{3} = \ln 1.3333$$
$$\mathbf{t = 0.288\ s}$$

Problem 4. A circuit consists of a resistor connected in series with a 0.5 μF capacitor and has a time constant of 12 ms. Determine (a) the value of the resistor and (b) the capacitor voltage 7 ms after connecting the circuit to a 10 V supply.

(a) The time constant $\tau = CR$, hence $R = \tau/C$, i.e.

$$R = \frac{12 \times 10^{-3}}{0.5 \times 10^{-6}} = 24 \times 10^3 = \mathbf{24\ k\Omega}$$

(b) The equation for the growth of capacitor voltage is $v_C = V(1 - e^{-t/\tau})$. Since $\tau = 12$ ms $= 12 \times 10^{-3}$ s, $V = 10$ V and $t = 7$ ms $= 7 \times 10^{-3}$ s, then

$$v_C = 10\left(1 - e^{-\frac{7\times10^{-3}}{12\times10^{-3}}}\right) = 10(1 - e^{-0.583})$$

$$= 10(1 - 0.558) \quad = \mathbf{4.42\ V}$$

Alternatively, the value of v_C when t is 7 ms may be determined using the growth characteristic as shown in Problem 1.

15.5 Discharging a capacitor

When a capacitor is charged (i.e. with the switch in position A in Fig. 15.7), and the switch is then moved to position B, the electrons stored in the capacitor keep the current flowing for a short time. Initially, at the instant of moving from A to B, the current flow is such that the capacitor voltage v_C is balanced by an equal and opposite voltage $v_R = iR$. Since initially $v_C = v_R = V$, then $i = I = V/R$. During the transient decay, by applying Kirchhoff's voltage law to Fig. 15.7, $v_C = v_R$. Finally the transients decay exponentially to zero, i.e. $v_C = v_R = 0$. The transient curves representing the voltages and current are as shown in Fig. 15.8.

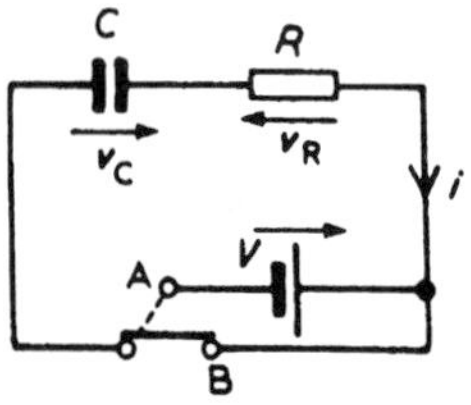

Figure 15.7

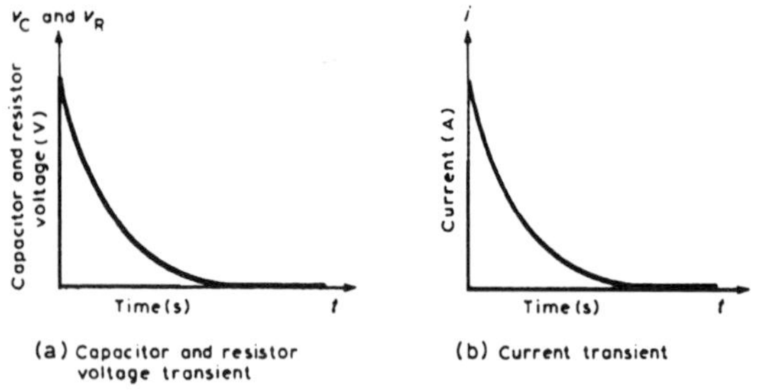

Figure 15.8

The equations representing the transient curves during the discharge period of a series connected C–R circuit are:

> decay of voltage, $v_C = v_R = Ve^{(-t/CR)} = Ve^{(-t/\tau)}$
>
> decay of current, $i = Ie^{(-t/CR)} = Ie^{(-t/\tau)}$

When a capacitor has been disconnected from the supply it may still be charged and it may retain this charge for some considerable time. Thus precautions must be taken to ensure that the capacitor is automatically discharged after the supply is switched off. This is done by connecting a high-value resistor across the capacitor terminals.

> Problem 5. A capacitor is charged to 100 V and then discharged through a 50 kΩ resistor. If the time constant of the circuit is 0.8 s, determine:
>
> (a) the value of the capacitor;
> (b) the time for the capacitor voltage to fall to 20 V;
> (c) the current flowing when the capacitor has been discharging for 0.5 s; and
> (d) the voltage drop across the resistor when the capacitor has been discharging for one second.

Parts (b), (c) and (d) of this problem may be solved graphically as shown in Problems 1 and 2 or by calculation as shown below.
$V = 100$ V, $\tau = 0.8$ s, $R = 50$ kΩ $= 50 \times 10^3$ Ω.

(a) Since time constant, $\tau = CR$, $C = \tau/R$, i.e.

$$C = \frac{0.8}{50 \times 10^3} = \mathbf{16\ \mu F}$$

(b) $v_C = Ve^{-t/\tau}$. $20 = 100e^{-t/0.8}$, i.e. $\frac{1}{5} = e^{-t/0.8}$. Thus $e^{t/0.8} = 5$ and taking natural logarithms of each side, gives $(t/0.8) = \ln 5$, i.e. $t = 0.8 \ln 5$. Hence $\boldsymbol{t} = \mathbf{1.29\ s}$.

(c) $i = Ie^{-t/\tau}$. The initial current flowing,

$$I = \frac{V}{R} = \frac{100}{50 \times 10^3} = 2\ \text{mA}$$

Working in mA units, $i = Ie^{-t/\tau} = 2e^{(-0.5/0.8)} = 2e^{-0.625} = 2 \times 0.535 = \mathbf{1.07\ mA}$.

(d) $v_R = v_C = Ve^{-t/\tau} = 100e^{-1/0.8} = 100e^{-1.25} = 100 \times 0.287 =$ **28.7 V**.

Problem 6. A 0.1 μF capacitor is charged to 200 V before being connected across a 4 kΩ resistor. Determine (a) the initial discharge current, (b) the time constant of the circuit, and (c) the minimum time required for the voltage across the capacitor to fall to less than 2 V.

(a) Initial discharge current,

$$i = \frac{V}{R} = \frac{200}{4 \times 10^3}$$

$$= \mathbf{0.05\ A} \text{ or } \mathbf{50\ mA}$$

(b) Time constant $\tau = CR = 0.1 \times 10^{-6} \times 4 \times 10^3 =$ **0.0004 s** or **0.4 ms**.

(c) The minimum time for the capacitor voltage to fall to less than 2 V, i.e. less than 2/200 or 1% of the initial value, is given by 5 τ: 5 $\tau = 5 \times 0.4 =$ **2 ms**.

In a d.c. circuit, a capacitor blocks the current except during the times that there are changes in the supply voltage.

15.6 Current growth in an *L–R* circuit

(a) The circuit diagram for a series connected *L–R* circuit is shown in Fig. 15.9. When switch S is closed, then by Kirchhoff's voltage law:

$$V = v_L + v_R \tag{15.3}$$

(b) The battery voltage V is constant. The voltage across the inductance is the induced voltage, i.e.

$$v_L = L \times \frac{\text{change of current}}{\text{change of time}}$$

shown as $L(di/dt)$. The voltage drop across R, v_R is given by iR. Hence, at all times:

$$V = L(di/dt) + iR \tag{15.4}$$

(c) At the instant of closing the switch, the rate of change of current is such that it induces an e.m.f. in the inductance which is equal and opposite to V, hence $V = v_L + 0$, i.e. $v_L = V$. From equation (15.3), because $v_L = V$, then $v_R = 0$ and $i = 0$.

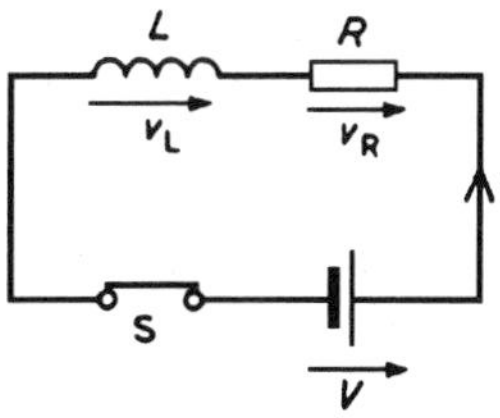

Figure 15.9

(d) A short time later at time t_1 seconds after closing S, current i_1 is flowing, since there is a rate of change of current initially, resulting in a voltage drop of i_1R across the resistor. Since V (constant) $= v_L + v_R$ the induced e.m.f. is reduced and equation (15.4) becomes:

$$V = L\frac{di_1}{dt_1} + i_1R$$

(e) A short time later still, say at time t_2 seconds after closing the switch, the current flowing is i_2, and the voltage drop across the resistor increases to i_2R. Since v_R increases, v_L decreases.

(f) Ultimately, a few seconds after closing S, the current flow is entirely limited by R, the rate of change of current is zero and hence v_L is zero. Thus $V = iR$. Under these conditions, steady-state current flows, usually signified by I. Thus, $I = V/R$, $v_R = IR$ and $v_L = 0$ at steady-state conditions.

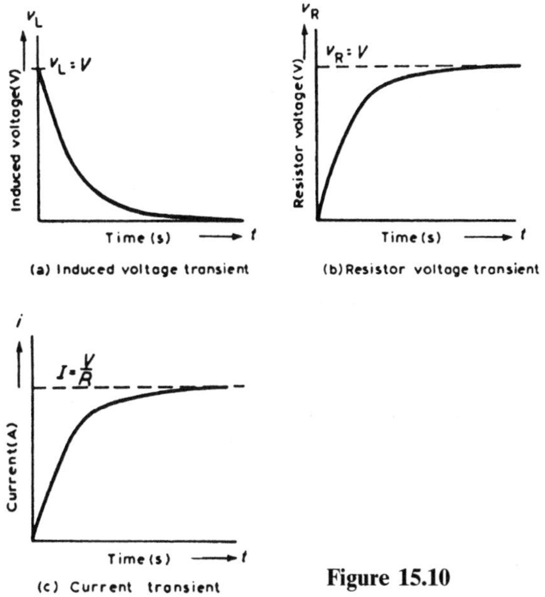

Figure 15.10

(g) Curves showing the changes in v_L, v_R and i with time are shown in Fig. 15.10 and indicate that v_L is a maximum value initially (i.e. equal to V), decaying exponentially to zero, whereas v_R and i grow exponentially from zero to their steady-state values of V and $I = V/R$, respectively.

15.7 Time constant for an L–R circuit

With reference to section 15.3, the time constant of a series connected L–R circuit is defined in the same way as the time constant for a series connected C–R circuit. Its value is given by:

time constant, $\tau = L/R$ seconds

15.8 Transient curves for an L–R circuit

Transient curves representing the induced voltage/time, resistor voltage/time and current/time characteristics may be drawn graphically, as outlined in section 15.4. The methods of construction are shown in Problems 7 and 8.

Each of the transient curves shown in Fig. 15.10 have mathematical equations, and these are:

decay of induced voltage,
$v_L = Ve^{(-Rt/L)} = Ve^{(-t/\tau)}$

growth of resistor voltage,
$v_R = V(1 - e^{-Rt/L}) = V(1 - e^{-t/\tau})$

growth of current flow,
$i = I(1 - e^{-Rt/L}) = I(1 - e^{-t/\tau})$

The application of these equations is shown in Problems 9 and 10.

Problem 7. A relay has an inductance of 100 mH and a resistance of 20 Ω. It is connected to a 60 V, d.c. supply. Use the 'initial slope and three-point' method to draw the current/time characteristic and hence determine the value of current flowing at a time equal to two time constants and the time for the current to grow to 1.5 A.

Before the current/time characteristic can be drawn, the time constant and steady-state value of the current have to be calculated.

$$\text{Time constant, } \tau = \frac{L}{R} = \frac{100 \times 10^{-3}}{20} = 5 \text{ ms}$$

Final value of current,

$$I = \frac{V}{R} = \frac{60}{20} = 3 \text{ A}$$

The method used to construct the characteristic is the same as that used in Problem 1.

(a) The scales should span at least five time constants (horizontally), i.e. 25 ms, and 3 A (vertically).
(b) With reference to Fig. 15.11, the initial slope is obtained by making AB equal to 1 time constant (5 ms), and joining OB.

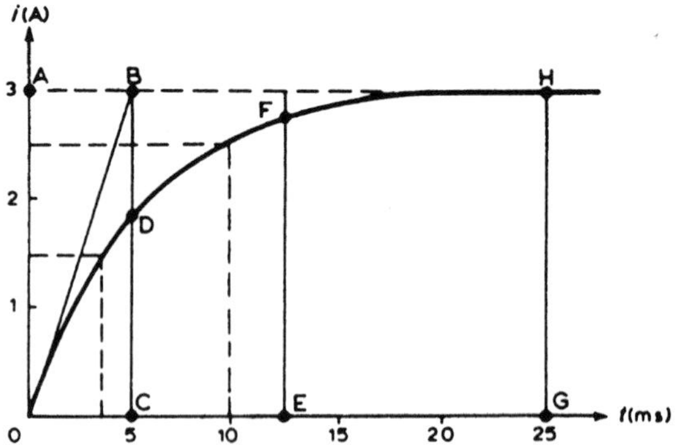

Figure 15.11

(c) At a time of 1 time constant, CD is 0.632 × I = 0.632 × 3 = 1.896 A.
At a time of 2.5 time constants, EF is 0.918 × I = 0.918 × 3 = 2.754 A. At a time of 5 time constants, GH is I = 3 A.
(d) A smooth curve is drawn through points O, D, F and H and this curve is the current/time characteristic.

From the characteristic, when $t = 2\tau$, $i \approx$ **2.6 A**. (This may be checked by calculation using $i = I(1 - e^{-t/\tau})$, where $I = 3$ and $t = 2\tau$, giving i = 2.59 A.) Also, when the current is 1.5 A, the corresponding time is about **3.6 ms**. (This may be checked by calculation, using $i = I(1 - e^{-t/\tau})$ where $i = 1.5$, $I = 3$ and $\tau = 5$ ms, giving t = 3.466 ms.)

Problem 8. A coil of inductance 0.04 H and resistance 10 Ω is connected to a 120 V, d.c. supply. Determine (a) the final value of current, (b) the time constant of the circuit, (c) the value of current after a time equal to the time constant from the instant the supply voltage is connected, (d) the expected time for the current to rise to within 1% of its final value.

(a) Final steady current, $I = (V/R) = 120/10 =$ **12 A**.
(b) Time constant of the circuit, $\tau = (L/R) = 0.04/10 =$ **0.004 s** or **4 ms**.
(c) In the time τ s the current rises to 63.2% of its final value of 12 A, i.e. in 4 ms the current rises to $0.632 \times 12 =$ **7.58 A**.
(d) The expected time for the current to rise to within 1% of its final value is given by 5τ s, i.e. $5 \times 4 =$ **20 ms**.

Problem 9. The winding of an electromagnet has an inductance of 3 H and a resistance of 15 Ω. When it is connected to a 120 V, d.c. supply, calculate: (a) the steady-state value of current flowing in the winding, (b) the time constant of the circuit, (c) the value of the induced e.m.f. after 0.1 s, (d) the time for the current to rise to 85% of its final value, and (e) the value of the current after 0.3 s.

(a) The steady-state value of current is $I = V/R$, i.e. $I = 120/15 =$ **8 A**.
(b) The time constant of the circuit, $\tau = L/R = 3/15 =$ **0.2 s**.

Parts (c), (d) and (e) of this problem may be determined by drawing the transients graphically, as shown in Problem 7 or by calculation as shown below.

(c) The induced e.m.f., v_L is given by $v_L = Ve^{-t/\tau}$. The d.c. voltage V is 120 V, t is 0.1 s and τ is 0.2 s, hence

$$V_L = 120e^{-0.1/0.2} = 120e^{-0.5} = 120 \times 0.6065$$

i.e.

$$v_L = \mathbf{72.78\ V}$$

(d) When the current is 85% of its final value, $i = 0.85I$. Also, $i = I(1 - e^{-t/\tau})$, thus

$$0.85I = I(1 - e^{-t/\tau})$$

$$0.85 = 1 - e^{-t/\tau} \text{ and } \tau = 0.2\text{, hence}$$

$$0.85 = 1 - e^{-t/0.2}$$

$$e^{-t/0.2} = 1 - 0.85 = 0.15$$

$$e^{t/0.2} = \frac{1}{0.15} = 6.\dot{6}$$

Taking natural logarithms of each side of this equation gives $\ln e^{t/0.2} = \ln 6.\dot{6}$ and by the laws of logarithms

$$\frac{t}{0.2} \ln e = \ln 6.\dot{6}$$

But $\ln e = 1$, hence $t = 0.2 \ln 6.\dot{6}$, i.e. $t =$ **0.379 s**.

(e) The current at any instant is given by $i = I(1 - e^{-t/\tau})$. When $I = 8$, $t = 0.3$ and $\tau = 0.2$, then

$$i = 8(1 - e^{-0.3/0.2}) = 8(1 - e^{-1.5})$$

$$= 8(1 - 0.2231) = 8 \times 0.7769, \text{ i.e. } i = \mathbf{6.215\ A}$$

15.9 Current decay in an *L–R* circuit

When a series-connected *L–R* circuit is connected to a d.c. supply as shown with S in position A of Fig. 15.12, a current $I = V/R$ flows after a short time, creating a magnetic field ($\Phi \propto I$) associated with the inductor. When S is moved to position B, the current value decreases, causing a decrease in the strength of the magnetic field. Flux linkages occur, generating a voltage v_L, equal to $L(di/dt)$. By Lenz's law, this voltage keeps current i flowing in the circuit, its value being limited by R. Thus $v_L = v_R$. The current decays exponentially to zero and since v_R is proportional to the current flowing, v_R decays exponentially to zero. Since $v_L = v_R$, v_L also decays exponentially to zero. The curves representing these transients are similar to those shown in Fig. 15.8.

The equations representing the decay transient curves are:

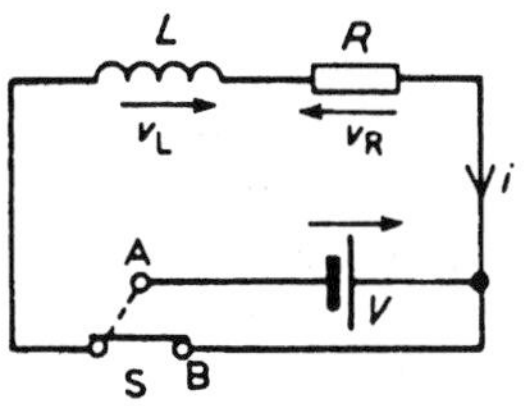

Figure 15.12

decay of voltages,
$v_L = v_R = Ve^{(-Rt/L)} = Ve^{(-t/\tau)}$

decay of current, $i = Ie^{(-Rt/L)} = Ie^{(-t/\tau)}$

Problem 10. The field winding of a 110 V, d.c. motor has a resistance of 15 Ω and a time constant of 2 s. Determine the inductance and use the tangential method to draw the current/time characteristic when the supply is removed and replaced by a shorting link. From the characteristic determine (a) the current flowing in the winding 3 s after being shorted-out, and (b) the time for the current to decay to 5 A.

Since the time constant, $\tau = (L/R)$, $L = R\tau$, i.e. inductance $L = 15 \times 2 =$ **30 H**.

The current/time characteristic is constructed in a similar way to that used in Problem 1.

(i) The scales should span at least five time constants horizontally, i.e. 10 s, and $I = V/R = 110/15 = 7.\dot{3}$ A vertically.
(ii) With reference to Fig. 15.3, the initial slope is obtained by making OB equal to 1 time constant (2 s), and joining AB.
(iii) At, say, $i = 6$ A, let C be the point on AB corresponding to a current of 6 A. Make DE equal to 1 time constant (2 s), and join CE.

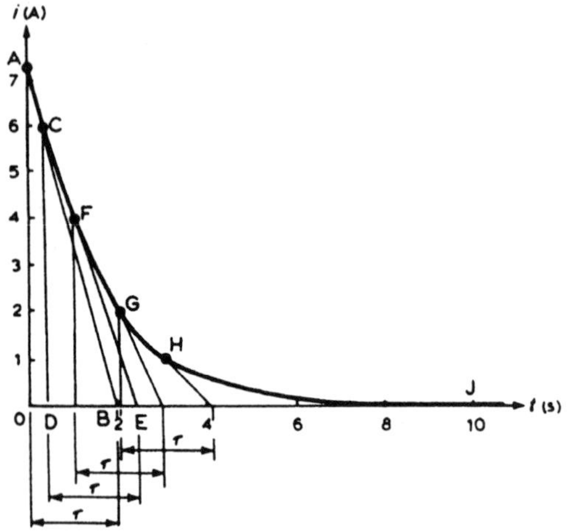

Figure 15.13

(iv) Repeat the procedure given in (iii) for current values of, say, 4 A, 2 A and 1 A, giving points F, G and H.
(v) Point J is at five time constants, when the value of current is zero.
(vi) Join points A, C, F, G, H and J with a smooth curve. This curve is the current/time characteristic.
(a) From the current/time characteristic, when $t =$ 3 s, $i =$ **1.2 A**. (This may be checked by calculation using $i = Ie^{-t/\tau}$, where $I = 7.3$, $t = 3$ and $\tau = 2$, giving $i = 1.64$ A.) The discrepancy between the two results is due to relatively few values, such as C, F, G and H, being taken.
(b) From the characteristic, when $i = 5$ A, $t =$ 0.65 s. (This may be checked by calculation using $i = Ie^{-t/\tau}$, where $i = 5$, $I = 7.3$, $\tau = 2$, giving $t = 0.766$ s.) The discrepancy between the graphical and calculated values is due to relatively few values such as C, F, G and H being taken.

Problem 11 A coil having an inductance of 6 H and a resistance of R Ω is connected in series with a resistor of 10 Ω to a 120 V, d.c. supply. The time constant of the circuit is 300 ms. When steady-state conditions have been reached, the supply is replaced instantaneously by a short-circuit. Determine:

(a) the resistance of the coil;
(b) the current flowing in the circuit one second after the shorting link has been placed in the circuit; and
(c) the time taken for the current to fall to 10% of its initial value.

(a) The time constant,

$$\tau = \frac{\text{circuit inductance}}{\text{total circuit resistance}} = \frac{L}{R + 10}$$

Thus

$$R = \frac{L}{\tau} - 10 = \frac{6}{0.3} - 10 = \mathbf{10\ \Omega}$$

Parts (b) and (c) may be determined graphically as shown in Problems 7 and 10 or by calculation as shown below.

(b) The steady-state current, $I = V/R = 120/(10 + 10) = 6$ A.
The transient current after 1 second, $i = Ie^{-t/\tau} = 6e^{-1/0.3}$

Thus

$$i = 6e^{-3.3} = 6 \times 0.035\,67$$

$$= \mathbf{0.214\ A}$$

(c) 10% of the initial value of the current is (10/100) × 6, i.e. 0.6 A. Using the equation $i = Ie^{-t/\tau}$ gives

$$0.6 = 6e^{-t/0.3}$$

i.e.

$$\frac{0.6}{6} = e^{-t/0.3} \text{ or } e^{t/0.3} = \frac{6}{0.6} = 10$$

Taking natural logarithms of each side of this equation gives:

$$\frac{t}{0.3} = \ln 10$$

$$t = 0.3 \ln 10 = \mathbf{0.691\ s}$$

Problem 10. An inductor has a negligible resistance and an inductance of 200 mH and is connected in series with a 1 kΩ resistor to a 24 V, d.c. supply. Determine the time constant of the circuit and the steady-state value of the current flowing in the circuit. Find (a) the current flowing in the circuit at a time equal to one time constant, (b) the voltage drop across the inductor at a time equal to two time constants, and (c) the voltage drop across the resistor after a time equal to three time constants.

The time constant, $\tau = (L/R) = 0.2/1000 = \mathbf{0.2\ ms}$. The steady-state current $I = (V/R) = 24/1000 = \mathbf{24\ mA}$.

(a) The transient current, $i = I(1 - e^{-t/\tau})$ and $t = 1\tau$. Working in mA units gives,

$$i = 24(1 - e^{-(1\tau/\tau)}) = 24(1 - e^{-1})$$

$$= 24(1 - 0.368) = \mathbf{15.17\ mA}$$

(b) The voltage drop across the inductor,

$$v_L = Ve^{-t/\tau}$$

$$= 24e^{-2\tau/\tau} = 24e^{-2}$$

$$= \mathbf{3.248\ V}$$

(c) The voltage drop across the resistor,

$$v_R = V(1 - e^{-t/\tau})$$

$$= 24(1 - e^{-3\tau/\tau}) = 24(1 - e^{-3})$$

$$= \mathbf{22.81\ V}$$

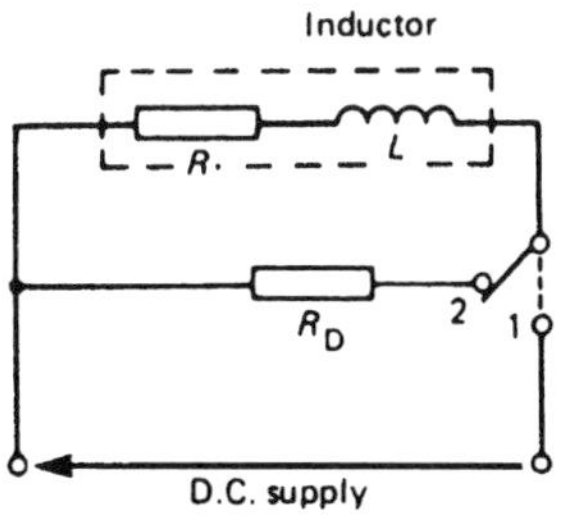

Figure 15.14

15.10 Switching inductive circuits

Energy stored in the magnetic field of an inductor exists because a current provides the magnetic field. When the d.c. supply is switched off the current falls rapidly, the magnetic field collapses causing a large induced e.m.f. which will either cause an arc across the switch contacts or will break down the insulation between adjacent turns of the coil. The high induced e.m.f. acts in a direction which tends to keep the current flowing, i.e. in the same direction as the applied voltage. The energy from the magnetic field will thus be aided by the supply voltage in maintaining an arc, which could cause severe damage to the switch. To reduce the induced e.m.f. when the supply switch is opened, a discharge resistor R_D is connected in parallel with the inductor as shown in Fig. 15.14. The magnetic field energy is dissipated as heat in R_D and R and arcing at the switch contacts is avoided.

15.11 The effects of time constant on a rectangular waveform

Integrator circuit

By varying the value of either C or R in a series-connected C–R circuit, the time constant ($\tau = CR$) of a circuit can be varied. If a rectangular waveform varying from $+E$ to $-E$ is applied to a C–R circuit as shown in Fig. 15.15, output

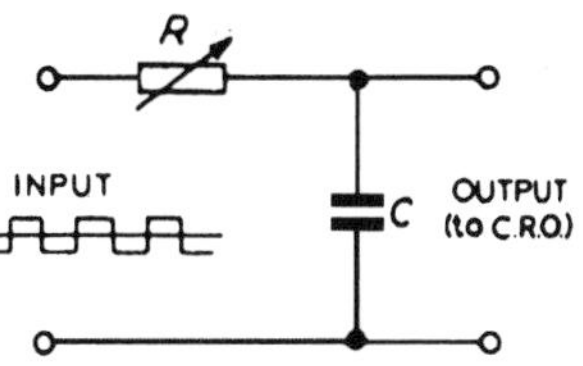

Figure 15.15

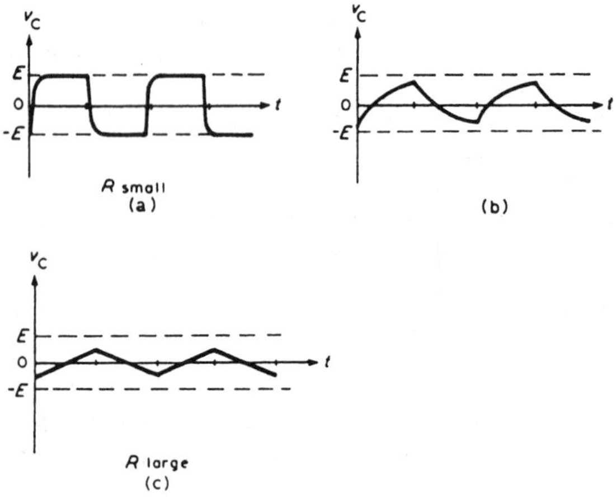

Figure 15.16

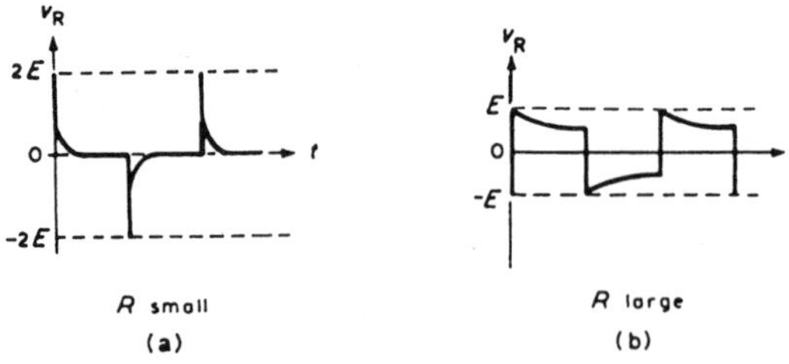

Figure 15.18

waveforms of the capacitor voltage have various shapes, depending on the value of R. When R is small, $\tau = CR$ is small and an output waveform such as that shown in Fig. 15.16(a) is obtained. As the value of R is increased, the waveform changes to that shown in Fig. 15.16(b). When R is large, the waveform is as shown in Fig. 15.16(c), the circuit then being described as an **integrator circuit**.

Differentiator circuit

If a rectangular waveform varying from $+E$ to $-E$ is applied to a series-connected C–R circuit and the waveform of the voltage drop across the resistor is observed, as shown in Fig. 15.17, the output waveform alters as R is varied due to the time constant ($\tau = CR$), altering. When R is small, the waveform is as shown in Fig. 15.18(a), the voltage being generated across R by the capacitor discharging fairly quickly. Since the change in capacitor voltage is from $+E$ to $-E$, the change in discharge current is $2E/R$, resulting in a change in voltage across the resistor of $2E$. This circuit is called a **differentiator circuit**. When R is large, the waveform is as shown in Fig. 15.18(b).

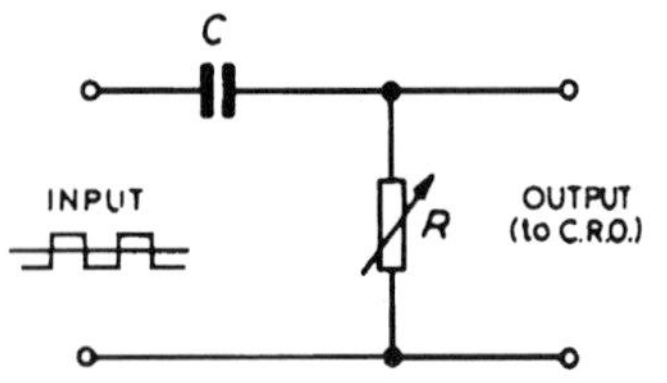

Figure 15.17

15.12 Multi-choice questions on d.c. transients

(Answers on page 284.)

An uncharged 2 μF capacitor is connected in series with a 5 MΩ resistor to a 100 V, constant voltage, d.c. supply. In questions 1 to 7, use this data to select the correct answer from those given below.

(a) 10 ms	(b) 100 V	(c) 10 s	(d) 10 V
(e) 20 μA	(f) 1 s	(g) 0 V	(h) 50 V
(i) 1 ms	(j) 50 μA	(k) 20 mA	(l) 0 A

1. Determine the time constant of the circuit.
2. Determine the final voltage across the capacitor.
3. Determine the initial voltage across the resistor.
4. Determine the final voltage across the resistor.
5. Determine the initial voltage across the capacitor.
6. Determine the initial current flowing in the circuit.
7. Determine the final current flowing in the circuit.

In questions 8 and 9, a series connected C–R circuit is suddenly connected to a d.c. source of V volts. Which of the statements is false?

8. (a) The initial current flowing is given by V/R.
   (b) The time constant of the circuit is given by CR.
   (c) The current grows exponentially.
   (d) The final value of the current is zero.

9. (a) The capacitor voltage is equal to the voltage drop across the resistor.
   (b) The voltage drop across the resistor decays exponentially.
   (c) The initial capacitor voltage is zero.
   (d) The initial voltage drop across the resistor is IR, where I is the steady-state current.
10. A capacitor which is charged to V volts is discharged through a resistor of R ohms. Which of the following statements is false?
   (a) The initial current flowing is V/R amperes.
   (b) The voltage drop across the resistor is equal to the capacitor voltage.
   (c) The time constant of the circuit is CR seconds.
   (d) The current grows exponentially to a final value of V/R amperes.

An inductor of inductance 0.1 H and negligible resistance is connected in series with a 50 Ω resistor to a 20 V, d.c. supply. In questions 11 to 15, use this data to determine the value required, selecting your answer from those given below.

(a) 5 ms (b) 12.6 V (c) 0.4 A (d) 500 ms
(e) 7.4 V (f) 2.5 A (g) 2 ms (h) 0 V
(i) 0 A (j) 20 V

11. The value of the time constant of the circuit.
12. The approximate value of the voltage across the resistor after a time equal to the time constant.
13. The final value of the current flowing in the circuit.
14. The initial value of the voltage across the inductor.
15. The final value of the steady-state voltage across the inductor.

15.13 Short answer questions on d.c. transients

A capacitor of capacitance C farads is connected in series with a resistor of R ohms and is switched across a constant voltage d.c. supply of V volts. After a time of t seconds, the current flowing is i amperes. Use this data to answer questions 1 to 10.

1. The voltage drop across the resistor at time t seconds is v_R = ...
2. The capacitor voltage at time t seconds is v_C = ...
3. The voltage equation for the circuit is V = ...
4. The time constant for the circuit is τ = ...
5. The final value of the current flowing is ....
6. The initial value of the current flowing is I = ...
7. The final value of capacitor voltage is ....
8. The initial value of capacitor voltage is ....
9. The final value of the voltage drop across the resistor is ....
10. The initial value of the voltage drop across the resistor is ....

A capacitor charged to V volts is disconnected from the supply and discharged through a resistor of R ohms. Use this data to answer questions 11 to 15.

11. The initial value of current flowing is I = ...
12. The approximate time for the current to fall to zero in terms of C and R is .... seconds.
13. If the value of resistance R is doubled, the time for the current to fall to zero is .... when compared with the time in question 12 above.
14. The approximate fall in the value of the capacitor voltage in a time equal to one time constant is .... %.
15. The time constant of the circuit is given by .... seconds.

An inductor of inductance L henrys and negligible resistance is connected in series with a resistor of resistance R ohms and is switched across a constant voltage d.c. supply of V volts. After a time interval of t seconds, the transient current flowing is i amperes. Use this data to answer questions 16 to 25.

16. The induced e.m.f., v_L, opposing the current flow when $t = 0$ is ....
17. The voltage drop across the resistor when $t = 0$ is v_R = ...
18. The current flowing when $t = 0$ is ....
19. V, v_R and v_L are related by the equation V = ...
20. The time constant of the circuit in terms of L and R is ....
21. The steady-state value of the current is reached in practice in a time equal to .... seconds.
22. The steady-state voltage across the inductor is .... volts.
23. The final value of the current flowing is .... amperes.
24. The steady-state resistor voltage is .... volts.

25. The e.m.f. induced in the inductor during the transient in terms of current, time and inductance is .... volts.

A series connected L–R circuit carrying a current of I amperes is suddenly short-circuited to allow the current to decay exponentially. Use this data to answer questions 26 to 30.

26. The current will fall to ....% of its initial value in a time equal to the time constant.
27. The voltage equation of the circuit is .... = ...
28. The time constant of the circuit in terms of L and R is ....
29. The current reaches zero in a time equal to .... seconds.
30. If the value of R is halved, the time for the current to fall to zero is .... when compared with the time in question 29.

15.14 Further questions on d.c. transients

Transients in series connected C–R circuits

1. An uncharged capacitor of 0.2 μF is connected to a 100 V, d.c. supply through a resistor of 100 kΩ. Determine, either graphically or by calculation, the capacitor voltage 10 ms after the voltage has been applied. [39.35 V]
2. A circuit consists of an uncharged capacitor connected in series with a 50 kΩ resistor and has a time constant of 15 ms. Determine either graphically or by calculation (a) the capacitance of the capacitor and (b) the voltage drop across the resistor 5 ms after connecting the circuit to a 20 V, d.c. supply. [(a) 0.3 μF (b) 14.33 V]
3. A 10 μF capacitor is charged to 120 V and then discharged through a 1.5 mΩ resistor. Determine either graphically or by calculation the capacitor voltage 2 s after discharging has commenced. Also find how long it takes for the voltage to fall to 25 V. [105.0 V, 23.53 s]
4. A capacitor is connected in series with a voltmeter of resistance 750 kΩ and a battery. When the voltmeter reading is steady the battery is replaced with a shorting link. If it takes 17 s for the voltmeter reading to fall to two thirds of its original value, determine the capacitance of the capacitor. [55.9 μF]
5. When a 3 μF charged capacitor is connected to a resistor, the voltage falls by 70% in 3.9 s. Determine the value of the resistor. [1.08 MΩ]
6. A circuit consists of a capacitor and resistor connected in series. When connected to a battery, sketch curves showing the variation of capacitor voltage, voltage drop across the resistor and current flow over the transient period. [See Fig. 15.2]
7. An uncharged capacitor is connected in series with a resistor across a d.c. supply. On the same axes sketch curves to show how (a) the current and (b) the capacitor voltage vary with time. [See Fig. 15.2]
8. Explain the meaning of the term 'time constant' of a circuit comprising a capacitor and resistor connected in series with a constant voltage d.c. supply. [See section 15.3]
9. A 50 μF uncharged capacitor is connected in series with a 1 kΩ resistor and the circuit is switched to a 100 V, d.c. supply. Determine:
   (a) the initial current flowing in the circuit;
   (b) the time constant;
   (c) the value of current when t is 50 ms; and
   (d) the voltage across the resistor 60 ms after closing the switch.
   [(a) 0.1 A (b) 50 ms (c) 36.8 mA (d) 30.1 V]
10. An uncharged 5 μF capacitor is connected in series with a 30 kΩ resistor across a 110 V, d.c. supply. Determine the time constant of the circuit and the initial charging current. Use a graphical method to draw the current/time characteristic of the circuit and hence determine the current flowing 120 ms after connecting to the supply. [150 ms, $3\frac{2}{3}$ mA, 1.65 mA]
11. An uncharged 80 μF capacitor is connected in series with a 1 kΩ resistor and is switched across a 110 V supply. Determine the time constant of the circuit and the initial value of current flowing. Derive graphically the current/time characteristic for the transient condition and hence determine the value of current flowing after (a) 40 ms, and (b) 80 ms. [80 ms, 0.11 A (a) 66.7 mA (b) 40.5 mA]
12. A resistor of 0.5 MΩ is connected in series with a 20 μF capacitor and the capacitor is charged to 200 V. The battery is replaced instantaneously by a conducting link. Draw a graph showing the variation of capacitor voltage with time over a period of at least 6 time constants. Determine from the graph

the approximate time for the capacitor voltage to fall to 75 V. [9.8 s]

Transients in series connected *L–R* circuits

13. A coil has an inductance of 1.2 H and a resistance of 40 Ω and is connected to a 200 V, d.c. supply. Draw the current/time characteristic and hence determine the approximate value of the current flowing 60 ms after connecting the coil to the supply. [4.3 A]
14. A 25 V d.c. supply is connected to a coil of inductance 1 H and resistance 5 Ω. Use a graphical method to draw the exponential growth curve of current and hence determine the approximate value of the current flowing 100 ms after being connected to the supply. [2 A]
15. An inductor has a resistance of 20 Ω and an inductance of 4 H. It is connected to a 50 V, d.c. supply. By drawing the appropriate characteristic find (a) the approximate value of current flowing after 0.1 s, and (b) the time for the current to grow to 1.5 A. [(a) 1 A (b) 0.18 s]
16. A direct voltage is suddenly applied to a coil of resistance *R* ohms and inductance *L* henrys. Explain briefly why current does not rise immediately to its steady-state value. [See section 15.6]
17. Explain briefly what you understand by the expression 'the time constant of a series connected *L–R* circuit'. Draw a graph to explain the significance of the term 'constant' and state the value of time constant in terms of *L* and *R*. [See section 15.7 and worked Problem 1]
18. The field winding of a 200 V, d.c. machine has a resistance of 20 Ω and an inductance of 500 mH. Calculate:
    (a) the time constant of the field winding;
    (b) the value of current flow one time constant after being connected to the supply; and
    (c) the current flowing 50 ms after the supply has been switched on.
    [(a) 25 ms (b) 6.32 A (c) 8.65 A]

The effects of circuit time constant on a rectangular wave

19. With the aid of a circuit diagram, explain briefly the effects on the waveform of the capacitor voltage of altering the value of resistance in a series connected *C–R* circuit when a rectangular wave is applied to the circuit. What do you understand by the term 'integrator circuit'? [See section 15.11]
20. With reference to a rectangular wave applied to a series connected *C–R* circuit, explain briefly the shape of the waveform when *R* is small and hence what you understand by the term 'differentiator circuit'. [See section 15.11]

Part 3

Electrical Technology

16

Three-phase systems

At the end of this chapter you should be able to:

- describe a single-phase supply
- describe a three-phase supply
- understand a star connection, and recognize that $I_L = I_p$ and $V_L = \sqrt{3}V_p$
- draw a complete phasor diagram for a balanced, star-connected load
- understand a delta connection, and recognize that $V_L = V_p$ and $I_L = \sqrt{3}I_p$
- draw a phasor diagram for a balanced, delta-connected load
- calculate power in three-phase systems using $P = \sqrt{3}\, V_L I_L \cos \phi$
- appreciate how power is measured in a three-phase system, by the one, two and three-wattmeter methods
- compare star and delta connections
- appreciate the advantages of three-phase systems

16.1 Introduction

Generation, transmission and distribution of electricity via the National Grid system is accomplished by three-phase alternating currents.

The voltage induced by a single coil when rotated in a uniform magnetic field is shown in Fig. 16.1 and is known as a **single-phase voltage**. Most consumers are fed by means of a single-phase a.c. supply. Two wires are used, one called the live conductor (usually coloured red) and the other is called the neutral conductor (usually coloured black). The neutral is usually connected via protective gear to earth, the earth wire being coloured green. The standard voltage for a single-phase a.c. supply is 240 V. The majority of single-phase supplies are obtained by connection to a three-phase supply (see Fig. 16.5).

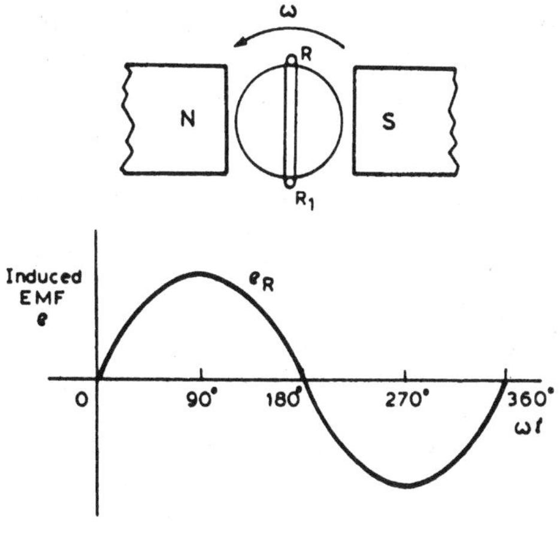

Figure 16.1

16.2 Three-phase supply

A **three-phase supply** is generated when three coils are placed 120° apart and the whole rotated in a uniform magnetic field as shown in Fig. 16.2(a). The result is three independent supplies of equal voltages which are each displaced by 120° from each other as shown in Fig. 16.2(b).

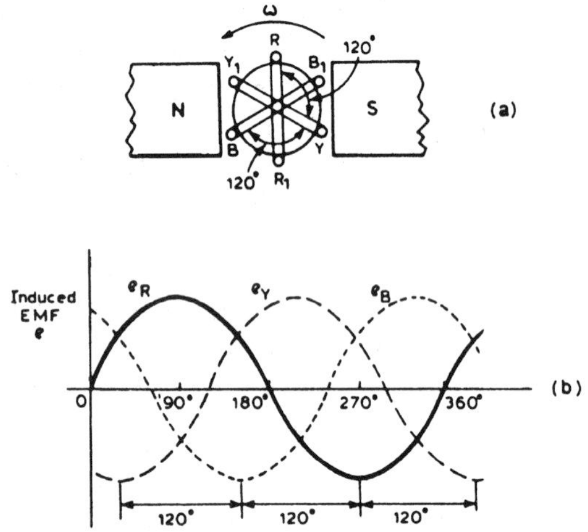

Figure 16.2

(i) The convention adopted to identify each of the phase voltages is: R-red, Y-yellow, and B-blue, as shown in Fig. 16.2.
(ii) The **phase-sequence** is given by the sequence in which the conductors pass the point initially taken by the red conductor. The national standard phase sequence is R, Y, B.

A three-phase a.c. supply is carried by three conductors, called '**lines**', which are coloured red, yellow and blue. The currents in these conductors are known as line currents (I_L) and the p.d.s between them are known as line voltages (V_L). A fourth conductor, called the **neutral** (coloured black, and connected through protective devices to earth), is often used with a three-phase supply. If the three-phase windings shown in Fig. 16.2 are kept independent then six wires are needed to connect a supply source (such as a generator) to a load (such as a motor). To reduce the number of wires it is usual to interconnect the three phases. There are two ways in which this can be done, these being (a) a star connection, and (b) a delta, or mesh, connection. Sources of three-phase supplies, i.e. alternators, are usually connected in star, whereas three-phase transformer windings, motors and other loads may be connected either in star or delta.

16.3 Star connection

(i) A **star-connected load** is shown in Fig. 16.3 where the three line conductors are each connected to a load and the outlets from the loads are joined together at N to form what is termed the **neutral point** or the **star point**.
(ii) The voltages V_R, V_Y and V_B are called **phase voltages** or line to neutral voltages. Phase voltages are generally denoted by V_p.
(iii) The voltages V_{RY}, V_{YB} and V_{BR} are called **line voltages**.

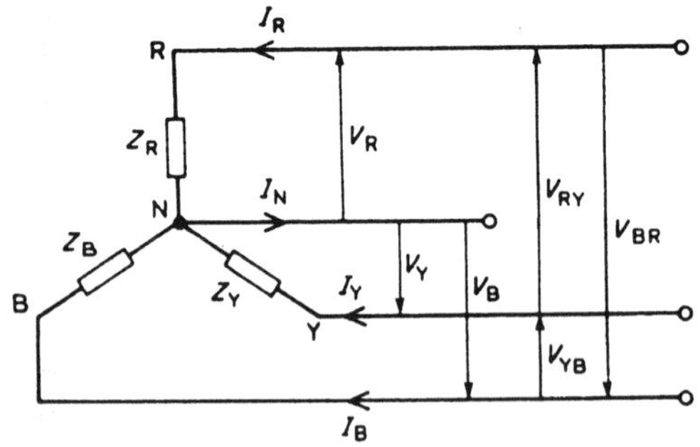

Figure 16.3

(iv) From Fig. 16.3 it can be seen that the phase currents (generally denoted by I_p) are equal to their respective line currents I_R, I_Y and I_B, i.e. for a star connection:

$$\boldsymbol{I_L = I_p}$$

(v) For a balanced system:

$I_R = I_Y = I_B$ $\quad$ $V_R = V_Y = V_B$

$V_{RY} = V_{YB} = V_{BR}$ $\quad$ $Z_R = Z_Y = Z_B$

and the current in the neutral conductor, $I_N = 0$.
When a star-connected system is balanced, then the neutral conductor is unnecessary and is often omitted.
(vi) The line voltage V_{RY}, shown in Fig. 16.4(a), is given by $V_{RY} = V_R - V_Y$ (V_Y is negative

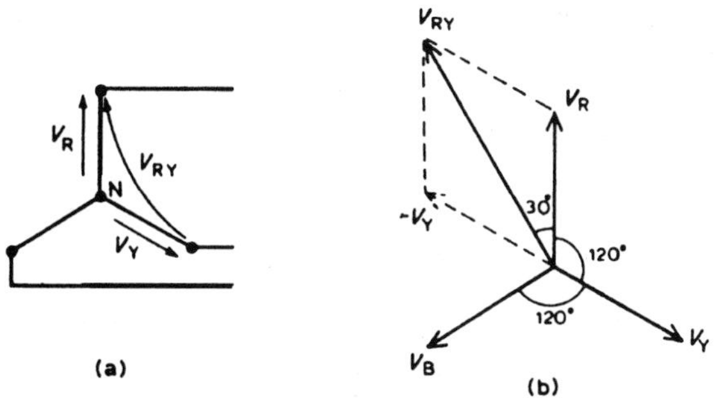

Figure 16.4

since it is in the opposite direction to V_{RY}). In the phasor diagram of Fig. 16.4(b), phasor V_Y is reversed (shown by the broken line) and then added phasorially to V_R (i.e. $V_{RY} = V_R + (-V_Y)$). By trigonometry, or by measurement, $V_{RY} = \sqrt{3}V_R$, i.e. for a balanced star connection:

$$\mathbf{V_L = \sqrt{3}V_p}$$

(See Problem 3 following for a complete phasor diagram of a star-connected system.)

(vii) The star connection of the three phases of a supply, together with a neutral conductor, allows the use of two voltages – the phase voltage and the line voltage. A 4-wire system is also used when the load is not balanced. The standard electricity supply to consumers in Great Britain is 415/240 V, 50 Hz, 3-phase, 4-wire alternating current, and a diagram of connections is shown in Fig. 16.5.

> Problem 1. Three loads, each of resistance 30 Ω, are connected in star to a 415 V, 3-phase supply. Determine (a) the system phase voltage, (b) the phase current and (c) the line current.

A '415 V, 3-phase supply' means that 415 V is the line voltage, V_L.

(a) For a star connection, $V_L = \sqrt{3}V_p$, hence phase voltage, $V_p = (V_L/\sqrt{3}) = 415/\sqrt{3} =$ **239.6 V** or **240 V** correct to 3 significant figures.

(b) Phase current, $I_p = V_p/R_p = 240/30 =$ **8 A**.

(c) For a star connection, $I_p = I_L$, hence the line current, $I_L =$ **8 A**.

> Problem 2. A star-connected load consists of three identical coils each of resistance 30 Ω and inductance 127.3 mH. If the line current is 5.08 A, calculate the line voltage if the supply frequency is 50 Hz.

Inductive reactance $X_L = 2\pi fL = 2\pi(50)(127.3 \times 10^{-3}) = 40\ \Omega$.

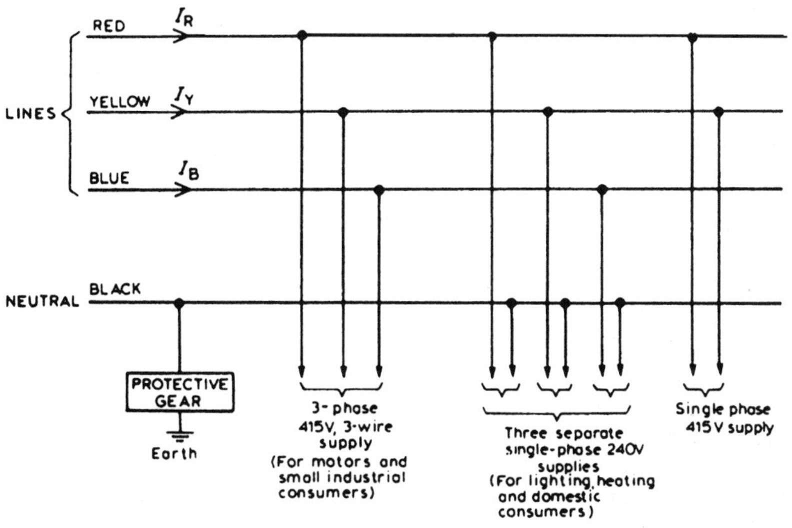

Figure 16.5

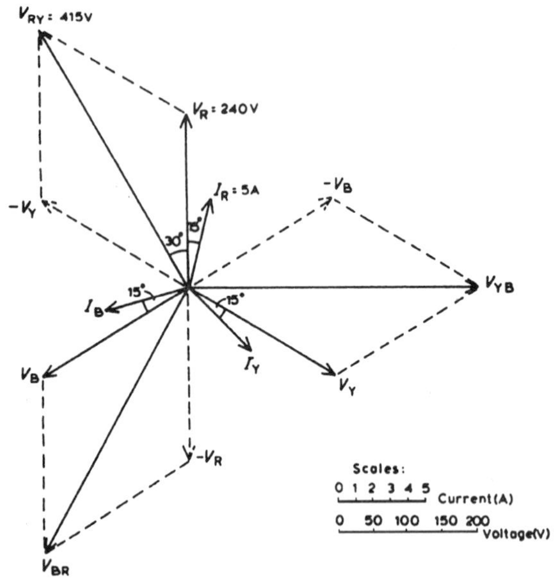

Figure 16.6

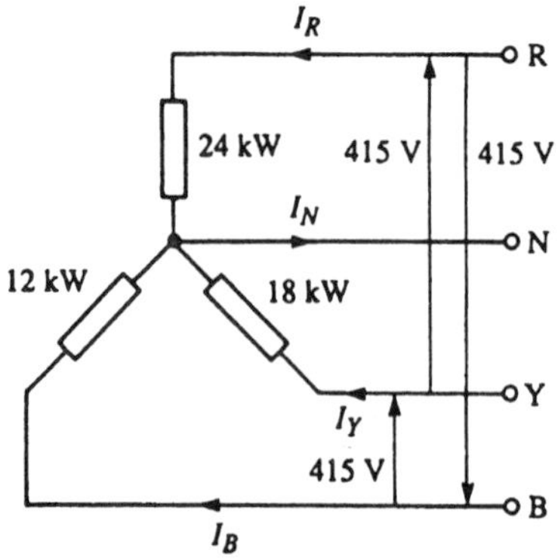

Figure 16.7

Impedance of each phase $Z_p = \sqrt{(R^2 + X_L{}^2)} = \sqrt{(30^2 + 40^2)} = 50\ \Omega$.
For a star connection $I_L = I_p = V_p/Z_p$. Hence phase voltage $V_p = I_pZ_p = (5.08)(50) = 254$ V, and line voltage $V_L = \sqrt{3}V_p = \sqrt{3}(254) = \mathbf{440\ V}$.

> Problem 3. A balanced, three-wire, star-connected, 3-phase load has a phase voltage of 240 V, a line current of 5 A and a lagging power factor of 0.966. Draw the complete phasor diagram.

The phasor diagram is shown in Fig. 16.6.
Procedure to construct the phasor diagram:

(i) Draw $V_R = V_Y = V_B = 240$ V and spaced 120° apart. (Note that V_R is shown vertically upwards – this however is immaterial for it may be drawn in any direction.)

(ii) Power factor = cos ϕ = 0.966 lagging. Hence the load phase angle is given by arccos 0.966, i.e. 15° lagging. Hence $I_R = I_Y = I_B = 5$ A, lagging V_R, V_Y and V_B respectively, by 15°.

(iii) $V_{RY} = V_R - V_Y$ (phasorially). Hence V_Y is reversed and added phasorially to V_R. By measurement, $V_{RY} = 415$ V (i.e. $\sqrt{3}(240)$) and leads V_R by 30°. Similarly, $V_{YB} = V_Y - V_B$ and $V_{BR} = V_B - V_R$.

> Problem 4. A 415 V, 3-phase, 4-wire, star-connected system supplies three resistive loads as shown in Fig. 16.7. Determine (a) the current in each line and (b) the current in the neutral conductor.

(a) For a star-connected system $V_L = \sqrt{3}V_p$, hence

$$V_p = \frac{V_L}{\sqrt{3}} = \frac{415}{\sqrt{3}} = 240\ \text{V}$$

Since current I = (Power P)/(Voltage V) for a resistive load then

$$I_R = \frac{P_R}{V_R} = \frac{24\,000}{240} = \mathbf{100\ A}$$

$$I_Y = \frac{P_Y}{V_Y} = \frac{18\,000}{240} = \mathbf{75\ A}$$

and

$$I_B = \frac{P_B}{V_B} = \frac{12\,000}{240} = \mathbf{50\ A}$$

(b) The three line currents are shown in the phasor diagram of Fig. 16.8. Since each load is resistive the currents are in phase with the phase voltages and are hence mutually displaced by 120°. The current in the neutral conductor is given by $I_N = I_R + I_Y + I_B$ phasorially.

Figure 16.9 shows the three line currents added phasorially. oa represents I_R in magnitude and direction. From the nose of oa, ab is drawn representing I_Y in magnitude and direction. From the nose of ab, bc is drawn representing I_B in magnitude and direction. oc represents the resultant, I_N. By measurement, $I_N = \mathbf{43\ A}$.

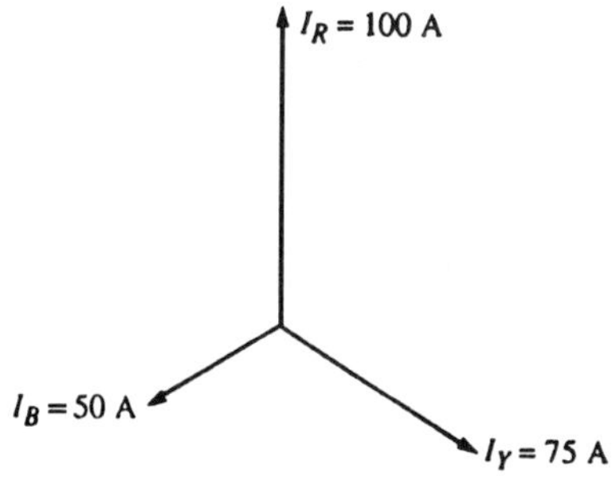

Figure 16.8

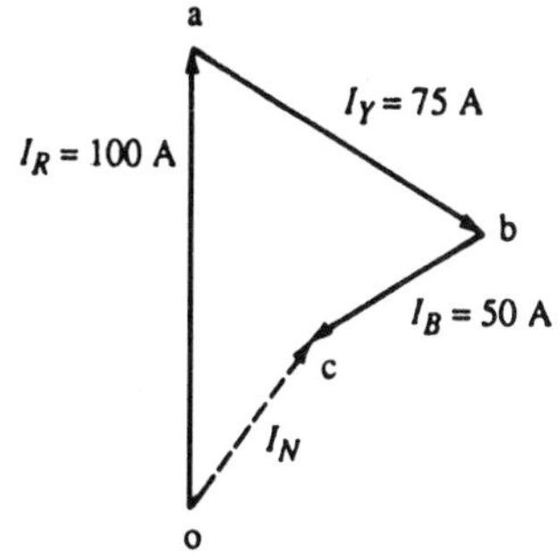

Figure 16.9

(Alternatively, by calculation, considering I_R at 90°, I_B at 210° and I_Y at 330°:
total horizontal component = 100 cos 90° + 75 cos 330° + 50 cos 210° = 21.65;
total vertical component = 100 sin 90° + 75 sin 330° + 50 sin 210° = 37.50.
Hence magnitude of $I_N = \sqrt{(21.65^2 + 37.50^2)}$ = **43.3 A**.)

16.4 Delta connection

(i) A **delta** (or **mesh**) **connected load** is shown in Fig. 16.10 where the end of one load is connected to the start of the next load.

(ii) From Fig. 16.10, it can be seen that the line voltages V_{RY}, V_{YB} and V_{BR} are the respective phase voltages, i.e. for a delta connection:

$$V_L = V_p$$

(iii) Using Kirchhoff's current law in Fig. 16.10, $I_R = I_{RY} - I_{BR}\ I_{RY} + (-I_{BR})$. From the phasor diagram shown in Fig. 16.11, by trigonometry or by measurement, $I_R = \sqrt{3}I_{RY}$, i.e. for a delta connection:

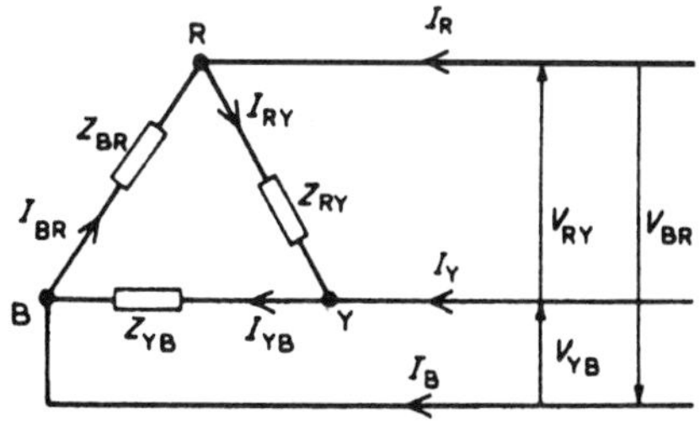

Figure 16.10

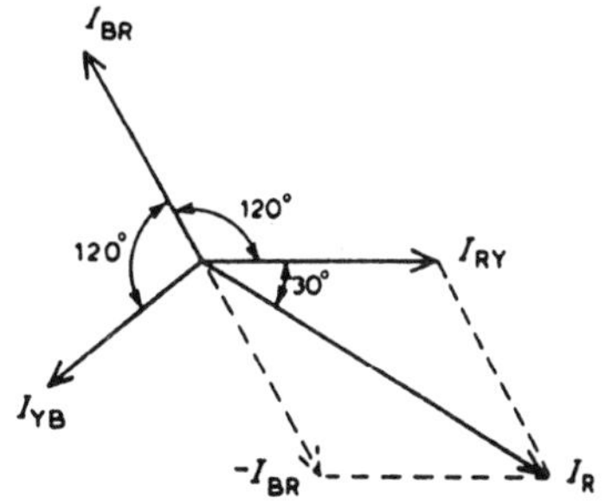

Figure 16.11

$$I_L = \sqrt{3}I_p$$

> Problem 5. Three identical coils each of resistance 30 Ω and inductance 127.3 mH are connected in delta to the 440 V, 50 Hz, 3-phase supply. Determine (a) the phase current and (b) the line current.

Phase impedance, Z_p = 50 Ω (from Problem 2) and for a delta connection, $V_p = V_L$.

(a) Phase current,

$$I_p = \frac{V_p}{Z_p} = \frac{V_L}{Z_p} = \frac{440}{50} = \mathbf{8.8\ A}$$

(b) For a delta connection, $I_L = \sqrt{3}I_p = \sqrt{3}(8.8)$ = **15.24 A**.

Thus when the load is connected in delta, three times the line current is taken from the supply than is taken if connected in star.

> Problem 6. Three identical capacitors are connected in delta to a 415 V, 50 Hz, 3-phase supply. If the line current is 15 A, determine the capacitance of each of the capacitors.

For a delta connection $I_L = \sqrt{3}I_p$. Hence phase current $I_p = I_L/\sqrt{3} = 15/\sqrt{3} = 8.66$ A.
Capacitive reactance per phase, $X_C = (V_p/I_p) = (V_L/I_p)$ (since for a delta connection $V_L = V_p$).
Hence $X_C = 415/8.66 = 47.92\ \Omega$.
$X_C = (1/2\pi fC)$, from which

$$\text{capacitance, } C = \frac{1}{2\pi f X_C} = \frac{1}{2\pi(50)(47.92)}\text{ F}$$
$$= \mathbf{66.43\ \mu F}$$

> Problem 7. Three coils each having resistance 3 Ω and inductive reactance 4 Ω are connected (i) in star, and (ii) in delta to a 415 V, 3-phase supply. Calculate for each connection (a) the line and phase voltages and (b) the phase and line currents.

(i) **For a star connection**: $I_L = I_p$ and $V_L = \sqrt{3}V_p$.
  (a) A 415 V, 3-phase supply means that the line voltage, V_L = **415 V**
  Phase voltage, $V_p = V_L/\sqrt{3} = 415/\sqrt{3}$ = **240 V**.
  (b) Impedance per phase, $Z_p = \sqrt{(R^2 + X_L^2)} = \sqrt{(3^2 + 4^2)} = 5\ \Omega$.
  Phase current, $I_p = (V_p/Z_p) = 240/5$ = **48 A**.
  Line current, $I_L = I_p$ = **48 A**.
(ii) **For a delta connection**: $V_L = V_p$ and $I_L = \sqrt{3}I_p$.
  (a) Line voltage, V_L = 415 V.
  Phase voltage, $V_p = V_L$ = **415 V**.
  (b) Phase current, $I_p = V_p/Z_p = 415/5$ = **83 A**.
  Line current, $I_L = \sqrt{3}I_p = \sqrt{3}(83)$ = **144 A**.

16.5 Power in three-phase systems

The power dissipated in a three-phase load is given by the sum of the power dissipated in each phase. If a load is balanced then the total power P is given by $P = 3 \times$ power consumed by one phase.

The power consumed in one phase $= I_p^2R_p$ or $V_pI_p \cos\phi$ (where ϕ is the phase angle between V_p and I_p).

For a star connection, $V_p = (V_L/\sqrt{3})$ and $I_p = I_L$ hence $P = 3(V_L/\sqrt{3})I_L \cos\phi = \sqrt{3}V_LI_L \cos\phi$.

For a delta connection, $V_p = V_L$ and $I_p = I_L/\sqrt{3}$, hence $P = 3V_L(I_L/\sqrt{3}) \cos\phi = \sqrt{3}V_LI_L \cos\phi$.
Hence for either a star or a delta balanced connection the total power P is given by:

> $\mathbf{P = \sqrt{3}V_LI_L \cos\phi}$ **watts or**
> $\mathbf{P = 3I_p^2R_p}$ **watts**

Total volt-amperes

> $\mathbf{S = \sqrt{3}V_LI_L}$ **volt-amperes**

> Problem 8. Three 12 Ω resistors are connected in star to a 415 V, 3-phase supply. Determine the total power dissipated by the resistors.

Power dissipated, $P = \sqrt{3}V_LI_L \cos\phi$ or $P = 3I_p^2R_p$.
Line voltage, $V_L = 415$ V and phase voltage $V_p = 415/\sqrt{3} = 240$ V (since the resistors are star connected).
Phase current, $I_p = V_p/Z_p = V_p/R_p = 240/12 = 20$ A.
For a star connection $I_L = I_p = 20$ A.
For a purely resistive load, the power factor $= \cos\phi = 1$. Hence power $P = \sqrt{3}V_LI_L \cos\phi = \sqrt{3}(415)(20)(1)$ = **14.4 kW**, or power $P = 3I_p^2R_p = 3(20)^2(12)$ = **14.4 kW**.

> Problem 9. The input power to a 3-phase a.c. motor is measured as 5 kW. If the voltage and current to the motor are 400 V and 8.6 A, respectively, determine the power factor of the system.

Power, $P = 5000$ W; line voltage, $V_L = 400$ V; line current, $I_L = 8.6$ A. Power, $P = \sqrt{3}V_LI_L \cos\phi$. Hence

$$\text{power factor} = \cos\phi = \frac{P}{\sqrt{3}V_LI_L}$$
$$= \frac{5000}{\sqrt{3}(400)(8.6)} = \mathbf{0.839}$$

> Problem 10. Three identical coils, each of resistance 10 Ω and inductance 42 mH, are connected (a) in star, and (b) in delta to a 415 V, 50 Hz, 3-phase supply. Determine the total power dissipated in each case.

(a) **Star connection**
  Inductive reactance $X_L = 2\pi fL = 2\pi(50)(42 \times 10^{-3}) = 13.19\ \Omega$.
  Phase impedance $Z_p = \sqrt{(R^2 + X_L^2)} = \sqrt{(10^2 + 13.19^2)} = 16.55\ \Omega$.
  Line voltage $V_L = 415$ V and phase voltage, $V_p = V_L/\sqrt{3} = 415/\sqrt{3} = 240$ V.

Phase current, $I_p = (V_p/Z_p) = 240/16.55 = 14.50$ A.
Line current, $I_L = I_p = 14.50$ A.
Power factor $= \cos\phi = (R_p/Z_p) = 10/16.55 = 0.6042$ lagging.
Power dissipated, $P = \sqrt{3}V_L I_L \cos\phi = \sqrt{3}(415)(14.50)(0.6042) =$ **6.3 kW**.
(Alternatively, $P = 3I_p^2 R_p = 3(14.50)^2(10) =$ **6.3 kW**.)

(b) **Delta connection**
$V_L = V_p = 415$ V, $Z_p = 16.55\ \Omega$, $\cos\phi = 0.6042$ lagging (from above).
Phase current, $I_p = (V_p/Z_p) = 415/16.55 = 25.08$ A.
Line current, $I_L = \sqrt{3}I_p = \sqrt{3}(25.08) = 43.44$ A.
Power dissipated, $P = \sqrt{3}V_L I_L \cos\phi = \sqrt{3}(415)(43.44)(0.6042) =$ **18.87 kW**.
(Alternatively, $P = 3I_p^2 R_p = 3(25.08)^2(10) =$ **18.87 kW**.)

Hence loads connected in delta dissipate three times the power than when connected in star and also take a line current three times greater.

> Problem 11. A 415 V, 3-phase a.c. motor has a power output of 12.75 kW and operates at a power factor of 0.77 lagging and with an efficiency of 85%. If the motor is delta-connected, determine (a) the power input, (b) the line current, and (c) the phase current.

(a) Efficiency = (power output)/(power input). Hence 85/100 = (12 750/power input) from which,

$$\text{power input} = \frac{12\,750 \times 100}{85} = \mathbf{15\,000\ W} \text{ or } \mathbf{15\ kW}$$

(b) Power, $P = \sqrt{3}V_L I_L \cos\phi$, hence line current,

$$I_L = \frac{P}{\sqrt{3}V_L \cos\phi} = \frac{15\,000}{\sqrt{3}(415)(0.77)} = \mathbf{27.10\ A}$$

(c) For a delta connection, $I_L = \sqrt{3}I_p$, hence phase current,

$$I_p = \frac{I_L}{\sqrt{3}} = \frac{27.10}{\sqrt{3}} = \mathbf{15.65\ A}$$

16.6 Measurement of power in three-phase systems

Power in three-phase loads may be measured by the following methods:

(i) **One-wattmeter method for a balanced load**
Wattmeter connections for both star and delta are shown in Fig. 16.12.

> Total power = 3 × wattmeter reading

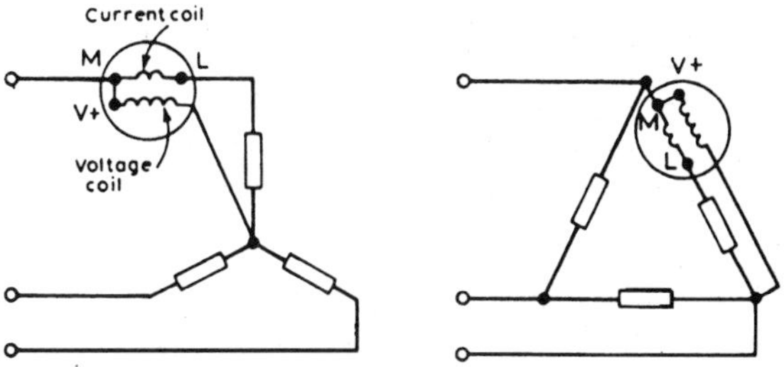

Figure 16.12

(ii) **Two-wattmeter method for balanced or unbalanced loads**
A connection diagram for this method is shown in Fig. 16.13 for a star-connected load. Similar connections are made for a delta-connected load.

> Total power = sum of wattmeter readings $= P_1 + P_2$

The power factor may be determined from:

$$\tan\phi = \sqrt{3}\left(\frac{P_1 - P_2}{P_1 + P_2}\right)$$

(see Problems 12 and 15 to 18).

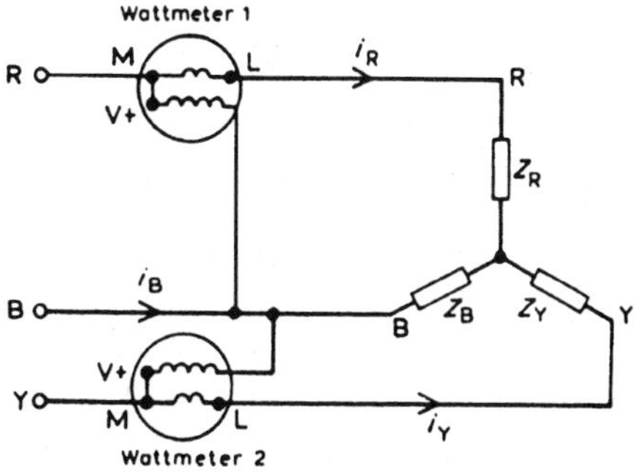

Figure 16.13

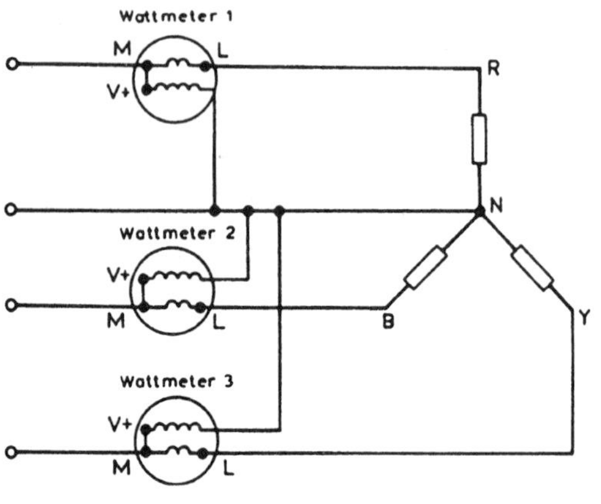

Figure 16.14

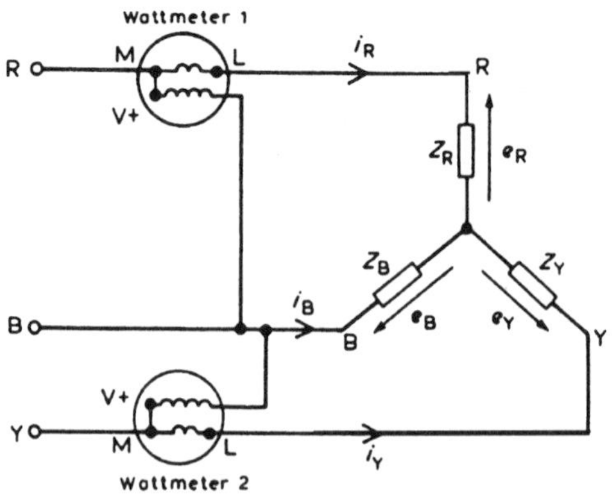

Figure 16.15

It is possible, depending on the load power factor, for one wattmeter to have to be 'reversed' to obtain a reading. In this case it is taken as a negative reading (see Problem 17).

(iii) **Three-wattmeter method for a 3-phase, 4-wire system for balanced and unbalanced loads** (see Fig. 16.14).

Total power $= P_1 + P_2 + P_3$

Problem 12. (a) Show that the total power in a 3-phase, 3-wire system using the two-wattmeter method of measurement is given by the sum of the wattmeter readings. Draw a connection diagram. (b) Draw a phasor diagram for the two-wattmeter method for a balanced load. (c) Use the phasor diagram of part (b) to derive a formula from which the power factor of a 3-phase system may be determined using only the wattmeter readings.

(a) A connection diagram for the two-wattmeter method of a power measurement is shown in Fig. 16.15 for a star-connected load. Total instantaneous power, $p = e_R i_R + e_Y i_Y + e_B i_B$ and in any 3-phase system $i_R + i_Y + i_B = 0$. Hence $i_B = -i_R - i_Y$. Thus,

$$p = e_R i_R + e_Y i_Y + e_B(-i_R - i_Y)$$
$$= (e_R - e_B)i_R + (e_Y - e_B)i_Y$$

However, $(e_R - e_B)$ is the p.d. across wattmeter 1 in Fig. 16.15 and $(e_Y - e_B)$ is the p.d. across wattmeter 2. Hence total instantaneous power, P = (wattmeter 1 reading) + (wattmeter 2 reading) $= P_1 + P_2$.

The moving systems of the wattmeters are unable to follow the variations which take place at normal frequencies and they indicate the mean power taken over a cycle. Hence the total power $P = P_1 + P_2$ for balanced or unbalanced loads.

(b) The phasor diagram for the two-wattmeter method for a balanced load having a lagging current is shown in Fig. 16.16, where $V_{RB} = V_R - V_B$ and $V_{YB} = V_Y - V_B$ (phasorially).

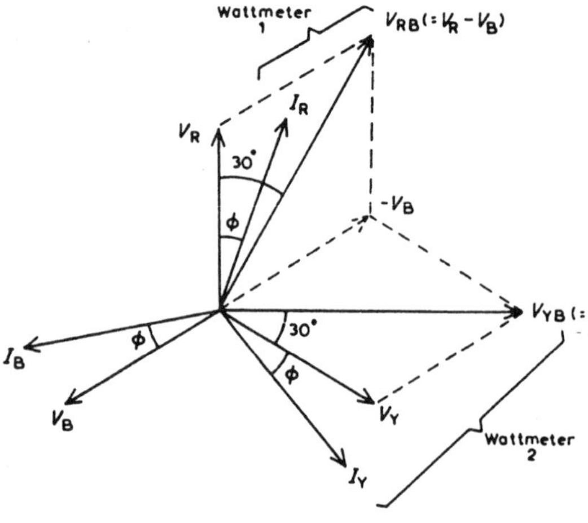

Figure 16.16

(c) Wattmeter 1 reads $V_{RB}I_R \cos(30° - \phi) = P_1$.
Wattmeter 2 reads $V_{YB}I_Y \cos(30° + \phi) = P_2$.

$$\frac{P_1}{P_2} = \frac{V_{RB}I_R \cos(30° - \phi)}{V_{YB}I_Y \cos(30° + \phi)}$$

$$= \frac{\cos(30° - \phi)}{\cos(30° + \phi)}$$

since $I_R = I_Y$ and $V_{RB} = V_{YB}$ for a balanced load. Hence

$$\frac{P_1}{P_2} = \frac{\cos 30° \cos\phi + \sin 30° \sin\phi}{\cos 30° \cos\phi - \sin 30° \sin\phi}$$

(from compound angle formulae, see *Engineering Mathematics*).
Dividing throughout by $\cos 30° \cos\phi$ gives:

$$\frac{P_1}{P_2} = \frac{1 + \tan 30° \tan\phi}{1 - \tan 30° \tan\phi}$$

$$= \frac{1 + \frac{1}{\sqrt{3}}\tan\phi}{1 - \frac{1}{\sqrt{3}}\tan\phi}, \text{ (since } \frac{\sin\phi}{\cos\phi} = \tan\phi)$$

Cross-multiplying gives:

$$P_1 - \frac{P_1}{\sqrt{3}}\tan\phi = P_2 + \frac{P_2}{\sqrt{3}}\tan\phi$$

Hence

$$P_1 - P_2 = (P_1 + P_2)\frac{\tan\phi}{\sqrt{3}}$$

from which

$$\mathbf{\tan\phi = \sqrt{3}\left(\frac{P_1 - P_2}{P_1 + P_2}\right)}$$

ϕ, $\cos\phi$ and thus power factor can be determined from this formula.

Problem 13. A 400 V, 3-phase star-connected alternator supplies a delta-connected load, each phase of which has a resistance of 30 Ω and inductive reactance 40 Ω. Calculate (a) the current supplied by the alternator, and (b) the output power and the kVA of the alternator, neglecting losses in the line between the alternator and load.

A circuit diagram of the alternator and load is shown in Fig. 16.17.

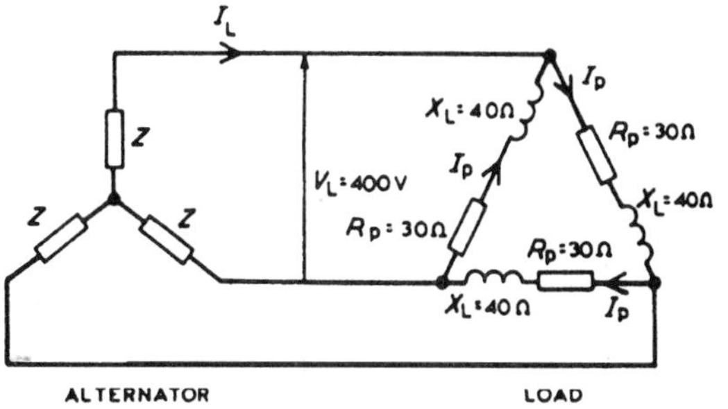

Figure 16.17

(a) Considering the load: Phase current, $I_p = V_p/Z_p$. $V_p = V_L$ for a delta connection. Hence $V_p = 400$ V.
Phase impedance, $Z_p = \sqrt{(R_p^2 + X_L^2)} = \sqrt{(30^2 + 40^2)} = 50\,\Omega$. Hence $I_p = (V_p/Z_p) = 400/50 = 8$ A.
For a delta connection, line current, $I_L = \sqrt{3}I_p = \sqrt{3}(8) = 13.86$ A.
Hence 13.86 A is the current supplied by the alternator.

(b) Alternator output power is equal to the power dissipated by the load, i.e. $P = \sqrt{3}V_LI_L \cos\phi$, where $\cos\phi = (R_p/Z_p) = 30/50 = 0.6$. Hence $P = \sqrt{3}(400)(13.86)(0.6) =$ **5.76 kW**.
Alternator output kVA, $S = \sqrt{3}V_LI_L = \sqrt{3}(400)(13.86) =$ **9.60 kVA**.

Problem 14. Each phase of a delta-connected load comprises a resistance of 30 Ω and an 80 μF capacitor in series. The load is connected to a 400 V, 50 Hz, 3-phase supply. Calculate (a) the phase current, (b) the line current, (c) the total power dissipated, and (d) the kVA rating of the load. Draw the complete phasor diagram for the load.

(a) Capacitive reactance,

$$X_C = \frac{1}{2\pi fC} = \frac{1}{2\pi(50)(80 \times 10^{-6})} = 39.79\,\Omega$$

Phase impedance, $Z_p = \sqrt{(R_p^2 + X_C^2)} = \sqrt{(30^2 + 39.79^2)} = 49.83\,\Omega$.
Power factor $= \cos\phi = (R_p/Z_p) = 30/49.83 = 0.602$. Hence $\phi = \arccos 0.602 = 52°59'$ leading.
Phase current, $I_p = V_p/Z_p$ and $V_p = V_L$ for a delta connection. Hence, $I_p = 400/49.83 =$ **8.027 A**.

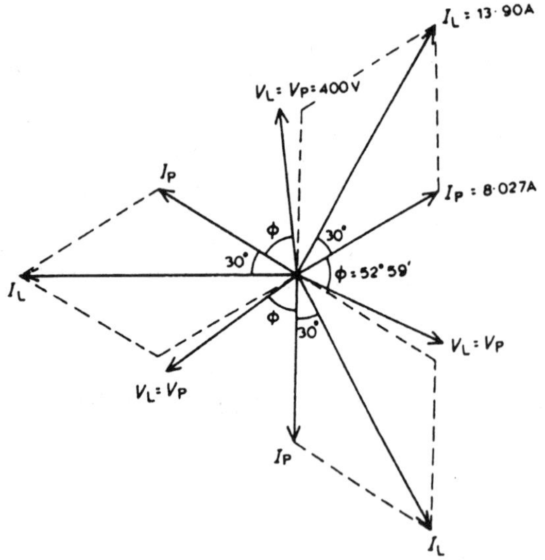

Figure 16.18

(b) Line current $I_L = \sqrt{3}I_p$ for a delta connection. Hence $I_L = \sqrt{3}(8.027) =$ **13.90 A**.
(c) Total power dissipated, $P = \sqrt{3}V_L I_L \cos\phi = \sqrt{3}(400)(13.90)(0.602) =$ **5.797 kW**.
(d) Total kVA, $S = \sqrt{3}V_L I_L = \sqrt{3}(400)(13.90) =$ **9.630 kVA**.
The phasor diagram for the load is shown in Fig. 16.18.

> Problem 15. Two wattmeters are connected to measure the input power to a balanced 3-phase load by the two-wattmeter method. If the instrument readings are 8 kW and 4 kW, determine (a) the total power input, and (b) the load power factor.

(a) Total input power, $P = P_1 + P_2 = 8 + 4 =$ **12 kW**.

(b) $\tan\phi = \sqrt{3}\left(\dfrac{P_1 - P_2}{P_1 + P_2}\right) = \sqrt{3}\left(\dfrac{8-4}{8+4}\right)$

$$= \sqrt{3}\left(\frac{4}{12}\right) = \sqrt{3}\left(\frac{1}{3}\right) = \frac{1}{\sqrt{3}}$$

Hence $\phi = \arctan(1/\sqrt{3}) = 30°$.
Power factor $= \cos\phi = \cos 30° =$ **0.866**.

> Problem 16. Two wattmeters connected to a 3-phase motor indicate the total power input to be 12 kW. The power factor is 0.6. Determine the readings of each wattmeter.

If the two wattmeters indicate P_1 and P_2, respectively, then $P_1 + P_2 = 12$ kW. (1)

$$\tan\phi = \sqrt{3}\left(\frac{P_1 - P_2}{P_1 + P_2}\right)$$

and power factor = 0.6 = $\cos\phi$. Angle $\phi = \arccos 0.6 = 53°8'$ and $\tan 53°8' = 1.3333$. Hence $1.3333 = \sqrt{3}(P_1 - P_2)/12$, from which

$$P_1 - P_2 = \frac{12(1.3333)}{\sqrt{3}}$$

i.e. $P_1 - P_2 = 9.237$ kW (2)

Adding equations (1) and (2) gives:

$$2P_1 = 21.237$$

i.e. $$P_1 = \frac{21.237}{2} = 10.62 \text{ kW}$$

Hence **wattmeter 1 reads 10.62 kW**.
From equation (1), **wattmeter 2 reads (12 – 10.62) = 1.38 kW**.

> Problem 17. Two wattmeters indicate 10 kW and 3 kW, respectively, when connected to measure the input power to a 3-phase balanced load, the reverse switch being operated on the meter indicating the 3 kW reading. Determine (a) the input power, and (b) the load power factor.

Since the reversing switch on the wattmeter had to be operated the 3 kW reading is taken as –3 kW.

(a) Total input power, $P = P_1 + P_2 = 10 + (-3) =$ **7 kW**.

(b) $\tan\phi = \sqrt{3}\left(\dfrac{P_1 - P_2}{P_1 + P_2}\right) = \sqrt{3}\left(\dfrac{10-(-3)}{10+(-3)}\right)$

$$= \sqrt{3}\left(\frac{13}{7}\right) = 3.2167$$

Angle $\phi = \arctan 3.2167 = 72°44'$.
Power factor $= \cos\phi = \cos 72°44' =$ **0.297**.

Problem 18. Three similar coils, each having a resistance of 8 Ω and an inductive reactance of 8 Ω, are connected (a) in star, and (b) in delta, across a 415 V, 3-phase supply. Calculate for each connection the readings on each of two wattmeters connected to measure the power by the two-wattmeter method.

(a) **Star connection**: $V_L = \sqrt{3}V_p$ and $I_L = I_p$.
Phase voltage, $V_p = V_L/\sqrt{3} = 415/\sqrt{3}$ and phase impedance,

$$Z_p = \sqrt{(R_p^2 + X_L^2)}$$
$$= \sqrt{(8^2 + 8^2)}$$
$$= 11.31\ \Omega$$

Hence phase current, $I_p = (V_p/Z_p) = (415/\sqrt{3})/11.31 = 21.18$ A.
Total power, $P = 3I_p^2R_p = 3(21.18)^2(8) = 10\,766$ W.
If wattmeter readings are P_1 and P_2 then $P_1 + P_2 = 10\,766$ W. (1)
Since $R_p = 8\ \Omega$ and $X_L = 8\ \Omega$, then phase angle $\phi = 45°$ (from impedance triangle):

$$\tan\phi = \sqrt{3}\left(\frac{P_1 - P_2}{P_1 + P_2}\right)$$

hence

$$\tan 45° = \frac{\sqrt{3}(P_1 - P_2)}{10\,766}$$

from which,

$$P_1 - P_2 = \frac{10\,766(1)}{\sqrt{3}} = 6216\text{ W} \qquad (16.2)$$

Adding equations (1) and (2) gives: $2P_1 = 10\,766 + 6216 = 16\,982$ W. Hence $P_1 = 8491$ W.
From equation (1), $P_2 = 10\,766 - 8491 = 2275$ W.
When the coils are star connected the wattmeter readings are thus 8.491 kW and 2.275 kW.

(b) **Delta connection**: $V_L = V_p$ and $I_L = \sqrt{3}I_p$.
Phase current, $I_p = V_p/Z_p = 415/11.31 = 36.69$ A.
Total power, $P = 3I_p^2R_p = 3(36.69)^2(8) = 32\,310$ W. Hence

$$P_1 + P_2 = 32\,310\text{ W} \qquad (3)$$

$$\tan\phi = \sqrt{3}\left(\frac{P_1 - P_2}{P_1 + P_2}\right) \text{ thus } 1 = \frac{\sqrt{3}(P_1 - P_2)}{32\,310}$$

from which,

$$P_1 - P_2 = \frac{32\,310}{\sqrt{3}} = 18\,650\text{ W} \qquad (4)$$

Adding equations (3) and (4) gives $2P_1 = 50\,960$, from which $P_1 = 25\,480$ W.
From equation (3), $P_2 = 32\,310 - 25\,480 = 6830$ W.
When the coils are delta connected the wattmeter readings are thus 25.48 kW and 6.83 kW.

16.7 Comparison of star and delta connections

(i) Loads connected in delta dissipate three times more power than when connected in star to the same supply.
(ii) For the same power, the phase currents must be the same for both delta and star connections (since power $= 3I_p^2R_p$), hence the line current in the delta-connected system is greater than the line current in the corresponding star-connected system. To achieve the same phase current in a star-connected system as in a delta-connected system, the line voltage in the star system is 3 times the line voltage in the delta system. Thus for a given power transfer, a delta system is associated with larger line currents (and thus larger conductor cross-sectional area) and a star system is associated with a larger line voltage (and thus greater insulation).

16.8 Advantages of three-phase systems

Advantages of three-phase systems over single-phase supplies include:

(i) For a given amount of power transmitted through a system, the three-phase system requires conductors with a smaller cross-sectional area. This means a saving of copper (or aluminium) and thus the original installation costs are less.
(ii) Two voltages are available (see section 16.3).
(iii) Three-phase motors are very robust, relatively cheap, generally smaller, have self-starting properties, provide a steadier output and require little maintenance compared with single-phase motors.

16.9 Multi-choice questions on three-phase systems

(Answers on page 284.)

Three loads, each of 10 Ω resistance, are connected in star to a 400 V, 3-phase supply. Determine the quantities stated in questions 1 to 5, selecting the correct answers from the following list.

(a) $\frac{40}{\sqrt{3}}$ A (b) $\sqrt{3}(16)$ kW (c) $\frac{400}{\sqrt{3}}$ V
(d) $\sqrt{3}(40)$ A (e) $\sqrt{3}(400)$ V (f) 16 kW
(g) 400 V (h) 48 kW (i) 40 A

1. Line voltage.
2. Phase voltage.
3. Phase current.
4. Line current.
5. Total power dissipated in the load.
6. Which of the following statements is false?
   (a) For the same power, loads connected in delta have a higher line voltage and a smaller line current than loads connected in star.
   (b) When using the two-wattmeter method of power measurement the power factor is unity when the wattmeter readings are the same.
   (c) A.c. may be distributed using a single-phase system with two wires, a three-phase system with three wires or a three-phase system with four wires.
   (d) The national standard phase sequence for a three-phase supply is R, Y, B.

Three loads, each of resistance 16 Ω and inductive reactance 12 Ω, are connected in delta to a 400 V, 3-phase supply. Determine the quantities stated in questions 7 to 12, selecting the correct answer from the following list.

(a) 4 Ω (b) $\sqrt{3}(400)$ V (c) $\sqrt{3}(6.4)$ kW
(d) 20 A (e) 6.4 kW (f) $\sqrt{3}(20)$ A
(g) 20 Ω (h) $\frac{20}{\sqrt{3}}$ V (i) $\frac{400}{\sqrt{3}}$ V
(j) 19.2 kW (k) 100 A (l) 400 V (m) 28 Ω

7. Phase impedance.
8. Line voltage.
9. Phase voltage.
10. Phase current.
11. Line current.
12. Total power dissipated in the load.

16.10 Short answer questions on three-phase systems

1. Explain briefly how a three-phase supply is generated.
2. State the national standard phase sequence for a three-phase supply.
3. State the two ways in which phases of a three-phase supply can be interconnected to reduce the number of conductors used compared with three single-phase systems.
4. State the relationships between line and phase currents and line and phase voltages for a star-connected system.
5. When may the neutral conductor of a star-connected system be omitted?
6. State the relationships between line and phase currents and line and phase voltages for a delta-connected system.
7. What is the standard electricity supply to consumers in Great Britain?
8. State two formulae for determining the power dissipated in the load of a three-phase balanced system.
9. By what methods may power be measured in a three-phase system?
10. State a formula from which power factor may be determined for a balanced system when using the two-wattmeter method of power measurement.
11. Loads connected in star dissipate .... the power dissipated when connected in delta and fed from the same supply.
12. Name three advantages of three-phase systems over single-phase systems.

16.11 Further questions on three-phase systems

1. Three loads, each of resistance 50 Ω, are connected in star to a 400 V, 3-phase supply. Determine (a) the phase voltage, (b) the phase current, and (c) the line current.
   [(a) 231 V (b) 4.62 A (c) 4.62 A]
2. If the loads in question 1 are connected in delta to the same supply determine (a) the phase voltage, (b) the phase current, and (c) the line current.
   [(a) 400 V (b) 8 A (c) 13.86 A]
3. A star-connected load consists of three identical coils, each of inductance 159.2 mH

and resistance 50 Ω. If the supply frequency is 50 Hz and the line current is 3 A determine (a) the phase voltage, and (b) the line voltage. [(a) 212 V (b) 367 V]

4. Obtain a relationship between the line and phase voltages and line and phase current for a delta-connected system. Three inductive loads each of resistance 75 Ω and inductance 318.4 mH are connected in delta to a 415 V, 50 Hz, 3-phase supply. Determine (a) the phase voltage, (b) the phase current, and (c) the line current.
[(a) 415 V (b) 3.32 A (c) 5.75 A]
5. Three identical capacitors are connected (a) in star, (b) in delta to a 400 V, 50 Hz, 3-phase supply. If the line current is 12 A determine in each case the capacitance of each of the capacitors.
[(a) 165.4 μF (b) 55.13 μF]
6. Three coils each having resistance 6 Ω and inductance L H are connected (a) in star, and (b) in delta, to a 415 V, 50 Hz, 3-phase supply. If the line current is 30 A, find for each connection the value of L.
[(a) 16.78 mH (b) 73.84 mH]
7. A 400 V, 3-phase, 4-wire, star-connected system supplies three resistive loads of 15 kW, 20 kW and 25 kW in the red, yellow and blue phases, respectively. Determine the current flowing in each of the four conductors.
[I_R = 64.95 A, I_Y = 86.60 A
I_B = 108.25 A, I_N = 37.50 A]
8. Determine the total power dissipated by three 20 Ω resistors when connected (a) in star, and (b) in delta to a 440 V, 3-phase supply. [(a) 9.68 kW (b) 29.04 kW]
9. Determine the power dissipated in the circuit of question 3. [1.35 kW]
10. A balanced delta-connected load has a line voltage of 400 V, a line current of 8 A and a lagging power factor of 0.94. Draw a complete phasor diagram of the load. What is the total power dissipated by the load?
[5.21 kW]
11. A 3-phase, star-connected alternator delivers a line current of 65 A to a balanced delta-connected load at a line voltage of 380 V. Calculate (a) the phase voltage of the alternator, (b) the alternator phase current, and (c) the load phase current.
[(a) 219.4 V (b) 65 A (c) 37.53 A]
12. Three inductive loads, each of resistance 4 Ω and reactance 9 Ω, are connected in delta. When connected to a 3-phase supply the loads consume 1.2 kW. Calculate (a) the power factor of the load, (b) the phase current, (c) the line current, and (d) the supply voltage.
[(a) 0.406 (b) 10 A (c) 17.32 A (d) 98.49 V]
13. The input voltage, current and power to a motor is measured as 415 V, 16.4 A and 6 kW, respectively. Determine the power factor of the system. [0.509]
14. A 440 V, 3-phase, a.c. motor has a power output of 11.25 kW and operates at a power factor of 0.8 lagging and with an efficiency of 84%. If the motor is delta connected determine (a) the power input, (b) the line current, and (c) the phase current.
[(a) 13.39 kW (b) 21.96 A (c) 12.68 A]
15. Two wattmeters are connected to measure the input power to a balanced 3-phase load. If the wattmeter readings are 9.3 kW and 5.4 kW determine (a) the total output power, and (b) the load power factor.
[(a) 14.7 kW (b) 0.909]
16. 8 kW is found by the two-wattmeter method to be the power input to a 3-phase motor. Determine the reading of each wattmeter if the power factor of the system is 0.85.
[5.431 kW, 2.569 kW]
17. Show that the power in a three-phase balanced system can be measured by two wattmeters and deduce an expression for the total power in terms of the wattmeter readings. When the two-wattmeter method is used to measure the input power of a balanced load, the readings on the wattmeters are 7.5 kW and 2.5 kW, the connections to one of the coils on the meter reading 2.5 kW having to be reversed. Determine (a) the total input power, and (b) the load power factor.
[(a) 5 kW (b) 0.277]
18. Three similar coils, each having a resistance of 4.0 Ω and an inductive reactance of 3.46 Ω are connected (a) in star, and (b) in delta across a 400 V, 3-phase supply. Calculate for each connection the readings on each of two wattmeters connected to measure the power by the two-wattmeter method.
[(a) 17.15 kW, 5.73 kW (b) 51.46 kW, 17.18 kW]
19. A 3-phase, star-connected alternator supplies a delta-connected load, each phase of which has a resistance of 15 Ω and inductive reactance 20 Ω. If the line voltage is 400 V,

calculate (a) the current supplied by the alternator, and (b) the output power and kVA rating of the alternator, neglecting any losses in the line between the alternator and the load.
[(a) 27.71 A (b) 11.52 kW, 19.2 kVA]

20. Each phase of a delta-connected load comprises a resistance of 40 Ω and a 40 μF capacitor in series. Determine, when connected to a 415 V, 50 Hz, 3-phase supply, (a) the phase current, (b) the line current, (c) the total power dissipated, and (d) the kVA rating of the load.
[(a) 4.66 A (b) 8.07 A (c) 2.605 kW (d) 5.80 kVA]

21. (a) State the advantages of three-phase supplies.
(b) Three 24 μF capacitors are connected in star across a 400 V, 50 Hz, 3-phase supply. What value of capacitance must be connected in delta in order to take the same line current? [8 μF]

17

Transformers

At the end of this chapter you should be able to:

- understand the principle of operation of a transformer
- understand the term 'rating' of a transformer
- use $V_1/V_2 = N_1/N_2 = I_2/I_1$ in calculations on transformers
- construct a transformer no-load phasor diagram and calculate magnetizing and core loss components of the no-load current
- state the e.m.f. equation for a transformer $E = 4.44 f \Phi_m N$ and use it in calculations
- construct a transformer on-load phasor diagram for an inductive circuit assuming the volt drop in the windings is negligible
- describe transformer construction
- derive the equivalent resistance, reactance and impedance referred to the primary of a transformer
- understand voltage regulation
- describe losses in transformers and calculate efficiency
- appreciate the concept of resistance matching and how it may be achieved
- perform calculations using $R_1 = (N_1/N_2)^2 R_L$
- describe an auto transformer, its advantages/disadvantages and uses
- describe an isolating transformer, stating uses
- describe a three-phase transformer
- describe current and voltage transformers

17.1 Introduction

A transformer is a device which uses the phenomenon of mutual induction (see Chapter 9) to change the values of alternating voltages and currents. In fact, one of the main advantages of a.c. transmission and distribution is the ease with which an alternating voltage can be increased or decreased by transformers.

Losses in transformers are generally low and thus efficiency is high. Being static they have a long life and are very stable.

Transformers range in size from the miniature units used in electronic applications to the large power transformers used in power stations. The principle of operation is the same for each.

A transformer is represented in Fig. 17.1(a) as consisting of two electrical circuits linked by a common ferromagnetic core. One coil is termed

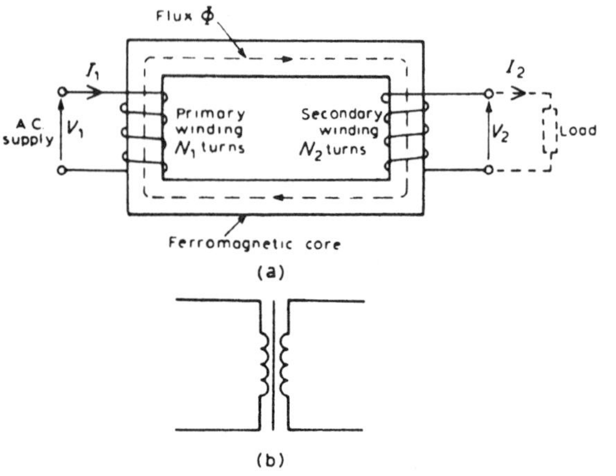

Figure 17.1

the **primary winding** which is connected to the supply of electricity, and the other the **secondary winding**, which may be connected to a load. A circuit diagram symbol for a transformer is shown in Fig. 17.1(b).

17.2 Transformer principle of operation

When the secondary is an open-circuit and an alternating voltage V_1 is applied to the primary winding, a small current – called the no-load current I_0 – flows, which sets up a magnetic flux in the core. This alternating flux links with both primary and secondary coils and induces in them e.m.f.s of E_1 and E_2, respectively, by mutual induction.

The induced e.m.f. E in a coil of N turns is given by $E = N(\Delta\Phi/t)$ volts, where $(\Delta\Phi/t)$ is the rate of change of flux. In an ideal transformer, the rate of change of flux is the same for both primary and secondary and thus $(E_1/N_1) = (E_2/N_2)$, i.e. **the induced e.m.f. per turn is constant.**

Assuming no losses, $E_1 = V_1$ and $E_2 = V_2$, hence

$$\frac{V_1}{N_1} = \frac{V_2}{N_2} \text{ or } \frac{V_1}{V_2} = \frac{N_1}{N_2} \qquad (17.1)$$

(V_1/V_2) is called the voltage ratio and (N_1/N_2) the turns ratio, or the **transformation ratio** of the transformer. If N_2 is less than N_1 then V_2 is less than V_1 and the device is termed a **step-down transformer**. If N_2 is greater than N_1 then V_2 is greater than V_1 and the device is termed a **step-up transformer**.

When a load is connected across the secondary winding, a current I_2 flows. In an ideal transformer losses are neglected and a transformer is considered to be 100% efficient. Hence input power = output power, or $V_1I_1 = V_2I_2$, i.e. in an ideal transformer, the **primary and secondary ampere-turns are equal**. Thus

$$\frac{V_1}{V_2} = \frac{I_2}{I_1} \qquad (17.2)$$

Combining equations (17.1) and (17.2) gives:

$$\boxed{\frac{V_1}{V_2} = \frac{N_1}{N_2} = \frac{I_2}{I_1}} \qquad (17.3)$$

The **rating** of a transformer is stated in terms of the volt-amperes that it can transform without overheating. With reference to Fig. 17.1(a), the transformer rating is either V_1I_1 or V_2I_2, where I_2 is the full-load secondary current.

> Problem 1. A transformer has 500 primary turns and 3000 secondary turns. If the primary voltage is 240 V, determine the secondary voltage, assuming an ideal transformer.

For an ideal transformer, voltage ratio = turns ratio, i.e.

$$\frac{V_1}{V_2} = \frac{N_1}{N_2}, \text{ hence } \frac{240}{V_2} = \frac{500}{3000}$$

Thus

$$\text{secondary voltage, } V_2 = \frac{(3000)(240)}{(500)}$$

= **1440 V** or **1.44 kV**

> Problem 2. An ideal transformer with a turns ratio of 2 : 7 is fed from a 240 V supply. Determine its output voltage.

A turns ratio of 2 : 7 means that the transformer has 2 turns on the primary for every 7 turns on the secondary (i.e. a step-up transformer). Thus, $(N_1/N_2) = 2/7$.

For an ideal transformer, $N_1/N_2 = V_1/V_2$, hence $2/7 = 240/V_2$. Thus

$$\text{secondary voltage, } V_2 = \frac{(240)(7)}{(2)} = \mathbf{840\ V}$$

> Problem 3. An ideal transformer has a turns ratio of 8 : 1 and the primary current is 3 A when it is supplied at 240 V. Calculate the secondary voltage and current.

A turns ratio of 8 : 1 means $(N_1/N_2) = (8/1)$, i.e. a step-down transformer.
$N_1/N_2 = V_1/V_2$, or secondary voltage $V_2 = V_1(N_2/N_1) = 240(1/8) =$ **30 volts**.
Also $N_1/N_2 = I_2/I_1$, hence secondary current $I_2 = I_1(N_1/N_2) = 3(8/1) =$ **24 A**.

> Problem 4. An ideal transformer, connected to a 240 V mains, supplies a 12 V, 150 W lamp. Calculate the transformer turns ratio and the current taken from the supply.

$V_1 = 240$ V, $V_2 = 12$ V, $I_2 = (P/V_2) = 150/12 = 12.5$ A.
Turns ratio $= N_1/N_2 = V_1/V_2 = 240/12 =$ **20**.
$V_1/V_2 = I_2/I_1$, from which $I_1 = I_2(V_2/V_1) = 12.5(12/240)$, hence current taken from the supply, $I_2 = 12.5/20 =$ **0.625 A**.

> Problem 5. A 12 Ω resistor is connected across the secondary winding of an ideal transformer whose secondary voltage is 120 V. Determine the primary voltage if the supply current is 4 A.

Secondary current $I_2 = (V_2/R_2) = (120/12) =$ **10 A**.
$(V_1/V_2) = (I_2/I_1)$, from which

$$\text{primary voltage } V_1 = V_2\left(\frac{I_2}{I_1}\right) = 120\left(\frac{10}{4}\right) = \mathbf{300\ volts}$$

> Problem 6. A 5 kVA single-phase transformer has a turns ratio of 10 : 1 and is fed from a 2.5 kV supply. Neglecting losses, determine (a) the full-load secondary current, (b) the minimum load resistance which can be connected across the secondary winding to give full load kVA, and (c) the primary current at full load kVA.

(a) $N_1/N_2 = 10/1$, $V_1 = 2.5$ kV $= 2500$ V.
Since $N_1/N_2 = V_1/V_2$, secondary voltage $V_2 = V_1(N_2/N_1) = 2500(1/10) = 250$ V.
The transformer rating in volt-amperes $= V_2I_2$ (at full load), i.e. $5000 = 250I_2$. Hence full load secondary current, $I_2 = 5000/250 =$ **20 A**.

(b) Minimum value of load resistance, $R_L = V_2/I_2 = 250/20 =$ **12.5 Ω**.

(c) $N_1/N_2 = I_2/I_1$, from which primary current $I_1 = I_2(N_2/N_1) = 20(1/10) =$ **2 A**.

17.3 Transformer no-load phasor diagram

(i) The core flux is common to both primary and secondary windings in a transformer and is thus taken as the reference phasor in a phasor diagram. On no-load the primary winding takes a small no-load current I_0 and since, with losses neglected, the primary winding is a pure inductor, this current lags the applied voltage V_1 by 90°. In the phasor diagram assuming no losses, shown in Fig. 17.2(a), current I_o produces the flux and is drawn in phase with the flux. The primary induced e.m.f. E_1 is in phase opposition to V_1 (by Lenz's law) and is shown 180° out of phase with V_1 and equal in magnitude. The secondary induced e.m.f. is shown for a 2 : 1 turns ratio transformer.

(ii) A no-load phasor diagram for a practical transformer is shown in Fig. 17.2(b). If current flows then losses will occur. When losses are considered then the no-load current I_0 is the phasor sum of two

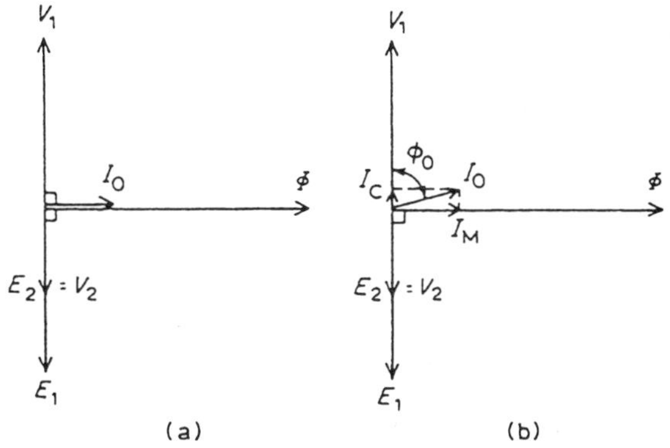

Figure 17.2

components – (i) $\boldsymbol{I_M}$, **the magnetizing component**, in phase with the flux, and (ii) $\boldsymbol{I_c}$, **the core loss component** (supplying the hysteresis and eddy current losses). From Fig. 17.2(b):
No-load current,

$$I_0 = \sqrt{(I_M^2 + I_c^2)}, \text{ where}$$
$$I_M = I_0 \sin \phi_0 \text{ and } I_c = I_0 \cos \phi_0$$

Power factor on no-load = $\cos \phi_0 = I_c/I_0$.
The total core losses (i.e. iron losses) = $V_1 I_0 \cos \phi_0$.

> Problem 7. A 2400/400 V single-phase transformer takes a no-load current of 0.5 A and the core loss is 400 W. Determine the values of the magnetizing and core loss components of the no-load current. Draw to scale the no-load phasor diagram for the transformer.

$V_1 = 2400$ V, $V_2 = 400$ V, $I_0 = 0.5$ A.
Core loss (i.e. iron loss) = 400 = $V_1 I_0 \cos \phi_0$ i.e. 400 = (2400)(0.5) $\cos \phi_0$.

$$\cos \phi_0 = \frac{400}{(2400)(0.5)} = 0.3333$$

$$\phi_0 = \arccos 0.3333 = 70.53°$$

The no-load phasor diagram is shown in Fig. 17.3.

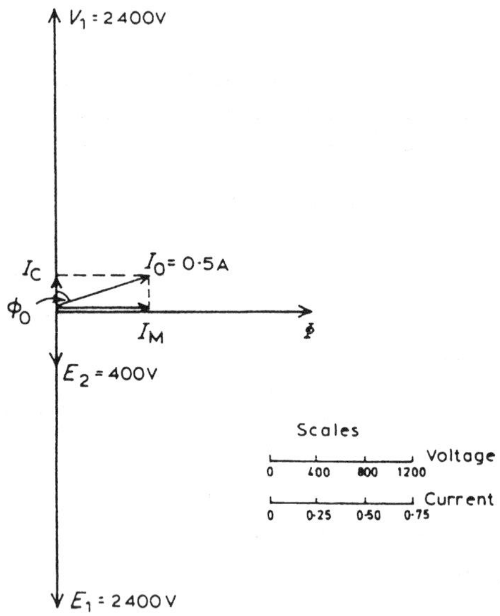

Figure 17.3

Magnetizing component, $I_M = I_0 \sin \phi_0 = 0.5 \sin 70.53° =$ **0.471 A**.
Core loss component, $I_C = I_0 \cos \phi_0 = 0.5 \cos 70.53° =$ **0.167 A**.

> Problem 8. A transformer takes a current of 0.8 A when its primary is connected to a 240 V, 50 Hz supply, the secondary being on open-circuit. If the power absorbed is 72 W, determine (a) the iron loss current, (b) the power factor on no-load, and (c) the magnetizing current.

$I_0 = 0.8$ A, $V = 240$ V.

(a) Power absorbed = total core loss = 72 = $V_1 I_0 \cos \phi_0$, hence $72 = 240 I_0 \cos \phi_0$ and iron loss current, $I_c = I_0 \cos \phi_0 = 72/240 =$ **0.30 A**.

(b) Power factor at no-load, $\cos \phi_0 = (I_c/I_0) = 0.30/0.80 =$ **0.375**.

(c) From the right-angled triangle in Fig. 17.2(b) and using Pythagoras' theorem, $I_0^2 = I_c^2 + I_M^2$, from which,

magnetizing current, I_M

$$= \sqrt{(I_0^2 - I_c^2)}$$
$$= \sqrt{(0.80^2 - 0.30^2)} = \mathbf{0.74\ A}$$

17.4 E.m.f. equation of a transformer

The magnetic flux Φ set up in the core of a transformer when an alternating voltage is applied to its primary winding is also alternating and is sinusoidal.

Let Φ_m be the maximum value of the flux and f be the frequency of the supply. The time for 1 cycle of the alternating flux is the periodic time T, where $T = (1/f)$ seconds.

The flux rises sinusoidally from zero to its maximum value in 1/4 cycle, and the time for 1/4 cycle is $1/4f$ seconds. Hence the average rate of change of flux is given by

$$\frac{\Phi_m}{1/4f} = 4f\Phi_m \text{ Wb/s}$$

and since 1 Wb/s = 1 volt, the average e.m.f. induced in each turn = $4f\Phi_m$ volts.

As the flux Φ varies sinusoidally, then a sinusoidal e.m.f. will be induced in each turn of both primary and secondary windings.

For a sine wave, form factor = (r.m.s. value/average value) = 1.11 (see Chapter 12). Hence r.m.s. value = form factor × average value = 1.11 × average value. Thus r.m.s. e.m.f. induced in each turn = 1.11 × $4f\Phi_m$ volts = $4.44f\Phi_m$ volts. Therefore, r.m.s. value of e.m.f. induced in primary is given by

$$\boldsymbol{E_1 = 4.44f\Phi_m N_1} \textbf{ volts} \qquad (17.4)$$

and r.m.s. value of e.m.f. induced in secondary is given by

$$\boldsymbol{E_2 = 4.44f\Phi_m N_2} \textbf{ volts} \qquad (17.5)$$

Dividing equation (17.4) by equation (17.5) gives $E_1/E_2 = N_1/N_2$, as previously obtained in section 17.2.

Problem 9. A 100 kVA, 4000 V/200 V, 50 Hz single-phase transformer has 100 secondary turns. Determine (a) the primary and secondary current, (b) the number of primary turns, and (c) the maximum value of the flux.

$V_1 = 4000$ V, $V_2 = 200$ V, $f = 50$ Hz, $N_2 = 100$ turns.

(a) Transformer rating = V_1I_1 = V_2I_2 = 100 000 VA.
Hence primary current
$I_1 = 100\,000/V_1 = 100\,000/4000 =$ **25 A**
and secondary current
$I_2 = 100\,000/V_2 = 100\,000/200 =$ **500 A**.

(b) From equation (17.3), $V_1/V_2 = N_1/N_2$ from which, primary turns, $N_1 = (V_1/V_2)(N_2) = (4000/200)(100)$, i.e. $\boldsymbol{N_1 = 2000}$ **turns**.

(c) From equation (17.5), $E_2 = 4.44f\Phi_m N_2$, from which,

$$\text{maximum flux } \Phi_m = \frac{E_2}{4.44fN_2}$$

$$= \frac{200}{4.44(50)(100)} \text{ (assuming } E_2 = V_2)$$

$$= \mathbf{9.01 \times 10^{-3}\ Wb} \text{ or } \mathbf{9.01\ mWb}$$

(Alternatively, equation (17.4) could have been used, where $E_1 = 4.44f\Phi_m N_1$, from which,

$$\Phi_m = \frac{E_1}{4.44fN_1} = \frac{4000}{4.44(50)(200)}$$

(assuming $E_1 = V_1$)

= 9.01 mWb, as above.)

Problem 10. A single-phase, 50 Hz transformer has 25 primary turns and 300 secondary turns. The cross-sectional area of the core is 300 cm². When the primary winding is connected to a 250 V supply, determine (a) the maximum value of the flux density in the core, and (b) the voltage induced in the secondary winding.

(a) From equation (17.4), e.m.f. $E_1 = 4.44f\Phi_m N_1$ volts, i.e. $250 = 4.44(50)\Phi_m(25)$, from which, maximum flux density,

$$\Phi_m = \frac{250}{(4.44)(50)(25)} \text{ Wb}$$

$$= 0.045\,05 \text{ Wb}$$

However, $\Phi_m = B_m \times A$, where B_m = maximum flux density in the core and A = cross-sectional area of the core (see Chapter 7). Hence $B_m \times 300 \times 10^{-4} = 0.045\,05$, from which,

$$\textbf{maximum flux density, } \boldsymbol{B_m} = \frac{0.045\,05}{300 \times 10^{-4}}$$

$$= \mathbf{1.50\ T}$$

(b) $V_1/V_2 = N_1/N_2$, from which, $V_2 = V_1(N_2/N_1)$, i.e. voltage induced in the secondary winding,

$$V_2 = (250)\left(\frac{300}{25}\right) = \mathbf{3000\ V} \text{ or } \mathbf{3\ kV}$$

Problem 11. A single phase 500 V/100 V, 50 Hz transformer has a maximum core flux density of 1.5 T and an effective core cross-sectional area of 50 cm². Determine the number of primary and secondary turns.

The e.m.f. equation for a transformer is $E = 4.44f\Phi_m N$,
and maximum flux,
$\Phi_m = B \times A = (1.5)(50 \times 10^{-4}) = 75 \times 10^{-4}$ Wb.
Since $E_1 = 4.44f\Phi_m N_1$, then primary turns,

$$N_1 = \frac{E_1}{4.44f\Phi_m} = \frac{500}{4.44(50)(75 \times 10^{-4})}$$

$$= \mathbf{300\ turns}$$

Since $E_2 = 4.4f\Phi_m N_2$, then secondary turns,

$$N_2 = \frac{E_2}{4.44 f \Phi_m} = \frac{100}{4.44(50)(75 \times 10^{-4})}$$
$$= \mathbf{60\ turns}$$

> Problem 12. A 4500 V/225 V, 50 Hz single-phase transformer is to have an approximate e.m.f. per turn of 15 V and operate with a maximum flux of 1.4 T. Calculate (a) the number of primary and secondary turns and (b) the cross-sectional area of the core.

(a) E.m.f. per turn = $E_1/N_1 = E_2/N_2 = 15$. Hence primary turns, $N_1 = E_1/15 = 4500/15 =$ **300**, and secondary turns, $N_2 = E_2/15 = 225/15 =$ **15**.

(b) E.m.f. $E_1 = 4.44 f \Phi_m N_1$, from which,

$$\Phi_m = \frac{E_1}{4.44 f N_1} = \frac{4500}{4.44(50)(300)} = 0.0676 \text{ Wb}$$

Now flux $\Phi_m = B_m \times A$, where A is the cross-sectional area of the core; hence

$$\text{area } A = \frac{\Phi_m}{B_m} = \frac{0.0676}{1.4} = \mathbf{0.0483\ m^2} \text{ or } \mathbf{483\ cm^2}$$

17.5 Transformer on-load phasor diagram

If the voltage drop in the windings of a transformer are assumed negligible, then the terminal voltage V_2 is the same as the induced e.m.f. E_2 in the secondary. Similarly, $V_1 = E_1$. Assuming an equal number of turns on primary and secondary windings, then $E_1 = E_2$, and let the load have a lagging phase angle ϕ_2.

In the phasor diagram of Fig. 17.4, current I_2 lags V_2 by angle ϕ_2. When a load is connected across the secondary winding a current I_2 flows in the secondary winding. The resulting secondary e.m.f. acts so as to tend to reduce the core flux. However this does not happen since reduction of the core flux reduces E_1, hence a reflected increase in primary current I_1' occurs which provides a restoring m.m.f. Hence at all loads primary and secondary m.m.f.s are equal, but in opposition, and the core flux remains constant. I_1' is sometimes called the 'balancing' current and is equal, but in the opposite direction, to current I_2 as shown in Fig. 17.4. I_0, shown at a phase angle ϕ_0 to V_1, is the no-load current of the transformer (see section 17.3).

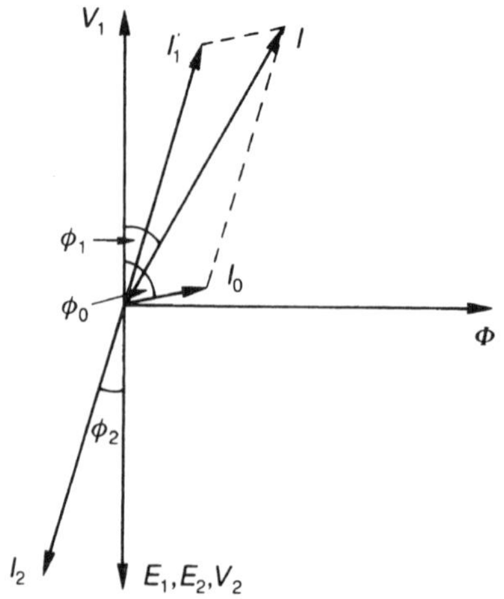

Figure 17.4

The phasor sum of I_1' and I_0 gives the supply current I_1 and the phase angle between V_1 and I_1 is shown as ϕ_1.

> Problem 13. A single-phase transformer has 2000 turns on the primary and 800 turns on the secondary. Its no-load current is 5 A at a power factor of 0.20 lagging. Assuming the volt drop in the windings is negligible, determine the primary current and power factor when the secondary current is 100 A at a power factor of 0.85 lagging.

Let I_1' be the component of the primary current which provides the restoring m.m.f. Then

$$I_1' N_1 = I_2 N_2$$

i.e.

$$I_1'(2000) = (100)(800)$$

from which,

$$I_1' = \frac{(100)(800)}{2000} = 40 \text{ A}$$

If the power factor of the secondary is 0.85, then $\cos\phi_2 = 0.85$, from which, $\phi_2 = \arccos 0.85 = 31.8°$. If the power factor on no-load is 0.20, then $\cos\phi_0 = 0.2$ and $\phi_0 = \arccos 0.2 = 78.5°$.

In the phasor diagram shown in Fig. 17.5, $I_2 = 100$ A is shown at an angle of 31.8° to V_2 and $I_1' = 40$ A is shown in anti-phase to I_2.

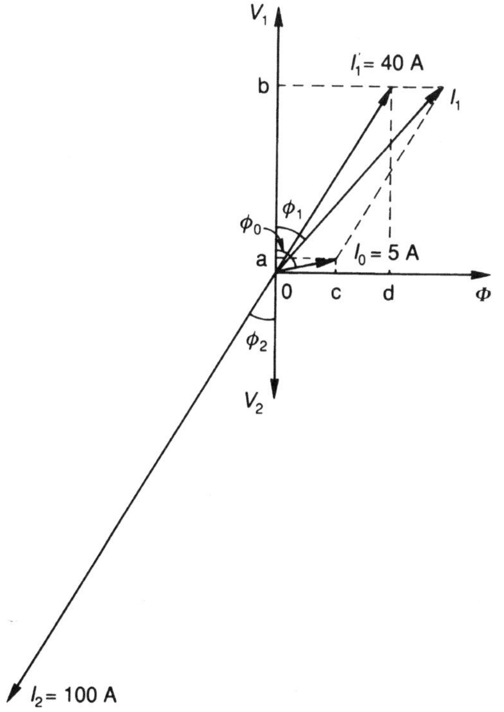

Figure 17.5

The no-load current I_0 = 5 A is shown at an angle of 78.5° to V_1. Current I_1 is the phasor sum of I_1' and I_0 and by drawing to scale, I_1 = 44 A and angle ϕ_1 = 37°.

By calculation,

$$I_1 \cos \phi_1 = \text{oa} + \text{ob}$$
$$= I_0 \cos \phi_0 + I_1' \cos \phi_2$$
$$= (5)(0.2) + (40)(0.85)$$
$$= 35.0 \text{ A}$$

and

$$I_1 \sin \phi_1 = \text{oc} + \text{od}$$
$$= I_0 \sin \phi_0 + I_1' \sin \phi_2$$
$$= (5) \sin 78.5° + 40 \sin 31.8°$$
$$= 25.98 \text{ A}$$

Hence the magnitude of $I_1 = \sqrt{(35.0^2 + 25.98^2)}$ = **43.59 A** and

$$\tan \phi_1 = \left(\frac{25.98}{35.0}\right)$$

from which,

$$\phi_1 = \arctan\left(\frac{25.98}{35.0}\right)$$
$$= \mathbf{36.59°}$$

Hence the power factor of the primary = $\cos \phi_1$ = cos 36.59° = **0.80**.

17.6 Transformer construction

(i) There are broadly two types of single-phase double-wound transformer constructions – the **core type** and the **shell type**, as shown in Fig. 17.6. The low and high voltage windings are wound as shown to reduce leakage flux.

(ii) For **power transformers**, rated possibly at several MVA and operating at a frequency of 50 Hz in Great Britain, the core material used is usually laminated silicon steel or stalloy, the laminations reducing eddy currents and the silicon steel keeping hysteresis loss to a minimum.
Large power transformers are used in the main distribution system and in industrial supply circuits. Small power transformers have many applications, examples including welding and rectifier supplies, domestic bell circuits, imported washing machines, and so on.

(iii) For **audio frequency (a.f.) transformers**, rated from a few mVA to no more than 20 VA, and operating at frequencies up to about 15 kHz, the small core is also made of laminated silicon steel. A typical application

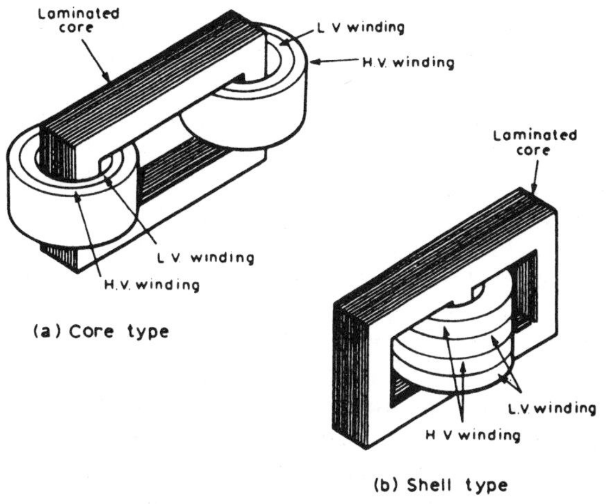

Figure 17.6

of a.f. transformers is in an audio amplifier system.

(iv) **Radio frequency (r.f.) transformers**, operating in the MHz frequency region have either an air core, a ferrite core or a dust core. Ferrite is a ceramic material having magnetic properties similar to silicon steel, but having a high resistivity. Dust cores consist of fine particules of carbonyl iron or permalloy (i.e. nickel and iron), each particle of which is insulated from its neighbour. Applications of r.f. transformers are found in radio and television receivers.

(v) Transformer **windings** are usually of enamel-insulated copper or aluminium.

(vi) **Cooling** is achieved by air in small transformers and oil in large transformers.

17.7 Equivalent circuit of a transformer

Figure 17.7 shows an equivalent circuit of a transformer. R_1 and R_2 represent the resistances of the primary and secondary windings and X_1 and X_2 represent the reactances of the primary and secondary windings, due to leakage flux.

The core losses due to hysteresis and eddy currents are allowed for by resistance R which takes a current I_c, the core loss component of the primary current. Reactance X takes the magnetizing component I_M. In a simplified equivalent circuit shown in Fig. 17.8, R and X are omitted since the no-load current I_0 is normally only about 3–5% of the full load primary current.

It is often convenient to assume all of the resistance and reactance as being on one side of the transformer. Resistance R_2 in Fig. 17.8 can be replaced by inserting an additional resistance R_2' in the primary circuit such that the power absorbed in R_2' when carrying the primary current is equal to that in R_2 due to the secondary current, i.e.

$$I_1^2 R_2' = I_2^2 R_2$$

from which,

$$R_2' = R_2 \left(\frac{I_2}{I_1}\right)^2 = R_2 \left(\frac{V_1}{V_2}\right)^2$$

Then the total equivalent resistance in the primary circuit R_e is equal to the primary and secondary resistances of the actual transformer. Hence $R_e = R_1 + R_2'$, i.e.

$$\mathbf{R_e = R_1 + R_2 \left(\frac{V_1}{V_2}\right)^2} \qquad (17.6)$$

By similar reasoning, the equivalent reactance in the primary circuit is given by $X_e = X_1 + X_2'$, i.e.

$$\mathbf{X_e = X_1 + X_2 \left(\frac{V_1}{V_2}\right)^2} \qquad (17.7)$$

The equivalent impedance Z_e of the primary and secondary windings referred to the primary is given by

$$\mathbf{Z_e = \sqrt{(R_e^2 + X_e^2)}} \qquad (17.8)$$

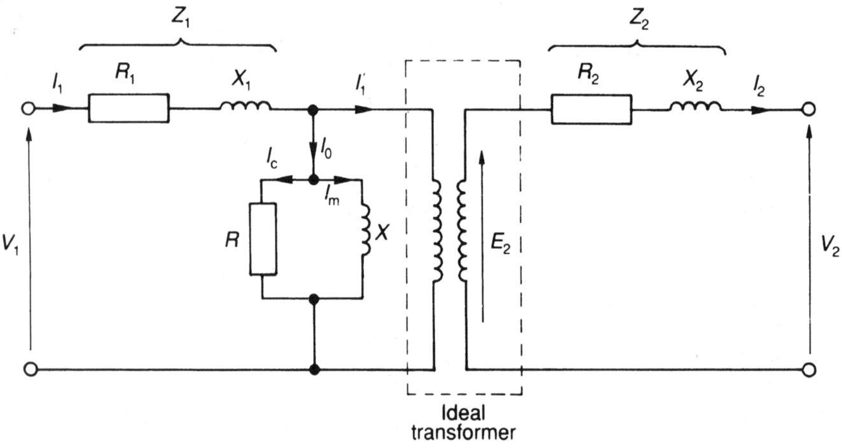

Figure 17.7

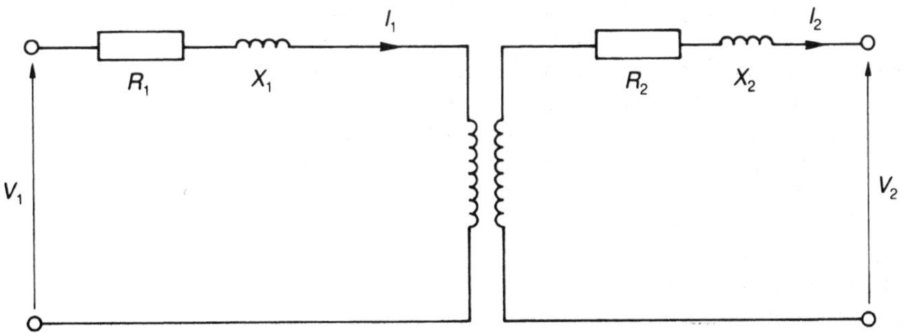

Figure 17.8

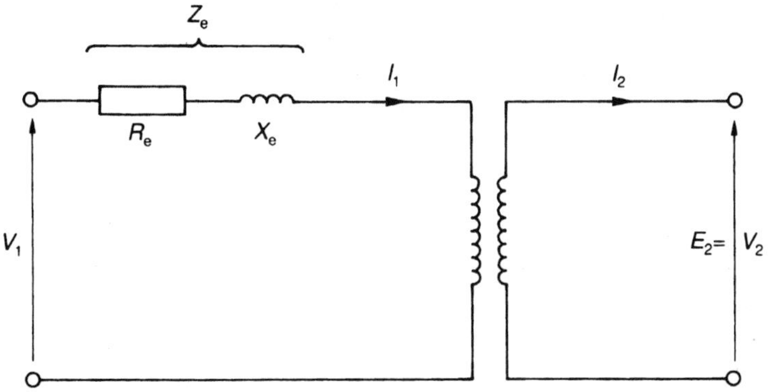

Figure 17.9

If ϕ_e is the phase angle between I_1 and the volt drop I_1Z_e then

$$\cos\phi_e = \frac{R_e}{Z_e} \qquad (17.9)$$

The simplified equivalent circuit of a transformer is shown in Fig. 17.9.

> Problem 14. A transformer has 600 primary turns and 150 secondary turns. The primary and secondary resistances are 0.25 Ω and 0.01 Ω, respectively, and the corresponding leakage reactances are 1.0 Ω and 0.04 Ω, respectively. Determine (a) the equivalent resistance referred to the primary winding, (b) the equivalent reactance referred to the primary winding, (c) the equivalent impedance referred to the primary winding, and (d) the phase angle of the impedance.

(a) From equation (17.6),

equivalent resistance $R_e = R_1 + R_2\left(\frac{V_1}{V_2}\right)^2$

i.e. $R_e = 0.25 + 0.01\left(\frac{600}{150}\right)^2$ since $\frac{V_1}{V_2} = \frac{N_1}{N_2}$

$= \mathbf{0.41\ \Omega}$

(b) From equation (17.7),

equivalent reactance, $X_e = X_1 + X_2\left(\frac{V_1}{V_2}\right)^2$

i.e. $X_e = 1.0 + 0.04\left(\frac{600}{150}\right)^2$

$= \mathbf{1.64\ \Omega}$

(c) From equation (17.8),

equivalent impedance, $Z_e = \sqrt{(R_e^2 + X_e^2)}$

$= \sqrt{(0.41^2 + 1.64^2)}$

$= \mathbf{1.69\ \Omega}$

(d) From equation (17.9),

$$\cos \phi_e = \frac{R_e}{Z_e} = \frac{0.41}{1.69}$$

Hence

$$\phi_e = \arccos\left(\frac{0.41}{1.69}\right) = \mathbf{13.64°}$$

17.8 Regulation of a transformer

When the secondary of a transformer is loaded, the secondary terminal voltage, V_2, falls. As the power factor decreases, this voltage drop increases. This is called the **regulation of the transformer** and it is usually expressed as a percentage of the secondary no-load voltage, E_2.

For full-load conditions:

$$\textbf{Regulation} = \left(\frac{E_2 - V_2}{E_2}\right) \times \mathbf{100\%} \quad (17.10)$$

The fall in voltage $(E_2 - V_2)$, is caused by the resistance and reactance of the windings.

Typical values of voltage regulation are about 3% in small transformers and about 1% in large transformers.

Problem 15. A 5 kVA, 200 V/400 V, single-phase transformer has a secondary terminal voltage of 387.6 volts when loaded. Determine the regulation of the transformer.

From equation (17.10):

regulation =

$$\frac{\text{(No-load secondary voltage – terminal voltage on load)}}{\text{no-load secondary voltage}} \times 100\%$$

$$= \left(\frac{400 - 387.6}{400}\right) \times 100\%$$

$$= \left(\frac{12.4}{400}\right) \times 100\% = \mathbf{3.1\%}$$

Problem 16. The open-circuit voltage of a transformer is 240 V. A tap-changing device is set to operate when the percentage regulation drops below 2.5%. Determine the load voltage at which the mechanism operates.

Regulation =

$$\frac{\text{(no load voltage – terminal load voltage)}}{\text{no load voltage}} \times 100\%$$

Hence

$$2.5 = \left(\frac{240 - V_2}{240}\right) 100\%$$

Therefore

$$\frac{(2.5)(240)}{100} = 240 - V_2$$

$$6 = 240 - V_2$$

from which,

load voltage, $V_2 = 240 - 6 =$ **234 volts**

17.9 Transformer losses and efficiency

There are broadly two sources of **losses in transformers** on load, these being copper losses and iron losses.

(a) **Copper losses** are variable and result in a heating of the conductors, due to the fact that they possess resistance. If R_1 and R_2 are the primary and secondary winding resistances then the total copper loss is $I_1^2R_1 + I_2^2R_2$.

(b) **Iron losses** are constant for a given value of frequency and flux density and are of two types – hysteresis loss and eddy current loss.

(i) **Hysteresis loss** is the heating of the core as a result of the internal molecular structure reversals which occur as the magnetic flux alternates. The loss is proportional to the area of the hysteresis loop and thus low-loss nickel iron alloys are used for the core since their hysteresis loops have small areas (see Chapter 7).

(ii) **Eddy current loss** is the heating of the core due to e.m.f.s being induced not only in the transformer windings but also in the core. These induced e.m.f.s set up circulating currents, called eddy currents. Owing to the low resistance of the core, eddy currents can be quite considerable and can cause a large power loss and excessive heating of the core. Eddy current losses can be reduced by increasing the resistivity of the core material or, more usually, by laminating the core (i.e. splitting it into layers or leaves) when very thin layers of insulating material can be inserted between each pair of laminations. This increases the resistance of the eddy current path, and reduces the value of the eddy current.

Transformer efficiency,

$$\eta = \frac{\text{output power}}{\text{input power}} = \frac{\text{input power} - \text{losses}}{\text{input power}}$$

i.e.

$$\boldsymbol{\eta = 1 - \frac{\textbf{losses}}{\textbf{input power}}} \qquad (17.11)$$

and is usually expressed as a percentage. It is not uncommon for power transformers to have efficiencies of between 95% and 98%.

Output power $= V_2 I_2 \cos\phi_2$; total losses = copper loss + iron losses, and input power = output power + losses.

Problem 17. A 200 kVA rated transformer has a full-load copper loss of 1.5 kW and an iron loss of 1 kW. Determine the transformer efficiency at full load and 0.85 power factor.

$$\text{Efficiency, } \eta = \frac{\text{output power}}{\text{input power}} = \frac{\text{input power} - \text{losses}}{\text{input power}} = 1 - \frac{\text{losses}}{\text{input power}}$$

Full-load output power $= VI\cos\phi = (200)(0.85)$ = 170 kW.
Total losses = 1.5 + 1.0 = 2.5 kW.

Input power = output power + losses = 170 + 2.5 = 172.5 kW. Hence

$$\text{efficiency} = \left(1 - \frac{2.5}{172.5}\right) = 1 - 0.014\,49$$
$$= 0.9855 \text{ or } \mathbf{98.55\%}$$

Problem 18. Determine the efficiency of the transformer in Problem 17 at half full-load and 0.85 power factor.

Half full-load power output $= \frac{1}{2}(200)(0.85) =$ 85 kW.
Copper loss (or I^2R loss) is proportional to current squared. Hence the copper loss at half full-load is $(\frac{1}{2})^2(1500) = 375$ W.
Iron loss = 1000 W (constant).
Total losses = 375 + 1000 = 1375 W or 1.375 kW.
Input power at half full-load = output power at half full-load + losses = 85 + 1.375 = 86.375 kW. Hence

$$\text{efficiency} = \left(1 - \frac{\text{losses}}{\text{input power}}\right)$$
$$= \left(1 - \frac{1.375}{86.375}\right) = 1 - 0.015\,92$$
$$= 0.9841 \text{ or } \mathbf{98.41\%}$$

Problem 19. A 400 kVA transformer has a primary winding resistance of 0.5 Ω and a secondary winding resistance of 0.001 Ω. The iron loss is 2.5 kW and the primary and secondary voltages are 5 kV and 320 V respectively. If the power factor of the load is 0.85, determine the efficiency of the transformer (a) on full-load, and (b) on half load.

(a) Rating = 400 kVA $= V_1 I_1 = V_2 I_2$. Hence primary current,

$$I_1 = \frac{400 \times 10^3}{V_1} = \frac{400 \times 10^3}{5000} = 80 \text{ A}$$

and secondary current,

$$I_2 = \frac{400 \times 10^3}{V_2} = \frac{400 \times 10^3}{320} = 1250 \text{ A}$$

Total copper loss $= I_1^2 R_1 + I_2^2 R_2$ (where R_1 = 0.5 Ω and R_2 = 0.001 Ω)

$$= (80)^2(0.5) + (1250)^2(0.001)$$
$$= 3200 + 1562.5 = 4762.5 \text{ watts}$$

On full-load, total loss = copper loss + iron loss

$= 4762.5 + 2500$

$= 7262.5\text{ W} = 7.2625\text{ kW}$

Total output power on full-load
$= V_2 I_2 \cos\phi_2$

$= (400 \times 10^3)(0.85)$

$= 340\text{ kW}$

Input power = output power + losses = 340 kW + 7.2625 kW, i.e. ≠ 347.2625 kW.

$$\text{Efficiency, } \eta = \left(1 - \frac{\text{losses}}{\text{input power}}\right) \times 100\%$$

$$= \left(1 - \frac{7.2625}{347.2625}\right) \times 100\% = \mathbf{97.91\%}$$

(b) Since the copper loss varies as the square of the current, then total copper loss on half-load $= 4762.5 \times (\frac{1}{2})^2 = 1190.625$ W. Hence total loss on half-load = 1190.625 + 2500

$= 3690.625\text{ W or } 3.691\text{ kW}$

Output power on half full-load $= \frac{1}{2}(340) =$ 170 kW.
Input power on half full-load = output power + losses

$= 170\text{ kW} + 3.691\text{ kW}$

$= 173.691\text{ kW}$

Hence efficiency at half full-load is given by

$$= \left(1 - \frac{\text{losses}}{\text{input power}}\right) \times 100\%$$

$$= \left(1 - \frac{3.691}{173.691}\right) \times 100\% = \mathbf{97.88\%}$$

Maximum efficiency

It may be shown that the efficiency of a transformer is a maximum when the variable copper loss (i.e. $I_1^2R_1 + I_2^2R_2$) is equal to the constant iron losses.

> Problem 20. A 500 kVA transformer has a full-load copper loss of 4 kW and an iron loss of 2.5 kW. Determine (a) the output kVA at which the efficiency of the transformer is a maximum, and (b) the maximum efficiency, assuming the power factor of the load is 0.75.

(a) Let x be the fraction of full-load kVA at which the efficiency is a maximum. The corresponding total copper loss = (4 kW)(x^2). At maximum efficiency, copper loss = iron loss. Hence $4x^2 = 2.5$, from which $x^2 = (2.5/4)$ and $x = \sqrt{(2.5/4)} = 0.791$.
Hence **the output kVA at maximum efficiency** = 0.791 × 500.

$= \mathbf{395.5\text{ kVA}}$

(b) Total loss at maximum efficiency = 2 × 2.5 = 5 kW.
Output power = 395.5 kVA × p.f. = 395.5 × 0.75, i.e. 296.625 kW.
Input power = output power + losses, i.e. 296.625 + 5 = 301.625 kW.

Maximum efficiency, η

$$= \left(1 - \frac{\text{losses}}{\text{input power}}\right) \times 100\%$$

$$= \left(1 - \frac{5}{301.625}\right) \times 100\% = \mathbf{98.34\%}$$

17.10 Resistance matching

Varying a load resistance to be equal, or almost equal, to the source internal resistance is called **matching**. Examples where resistance matching is important include coupling an aerial to a transmitter or receiver, or in coupling a loudspeaker to an amplifier where coupling transformers may be used to give maximum power transfer.

With d.c. generators or secondary cells, the internal resistance is usually very small. In such cases, if an attempt is made to make the load resistance as small as the source internal resistance, overloading of the source results.

A method of achieving maximum power transfer between a source and a load (see section 11.9), is to adjust the value of the load resistance to 'match' the source internal resistance. A transformer may be used as a **resistance matching device** by connecting it between the load and the source.

The reason why a transformer can be used for this is shown below. With reference to Fig. 17.10:

$$R_L = \frac{V_2}{I_2} \text{ and } R_1 = \frac{V_1}{I_1}$$

For an ideal transformer,

$$V_1 = \left(\frac{N_1}{N_2}\right) V_2 \text{ and } I_1 = \left(\frac{N_2}{N_1}\right) I_2$$

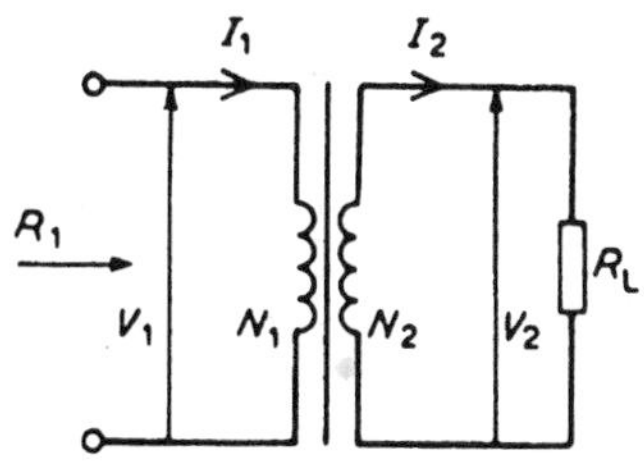

Figure 17.10

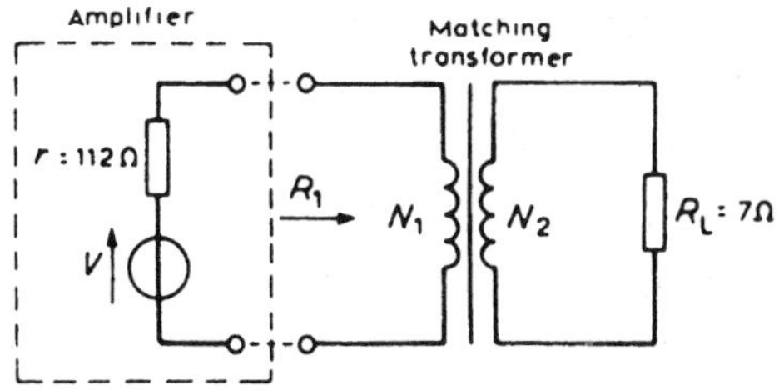

Figure 17.11

Thus the equivalent input resistance R_1 of the transformer is given by:

$$R_1 = \frac{V_1}{I_1} = \frac{\left(\frac{N_1}{N_2}\right)V_2}{\left(\frac{N_2}{N_1}\right)I_2} = \left(\frac{N_1}{N_2}\right)^2\left(\frac{V_2}{I_2}\right) = \left(\frac{N_1}{N_2}\right)^2 R_L$$

i.e.

$$\boxed{R_1 = \left(\frac{N_1}{N_2}\right)^2 R_L}$$

Hence by varying the value of the turns ratio, the equivalent input resistance of a transformer can be 'matched' to the internal resistance of a load to achieve maximum power transfer.

> Problem 21. A transformer having a turns ratio of 4 : 1 supplies a load of resistance 100 Ω. Determine the equivalent input resistance of the transformer.

From above, the equivalent input resistance is given by

$$R_1 = \left(\frac{N_1}{N_2}\right)^2 R_L$$

$$= \left(\frac{4}{1}\right)^2 (100) = \mathbf{1600\ \Omega}$$

> Problem 22. The output stage of an amplifier has an output resistance of 112 Ω. Calculate the optimum turns ratio of a transformer which would match a load resistance of 7 Ω to the output resistance of the amplifier.

The circuit is shown in Fig. 17.11.

The equivalent input resistance, R_1 of the transformer needs to be 112 Ω for maximum power transfer.

$$R_1 = \left(\frac{N_1}{N_2}\right)^2 R_L$$

Hence

$$\left(\frac{N_1}{N_2}\right)^2 = \frac{R_1}{R_L} = \frac{112}{7} = 16$$

i.e. $$\frac{N_1}{N_2} = \sqrt{(16)} = 4$$

Hence the optimum turns ratio is 4 : 1.

> Problem 23. Determine the optimum value of load resistance for maximum power transfer if the load is connected to an amplifier of output resistance 150 Ω through a transformer with a turns ratio of 5 : 1.

The equivalent input resistance R_1 of the transformer needs to be 150 Ω for maximum power transfer.

$R_1 = (N_1/N_2)^2 R_L$, from which

$$R_L = R_1\left(\frac{N_2}{N_1}\right)^2 = 150\left(\frac{1}{5}\right)^2 = \mathbf{6\ \Omega}$$

> Problem 24. A single-phase, 220 V/1760 V ideal transformer is supplied from a 220 V source through a cable of resistance 2 Ω. If the load across the secondary winding is 1.28 kΩ determine (a) the primary current flowing, and (b) the power dissipated in the load resistor.

The circuit diagram is shown in Fig. 17.12.

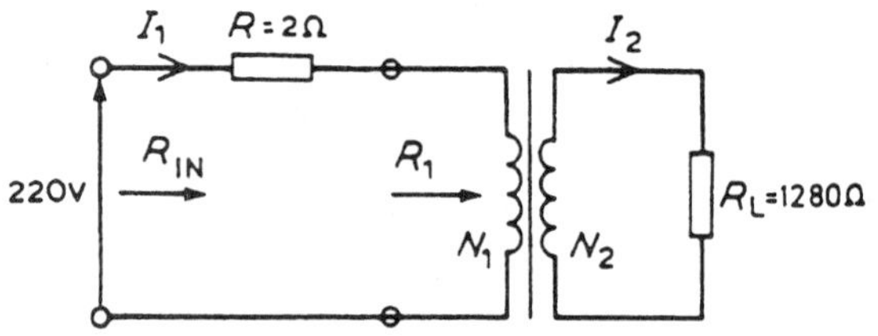

Figure 17.12

(a) Turns ratio $N_1/N_2 = V_1/V_2 = 220/1760 = 1/8$. Equivalent input resistance of the transformer,

$$R_1 = \left(\frac{N_1}{N_2}\right)^2 R_L$$

$$= \left(\frac{1}{8}\right)^2 (1.28 \times 10^3) = 20\ \Omega$$

Total input resistance, $R_{IN} = R + R_1 = 2 + 20 = 22\ \Omega$.

Primary current, $I_1 = \dfrac{V_1}{R_{IN}} = \dfrac{220}{22} = \mathbf{10\ A}$

(b) For an ideal transformer $V_1/V_2 = I_2/I_1$, from which

$$I_2 = I_1 \left(\frac{V_1}{V_2}\right)$$

$$= 10 \left(\frac{220}{1760}\right) = 1.25\ \text{A}$$

Power dissipated in load resistor R_L is given by

$$P = I_2^2 R_L = (1.25)^2 (1.28 \times 10^3)$$

$$= \mathbf{2000\ watts}\ \text{or}\ \mathbf{2\ kW}$$

> Problem 25. An a.c. source of 24 V and internal resistance 15 kΩ is matched to a load by a 25 : 1 ideal transformer. Determine (a) the value of the load resistance, and (b) the power dissipated in the load.

The circuit diagram is shown in Fig. 17.13.

(a) For maximum power transfer R_1 needs to be equal to 15 kΩ. $R_1 = (N_1/N_2)^2 R_L$, from which

load resistance, $R_L = R_1 \left(\dfrac{N_2}{N_1}\right)^2$

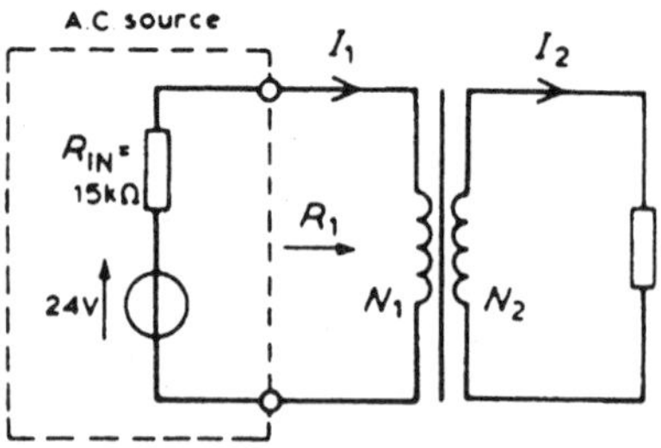

Figure 17.13

$$= (15\,000) \left(\frac{1}{25}\right)^2 = \mathbf{24\ \Omega}$$

(b) The total input resistance when the source is connected to the matching transformer is $R_{IN} + R_1$, i.e. 15 kΩ + 15 kΩ = 30 kΩ. Primary current, $I_1 = V/30\,000 = 24/30\,000 = 0.8$ mA. $N_1/N_2 = I_2/I_1$, from which

$$I_2 = I_1 \left(\frac{N_1}{N_2}\right) = (0.8 \times 10^{-3}) \left(\frac{25}{1}\right)$$

$$= 20 \times 10^{-3}\ \text{A}$$

Power dissipated in the load R_L is given by

$$P = I_2^2 R_L = (20 \times 10^{-3})^2 (24)$$

$$= 9600 \times 10^{-6}\ \text{W} = \mathbf{9.6\ mW}$$

17.11 Auto transformers

An auto transformer is a transformer which has part of its winding common to the primary and secondary circuits. Fig. 17.14(a) shows the circuit for a double-wound transformer and Fig. 17.14(b) that for an auto transformer. The latter shows that the secondary is actually part of the primary, the current in the secondary being $(I_2 - I_1)$. Since the current is less in this section, the cross-sectional area of the winding can be reduced, which reduces the amount of material necessary.

Figure 17.15 shows the circuit diagram symbol for an auto transformer.

> Problem 26. A single-phase auto transformer has a voltage ratio 320 V : 250 V and supplies a load of 200 kVA at 250 V. Assuming an ideal transformer, determine the current in each section of the winding.

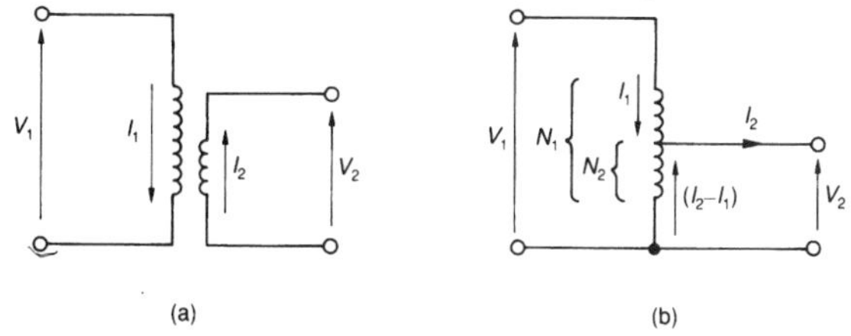

Figure 17.14

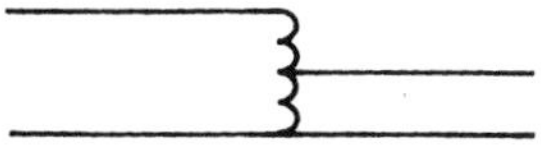

Figure 17.15

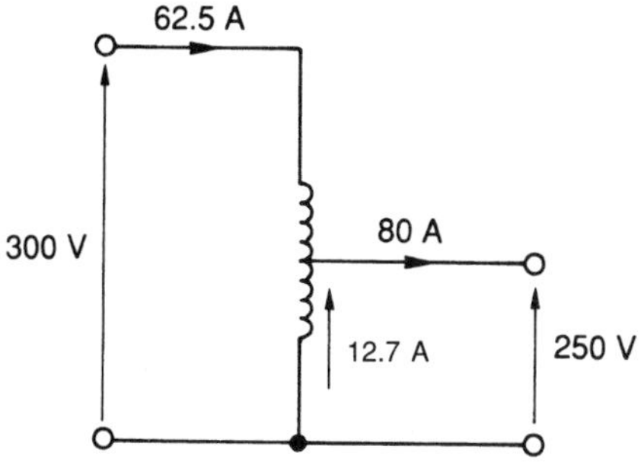

Figure 17.16

Rating = 20 kVA = $V_1I_1 = V_2I_2$. Hence primary current,

$$I_1 = \frac{20 \times 10^3}{V_1} = \frac{20 \times 10^3}{320} = \mathbf{62.5\ A}$$

and secondary current,

$$I_2 = \frac{20 \times 10^3}{V_2} = \frac{20 \times 10^3}{250} = \mathbf{80\ A}$$

Hence current in common part of the winding = 80 – 62.5 = **17.5 A.**

The current flowing in each section of the transformer is shown in Fig. 17.16.

Saving of copper in an auto transformer

For the same output and voltage ratio, the auto transformer requires less copper than an ordinary double-wound transformer. This is explained below.

The volume, and hence weight, of copper required in a winding is proportional to the number of turns and to the cross-sectional area of the wire. In turn this is proportional to the current to be carried, i.e. volume of copper is proportional to NI.

Volume of copper in an auto transformer

$$\propto (N_1 - N_2)I_1 + N_2(I_2 - I_1)$$

$$\propto N_1I_1 - N_2I_1 + N_2I_2 - N_2I_1$$

$$\propto N_1I_1 + N_2I_2 - 2N_2I_1$$

$$\propto 2N_1I_1 - 2N_2I_1 \text{ (since } N_2I_2 = N_1I_1)$$

Volume of copper in a double-wound transformer

$$\propto N_1I_1 + N_2I_2$$

$$\propto 2N_1I_1 \text{ (again, since } N_2I_2 = N_1I_1)$$

Hence

$$\frac{\text{volume of copper in an auto transformer}}{\text{volume of copper in a double-wound transformer}}$$

$$= \frac{2N_1I_1 - 2N_2I_1}{2N_1I_1}$$

$$= \frac{2N_1I_1}{2N_1I_1} - \frac{2N_2I_1}{2N_1I_1} = 1 - \frac{N_2}{N_1}$$

If $(N_2/N_1) = x$ then

(volume of copper in an auto transformer) = (1 – x)(volume of copper in a double-wound transformer) (17.12)

If, say, $x = \frac{4}{5}$ then

(volume of copper in auto transformer)

$$= \left(1 - \frac{4}{5}\right)\left(\begin{array}{c}\text{volume of copper in a}\\ \text{double-wound transformer}\end{array}\right)$$

$$= \frac{1}{5}\text{(volume in double-wound transformer)}$$

i.e. a saving of 80%.

Similarly, if $x = \frac{1}{4}$, the saving is 25%, and so on. The closer N_2 is to N_1, the greater the saving in copper.

Problem 27. Determine the saving in the volume of copper used in an auto transformer compared with a double-wound transformer for (a) a 200 V : 150 V transformer, and (b) a 500 V : 100 V transformer.

(a) For a 200 V : 150 V transformer,

$$x = \frac{V_2}{V_1} = \frac{150}{200} = 0.75$$

Hence from equation (17.12),

(volume of copper in auto transformer)

= (1 – 0.75) (volume of copper in double-wound transformer

= (0.25) (volume of copper in double-wound transformer)

or 25% of copper in a double-wound transformer. **Hence the saving is 75%.**

(b) For a 500 V : 100 V transformer,

$$x = \frac{V_2}{V_1} = \frac{100}{500} = 0.2$$

Hence

(volume of copper in auto transformer)

= (1 – 0.2) (volume of copper in double-wound transformer)

= (0.8) (volume in double-wound transformer)

or 80% of copper in a double-wound transformer. **Hence the saving is 20%.**

Advantages of auto transformers

The advantages of auto transformers over double-wound transformers include:

1. a saving in cost since less copper is needed (see above);
2. less volume, hence less weight;
3. a higher efficiency, resulting from lower I^2R losses;
4. a continuously variable output voltage is achievable if a sliding contact is used;
5. a smaller percentage voltage regulation.

Disadvantages of auto transformers

The primary and secondary windings are not electrically separate, hence if an open-circuit occurs in the secondary winding the full primary voltage appears across the secondary.

Uses of auto transformers

Auto transformers are used for reducing the voltage when starting induction motors (see Chapter 19) and for interconnecting systems that are operating at approximately the same voltage.

17.12 Isolating transformers

Transformers not only enable current or voltage to be transformed to some different magnitude but provide a means of isolating electrically one part of a circuit from another when there is no electrical connection between primary and secondary windings. An **isolating transformer** is a

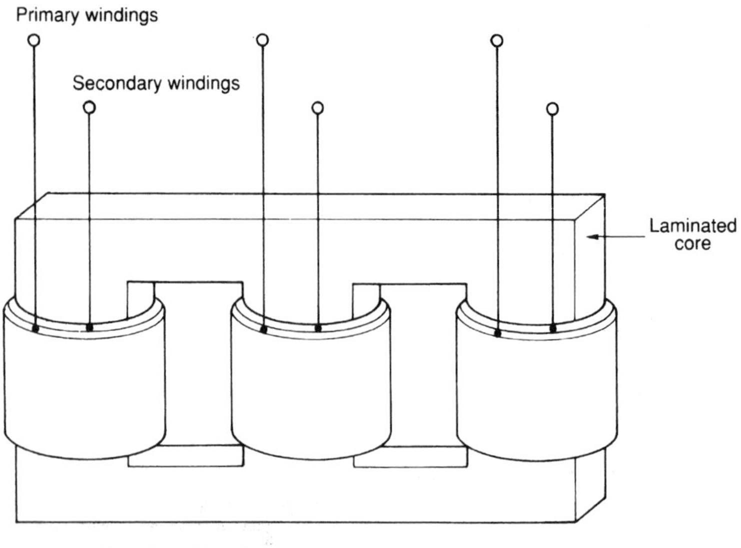

Figure 17.17

1 : 1 ratio transformer with several important applications, including bathroom shaver-sockets, portable electric tools, model railways, and so on.

17.13 Three-phase transformers

Three-phase double-wound transformers are mainly used in power transmission and are usually of the core type. They basically consist of three pairs of single-phase windings mounted on one core, as shown in Fig. 17.17, which gives a considerable saving in the amount of iron used. The primary and secondary windings in Fig. 17.17 are wound on top of each other in the form of concentric cylinders, similar to that shown in Fig. 17.6(a). The windings may be with the primary delta connected and the secondary star connected, or star–delta, star–star or delta–delta, depending on its use.

A delta connection is shown in Fig. 17.18(a) and a star connection in Fig. 17.18(b).

> Problem 28. A three-phase transformer has 500 primary turns and 50 secondary turns. If the supply voltage is 2.4 kV find the secondary line voltage on no-load when the windings are connected (a) star–delta, (b) delta–star.

(a) For a star connection, $V_L = \sqrt{3}V_p$ (see Chapter 16).
Primary phase voltage,

$$V_p = \frac{V_{L1}}{\sqrt{3}} = \frac{2400}{\sqrt{3}} = 1385.64 \text{ volts}$$

For a delta connection, $V_L = V_p$. $(N_1/N_2) = (V_1/V_2)$, from which,

secondary phase voltage, $V_{p2} = V_1\left(\frac{N_2}{N_1}\right)$

$= (1385.64)\left(\frac{50}{500}\right)$

$=$ **138.6 volts**

(b) For a delta connection, $V_L = V_p$, hence primary phase voltage = 2.4 kV = 2400 volts.

Secondary phase voltage, $V_{p2} = V_1\left(\frac{N_2}{N_1}\right)$

$= (2400)\left(\frac{50}{500}\right)$

$= 240$ volts

For a star connection, $V_L = \sqrt{3}V_p$, hence the secondary line voltage $= \sqrt{3}(240) =$ **416 volts.**

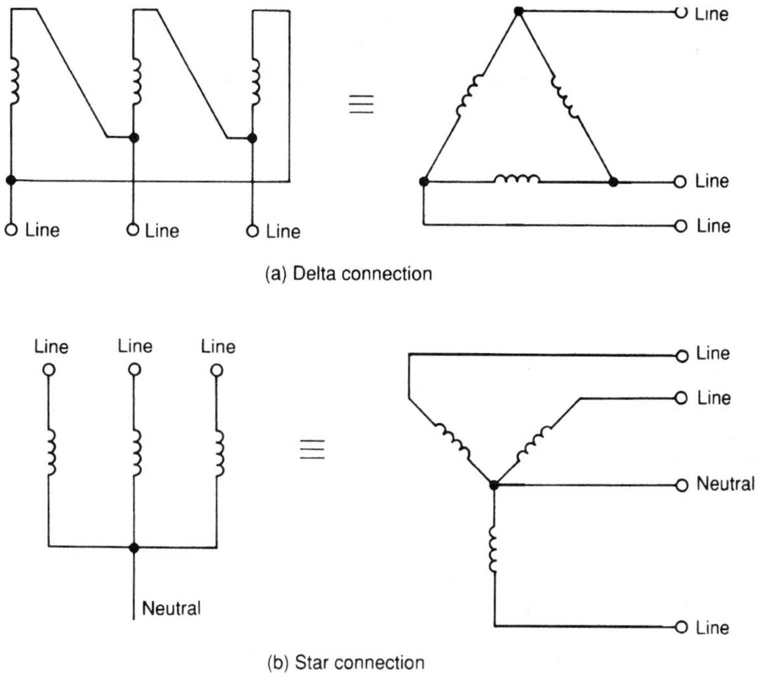

Figure 17.18

17.14 Current transformers

For measuring currents in excess of about 100 A a current transformer is normally used. With a d.c. moving-coil ammeter the current required to give full-scale deflection is very small – typically a few milliamperes. When larger currents are to be measured a shunt resistor is added to the circuit (see Chapter 10). However, even with shunt resistors added it is not possible to measure very large currents. When a.c. is being measured a shunt cannot be used since the proportion of the current which flows in the meter will depend on its impedance, which varies with frequency.

In a double-wound transformer:

$$\frac{I_1}{I_2} = \frac{N_2}{N_1}$$

from which,

$$\textbf{secondary current } \boldsymbol{I_2 = I_1 \left(\frac{N_1}{N_2}\right)}$$

In current transformers the primary usually consists of one or two turns while the secondary can have several hundred turns. A typical arrangement is shown in Fig. 17.19.

If, for example, the primary has 2 turns and the secondary 200 turns, then if the primary current is 500 A,

$$\text{secondary current, } I_2 = I_1 \left(\frac{N_1}{N_2}\right)$$

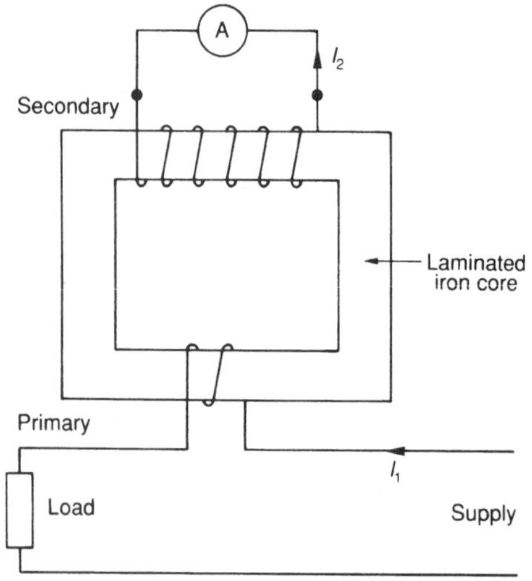

Figure 17.19

$$= (500)\left(\frac{2}{200}\right) = 5 \text{ A}$$

Current transformers isolate the ammeter from the main circuit and allow the use of a standard range of ammeters giving full-scale deflections of 1 A, 2 A or 5 A.

For very large currents the transformer core can be mounted around the conductor or bus-bar. Thus the primary then has just one turn.

It is very important to short-circuit the secondary winding before removing the ammeter. This is because if current is flowing in the primary, dangerously high voltages could be induced in the secondary should it be open-circuited.

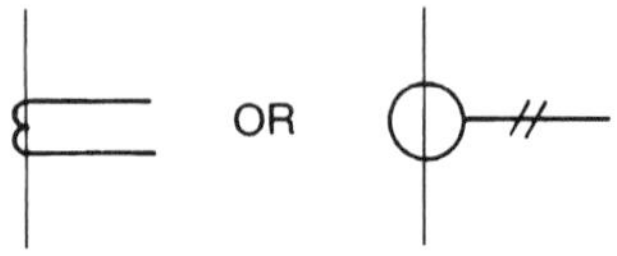

Figure 17.20

Current transformer circuit diagram symbols are shown in Fig. 17.20.

> Problem 29. A current transformer has a single turn on the primary winding and a secondary winding of 60 turns. The secondary winding is connected to an ammeter with a resistance of 0.15 Ω. The resistance of the secondary winding is 0.25 Ω. If the current in the primary winding is 300 A, determine (a) the reading on the ammeter, (b) the potential difference across the ammeter, and (c) the total load (in VA) on the secondary.

(a) Reading on the ammeter, $I_2 = I_1(N_1/N_2) = 300(1/60) =$ **5 A**.

(b) P.d. across the ammeter $= I_2R_A$, where R_A is the ammeter resistance, i.e. $(5)(0.15) =$ **0.75 volts**.

(c) Total resistance of secondary circuit $= 0.15 + 0.25 = 0.40\ \Omega$.

Induced e.m.f. in secondary $= (5)(0.40) = 2.0$ V.

Total load on secondary $= (2.0)(5) =$ **10 VA**.

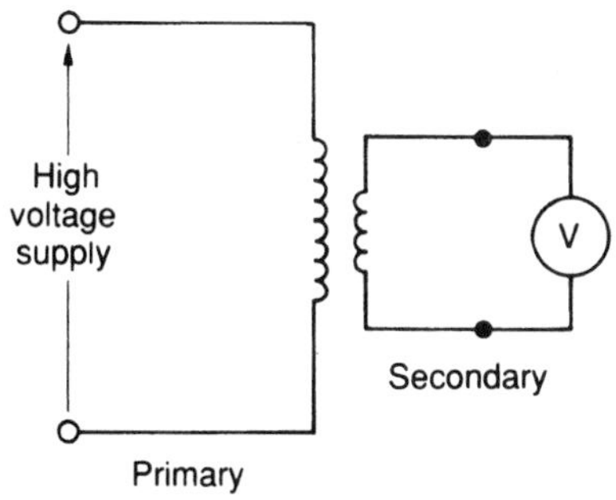

Figure 17.21

17.15 Voltage transformers

For measuring voltages in excess of about 500 V it is often safer to use a voltage transformer. These are normal double-wound transformers with a large number of turns on the primary, which is connected to a high voltage supply, and a small number of turns on the secondary. A typical arrangement is shown in Fig. 17.21.

Since $V_1/V_2 = N_1/N_2$

$$\textbf{secondary voltage, } V_2 = V_1\left(\frac{N_2}{N_1}\right)$$

Thus if the arrangement in Fig. 17.21 has 4000 primary turns and 20 secondary turns then for a voltage of 22 kV on the primary, the voltmeter on the secondary,

$$V_2 = V_1\left(\frac{N_2}{N_1}\right) = 22\,000\left(\frac{20}{4000}\right) = \textbf{110 volts}$$

17.16 Multi-choice questions on transformers

(Answers on page 284.)

1. A transformer has 800 primary turns and 100 secondary turns. To obtain 40 V from the secondary winding the voltage applied to the primary winding must be:
   (a) 5 V; (b) 320 V; (c) 2.5 V; (d) 20 V.

A 100 kVA, 250 V/10 kV, single-phase transformer has a full-load copper loss of 800 W and an iron loss of 500 W. The primary winding contains 120 turns. For the statements in questions 2 to 8, select the correct answer from this list.

(a) 81.3 kW (b) 800 W (c) 97.32%
(d) 80 kW (e) 3 (f) 4800
(g) 1.3 kW (h) 98.40% (i) 100 kW
(j) 98.28% (k) 200 W (l) 101.3 kW
(m) 96.38% (n) 400 W

2. The total full-load losses.
3. The full-load output power at 0.8 power factor.
4. The full-load input power at 0.8 power factor.
5. The full-load efficiency at 0.8 power factor.
6. The half full-load copper loss.
7. The transformer efficiency at half full-load, 0.8 power factor.
8. The number of secondary winding turns.
9. Which of the following statements is false?
   (a) In an ideal transformer, the volts per turn are constant for a given value of primary voltage.
   (b) In a single-phase transformer, the hysteresis loss is proportional to frequency.
   (c) A transformer whose secondary current is greater than the primary current is a step-up transformer.
   (d) In transformers, eddy-current loss is reduced by laminating the core.
10. An ideal transformer has a turns ratio of 1 : 5 and is supplied at 200 V when the primary current is 3 A. Which of the following statements is false?
    (a) The turns ratio indicates a step-up transformer.
    (b) The secondary voltage is 40 V.
    (c) The secondary current is 15 A.
    (d) The transformer rating is 0.6 kVA.
    (e) The secondary voltage is 1 kV.
    (f) The secondary current is 0.6 A.
11. Iron losses in a transformer are due to
    (a) eddy currents only,
    (b) flux leakage,
    (c) both eddy current and hysteresis losses,
    (d) the resistance of the primary and secondary windings.
12. A load is to be matched to an amplifier having an effective internal resistance of 10 Ω via a coupling transformer having a turns ratio of 1 : 10. The value of the load resistance for maximum power transfer is:
    (a) 100 Ω; (b) 1 kΩ; (c) 100 mΩ; (d) 1 mΩ.

17.17 Short answer questions on transformers

1. What is a transformer?
2. Explain briefly how a voltage is induced in the secondary winding of a transformer.

3. Draw the circuit diagram symbol for a transformer.
4. State the relationship between turns and voltage ratios for a transformer.
5. How is a transformer rated?
6. Briefly describe the principle of operation of a transformer.
7. Draw a phasor diagram for an ideal transformer on no-load.
8. State the e.m.f. equation for a transformer.
9. Draw an on-load phasor diagram for an ideal transformer with an inductive load.
10. Name two types of transformer construction.
11. What core material is normally used for power transformers.
12. Name three core materials used in r.f. transformers.
13. State a typical application for (a) a.f. transformers; (b) r.f. transformers.
14. How is cooling achieved in a transformer?
15. State the expressions for the equivalent resistance and reactance of a transformer, referred to the primary.
16. Define regulation of a transformer.
17. Name two main sources of loss in a transformer.
18. What is hysteresis loss? How is it minimized in a transformer?
19. What are eddy currents? How may they be reduced in transformers?
20. How is efficiency of a transformer calculated?
21. What is the condition for maximum efficiency of a transformer?
22. What does 'resistance matching' mean?
23. State a practical source to which the maximum power transfer theorem is applicable.
24. Derive a formula for the equivalent resistance of a transformer having a turns ratio $N_1 : N_2$ and load resistance R_L.
25. What is an auto transformer?
26. State three advantages and one disadvantage of an auto transformer compared with a double-wound transformer.
27. In what applications are auto transformers used?
28. What is an isolating transformer? Give two applications.
29. Describe briefly the construction of a three-phase transformer.
30. For what reason are current transformers used?
31. Describe how a current transformer operates.
32. For what reason are voltage transformers used?
33. Describe how a voltage transformer operates.

17.18 Further questions on transformers

Principle of operation

1. A transformer has 600 primary turns connected to a 1.5 kV supply. Determine the number of secondary turns for a 240 V output voltage, assuming no losses. [96]
2. An ideal transformer with a turns ratio of 2 : 9 is fed from a 220 V supply. Determine its output voltage. [990 V]
3. A transformer has 800 primary turns and 2000 secondary turns. If the primary voltage is 160 V, determine the secondary voltage assuming an ideal transformer. [400 V]
4. An ideal transformer with a turns ratio of 3 : 8 is fed from a 240 V supply. Determine its output voltage. [640 V]
5. An ideal transformer has a turns ratio of 12 : 1 and is supplied at 192 V. Calculate the secondary voltage. [16 V]
6. A transformer primary winding connected across a 415 V supply has 750 turns. Determine how many turns must be wound on the secondary side if an output of 1.66 kV is required. [3000 turns]
7. (a) Describe the transformer principle of operation.
   (b) An ideal transformer has a turns ratio of 12 : 1 and is supplied at 180 V when the primary current is 4 A. Calculate the secondary voltage and current. [15 V, 48 A]
8. A step-down transformer having a turns ratio of 20 : 1 has a primary voltage of 4 kV and a load of 10 kW. Neglecting losses, calculate the value of the secondary current. [50 A]
9. A transformer has a primary to secondary turns ratio of 1 : 15. Calculate the primary voltage necessary to supply a 240 V load. If the load current is 3 A determine the primary current. Neglect any losses. [16 V, 45 A]
10. A 10 kVA, single-phase transformer has a turns ratio of 12 : 1 and is supplied from a 2.4 kV supply. Neglecting losses, determine (a) the full-load secondary current, (b) the minimum value of load resistance which can

be connected across the secondary winding without the kVA rating being exceeded, and (c) the primary current.

[(a) 50 A (b) 4 Ω (c) 4.17 A]

11. State the essential features of construction of power, a.f. and r.f. transformers. A 20 Ω resistance is connected across the secondary winding of a single-phase power transformer whose secondary voltage is 150 V. Calculate the primary voltage and the turns ratio if the supply current is 5 A, neglecting losses.

[225 V, 3 : 2]

No-load phasor diagram

12. (a) Draw the phasor diagram for an ideal transformer on no-load.
    (b) A 500 V/100 V, single-phase transformer takes a full-load primary current of 4 A. Neglecting losses, determine (a) the full-load secondary current, and (b) the rating of the transformer.

    [(a) 20 A (b) 2 kVA]

13. A 3300 V/440 V, single-phase transformer takes a no-load current of 0.8 A and the iron loss is 500 W. Draw the no-load phasor diagram and determine the values of the magnetizing and core loss components of the no-load current. [0.786 A, 0.152 A]
14. A transformer takes a current of 1 A when its primary is connected to a 300 V, 50 Hz supply, the secondary being on open-circuit. If the power absorbed is 120 W, calculate (a) the iron loss current, (b) the power factor on no-load, and (c) the magnetizing current.

    [(a) 0.4 A (b) 0.4 (c) 0.92 A]

E.m.f. equation

15. A 600 kVA, 1600 V/100 V, 50 Hz, single-phase transformer has 50 secondary windings. Calculate (a) the primary and secondary current, (b) the number of primary turns, and (c) the maximum value of the flux.

    [(a) 37.5 A, 600 A (b) 800 (c) 9.0 mWb]

16. A single-phase, 50 Hz transformer has 40 primary turns and 520 secondary turns. The cross-sectional area of the core is 270 cm^2. When the primary winding is connected to a 300 volt supply, determine (a) the maximum value of flux density in the core, and (b) the voltage induced in the secondary winding.

    [(a) 1.25 T (b) 3.90 kV]

17. A single-phase 800 V/125 V, 50 Hz transformer has a maximum core flux density of 1.294 T and an effective cross-sectional area of 60 cm^2. Calculate the number of turns on the primary and secondary windings.

    [464, 58]

18. A 3.3 kV/110 V, 50 Hz, single-phase transformer is to have an approximate e.m.f. per turn of 22 V and operate with a maximum flux of 1.25 T. Calculate (a) the number of primary and secondary turns, and (b) the cross-sectional area of the core.

    [(a) 150, 5 (b) 792.8 cm^2]

Transformer on-load

19. A single-phase transformer has 2400 turns on the primary and 600 turns on the secondary. Its no-load current is 4 A at a power factor of 0.25 lagging. Assuming the volt drop in the windings is negligible, calculate the primary current and power factor when the secondary current is 80 A at a power factor of 0.8 lagging. [23.26 A, 0.73]

Equivalent circuit of a transformer

20. A transformer has 1200 primary turns and 200 secondary turns. The primary and secondary resistances are 0.2 Ω and 0.02 Ω, respectively, and the corresponding leakage reactances are 1.2 Ω and 0.05 Ω, respectively. Calculate (a) the equivalent resistance, reactance and impedance referred to the primary winding, and (b) the phase angle of the impedance.

    [(a) 0.92 Ω, 3.0 Ω, 3.14 Ω (b) 72.96°]

Regulation

21. A 6 kVA, 100 V/500 V, single-phase transformer has a secondary terminal voltage of 487.5 volts when loaded. Determine the regulation of the transformer. [2.5%]
22. A transformer has an open-circuit voltage of 110 volts. A tap-changing device operates when the regulation falls below 3%. Calculate the load voltage at which the tap-changer operates. [106.7 volts]

Losses and efficiency

23. A single-phase transformer has a voltage ratio of 6 : 1 and the h.v. winding is supplied at 540 V. The secondary winding provides a full load current of 30 A at a power factor of 0.8 lagging. Neglecting losses, find (a) the rating of the transformer, (b) the power supplied to the load, and (c) the primary current.

    [(a) 2.7 kVA (b) 2.16 kW (c) 5 A]

24. A single-phase transformer is rated at 40 kVA. The transformer has full-load copper losses of 800 W and iron losses of 500 W. Determine the transformer efficiency at full load and 0.8 power factor. [96.10%]
25. Determine the efficiency of the transformer in question 24 at half full load and 0.8 power factor. [95.81%]
26. (a) Explain the choice of transformer core materials and construction to minimize core losses.
    (b) A 100 kVA, 2000 V/400 V, 50 Hz, single-phase transformer has an iron loss of 600 W and a full-load copper loss of 1600 W. Calculate its efficiency for a load of 60 kW at 0.8 power factor. [97.56%]
27. (a) What are eddy currents? State how their effect is reduced in transformers.
    (b) Determine the efficiency of a 15 kVA transformer for the following conditions:
    (i) full-load, unity power factor;
    (ii) 0.8 full-load, unity power factor;
    (iii) half full-load, 0.8 power factor.
    Assume that iron losses are 200 W and the full-load copper loss is 300 W.
    [(a) 96.77% (ii) 96.84% (iii) 95.62%]
28. A 300 kVA transformer has a primary winding resistance of 0.4 Ω and a secondary winding resistance of 0.0015 Ω. The iron loss is 2 kW and the primary and secondary voltages are 4 kV and 200 V, respectively. If the power factor of the load is 0.78, determine the efficiency of the transformer (a) on full load, and (b) on half load.
    [(a) 96.84% (b) 97.63%]
29. A 250 kVA transformer has a full-load copper loss of 3 kW and an iron loss of 2 kW. Calculate (a) the output kVA at which the efficiency of the transformer is a maximum, and (b) the maximum efficiency, assuming the power factor of the load is 0.80.
    [(a) 204.1 kVA (b) 97.61%]

Resistance matching

30. A transformer having a turns ratio of 8 : 1 supplies a load of resistance 50 Ω. Determine the equivalent input resistance of the transformer. [3.2 kΩ]
31. What ratio of transformer is required to make a load of resistance 30 Ω appear to have a resistance of 270 Ω? [3 : 1]
32. Determine the optimum value of load resistance for maximum power transfer if the load is connected to an amplifier of output resistance 147 Ω through a transformer with a turns ratio of 7 : 2. [12 Ω]
33. A single-phase, 240 V/2880 V ideal transformer is supplied from a 240 V source through a cable of resistance 3 Ω. If the load across the secondary winding is 720 Ω determine (a) the primary current flowing, and (b) the power dissipated in the load resistance.
    [(a) 30 A (b) 4.5 kW]
34. A load of resistance 768 Ω is to be matched to an amplifier which has an effective output resistance of 12 Ω. Determine the turns ratio of the coupling transformer. [1 : 8]
35. An a.c. source of 20 V and internal resistance 20 kΩ is matched to a load by a 16 : 1 single-phase transformer. Determine (a) the value of the load resistance, and (b) the power dissipated in the load.
    [(a) 78.13 Ω (b) 20 mW]

Auto-transformer

36. A single-phase auto transformer has a voltage ratio of 480 V : 300 V and supplies a load of 30 kVA at 300 V. Assuming an ideal transformer, calculate the current in each section of the winding.
    [$I_1 = 62.5$ A, $I_2 = 100$ A, $(I_2 - I_1) = 37.5$ A]
37. Calculate the saving in the volume of copper used in an auto transformer compared with a double-wound transformer for (a) a 300 V : 240 V transformer, and (b) a 400 V : 100 V transformer.
    [(a) 80% (b) 25%]

Three-phase transformer

38. A three-phase transformer has 600 primary turns and 150 secondary turns. If the supply voltage is 1.5 kV determine the secondary line voltage on no-load when the windings are connected (a) delta–star, (b) star–delta.
    [(a) 649.5 V (b) 216.5 V]

Current transformer

39. A current transformer has two turns on the primary winding and a secondary winding of 260 turns. The secondary winding is connected to an ammeter with a resistance of 0.2 Ω. The resistance of the secondary winding is 0.3 Ω. If the current in the primary winding is 650 A, determine (a) the reading on the ammeter, (b) the potential difference across the ammeter, and (c) the total load in VA on the secondary.
    [(a) 5 A (b) 1 V (c) 5 VA]

18

D.c. machines

At the end of this chapter you should be able to:

- distinguish between the function of a motor and a generator
- describe the action of a commutator
- describe the construction of a d.c. machine
- distinguish between wave and lap windings
- understand shunt, series and compound windings of d.c. machines
- understand armature reaction
- calculate generated e.m.f. in an armature winding using $E = 2p\Phi nZ/c$
- describe types of d.c. generator and their characteristics
- calculate generated e.m.f. for a generator using $E = V + I_a R_a$
- state typical applications of d.c. generators
- list d.c. machine losses and calculate efficiency
- calculate back e.m.f. for a d.c. motor using $E = V - I_a R_a$
- calculate the torque of a d.c. motor using

  $$T = \frac{EI_a}{2\pi n} \text{ and } T = \frac{p\Phi ZI_a}{\pi c}$$
- describe types of d.c. motor and their characteristics
- state typical applications of d.c. motors
- describe a d.c. motor starter
- describe methods of speed control of d.c. motors
- list types of enclosure for d.c. motors

18.1 Introduction

When the input to an electrical machine is electrical energy (seen as applying a voltage to the electrical terminals of the machine), and the output is mechanical energy (seen as a rotating shaft), the machine is called an electric **motor**. Thus an electric motor converts electrical energy into mechanical energy.

The principle of operation of a motor is explained in section 8.4, p. 79.

When the input to an electrical machine is mechanical energy (seen as, say, a diesel motor, coupled to the machine by a shaft), and the output is electrical energy (seen as a voltage appearing at the electrical terminals of the machine), the machine is called a **generator**. Thus, a generator converts mechanical energy to electrical energy.

The principle of operation of a generator is explained in section 9.2, p. 84.

18.2 The action of a commutator

In an electric motor, conductors rotate in a uniform magnetic field. A single-loop conductor mounted between permanent magnets is shown in Fig. 18.1. A voltage is applied at points A and B in Fig. 18.1(a).

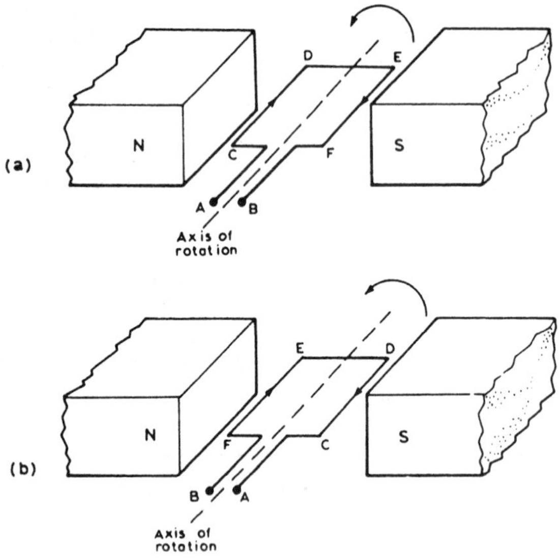

Figure 18.1

A force, *F*, acts on the loop due to the interaction of the magnetic field of the permanent magnets and the magnetic field created by the current flowing in the loop. This force is proportional to the flux density, *B*, the current flowing, *I*, and the effective length of the conductor, *l*, i.e. $F = BIl$. The force is made up of two parts, one acting vertically downwards due to the current flowing from C to D and the other acting vertically upwards due to the current flowing from E to F (from Fleming's left-hand rule). If the loop is free to rotate, then when it has rotated through 180°, the conductors are as shown in Fig. 18.1(b). For rotation to continue in the same direction, it is necessary for the current flow to be as shown in Fig. 18.1(b), i.e. from D to C and from F to E. This apparent reversal in the direction of current flow is achieved by a process called **commutation**. With reference to Fig. 18.2(a), when a direct voltage is applied at A and B, then as the single-loop conductor rotates, current flow will always be away from the commutator for the part of the conductor adjacent to the N-pole and towards the commutator for the part of the conductor adjacent to the S-pole. Thus the forces act to give continuous rotation in an anti-clockwise direction. The arrangement shown in Fig. 18.2(a) is called a 'two-segment' commutator and the voltage is applied to the rotating segments by stationary **brushes** (usually carbon blocks) which slide on the commutator material (usually copper) when rotation takes place.

In practice, there are many conductors on the rotating part of a d.c. machine and these are attached to many commutator segments. A schematic diagram of a multi-segment commutator is shown in Fig. 18.2(b).

Poor commutation results in sparking at the trailing edge of the brushes. This can be improved by using interpoles (situated between each pair of main poles), high resistance brushes, or using brushes spanning several commutator segments.

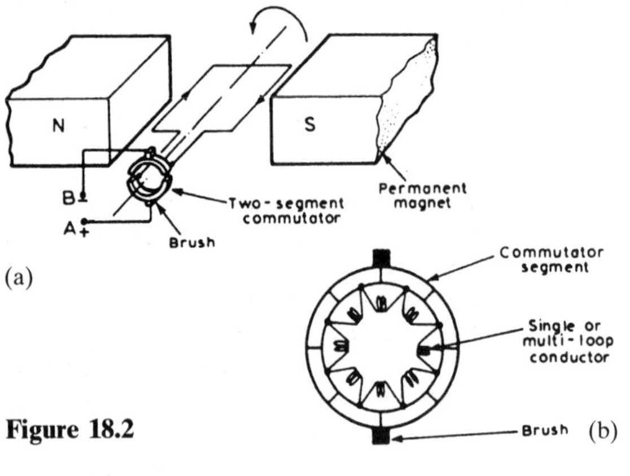

Figure 18.2

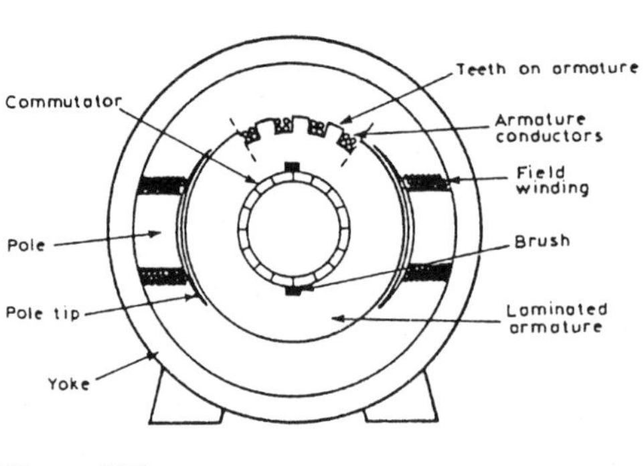

Figure 18.3

18.3 D.c. machine construction

The basic parts of any d.c. machine are shown in Fig. 18.3, and comprise:

(a) a stationary part called the **stator** having,
  (i) a steel ring called the **yoke**, to which are attached
  (ii) the magnetic **poles**, around which are the
  (iii) **field windings**, i.e. many turns of a conductor wound round the pole core. Current passing through this conductor creates an electromagnet (rather than the permanent magnets shown in Figs. 18.1 and 18.2).

(b) a rotating part called the **armature** mounted in bearings housed in the stator and having,
  (iv) a laminated cylinder of iron or steel called the **core**, on which teeth are cut to house the
  (v) **armature winding**, i.e. a single or multi-loop conductor system and
  (vi) the **commutator** (see section 18.2).

Armature windings can be divided into two groups, depending on how the wires are joined to the commutator. These are called **wave windings** and **lap windings**.

(a) In **wave windings** there are two paths in parallel irrespective of the number of poles, each path supplying half the total current output. Wave-wound generators produce high voltage, low current outputs.

(b) In **lap windings** there are as many paths in parallel as the machine has poles. The total current output divides equally between them. Lap-wound generators produce high current, low voltage output.

18.4. Shunt, series and compound windings

When the field winding of a d.c. machine is connected in parallel with the armature, as shown in Fig. 18.4(a), the machine is said to be **shunt** wound. If the field winding is connected in series with the armature, as shown in Fig. 18.4(b), then the machine is said to be **series** wound. A **compound** wound machine has a combination of series and shunt windings.

Depending on whether the electrical machine is series wound, shunt wound or compound wound, it behaves differently when a load is applied. The behaviour of a d.c. machine under various conditions is shown by means of graphs, called characteristic curves or just **characteristics**. The characteristics shown in the following sections are theoretical, since they neglect the effects of armature reaction.

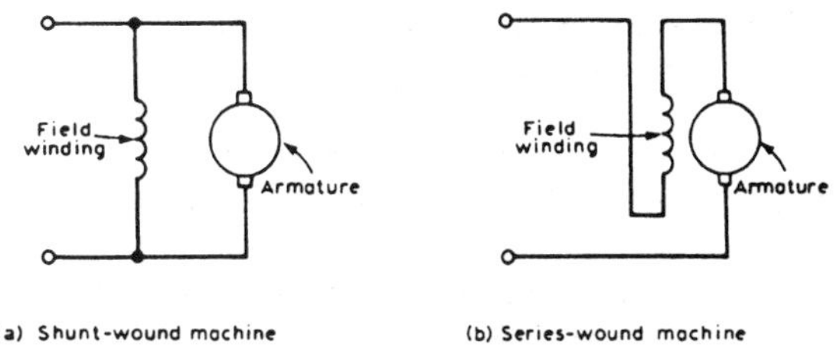

Figure 18.4

Armature reaction is the effect that the magnetic field produced by the armature current has on the magnetic field produced by the field system. In a generator, armature reaction results in a reduced output voltage, and in a motor, armature reaction results in increased speed.

A way of overcoming the effect of armature reaction is to fit compensating windings, located in slots in the pole face.

18.5 Em.f. generated in an armature winding

Let Z = number of armature conductors,
Φ = useful flux per pole, in webers,
p = number of **pairs** of poles, and
n = armature speed in rev/s.

The e.m.f. generated by the armature is equal to the e.m.f. generated by one of the parallel paths. Each conductor passes $2p$ poles per revolution and thus cuts $2p\Phi$ webers of magnetic flux per revolution. Hence flux cut by one conductor per second = $2p\Phi n$ Wb and so the average e.m.f. E generated per conductor is given by:

$$E = 2p\Phi n \text{ volts (since 1 volt = 1 Weber per second)}$$

Let c = number of parallel paths through the winding between positive and negative brushes

c = 2 for a wave winding

c = $2p$ for a lap winding

The number of conductors in series in each path = Z/c. The total e.m.f. between brushes is given by (average e.m.f./conductor)(number of conductors in series per path)

$$= 2p\Phi n \frac{Z}{c}$$

i.e.

generated e.m.f. $\boxed{E = \dfrac{2p\Phi nZ}{c} \text{ volts}}$ (18.1)

Since Z, p and c are constant for a given machine, then $E \propto \Phi n$. However, $2\pi n$ is the angular velocity ω in radians per second, hence the generated e.m.f. is proportional to Φ and ω; i.e.

generated e.m.f. $\boxed{E \propto \Phi\omega}$ (18.2)

Problem 1. An 8-pole, wave-connected armature has 600 conductors and is driven at 625 rev/min. If the flux per pole is 20 mWb, determine the generated e.m.f.

$Z = 600$, $c = 2$ (for a wave winding), $p = 4$ pairs, $n = (625/60)$ rev/s, $\Phi = 20 \times 10^{-3}$ Wb.

$$\text{Generated e.m.f., } E = \frac{2p\Phi nZ}{c}$$

$$= \frac{2(4)(20 \times 10^{-3})\left(\frac{625}{60}\right)(600)}{2} = \mathbf{500\ volts}$$

Problem 2. A 4-pole generator has a lap-wound armature with 50 slots with 16 conductors per slot. The useful flux per pole is 30 mWb. Determine the speed at which the machine must be driven to generate an e.m.f. of 240 V.

$E = 240$ V, $c = 2p$ (for a lap winding), $Z = 50 \times 16 = 800$, $\Phi = 30 \times 10^{-3}$ Wb.

$$\text{Generated e.m.f. } E = \frac{2p\Phi nZ}{c}$$

$$= \frac{2p\Phi nZ}{2p} = \Phi nZ$$

Rearranging gives,

$$\text{speed, } n = \frac{E}{\Phi Z} = \frac{240}{(30 \times 10^{-3})(800)}$$

$$= \mathbf{10\ rev/s} \text{ or } \mathbf{600\ rev/min}$$

Problem 3. An 8-pole, lap-wound armature has 1200 conductors and a flux per pole of 0.03 Wb. Determine the e.m.f. generated when running at 500 rev/min.

$$\text{Generated e.m.f., } E = \frac{2p\Phi nZ}{c} = \frac{2p\Phi nZ}{2p}$$

for a lap-wound machine, i.e.

$$E = \Phi nZ = (0.03)\left(\frac{500}{60}\right)(1200)$$

$$= \mathbf{300\ volts}$$

Problem 4. Determine the generated e.m.f. in Problem 3 if the armature is wave wound.

$$\text{Generated e.m.f., } E = \frac{2p\Phi nZ}{c} = \frac{2p\Phi nZ}{2}$$

$$= p\Phi nZ = (4)(\Phi nZ)$$

$$= (4)(300) \text{ from Problem 3}$$

$$= \mathbf{1200\ volts}$$

Problem 5. A d.c. shunt-wound generator running at constant speed generates a voltage of 150 V at a certain value of field current. Determine the change in the generated voltage when the field current is reduced by 20%, assuming the flux is proportional to the field current.

The generated e.m.f. E of a generator is proportional to $\Phi\omega$, i.e. is proportional to Φn, where Φ is the flux and n is the speed of rotation. It follows that $E = k\Phi n$, where k is a constant.
At speed n_1 and flux Φ_1, $E_1 = k\Phi_1 n_1$.
At speed n_2 and flux Φ_2, $E_2 = k\Phi_2 n_2$.
Thus, by division:

$$\frac{E_1}{E_2} = \frac{k\Phi_1 n_1}{k\Phi_2 n_2} = \frac{\Phi_1 n_1}{\Phi_2 n_2}$$

The initial conditions are $E_1 = 150$ V, $\Phi = \Phi_1$ and $n = n_1$. When the flux is reduced by 20%, the new value of flux is 80/100 or 0.8 of the initial value, i.e. $\Phi_2 = 0.8\Phi_1$. Since the generator is running at constant speed, $n_2 = n_1$. Thus

$$\frac{E_1}{E_2} = \frac{\Phi_1 n_1}{\Phi_2 n_2} = \frac{\Phi_1 n_1}{0.8\Phi_1 n_1} = \frac{1}{0.8}$$

that is, $E_2 = 150 \times 0.8 = 120$ V.
Thus, a reduction of 20% in the value of the flux **reduces the generated voltage to 120 V** at constant speed.

Problem 6. A d.c. generator running at 30 rev/s generates an e.m.f. of 200 V. Determine the percentage increase in the flux per pole required to generate 250 V at 20 rev/s.

From equation (18.2), generated e.m.f., $E \propto \Phi\omega$ and since $\omega = 2\pi n$, $E \propto \Phi n$.
Let $E_1 = 200$ V, $n_1 = 30$ rev/s and flux per pole at this speed be Φ_1.
Let $E_2 = 250$ V, $n_2 = 20$ rev/s and flux per pole at this speed be Φ_2.
Since $E \propto \Phi n$ then

$$\frac{E_1}{E_2} = \frac{\Phi_1 n_1}{\Phi_2 n_2}$$

Hence

$$\frac{200}{250} = \frac{\Phi_1(30)}{\Phi_2(20)}$$

from which,

$$\Phi_2 = \frac{\Phi_1(30)(250)}{(20)(200)} = 1.875\Phi_1$$

Hence the increase in flux per pole needs to be **87.5%**.

18.6 D.c. generators

D.c. generators are classified according to the method of their field excitation. These groupings are:

(i) **Separately excited generators**, where the field winding is connected to a source of supply other than the armature of its own machine.
(ii) **Self-excited generators**, where the field winding receives its supply from the armature of its own machine, and which are sub-divided into (a) shunt, (b) series, and (c) compound wound generators.

18.7 Types of d.c. generator and their characteristics

Separately excited generator

A typical separately excited generator circuit is shown in Fig. 18.5.

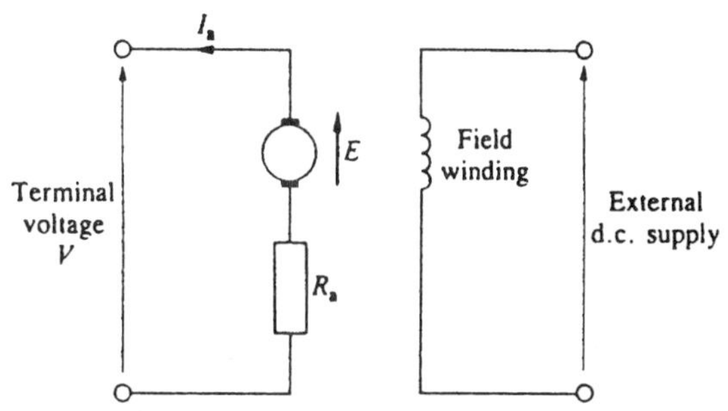

Figure 18.5

When a load is connected across the armature terminals, a load current I_a will flow. The terminal voltage V will fall from its open-circuit e.m.f. E due to a volt drop caused by current flowing through the armature resistance, shown as R_a, i.e.

terminal voltage $\boxed{V = E - I_a R_a}$

or

generated e.m.f. $\boxed{E = V + I_a R_a}$ (18.3)

Problem 7. Determine the terminal voltage of a generator which develops an e.m.f. of 200 V and has an armature current of 30 A on load. Assume the armature resistance is 0.30 Ω.

With reference to Fig. 18.5,

$$\text{terminal voltage, } V = E - I_a R_a = 200 - (30)(0.30) = 200 - 9 = \mathbf{191\ volts}$$

Problem 8. A generator is connected to a 60 Ω load and a current of 8 A flows. If the armature resistance is 1 Ω determine (a) the terminal voltage, and (b) the generated e.m.f.

(a) Terminal voltage, $V = I_a R_L = (8)(60) =$ **480 volts.**

(b) Generated e.m.f., $E = V + I_a R_a$

$= 480 + (8)(1)$

$= 480 + 8$

$=$ **488 volts**

Problem 9. A separately excited generator develops a no-load e.m.f. of 150 V at an armature speed of 20 rev/s and a flux per pole of 0.10 Wb. Determine the generated e.m.f. when (a) the speed increases to 25 rev/s and the pole flux remaining unchanged, (b) the speed remains at 20 rev/s and the pole flux is decreased to 0.08 Wb, and (c) the speed increases to 24 rev/s and the pole flux is 0.07 Wb.

(a) From section 18.5, generated e.m.f. $E \propto \Phi n$, from which,

$$\frac{E_1}{E_2} = \frac{\Phi_1 n_1}{\Phi_2 n_2}$$

Hence

$$\frac{150}{E_2} = \frac{(0.10)(20)}{(0.10)(25)}$$

from which,

$$E_2 = \frac{(150)(0.10)(25)}{(0.10)(20)}$$

$=$ **187.5 volts**

(b) $$\frac{150}{E_3} = \frac{(0.10)(20)}{(0.08)(20)}$$

from which,

$$\text{e.m.f., } E_3 = \frac{(150)(0.08)(20)}{(0.10)(20)} = \textbf{120 volts}$$

(c) $$\frac{150}{E_4} = \frac{(0.10)(20)}{(0.07)(24)}$$

from which,

$$\text{e.m.f., } E_4 = \frac{(150)(0.07)(24)}{(0.10)(20)} = \textbf{126 volts}$$

Characteristics

The two principal generator characteristics are the generated voltage/field current characteristics, called the **open-circuit characteristic** and the terminal voltage/load current characteristic, called the **load characteristic**.

A typical separately excited generator **open-circuit characteristic** is shown in Fig. 18.6(a) and a typical **load characteristic** is shown in Fig. 18.6(b).

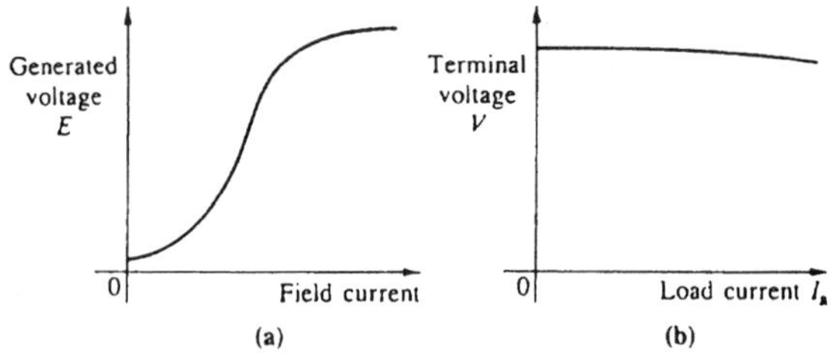

Figure 18.6

A separately excited generator is used only in special cases, such as when a wide variation in terminal p.d. is required, or when exact control of the field current is necessary. Its disadvantage lies in requiring a separate source of direct current.

Shunt-wound generator

In a shunt-wound generator the field winding is connected in parallel with the armature as shown

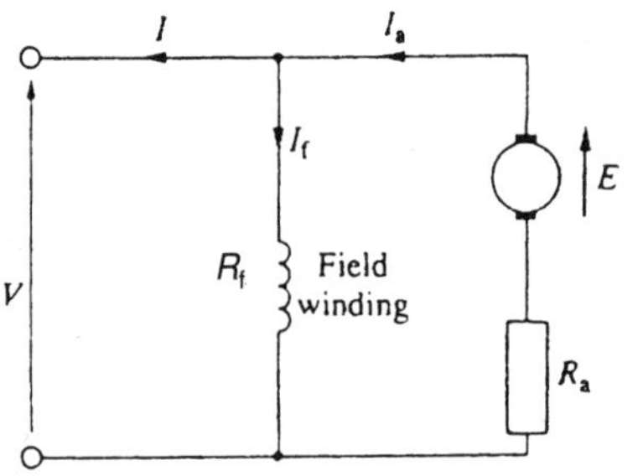

Figure 18.7

in Fig. 18.7. The field winding has a relatively high resistance and therefore the current carried is only a fraction of the armature current.

For the circuit shown in Fig. 18.7, terminal voltage $V = E - I_aR_a$, or generated e.m.f., $E = V + I_aR_a$.

$I_a = I_f + I$, from Kirchhoff's laws, where I_a = armature current, I_f = field current ($= V/R_f$), and I = load current.

Problem 10. A shunt generator supplies a 20 kW load at 200 V through cables of resistance, $R = 100\ \text{m}\Omega$. If the field winding resistance, $R_f = 50\ \Omega$ and the armature resistance, $R_a = 40\ \text{m}\Omega$, determine (a) the terminal voltage, and (b) the e.m.f. generated in the armature.

(a) The circuit is as shown in Fig. 18.7.

$$\text{Load current, } I = \frac{20\,000 \text{ watts}}{200 \text{ volts}} = 100 \text{ A}$$

Volt drop in circuit $= IR = (100)(100 \times 10^{-3}) = 10$ V. Hence terminal voltage, $V = 200 + 10 =$ **210 volts**.

(b) Armature current $I_a = I_f + I$.

$$\text{Field current, } I_f = \frac{V}{R_f} = \frac{210}{50} = 4.2 \text{ A}$$

Hence $I_a = I_f + I = 4.2 + 100 = 104.2$ A.

$$\begin{aligned}\text{Generated e.m.f., } E &= V + I_aR_a \\ &= 210 + (104.2)(40 \times 10^{-3}) \\ &= 210 + 4.168 \\ &= 214.17 \text{ volts}\end{aligned}$$

Characteristics

The generated e.m.f., E, is proportional to $\Phi\omega$, (see section 18.5), hence at constant speed, since $\omega = 2\pi n$, $E \propto \Phi$. Also the flux Φ is proportional to field current I_f until magnetic saturation of the iron circuit of the generator occurs. Hence the open-circuit characteristic is as shown in Fig. 18.8(a).

As the load current on a generator having constant field current and running at constant speed increases, the value of armature current increases, hence the armature volt drop, I_aR_a increases. The generated voltage E is larger than the terminal voltage V and the voltage equation for the armature circuit is $V = E - I_aR_a$. Since E

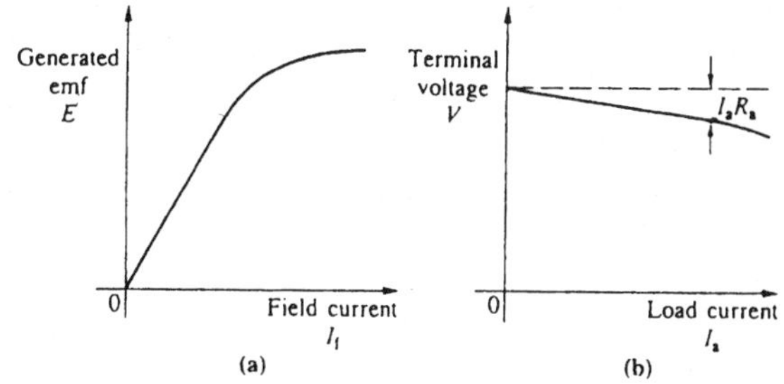

Figure 18.8

is constant, V decreases with increasing load. The load characteristic is as shown in Fig. 18.8(b). In practice, the fall in voltage is about 10% between no-load and full-load for many d.c. shunt-wound generators.

The shunt-wound generator is the type most used in practice, but the load current must be limited to a value that is well below the maximum value. This then avoids excessive variation of the terminal voltage. Typical applications are with battery charging and motor car generators.

Series-wound generator

In the series-wound generator the field winding is connected in series with the armature as shown in Fig. 18.9.

Characteristic

The load characteristic is the terminal voltage/current characteristic. The generated e.m.f., E, is proportional to $\Phi\omega$ and at constant

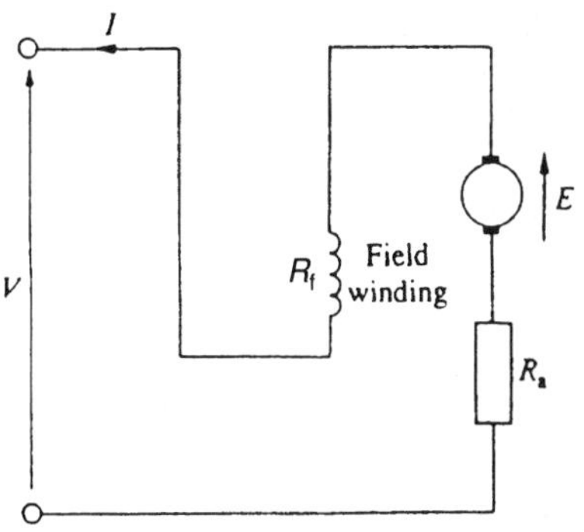

Figure 18.9

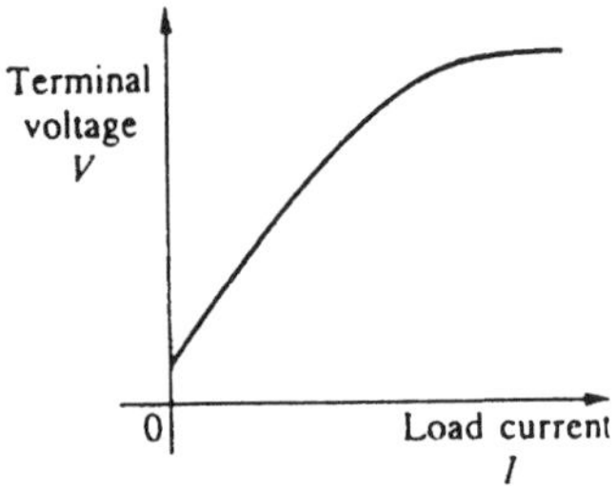

Figure 18.10

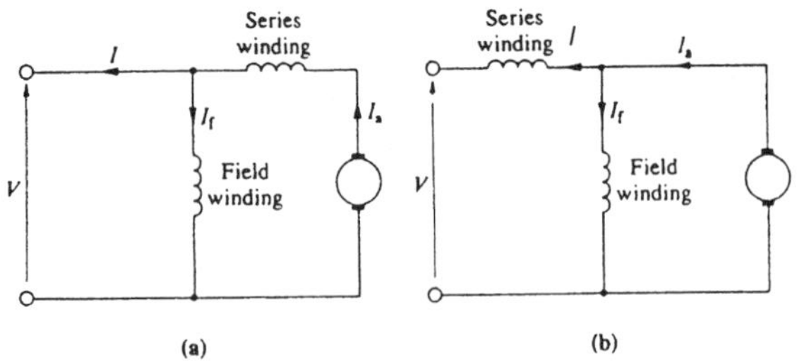

Figure 18.11

speed ω (= $2\pi n$) is a constant. Thus E is proportional to Φ. For values of current below magnetic saturation of the yoke, poles, air gaps and armature core, the flux Φ is proportional to the current, hence $E \propto I$. For values of current above those required for magnetic saturation, the generated e.m.f. is approximately constant. The values of field resistance and armature resistance in a series-wound machine are small, hence the terminal voltage V is very nearly equal to E. A typical load characteristic for a series generator is shown in Fig. 18.10.

In a series-wound generator, the field winding is in series with the armature and it is not possible to have a value of field current when the terminals are open-circuited, thus it is not possible to obtain an open-circuit characteristic.

Series-wound generators are rarely used in practice, but can be used as a 'booster' on d.c. transmission lines.

Compound-wound generator

In the compound-wound generator two methods of connection are used, both having a mixture of shunt and series windings, designed to combine the advantages of each. Fig. 18.11(a) shows what is termed a **long-shunt** compound generator, and Fig. 18.11(b) shows a **short-shunt** compound generator. The latter is the most generally used form of d.c. generator.

Problem 11. A short-shunt compound generator supplies 80 A at 200 V. If the field resistance, $R_f = 40\ \Omega$, the series resistance, $R_{Se} = 0.02\ \Omega$ and the armature resistance, $R_a = 0.04\ \Omega$, determine the e.m.f. generated.

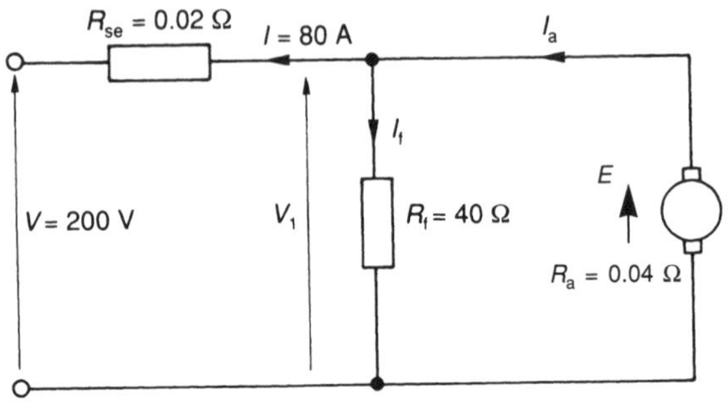

Figure 18.12

The circuit is shown in Fig. 18.12.

Volt drop in series winding = IR_{Se} = (80)(0.02) = 1.6 V.

P.d. across the field winding = p.d. across armature, i.e. $V_1 = 200 + 1.6 = 201.6$ V.

$$\text{Field current } I_f = \frac{V_1}{R_f} = \frac{201.6}{40} = 5.04 \text{ A}$$

Armature current, $I_a = I + I_f = 80 + 5.04 = 85.04$ A.

$$\begin{aligned}\text{Generated e.m.f., } E &= V_1 + I_aR_a \\ &= 201.6 + (85.04)(0.04) \\ &= 201.6 + 3.4016 \\ &= \mathbf{205\ volts}\end{aligned}$$

Characteristics

In cumulative-compound machines the magnetic flux produced by the series and shunt fields are additive. Included in this group are **over-compounded**, **level-compounded** and **under-compounded machines** – the degree of compounding obtained depending on the number of turns of wire on the series winding.

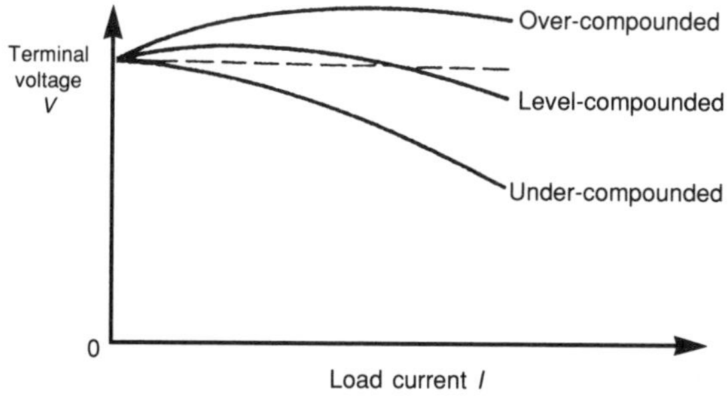

Figure 18.13

A large number of series winding turns results in an over-compounded characteristic, as shown in Fig. 18.13, in which the full-load terminal voltage exceeds the no-load voltage. A level-compound machine gives a full-load terminal voltage which is equal to the no-load voltage, as shown in Fig. 18.13. An under-compounded machine gives a full-load terminal voltage which is less than the no-load voltage, as shown in Fig. 18.13. However, even this latter characteristic is a little better than that for a shunt generator alone.

Compound-wound generators are used in electric arc welding, with lighting sets and with marine equipment.

18.8 D.c. machine losses

As stated in section 18.1, a generator is a machine for converting mechanical energy into electrical energy and a motor is a machine for converting electrical energy into mechanical energy. When such conversions take place, certain losses occur which are dissipated in the form of heat.

The principal **losses of machines** are:

(i) **Copper loss**, due to I^2R heat losses in the armature and field windings.

(ii) **Iron (or core) loss**, due to hysteresis and eddy-current losses in the armature. This loss can be reduced by constructing the armature of silicon steel laminations having a high resistivity and low hysteresis loss. At constant speed, the iron loss is assumed constant.

(iii) **Friction and windage losses**, due to bearing and brush contact friction and losses due to air resistance against moving parts (called windage). At constant speed, these losses are assumed to be constant.

(iv) **Brush contact loss** between the brushes and commutator. This loss is approximately proportional to the load current.

The total losses of a machine can be quite significant and operating efficiencies of between 80% and 90% are common.

18.9. Efficiency of a d.c. generator

The efficiency of an electrical machine is the ratio of the output power to the input power and is usually expressed as a percentage. The Greek letter 'η' (eta) is used to signify efficiency and since the units are power/power, then efficiency has no units. Thus

$$\textbf{efficiency, } \eta = \left(\frac{\textbf{output power}}{\textbf{input power}}\right) \times \mathbf{100\%}$$

If the total resistance of the armature circuit (including brush contact resistance) is R_a, then **the total loss in the armature circuit is $I_a^2R_a$.**

If the terminal voltage is V and the current in the shunt circuit is I_f, then **the loss in the shunt circuit is I_fV.**

If the sum of the iron, friction and windage losses is C then **the total loss is given by**:

$$I_a^2 R_a + I_f V + C$$

($I_a^2R_a + I_fV$ is, in fact, the 'copper loss'.)

If the output current is I, then **the output power is VI.**

Total input power $= VI + I_a^2R_a + I_fV + C$. Hence

$$\textbf{efficiency, } \eta = \frac{\textbf{output}}{\textbf{input}} = \left(\frac{VI}{VI + I_a^2 R_a + I_f V + C}\right) \times \mathbf{100\%} \qquad (18.4)$$

The **efficiency of a generator** is a maximum when the load is such that:

$$I_a^2R_a = VI_f + C$$

i.e. when the variable loss = the constant loss.

Problem 12. A 10 kW shunt generator having an armature circuit resistance of 0.75 Ω and a field resistance of 125 Ω, generates a terminal voltage of 250 V at full-load. Determine the efficiency of the generator at full load,assuming the iron, friction and windage losses amount to 600 W.

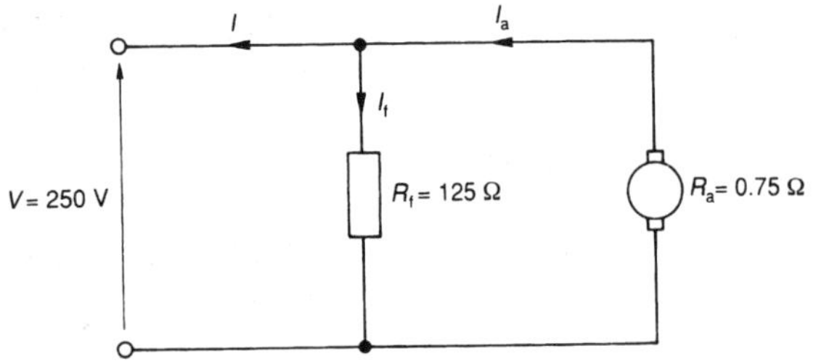

Figure 18.14

The circuit is shown in Fig. 18.14.

Output power = 10 000 W = VI, from which, load current $I = (10\,000/V) = 10\,000/250 = 40$ A.
Field current, $I_f = (V/R_f) = 250/125 = 2$ A.
Armature current, $I_a = I_f + I = 2 + 40 = 42$ A.

Efficiency, η

$$= \left(\frac{VI}{VI + I_a^2R_a + I_fV + C}\right) \times 100\%$$

$$= \left(\frac{10\,000}{10\,000 + (42)^2(0.75) + (2)(250) + 600}\right) \times 100\%$$

$$= \left(\frac{10\,000}{12\,423}\right) \times 100\% = \mathbf{80.50\%}$$

18.10 D.c. motors

The construction of a d.c. motor is the same as a d.c. generator. The only difference is that in a generator the generated e.m.f. is greater than the terminal voltage, whereas in a motor the generated e.m.f. is less than the terminal voltage.

D.c. motors are often used in power stations to drive emergency stand-by pump systems which come into operation to protect essential equipment and plant should the normal a.c. supplies or pumps fail.

Back e.m.f.

When a d.c. motor rotates, an e.m.f. is induced in the armature conductors. By Lenz's law this induced e.m.f. E opposes the supply voltage V and is called a **back e.m.f.**, and the supply voltage V is given by:

$$\boxed{V = E + I_aR_a} \quad \text{or} \quad \boxed{E = V - I_aR_a} \qquad (18.5)$$

> Problem 13. A d.c. motor operates from a 240 V supply. The armature resistance is 0.2 Ω. Determine the back e.m.f. when the armature current is 50 A.

For a motor, $V = E + I_aR_a$. Hence

$$\text{back e.m.f., } E = V - I_aR_a = 240 - (50)(0.2) = 240 - 10 = \mathbf{230\ volts}$$

> Problem 14. The armature of a d.c. machine has a resistance of 0.25 Ω and is connected to a 300 V supply. Calculate the e.m.f. generated when it is running:
>
> (a) as a generator giving 100 A; and
> (b) as a motor taking 80 A.

(a) As a generator,

generated e.m.f., E

$= V + I_aR_a$ (from equation (18.3)

$= 300 + (100)(0.25)$

$= 300 + 25 = \mathbf{325\ volts}$

(b) As a motor, generated e.m.f. (or back e.m.f.),

$E = V - I_aR_a$ (from equation (18.5))

$= 300 - (100)(0.25) = \mathbf{275\ volts}$

18.11 Torque of a d.c. motor

From equation (18.5), for a d.c. motor, the supply voltage V is given by

$$V = E + I_aR_a$$

Multiplying each term by current I_a gives:

$$VI_a = EI_a + I_a^2R_a$$

The term $\boldsymbol{VI_a}$ is the **total electrical power supplied to the armature**, the term $\boldsymbol{I_a^2R_a}$ is the **loss due to armature resistance**, and the term EI_a is the **mechanical power developed by the armature**.

If T is the torque, in newton metres, then the mechanical power developed is given by $T\omega$ watts (see Chapter 20, *Science for Engineering*). Hence

$$T\omega = 2\pi nT = EI_a$$

from which,

$$\textbf{torque } T = \frac{EI_a}{2\pi n} \textbf{ newton metres} \qquad (18.6)$$

From section 18.5, equation (18.1), the e.m.f. E generated is given by

$$E = \frac{2p\Phi nZ}{c}$$

Hence

$$2\pi nT = EI_a = \left(\frac{2p\Phi nZ}{c}\right) I_a$$

Hence

$$\text{torque } T = \frac{\left(\frac{2p\Phi nZ}{c}\right)}{2\pi n} I_a$$

i.e.

$$T = \frac{p\Phi ZI_a}{\pi c} \textbf{ newton metres} \qquad (18.7)$$

For a given machine, Z, c and p are fixed values, hence

$$\text{torque, } T \propto I_a\Phi \qquad (18.8)$$

Problem 15. An 8-pole d.c. motor has a wave-wound armature with 900 conductors. The useful flux per pole is 25 mWb. Determine the torque exerted when a current of 30 A flows in each armature conductor.

$p = 4$, $c = 2$ for a wave winding, $\Phi = 25 \times 10^{-3}$ Wb, $Z = 900$, $I_a = 30$ A.
From equation (18.7),

$$\text{torque } T = \frac{p\Phi ZI_a}{\pi c}$$

$$= \frac{(4)(25 \times 10^{-3})(900)(30)}{\pi(2)}$$

$$= \mathbf{429.7\ N\ m}$$

Problem 16. Determine the torque developed by a 350 V d.c. motor having an armature resistance of 0.5 Ω and running at 15 rev/s. The armature current is 60 A.

$V = 350$ V, $R_a = 0.5\ \Omega$, $n = 15$ rev/s, $I_a = 60$ A.
Back e.m.f. $E = V - I_aR_a = 350 - (60)(0.5) = 320$ V.
From equation (18.6),

$$\text{torque } T = \frac{EI_a}{2\pi n} = \frac{(320)(60)}{2\pi(15)} = \mathbf{203.7\ N\ m}$$

Problem 17. A six-pole lap wound motor is connected to a 250 V d.c. supply. The armature has 500 conductors and a resistance of 1 Ω. The flux per pole is 20 mWb. Calculate (a) the speed and (b) the torque developed when the armature current is 40 A.

$V = 250$ V, $Z = 500$, $R_a = 1\ \Omega$, $\Phi = 20 \times 10^{-3}$ Wb, $I_a = 40$ A, $c = 2p$ for a lap winding.

(a) Back e.m.f. $E = V - I_aR_a = 250 - (40)(1) = 210$ V.

$$E = \frac{2p\Phi nZ}{c}$$

i.e.

$$210 = \frac{2p(20 \times 10^{-3})n(500)}{2p}$$

Hence

$$\text{speed } n = \frac{210}{(20 \times 10^{-3})(500)} = \mathbf{21\ rev/s}$$

or $(21 \times 60) =$ **1260 rev/min**

(b) $$\text{Torque } T = \frac{EI_a}{2\pi n} = \frac{(210(40)}{2\pi(21)} = \mathbf{63.66\ N\ m}$$

Problem 18. The shaft torque of a diesel motor driving a 100 V d.c. shunt-wound generator is 25 N m. The armature current of the generator is 16 A at this value of torque. If the shunt field regulator is adjusted so that the flux is reduced by 15%, the torque increases to 35 N m. Determine the armature current at this new value of torque.

From equation (18.8), the shaft torque T of a generator is proportional to ΦI_a, where Φ is the flux and I_a is the armature current. Thus, $T = k\Phi I_a$, where k is a constant.

The torque at flux Φ_1 and armature current I_{a1} is $T_1 = k\Phi_1 I_{a1}$.
Similarly, $T_2 = k\Phi_2 I_{a2}$. By division

$$\frac{T_1}{T_2} = \frac{k\Phi_1 I_{a1}}{k\Phi_2 I_{a2}} = \frac{\Phi_1 I_{a1}}{\Phi_2 I_{a2}}$$

Hence

$$\frac{25}{35} = \frac{\Phi_1 \times 16}{0.85\Phi_1 \times I_{a2}}$$

i.e.

$$I_{a2} = \frac{16 \times 35}{0.85 \times 25} = 26.35 \text{ A}$$

That is, **the armature current at the new value of torque is 26.35 A**.

> Problem 19. A 100 V d.c. generator supplies a current of 15 A when running at 1500 rev/min. If the torque on the shaft driving the generator is 12 N m, determine (a) the efficiency of the generator, and (b) the power loss in the generator.

(a) From section 18.9, the efficiency of a generator = (output power/input power) × 100%. The output power is the electrical output, i.e. VI watts. The input power to a generator is the mechanical power in the shaft driving the generator, i.e. $T\omega$ or $T(2\pi n)$ watts, where T is the torque in N m and n is speed of rotation in rev/s. Hence, for a generator

$$\text{efficiency, } \eta = \frac{VI}{T(2\pi n)} \times 100\%$$

i.e.

$$\eta = \frac{(100)(15)(100)}{(12)(2\pi)\left(\frac{1500}{60}\right)}$$

i.e. **efficiency = 79.6%**.

(b) input power = output power + losses

Hence,

$T(2\pi n) = VI + \text{losses}$

i.e.

$\text{losses} = T(2\pi n) - VI$

$$= \left[(12)(2\pi)\left(\frac{1500}{60}\right)\right] - \left[(100)(15)\right]$$

i.e.

power loss = 1885 − 1500 = **385 W**

18.12 Types of d.c. motor and their characteristics

Shunt-wound motor

In the shunt-wound motor the field winding is in parallel with the armature across the supply as shown in Fig. 18.15.

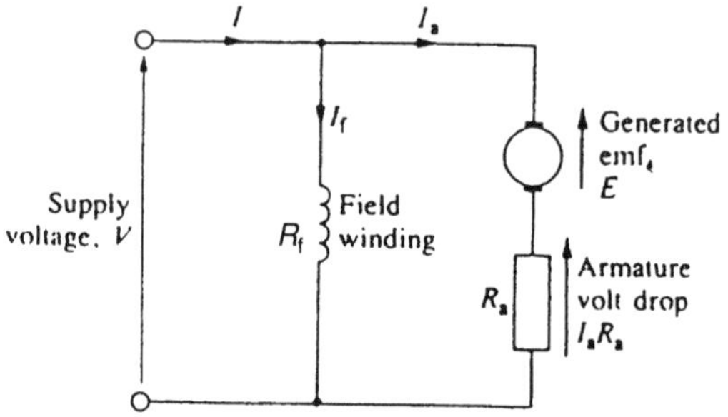

Figure 18.15

For the circuit shown in Fig. 18.15, supply voltage, $V = E + I_a R_a$, or generated e.m.f., $E = V - I_a R_a$, and supply current, $I = I_a + I_f$ (from Kirchhoff's laws).

> Problem 20. A 240 V shunt motor takes a total current of 30 A. If the field winding resistance $R_f = 150\ \Omega$ and the armature resistance $R_a = 0.4\ \Omega$, determine (a) the current in the armature, and (b) the back e.m.f.

(a) Field current $I_f = (V/R_f) = (240/150) = 1.6$ A. Supply current $I = I_a + I_f$. Hence armature current, $I_a = I - I_f = 30 - 1.6 =$ **28.4 A**.

(b) Back e.m.f., $E = V - I_a R_a$

$= 240 - (28.4)(0.4)$

= **228.64 volts**

Characteristics

The two principal characteristics are the torque/armature current and speed/armature

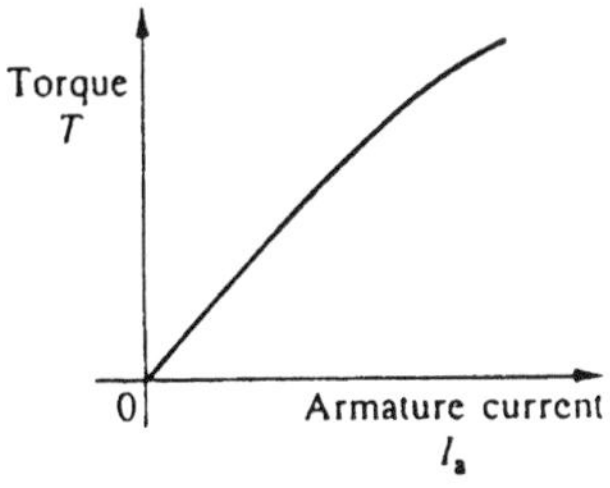

Figure 18.16

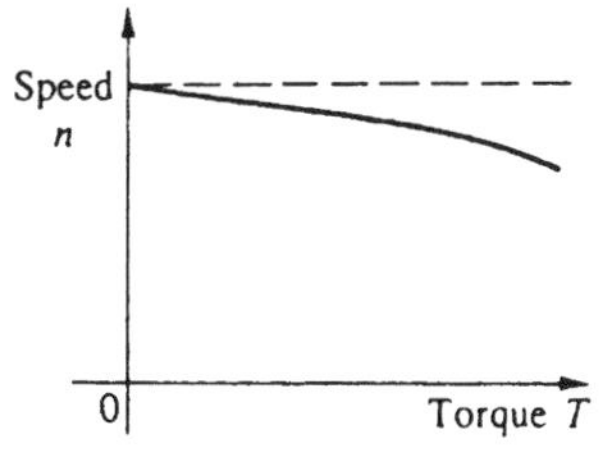

Figure 18.18

current relationships. From these, the torque/speed relationship can be derived.

(i) The theoretical torque/armature current characteristic can be derived from the expression $T \propto \Phi I_a$ (see section 18.11). For a shunt-wound motor, the field winding is connected in parallel with the armature circuit and thus the applied voltage gives a constant field current, i.e. a shunt-wound motor is a constant flux machine. Since Φ is constant, it follows that $T \propto I_a$, and the characteristic is as shown in Fig. 18.16.

(ii) The armature circuit of a d.c. motor has resistance due to the armature winding and brushes, R_a ohms, and when armature current I_a is flowing through it, there is a voltage drop of I_aR_a volts. In Fig. 18.15 the armature resistance is shown as a separate resistor in the armature circuit to help understanding. Also, even though the machine is a motor, because conductors are rotating in a magnetic field, a voltage, $E \propto \Phi\omega$, is generated by the armature conductors. From equation (18.5) $V = E + I_aR_a$, or $E = V - I_aR_a$. But from section 18.5, $E \propto \Phi n$, hence $n \propto E/\Phi$, i.e.

$$\text{speed of rotation, } n \propto \frac{E}{\Phi} \propto \frac{V - I_aR_a}{\Phi} \quad (18.9)$$

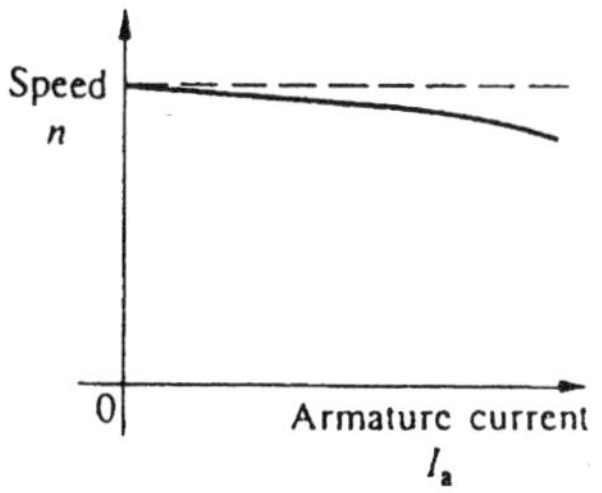

Figure 18.17

For a shunt motor, V, Φ and R_a are constants, hence as armature current I_a increases, I_aR_a increases and $V - I_aR_a$ decreases, and the speed is proportional to a quantity which is decreasing and is as shown in Fig. 18.17. As the load on the shaft of the motor increases, I_a increases and the speed drops slightly. In practice, the speed falls by about 10% between no-load and full-load on many d.c. shunt-wound motors. Due to this relatively small drop in speed, the d.c. shunt-wound motor is taken as basically being a constant-speed machine and may be used for driving lathes, lines of shafts, fans, conveyor belts, pumps, compressors, drilling machines and so on.

(iii) Since torque is proportional to armature current (see (i) above), the theoretical speed/torque characteristic is as shown in Fig. 18.18.

Problem 21. A 200 V d.c. shunt-wound motor has an armature resistance of 0.4 Ω and at a certain load has an armature current of 30 A and runs at 1350 rev/min. If the load on the shaft of the motor is increased so that the armature current increases to 45 A, determine the speed of the motor, assuming the flux remains constant.

The relationship $E \propto \Phi n$ applies to both generators and motors. For a motor, $E = V - I_aR_a$ (equation (18.5)). Hence

$$E_1 = 200 - 30 \times 0.4 = 188 \text{ V}$$

and

$$E_2 = 200 - 45 \times 0.4 = 182 \text{ V}$$

The relationship,

$$\frac{E_1}{E_2} = \frac{\Phi_1 n_1}{\Phi_2 n_2}$$

applies to both generators and motors. Since the flux is constant, $\Phi_1 = \Phi_2$. Hence

$$\frac{188}{182} = \frac{\Phi_1 \times \left(\frac{1350}{60}\right)}{\Phi_1 \times n_2}$$

i.e.

$$n_2 = \frac{22.5 \times 182}{188} = 21.78 \text{ rev/s}$$

Thus the speed of the motor when the armature current is 45 A is 21.78 × 60 rev/min, i.e. **1307 rev/min**.

> Problem 22. A 220 V, d.c. shunt-wound motor runs at 800 rev/min and the armature current is 30 A. The armature circuit resistance is 0.4 Ω. Determine (a) the maximum value of armature current if the flux is suddenly reduced by 10%, and (b) the steady-state value of the armature current at the new value of flux, assuming the shaft torque of the motor remains constant.

(a) For a d.c. shunt-wound motor, $E = V - I_a R_a$. Hence initial generated e.m.f., $E_1 = 220 - 30 \times 0.4 = 208$ V. The generated e.m.f. is also such that $E \propto \Phi n$, so at the instant the flux is reduced, the speed has not had time to change, and $E = 208 \times (90/100) = 187.2$ V. Hence, the voltage drop due to the armature resistance is 220 − 187.2, i.e. 32.8 V. The **instantaneous value of the current** is 32.8/0.4, i.e. **82 A**. This increase in current is about three times the initial value and causes an increase in torque ($T \propto \Phi I_a$). The motor accelerates because of the larger torque value until steady-state conditions are reached.

(b) $T \propto \Phi I_a$ and since the torque is constant, $\Phi_1 I_{a1} = \Phi_2 I_{a2}$. The flux Φ is reduced by 10%, hence $\Phi_2 = 0.9\Phi_1$. Thus, $\Phi_1 \times 30 = 0.9\Phi_1 \times I_{a2}$, i.e. the steady-state value of armature current, $I_{a2} = (30/0.9) = \mathbf{33\frac{1}{3}}$ **A**.

Series-wound motor

In the series-wound motor the field winding is in series with the armature across the supply as shown in Fig. 18.19.

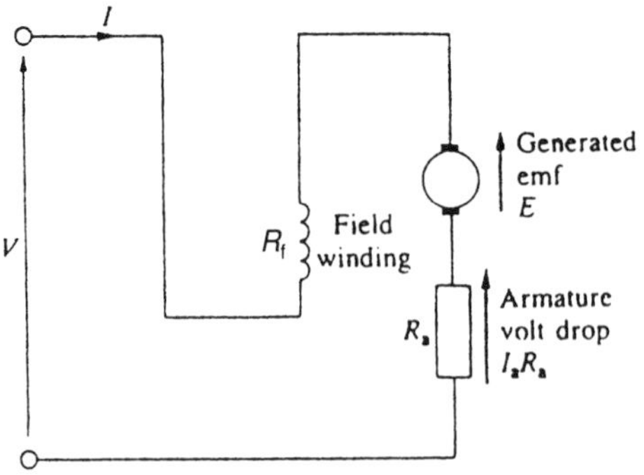

Figure 18.19

For the series motor shown in Fig. 18.19, supply voltage, $V = E + I(R_a + R_f)$, or generated e.m.f. $E = V - I(R_a + R_f)$.

Characteristics

In a series motor, the armature current flows in the field winding and is equal to the supply current, *I*.

(i) **The torque/current characteristic**
It is shown in section 18.11 that torque $T \propto \Phi I_a$. Since the armature and field currents are the same current, *I*, in a series machine, then $T \propto \Phi I$ over a limited range, before magnetic saturation of the magnetic circuit of the motor is reached (i.e. the linear portion of the *B–H* curve for the yoke, poles, air gap, brushes and armature in series). Thus $\Phi \propto I$ and $T \propto I^2$. After magnetic saturation, Φ almost becomes a constant and $T \propto I$. Thus the theoretical torque/current characteristic is as shown in Fig. 18.20.

(ii) **The speed/current characteristic**
It is shown in equation (18.9) that $n \propto (V - I_a R_a)/\Phi$. In a series motor, $I_a = I$ and below the magnetic saturation level, $\Phi \propto I$.

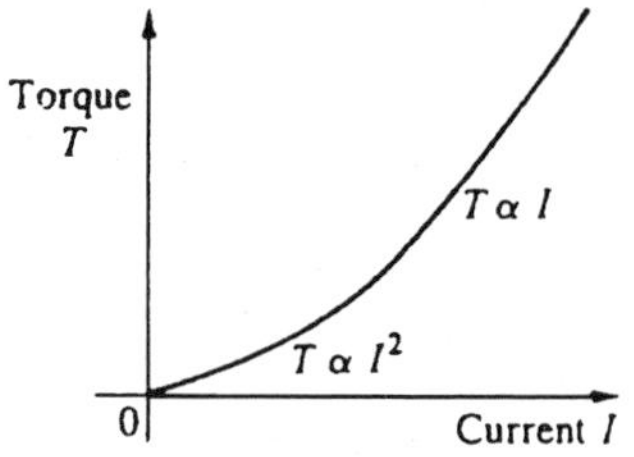

Figure 18.20

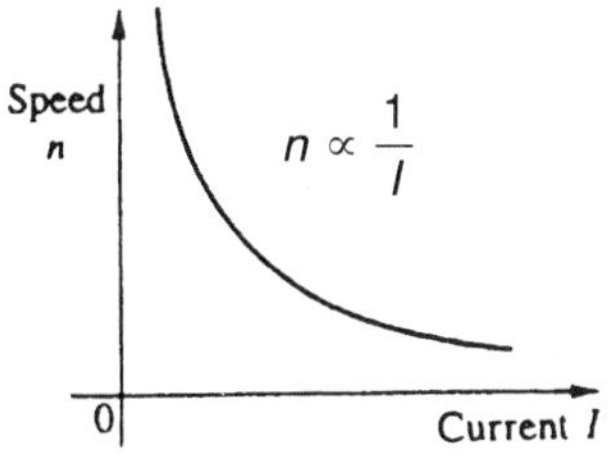

Figure 18.21

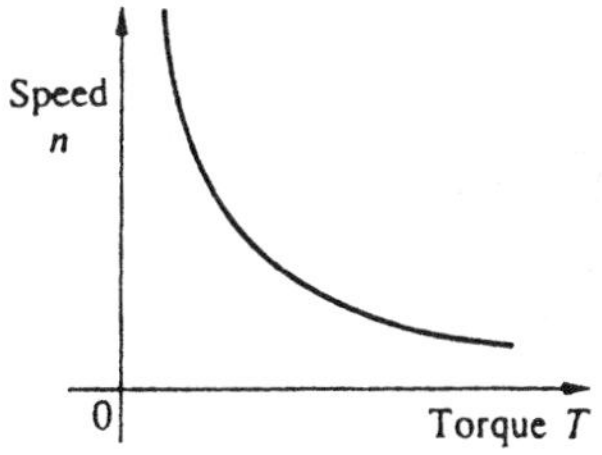

Figure 18.22

Thus $n \propto (V - IR)/I$ where R is the combined resistance of the series field and armature circuit. Since IR is small compared with V, then an approximate relationship for the speed is $n \propto 1/I$. Hence the theoretical speed/current characteristic is as shown in Fig. 18.21. The high speed at small values of current indicates that this type of motor must not be run on very light loads and invariably such motors are permanently coupled to their loads.

(iii) The theoretical **speed/torque characteristic** may be derived from (i) and (ii) above by obtaining the torque and speed for various values of current and plotting the coordinates on the speed/torque characteristics. A typical speed/torque characteristic is shown in Fig. 18.22.

A d.c. series motor takes a large current on starting and the characteristic shown in Fig. 18.20 shows that the series-wound motor has a large torque when the current is large. Hence these motors are used for traction (such as trains, milk delivery vehicles, etc.), driving fans and for cranes and hoists, where a large initial torque is required.

> Problem 23. A series motor has an armature resistance of 0.2 Ω and a series field resistance of 0.3 Ω. It is connected to a 240 V supply and at a particular load runs at 24 rev/s when drawing 15 A from the supply.
>
> (a) Determine the generated e.m.f. at this load.
> (b) Calculate the speed of the motor when the load is changed such that the current is increased to 30 A. Assume that this causes a doubling of the flux.

(a) With reference to Fig. 18.19, generated e.m.f., E, at initial load, is given by

$$E_1 = V - I_a(R_a + R_f)$$
$$= 240 - (15)(0.2 + 0.3) = 240 - 7.5$$
$$= \mathbf{232.5\ volts}$$

(b) When the current is increased to 30 A, the generated e.m.f. is given by

$$E_2 = V - I_2(R_a + R_f)$$
$$= 240 - (30)(0.2 + 0.3) = 240 - 15$$
$$= 225\ \text{volts}$$

Now e.m.f. $E \propto \Phi n$, thus

$$\frac{E_1}{E_2} = \frac{\Phi_1 n_1}{\Phi_2 n_2}$$

i.e.

$$\frac{232.5}{225} = \frac{\Phi_1(24)}{(2\Phi_1)(n_2)} \quad \text{since } \Phi_2 = 2\Phi_1$$

Hence

$$\textbf{speed of motor, } n_2 = \frac{(24)(225)}{(232.5)(2)} = \mathbf{11.6\ rev/s}$$

As the current has been increased from 15 A to 30 A, the speed has decreased from 24 rev/s to 11.6 rev/s. Its speed/current characteristic is similar to Fig. 18.21.

Compound-wound motor

There are two types of compound-wound motor:

(i) **Cumulative compound**, in which the series winding is so connected that the field due to it assists that due to the shunt winding.
(ii) **Differential compound**, in which the series winding is so connected that the field due to it opposes that due to the shunt winding.

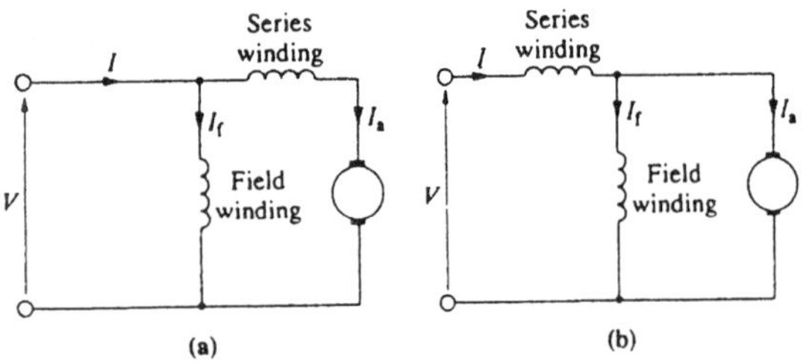

Figure 18.23

Figure 18.23(a) shows a **long-shunt** compound motor and Fig. 18.23(b) a **short-shunt** compound motor.

Characteristics

A compound-wound motor has both a series and a shunt field winding (i.e. one winding in series and one in parallel with the armature), and is usually wound to have a characteristic similar in shape to a series-wound motor (Figs. 18.20–18.22). A limited amount of shunt winding is present to restrict the no-load speed to a safe value. However, by varying the number of turns on the series and shunt windings and the directions of the magnetic fields produced by these windings (assisting or opposing), families of characteristics may be obtained to suit almost all applications. Generally, compound-wound motors are used for heavy duties, particularly in applications where sudden heavy loads may occur such as for driving plunger pumps, presses, geared lifts, conveyors, hoists and so on.

Typical compound motor torque and speed characteristics are shown in Fig. 18.24.

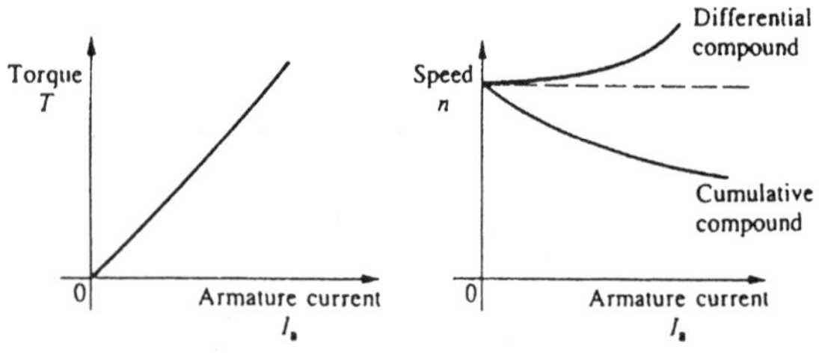

Figure 18.24

18.13 The efficiency of a d.c. motor

It was stated in section 18.9, that the efficiency of a d.c. machine is given by

$$\text{efficiency, } \eta = \frac{\text{output power}}{\text{input power}} \times 100\%$$

Also, the total losses $= I_a^2R_a + I_fV + C$ (for a shunt motor) where C is the sum of the iron, friction and windage losses.

For a motor, the input power $= VI$ and the output power $= VI -$ losses, i.e.

$$VI - I_a^2R_a - I_fV - C$$

Hence

$$\textbf{efficiency, } \boldsymbol{\eta} = \left(\frac{VI - I_a^2R_a - I_fV - C}{VI}\right) \times \mathbf{100\%} \qquad (18.10)$$

The efficiency of a motor is a maximum when the load is such that:

$$I_a^2R_a = I_fV + C$$

Problem 24. A 320 V shunt motor takes a total current of 80 A and runs at 1000 rev/min. If the iron, friction and windage losses amount to 1.5 kW, the shunt field resistance is 40 Ω and the armature resistance is 0.2 Ω, determine the overall efficiency of the motor.

The circuit is shown in Fig. 18.25.
Field current, $I_f = (V/R_f) = 320/40 = 8$ A.
Armature current $I_a = I - I_f = 80 - 8 = 72$ A.

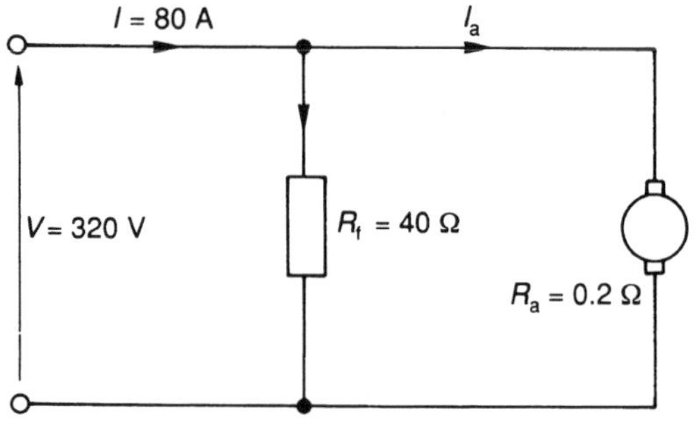

Figure 18.25

C = iron, friction and windage losses = 1500 W.
Efficiency, η

$$= \left(\frac{VI - I_a^2R_a - I_fV - C}{VI}\right) \times 100\%$$

$$= \left(\frac{(320)(80) - (72)^2(0.2) - (8)(320) - 1500}{(300)(80)}\right) \times 100\%$$

$$= \left(\frac{25\,600 - 1036.8 - 2560 - 1500}{25\,600}\right) \times 100\%$$

$$= \left(\frac{20\,503.2}{25\,600}\right) \times 100\%$$

= **80.1%**

> Problem 25. A 250 V series motor draws a current of 40 A. The armature resistance is 0.15 Ω and the field resistance is 0.05 Ω. Determine the maximum efficiency of the motor.

The circuit is as shown in Fig. 18.26.
From equation (18.10),

$$\text{efficiency, } \eta = \left(\frac{VI - I_a^2R_a - I_fV - C}{VI}\right) \times 100\%$$

However, for a series motor, $I_f = 0$ and the $I_a^2R_a$ loss needs to be $I^2(R_a + R_f)$. Hence

$$\text{efficiency, } \eta = \left(\frac{VI - I^2(R_a + R_p) - C}{VI}\right) \times 100\%$$

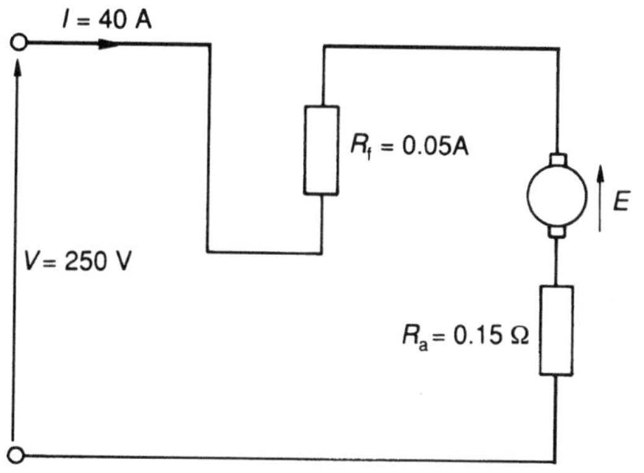

Figure 18.26

For maximum efficiency $I^2(R_a + R_f) = C$. Hence efficiency, η

$$= \left(\frac{VI - 2I^2(R_a + R_f)}{VI}\right) \times 100\%$$

$$= \left(\frac{(250)(40) - 2(40)^2(0.15 + 0.05)}{(250)(40)}\right) \times 100\%$$

$$= \left(\frac{10\,000 - 640}{10\,000}\right) \times 100\%$$

$$= \left(\frac{9360}{10\,000}\right) \times 100\% = \mathbf{93.6\%}$$

> Problem 26. A 200 V d.c. motor develops a shaft torque of 15 N m at 1200 rev/min. If the efficiency is 80%, determine the current supplied to the motor.

The efficiency of a motor = (output power/input power) × 100%.

The output power of a motor is the power available to do work at its shaft and is given by $T\omega$ or $T(2\pi n)$ watts, where T is the torque in N m and n is thespeed of rotation in rev/s. The input power is the electrical power in watts supplied to the motor, i.e. VI watts.

Thus for a motor,

$$\text{efficiency, } \eta = \frac{T(2\pi n)}{VI} \times 100\%$$

i.e.

$$80 = \left\{\frac{(15)(2\pi)\left(\frac{1200}{60}\right)}{(200)(I)}\right\}(100)$$

Thus the current supplied,

$$I = \frac{(15)(2\pi)(20)(100)}{(200)(80)} = \mathbf{11.8\ A}$$

> Problem 27. A d.c. series motor drives a load at 30 rev/s and takes a current of 10 A when the supply voltage is 400 V. If the total resistance of the motor is 2 Ω and the iron, friction and windage losses amount to 300 W, determine the efficiency of the motor.

Efficiency, η

$$= \left(\frac{VI - I^2R - C}{VI}\right) \times 100\%$$

$$= \left(\frac{(400)(10) - (10)^2(2) - 300}{(400)(10)}\right) \times 100\%$$

$$= \left(\frac{4000 - 200 - 300}{4000}\right) \times 100\%$$

$$= \left(\frac{3500}{4000}\right) \times 100\% = \mathbf{87.5\%}$$

18.14 D.c. motor starter

If a d.c. motor whose armature is stationary is switched directly to its supply voltage, it is likely that the fuses protecting the motor will burn out. This is because the armature resistance is small, frequently being less than one ohm. Thus, additional resistance must be added to the armature circuit at the instant of closing the switch to start the motor.

As the speed of the motor increases, the armature conductors are cutting flux and a generated voltage, acting in opposition to the applied voltage, is produced, which limits the flow of armature current. Thus the value of the additional armature resistance can then be reduced.

When at normal running speed, the generated e.m.f. is such that no additional resistance is required in the armature circuit. To achieve this varying resistance in the armature circuit on starting, a d.c. motor starter is used, as shown in Fig. 18.27.

The starting handle is moved **slowly** in a clockwise direction to start the motor. For a shunt-wound motor, the field winding is connected to stud 1 or to L via a sliding contact on the starting handle, to give maximum field current, hence maximum flux, hence maximum torque on starting, since $T \propto \Phi I_a$. A similar arrangement without the field connection is used for series motors.

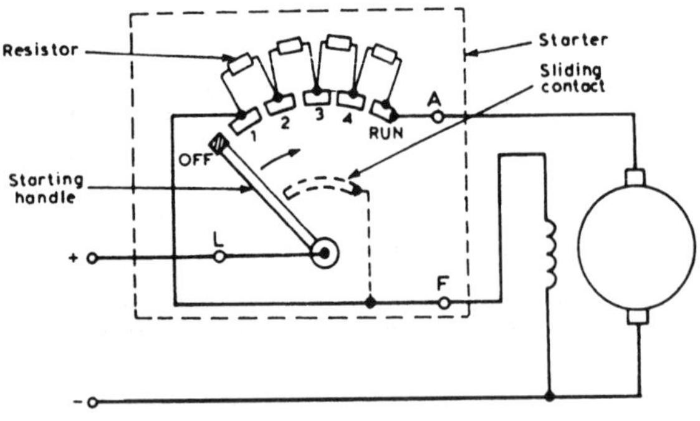

Figure 18.27

18.15 Speed control of d.c. motors

Shunt-wound motors

The speed of a shunt-wound d.c. motor, n, is proportional to $(V - I_aR_a)/\Phi$ (equation (18.9)). The speed is varied either by varying the value of flux, Φ, or by varying the value of R_a. The former is achieved by using a variable resistor in series with the field winding, as shown in Fig. 18.28(a), and such a resistor is called the **shunt field regulator**. As the value of resistance of the shunt field regulator is increased, the value of the field current, I_f, is decreased. This results in a decrease in the value of flux, Φ, and hence an increase in the speed, since $n \propto 1/\Phi$. Thus only speeds **above** that given without a shunt field regulator can be obtained by this method. Speeds below those given by $(V - I_aR_a)/\Phi$ are obtained by increasing the resistance in the armature circuit, as shown in Fig. 18.28(b), where

$$n \propto \frac{V - I_a(R_a + R)}{\Phi}$$

Since resistor R is in series with the armature, it carries the full armature current and results in a large power loss in large motors where a considerable speed reduction is required for long periods.

These methods of speed control are demonstrated in the following worked problem.

Problem 28. A 500 V shunt motor runs at its normal speed of 10 rev/s when the armature current is 120 A. The armature resistance is 0.2 Ω.

(a) Determine the speed when the current is 60 A and a resistance of 0.5 Ω is connected in series with the armature, the shunt field remaining constant.
(b) Determine the speed when the current is 60 A and the shunt field is reduced to 80% of its normal value by increasing resistance in the field circuit.

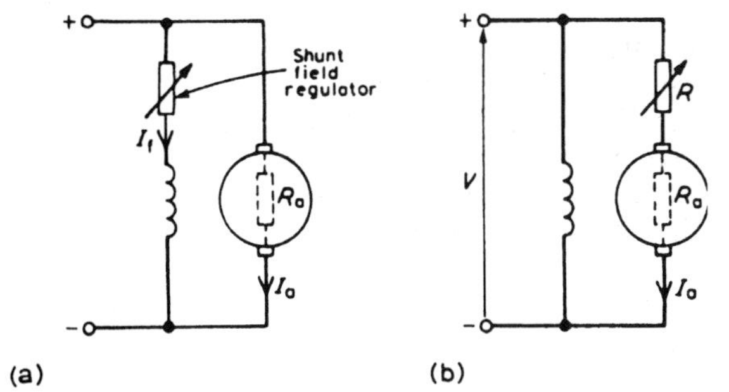

Figure 18.28

(a) With reference to Fig. 18.28(b), back e.m.f. at 120 A, $E_1 = V - I_aR_a = 500 - (120)(0.2)$, $500 - 24 = 476$ volts.
When $I_a = 60$ A,

$$E_2 = 500 - (60)(0.2 + 0.5)$$
$$= 500 - (60)(0.7)$$
$$= 500 - 42 = 458 \text{ volts}$$

Now

$$\frac{E_1}{E_2} = \frac{\Phi_1 n_1}{\Phi_2 n_2}$$

i.e.

$$\frac{476}{458} = \frac{\Phi_1(10)}{\Phi_1(n_2)} \text{ since } \Phi_2 = \Phi_1$$

from which,

$$\text{speed } n_2 = \frac{(10)(458)}{(476)} = \mathbf{9.62\ rev/s}$$

(b) Back e.m.f. when $I_a = 60$ A,

$$E_2 = 500 - (60)(0.2)$$
$$= 500 - 12 = 488 \text{ volts}$$

Now

$$\frac{E_1}{E_2} = \frac{\Phi_1 n_1}{\Phi_2 n_2}$$

i.e.

$$\frac{476}{488} = \frac{(\Phi_1)(10)}{(0.8\Phi_1)(n_3)}, \text{ since } \Phi_2 = 0.8\Phi_1,$$

from which,

$$\text{speed } n_3 = \frac{(10)(488)}{(0.8)(476)} = \mathbf{12.82\ rev/s}$$

Series-wound motors

The speed control of series-wound motors is achieved using either (a) field resistance, or (b) armature resistance techniques.

(a) The speed of a d.c. series-wound motor is given by

$$n = k\left(\frac{V - IR}{\Phi}\right)$$

where k is a constant, V is the terminal voltage, R is the combined resistance of the armature and series field and Φ is the flux.

Thus, a reduction in flux results in an increase in speed. This is achieved by putting a variable resistance in parallel with the field winding and reducing the field current, and hence flux, for a given value of supply current. A circuit diagram of this arrangement is shown in Fig. 18.29(a). A variable resistor connected in parallel with the series-wound field to control speed is called a **diverter**. Speeds above those given with no diverter are obtained by this method. Worked Problem 29 demonstrates this method.

(b) Speeds below normal are obtained by connecting a variable resistor in series with the field winding and armature circuit, as shown in Fig. 18.29(b). This effectively increases the value of R in the equation

$$n = k\left(\frac{V - IR}{\Phi}\right)$$

and thus reduces the speed. Since the additional resistor carries the full supply current, a large power loss is associated with large motors in which a considerable speed reduction is required for long periods. This method is demonstrated in worked Problem 30.

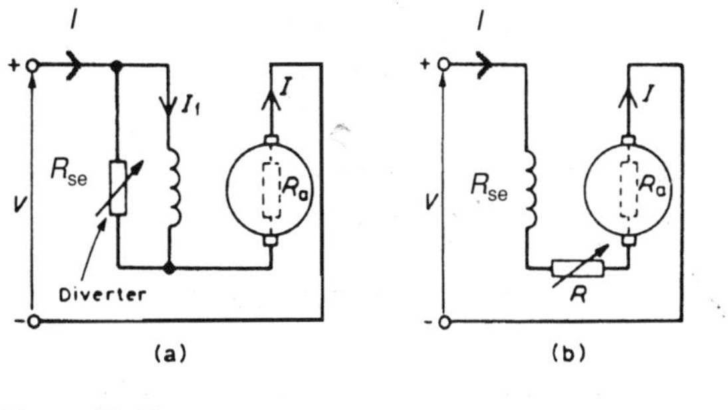

Figure 18.29

> Problem 29. On full-load a 300 V series motor takes 90 A and runs at 15 rev/s. The armature resistance is 0.1 Ω and the series winding resistance is 50 mΩ. Determine the speed when developing full-load torque but with a 0.2 Ω diverter in parallel with the field winding. (Assume that the flux is proportional to the field current.)

At 300 V,

$$\begin{aligned}\text{e.m.f. } E_1 &= V - IR\\ &= V - I(R_a + R_{se})\\ &= 300 - (90)(0.1 + 0.05)\\ &= 300 - (90)(0.15)\\ &= 300 - 13.5 = 286.5 \text{ volts}\end{aligned}$$

With the 0.2 Ω diverter in parallel with R_{se} (see Fig. 18.29(a)), the equivalent resistance,

$$R = \frac{(0.2)(0.05)}{(0.2) + (0.05)} = \frac{(0.2)(0.05)}{(0.25)} = 0.04\ \Omega$$

By current division,

current I_1 (in Fig. 18.29(a))

$$= \left(\frac{0.2}{0.2 + 0.05}\right) I = 0.8I$$

Torque, $T \propto I_a\Phi$ and for full-load torque, $I_{a1}\Phi_1 = I_{a2}\Phi_2$. Since flux is proportional to field current $\Phi_1 \propto I_{a1}$ and $\Phi_2 \propto 0.8I_{a2}$, then $(90)(90) = (I_{a2})(0.8I_{a2})$, from which, $I_{a2}^2 = (90)^2/0.8$ and $I_{a2} = 90/\surd(0.8) = 100.62$ A. Hence

$$\begin{aligned}\text{e.m.f. } E_2 &= V - I_{a2}(R_a + R)\\ &= 300 - (100.62)(0.1 + 0.04)\\ &= 300 - (100.62)(0.14)\\ &= 300 - 14.087 = 285.9 \text{ volts}\end{aligned}$$

Now e.m.f., $E \propto \Phi n$, from which,

$$\frac{E_1}{E_2} = \frac{\Phi_1 n_1}{\Phi_2 n_2} = \frac{I_{a1}n_1}{0.8I_{a2}n_2}$$

Hence

$$\frac{286.5}{285.9} = \frac{(90)(15)}{(0.8)(100.62)n_2}$$

and new speed,

$$n_2 = \frac{(285.9)(90)(150)}{(286.5)(0.8)(100.62)} = \mathbf{16.74\ rev/s}$$

Thus the speed of the motor has increased from 15 rev/s (900 rev/min) to 16.74 rev/s (1004 rev/min) by inserting a 0.2 Ω diverter resistance in parallel with the series winding.

> Problem 30. A series motor runs at 800 rev/min when the voltage is 400 V and the current is 25 A. The armature resistance is 0.4 Ω and the series field resistance is 0.2 Ω. Determine the resistance to be connected in series to reduce the speed to 600 rev/min with the same current.

With reference to Fig. 18.29(b), at 800 rev/min,

$$\begin{aligned}\text{e.m.f., } E_1 &= V - I(R_a + R_{se})\\ &= 400 - (25)(0.4 + 0.2)\\ &= 400 - (25)(0.6)\\ &= 400 - 15 = 385 \text{ volts}\end{aligned}$$

At 600 rev/min, since the current is unchanged, the flux is unchanged. Thus $E \propto \Phi n$, or $E \propto n$, and $(E_1/E_2) = (n_1/n_2)$. Hence $(385/E_2) = (800/600)$, from which,

$$E_2 = \frac{(385)(600)}{(800)} = 288.75 \text{ volts}$$

and $E_2 = V - I(R_a + R_{se} + R)$. Hence $288.75 = 400 - 25(0.4 + 0.2 + R)$. Rearranging gives:

$$0.6 + R = \frac{400 - 288.75}{25} = 4.45$$

from which extra series resistance,

$$R = 4.45 - 0.6$$

i.e.

$$\mathbf{R = 3.85\ \Omega}$$

The addition of a series resistance of 3.85 Ω has reduced the speed from 800 rev/min to 600 rev/min.

18.16 Motor cooling

Motors are often classified according to the type of enclosure used, the type depending on the conditions under which the motor is used and the degree of ventilation required.

The most common type of protection is the **screen-protected type**, where ventilation is achieved by fitting a fan internally, with the openings at the end of the motor fitted with wire mesh.

A **drip-proof type** is similar to the screen-protected type but has a cover over the screen to prevent drips of water entering the machine.

A **flame-proof type** is usually cooled by the conduction of heat through the motor casing.

With a **pipe-ventilated type** air is piped into the motor from a dust-free area, and an internally fitted fan ensures the circulation of this cool air.

18.17 Multi-choice questions on d.c. machines

(Answers on page 284.)

1. Which of the following statements is false?
   (a) A d.c. motor converts electrical energy to mechanical energy.
   (b) The efficiency of a d.c. motor is the ratio (input power/output power) × 100%.
   (c) A d.c. generator converts mechanical energy to electrical energy.
   (d) The efficiency of a d.c. generator is the ratio (output power/input power) × 100%.

A shunt-wound d.c. machine is running at n rev/s and has a shaft torque of T N m. The supply current is I A when connected to d.c. bus-bars of voltage V volts. The armature resistance of the machine is R_a ohms, the armature current is I_a A and the generated voltage is E volts. Use this data to find the formulae of the quantities stated in questions 2 to 9, selecting the correct answer from the list given below.

(a) $V - I_aR_a$ (b) $E + I_aR_a$ (c) VI
(d) $E - I_aR_a$ (e) $T(2\pi n)$ (f) $V + I_aR_a$

2. The input power when running as a generator.
3. The output power when running as a motor.
4. The input power when running as a motor.
5. The output power when running as a generator.
6. The generated voltage when running as a motor.
7. The terminal voltage when running as a generator.
8. The generated voltage when running as a generator.
9. The terminal voltage when running as a motor.
10. Which of the following statements is false?
    (a) A commutator is necessary as part of a d.c. motor to keep the armature rotating in the same direction.
    (b) A commutator is necessary as part of a d.c. generator to produce a uni-directional voltage at the terminals of the generator.
    (c) The field winding of a d.c. machine is housed in slots on the armature.
    (d) The brushes of a d.c. machine are usually made of carbon and do not rotate with the armature.
11. If the speed of a d.c. machine is doubled and the flux remains constant, the generated e.m.f.
    (a) remains the same, (b) is doubled, (c) is halved.
12. If the flux per pole of a shunt-wound d.c. generator is increased, and all other variables are kept the same, the speed
    (a) decreases, (b) stays the same, (c) increases.
13. If the flux per pole of a shunt-wound d.c. generator is halved, the generated e.m.f. at constant speed
    (a) is doubled, (b) is halved, (c) remains the same.
14. In a series-wound generator running at constant speed, as the load current increases, the terminal voltage
    (a) increases, (b) decreases, (c) stays the same.
15. Which of the following statements is false for a series-wound d.c. motor?
    (a) The speed decreases with increase of resistance in the armature circuit.
    (b) The speed increases as the flux decreases.
    (c) The speed can be controlled by a diverter.
    (d) The speed can be controlled by a shunt field regulator.
16. Which of the following statements is false?
    (a) A series-wound motor has a large starting torque.
    (b) A shunt-wound motor must be permanently connected to its load.
    (c) The speed of a series-wound motor drops considerably when load is applied.
    (d) A shunt-wound motor is essentially a constant-speed machine.
17. The speed of a d.c. motor may be increased by
    (a) increasing the armature current,
    (b) decreasing the field current,

(c) decreasing the applied voltage,
(d) increasing the field current.

18. The armature resistance of a d.c. motor is 0.5 Ω, the supply voltage is 200 V and the back e.m.f. is 196 V at full speed. The armature current is:
(a) 4 A; (b) 8 A; (c) 400 A; (d) 392 A.
19. In d.c. generators iron losses are made up of
(a) hysteresis and friction losses,
(b) hysteresis, eddy current and brush contact losses,
(c) hysteresis and eddy current losses,
(d) hysteresis, eddy current and copper losses.
20. The effect of inserting a resistance in series with the field winding of a shunt motor is to
(a) increase the magnetic field,
(b) increase the speed of the motor,
(c) decrease the armature current,
(d) reduce the speed of the motor.

18.18 Short answer questions on d.c. machines

1. A .... converts mechanical energy into electrical energy.
2. A .... converts electrical energy into mechanical energy.
3. What does 'commutation' achieve?
4. Poor commutation may cause sparking. How can this be improved?
5. State any five basic parts of a d.c. machine.
6. State the two groups armature windings can be divided into.
7. What is armature reaction? How can it be overcome?
8. The e.m.f. generated in an armature winding is given by
$$E = \frac{2p\Phi nZ}{c} \text{ volts}$$
State what p, Φ, n, Z and c represent.
9. In a series-wound d.c. machine, the field winding is in .... with the armature circuit.
10. In a d.c. generator, the relationship between the generated voltage, terminal voltage, current and armature resistance is given by E = ...
11. A d.c. machine has its field winding in parallel with the armature circuit. It is called a .... wound machine.
12. Sketch a typical open-circuit characteristic for (a) a separately excited generator, (b) a shunt generator, (c) a series generator.
13. Sketch a typical load characteristic for (a) a separately excited generator, (b) a shunt generator.
14. State one application for (a) a shunt generator, (b) a series generator, (c) a compound generator.
15. State the principle losses in d.c. machines.
16. The efficiency of a d.c. machine is given by the ratio (........) %.
17. The equation relating the generated e.m.f., terminal voltage, armature current and armature resistance for a d.c. motor is .... = ...
18. The torque T of a d.c. motor is given by
$$T = \frac{p\Phi ZI_a}{\pi c} \text{ newton metres.}$$
State what p, Φ, Z, I_a and c represent.
19. Complete the following. In a d.c. machine
(a) generated e.m.f. ∝ .... × ....,
(b) torque ∝ .... × ....
20. Sketch typical characteristics of torque/armature current for
(a) a shunt motor, (b) a series motor,
(c) a compound motor.
21. Sketch typical speed/torque characteristics for a shunt and series motor.
22. State two applications for each of the following motors
(a) shunt, (b) series, (c) compound.

In questions 23 to 26, an electrical machine runs at n rev/s, has a shaft torque of T and takes a current of I from a supply voltage V.

23. The power input to a generator is .... watts.
24. The power input to a motor is .... watts.
25. The power output from a generator is .... watts.
26. The power output from a motor is .... watts.
27. The generated e.m.f. of a d.c. machine is proportional to .... volts.
28. The torque produced by a d.c. motor is proportional to .... N m.
29. A starter is necessary for a d.c. motor because the generated e.m.f. is .... at low speeds.
30. The speed of a d.c. shunt-wound motor will .... if the value of resistance of the shunt field regulator is increased.
31. The speed of a d.c. motor will .... if the value of resistance in the armature circuit is increased.
32. The value of the speed of a d.c. shunt-wound motor .... as the value of the armature current increases.

33. At a large value of torque, the speed of a d.c. series-wound motor is ....
34. At a large value of field current, the generated e.m.f. of a d.c. shunt-wound generator is approximately ....
35. In a series-wound generator, the terminal voltage increases as the load current ....
36. Name three types of motor enclosure.

18.19 Further questions on d.c. machines

1. A 4-pole, wave-connected armature of a d.c. machine has 750 conductors and is driven at 720 rev/min. If the useful flux per pole is 15 mWb, determine the generated e.m.f.. [270 volts]
2. A 6-pole generator has a lap-wound armature with 40 slots with 20 conductors per slot. The flux per pole is 25 mWb. Calculate the speed at which the machine must be driven to generate an e.m.f. of 300 V. [15 rev/s or 900 rev/min]
3. A 4-pole armature of a d.c. machine has 1000 conductors and a flux per pole of 20 mWb. Determine the e.m.f. generated when running at 600 rev/min when the armature is (a) wave-wound, (b) lap-wound. [(a) 400 volts (b) 200 volts]
4. A d.c. generator running at 25 rev/s generates an e.m.f. of 150 V. Determine the percentage increase in the flux per pole required to generate 180 V at 20 rev/s [50%]
5. Determine the terminal voltage of a generator which develops an e.m.f. of 240 V and has an armature current of 50 A on load. Assume the armature resistance is 40 mΩ . [238 volts]
6. Explain the function of a commutator on the armature of a d.c. motor.
7. Draw a labelled diagram showing a cross-section of a 2-pole d.c. machine. Describe the functions performed by the field windings, the armature, the commutator and the brushes.
8. A generator is connected to a 50 Ω load and a current of 10 A flows. If the armature resistance is 0.5 Ω, determine (a) the terminal voltage, and (b) the generated e.m.f. [(a) 500 volts (b) 505 volts]
9. A separately excited generator develops a no-load e.m.f. of 180 V at an armature speed of 15 rev/s and a flux per pole of 0.20 Wb. Calculate the generated e.m.f. when
   (a) the speed increases to 20 rev/s and the flux per pole remaining unchanged,
   (b) the speed remains at 15 rev/s and the pole flux is decreased to 0.15 Wb, and
   (c) the speed increases to 25 rev/s and the pole flux is 0.18 Wb.
   [(a) 240 volts (b) 112.5 volts (c) 270 volts]
10. A shunt generator supplies a 50 kW load at 400 V through cables of resistance 0.2 Ω. If the field winding resistance is 50 Ω and the armature resistance is 0.05 Ω, determine (a) the terminal voltage, (b) the e.m.f. generated in the armature. [(a) 425 volts (b) 431.68 volts]
11. A short-shunt compound generator supplies 50 A at 300 V. If the field resistance is 30 Ω, the series resistance 0.03 Ω and the armature resistance 0.05 Ω, determine the e.m.f. generated. [304.5 volts]
12. A d.c. generator has a generated e.m.f. of 210 V when running at 700 rev/min and the flux per pole is 120 mWb. Determine the generated e.m.f. (a) at 1050 rev/min, assuming the flux remains constant, (b) if the flux is reduced by one-sixth at constant speed, and (c) at a speed of 1155 rev/min and a flux of 132 mWb. [(a) 315 V (b) 175 V (c) 381.2 V]
13. A 250 V d.c. shunt-wound generator has an armature resistance of 0.1 Ω. Determine the generated e.m.f. when the generator is supplying 50 kW, neglecting the field current of the generator. [270 V]
14. A 15 kW shunt generator having an armature circuit resistance of 0.4 Ω and a field resistance of 100 Ω, generates a terminal voltage of 240 V at full load. Determine the efficiency of the generator at full load, assuming the iron, friction and windage losses amount to 1 kW. [80.29%]
15. A d.c. motor operates from a 350 V supply. If the armature resistance is 0.4 Ω determine the back e.m.f. when the armature current is 60 A. [326 volts]
16. The armature of a d.c. machine has a resistance of 0.5 Ω and is connected to a 200 V supply. Calculate the e.m.f. generated when it is running (a) as a motor taking 50 A, and (b) as a generator giving 70 A. [(a) 175 volts (b) 235 volts]
17. (a) State the principal losses in d.c. machines.

(b) The shaft torque required to drive a d.c. generator is 18.7 N m when it is running at 1250 rev/min. If its efficiency is 87% under these conditions and the armature current is 17.3 A, determine the voltage at the terminals of the generator. [123.1 V]

18. A 220 V, d.c. generator supplies a load of 37.5 A and runs at 1550 rev/min. Determine the shaft torque of the diesel motor driving the generator, if the generator efficiency is 78%. [65.2 N m]
19. A 4-pole d.c. motor has a wave-wound armature with 800 conductors. The useful flux per pole is 20 mWb. Calculate the torque exerted when a current of 40 A flows in each armature conductor. [203.7 N m]
20. Calculate the torque developed by a 240 V d.c. motor whose armature current is 50 A, armature resistance is 0.6 Ω and is running at 10 rev/s. [167.1 N m]
21. Determine the generated e.m.f. of a d.c. machine if the armature resistance is 0.1 Ω and it (a) is running as a motor connected to a 230 V supply, the armature current being 60 A, and (b) is running as a generator with a terminal voltage of 230 V, the armature current being 80 A. [(a) 224 V (b) 238 V]
22. An 8-pole lap-wound d.c. motor has a 200 V supply. The armature has 800 conductors and a resistance of 0.8 Ω. If the useful flux per pole is 40 mWb and the armature current is 30 A, calculate (a) the speed, and (b) the torque developed. [(a) 5.5 rev/s or 330 rev/min (b) 152.8 N m]
23. A 150 V d.c. generator supplies a current of 25 A when running at 1200 rev/min. If the torque on the shaft driving the generator is 35.8 N m, determine (a) the efficiency of the generator, and (b) the power loss in the generator. [(a) 83.4% (b) 748.8 W]
24. A 240 V shunt motor takes a total current of 80 A. If the field winding resistance is 120 Ω and the armature resistance is 0.4 Ω, determine (a) the current in the armature, and (b) the back e.m.f. [(a) 86.5 A (b) 385.4 V]
25. A d.c. motor has a speed of 900 rev/min when connected to a 460 V supply. Find the approximate value of the speed of the motor when connected to a 200 V supply, assuming the flux decreases by 30% and neglecting the armature volt drop. [559 rev/min]
26. A series motor having a series field resistance of 0.25 Ω and an armature resistance of 0.15 Ω, is connected to a 220 V supply and at a particular load runs at 20 rev/s when drawing 20 A from the supply. Calculate the e.m.f. generated at this load. Determine also the speed of the motor when the load is changed such that the current increases to 25 A. Assume the flux increases by 25%. [212 V, 15.85 rev/s]
27. A 500 V shunt motor takes a total current of 100 A and runs at 1200 rev/min. If the shunt field resistance is 50 Ω, the armature resistance is 0.25 Ω and the iron, friction and windage losses amount to 2 kW, determine the overall efficiency of the motor. [81.95%]
28. A 250 V, series-wound motor is running at 500 rev/min and its shaft torque is 130 N m. If its efficiency at this load is 88%, find the current taken from the supply. [30.94 A]
29. In a test on a d.c. motor, the following data was obtained. Supply voltage: 500 V; current taken from the supply: 42.4 A; speed: 850 rev/min; shaft torque: 187 N m. Determine the efficiency of the motor correct to the nearest 0.5%. [78.5%]
30. A 300 V series motor draws a current of 50 A. The field resistance is 40 mΩ and the armature resistance is 0.2 Ω. Determine the maximum efficiency of the motor. [92%]
31. A series motor drives a load at 1500 rev/min and takes a current of 20 A when the supply voltage is 250 V. If the total resistance of the motor is 1.5 Ω and the iron, friction and windage losses amount to 400 W, determine the efficiency of the motor. [80%]
32. A series-wound motor is connected to a d.c. supply and develops full-load torque when the current is 30 A and speed is 1000 rev/min. If the flux per pole is proportional to the current flowing, find the current and speed at half full-load torque, when connected to the same supply. [212 A, 1415 rev/min]
33. Sketch the theoretical speed/torque characteristic for (a) a series-wound d.c. motor, and (b) a shunt-wound, d.c. motor. Use the e.m.f. and torque relationships to explain their shapes.
34. A 350 V shunt motor runs at its normal speed of 12 rev/s when the armature current is 90 A. The resistance of the armature is 0.3 Ω.
(a) Find the speed when the current is 45 A and a resistance of 0.4 Ω is connected in series with the armature, the shunt field remaining constant.

(b) Find the speed when the current is 45 A and the shunt field is reduced to 75% of its normal value by increasing resistance in the field circuit.
[(a) 11.83 rev/s (b) 16.67 rev/s]

35. A series motor runs at 900 rev/min when the voltage is 420 V and the current is 40 A. The armature resistance is 0.3 Ω and the series field resistance is 0.2 Ω. Calculate the resistance to be connected in series to reduce the speed to 720 rev/min with the same current. [2 Ω]

36. A 320 V series motor takes 80 A and runs at 1080 rev/min at full load. The armature resistance is 0.2 Ω and the series winding resistance is 0.05 Ω. Assuming the flux is proportional to the field current, calculate the speed when developing full-load torque, but with a 0.15 Ω diverter in parallel with the field winding. [1430 rev/min]

37. As the current supplied by a d.c. shunt-wound generator increases, the terminal voltage falls. Explain why the voltage falls and sketch the theoretical load characteristic for this generator.

38. Explain the effect of the generated e.m.f. of a d.c. motor and why a d.c. motor starter is necessary.

39. Explain why a d.c. shunt-wound motor needs a starter when connected to a constant-voltage supply and make a sketch of such a starter.

40. One type of d.c. motor uses resistance in series with the field winding to obtain speed variations and another type uses resistance in parallel with the field winding for the same purpose. Explain why these two distinct methods are used and why the field current plays a significant part in controlling the speed of a d.c. motor.

19

Three-phase induction motors

At the end of this chapter you should be able to:

- appreciate the merits of three-phase induction motors
- understand how a rotating magnetic field is produced
- state the synchronous speed, $n_s = (f/p)$ and use in calculations
- describe the principle of operation of a three-phase induction motor
- distinguish between squirrel cage and wound rotor types of motor
- understand how a torque is produced causing rotor movement
- understand and calculate slip
- derive expressions for rotor e.m.f., frequency, resistance, reactance, impedance, current and copper loss, and use them in calculations
- state the losses in an induction motor and calculate efficiency
- derive the torque equation for an induction motor, state the condition for maximum torque, and use in calculations
- describe torque–speed and torque–slip characteristics for an induction motor
- state and describe methods of starting induction motors
- state advantages of cage rotor and wound rotor types of induction motor
- describe the double cage induction motor
- state typical applications of three-phase induction motors

19.1 Introduction

In d.c. motors, introduced in Chapter 18, conductors on a rotating armature pass through a stationary magnetic field. In a **three-phase induction motor**, the magnetic field rotates and this has the advantage that no external electrical connections to the rotor need be made. Its name is derived from the fact that the current in the rotor is **induced** by the magnetic field instead of being supplied through electrical connections to the supply.

The result is a motor which: (i) is cheap and robust, (ii) is explosion proof, due to the absence of a commutator or slip-rings and brushes with their associated sparking, (iii) requires little or no skilled maintenance, and (iv) has self-starting properties when switched to a supply with no additional expenditure on auxiliary equipment. The principal disadvantage of a three-phase induction motor is that its speed cannot be readily adjusted.

19.2. Production of a rotating magnetic field

When a three-phase supply is connected to symmetrical three-phase windings, the currents flowing in the windings produce a magnetic field.

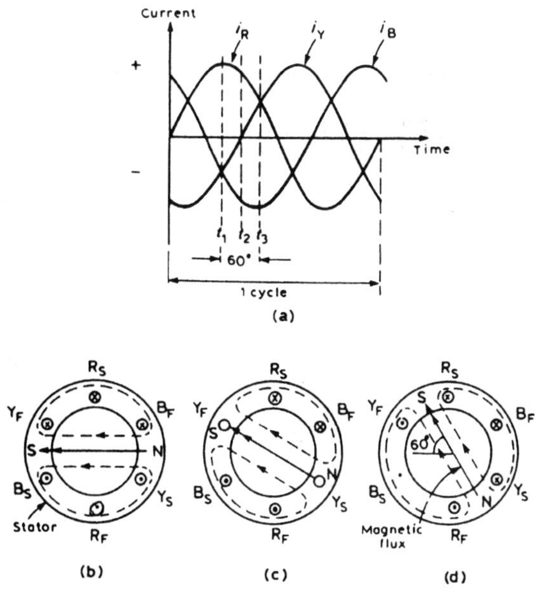

Figure 19.1

This magnetic field is constant in magnitude and rotates at constant speed as shown below, and is called the **synchronous speed**.

With reference to Fig. 19.1, the windings are represented by three single-loop conductors, one for each phase, marked R_SR_F, Y_SY_F and B_SB_F, the S and F signifying start and finish. In practice, each phase winding comprises many turns and is distributed around the stator (i.e. the stationary part); the single-loop approach is for clarity only.

When the stator windings are connected to a three-phase supply, the current flowing in each winding varies with time and is as shown in Fig. 19.1(a). If the value of current in a winding is positive, the assumption is made that it flows from start to finish of the winding, i.e. if it is the red phase, current flows from R_S to R_F, i.e. away from the viewer in R_S and towards the viewer in R_F. When the value of current is negative, the assumption is made that it flows from finish to start, i.e. towards the viewer in an 'S' winding and away from the viewer in an 'F' winding. At time, say t_1, shown in Fig. 19.1(a), the current flowing in the red phase is a maximum positive value. At the same time, t_1, the currents flowing in the yellow and blue phases are both 0.5 times the maximum value and are negative.

The current distribution in the stator windings is therefore as shown in Fig. 19.1(b), in which current flows away from the viewer (shown as ⊗) in R_S since it is positive, but towards the viewer (shown as ⊙) in Y_S and B_S, since these are negative. The resulting magnetic field is as shown, due to the 'solenoid' action and application of the corkscrew rule.

A short time later at time t_2, the current flowing in the red phase has fallen to about 0.87 times its maximum value and is positive, the

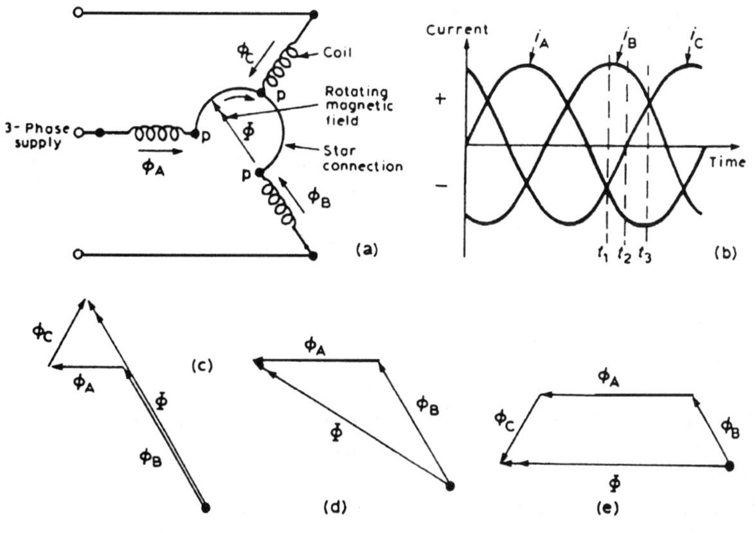

Figure 19.2

current in the yellow phase is zero and the current in the blue phase is about 0.87 times its maximum value and is negative. Hence the currents and resultant magnetic field are as shown in Fig. 19.1(c). At time t_3, the currents in the red and yellow phases are 0.5 of their maximum values and the current in the blue phase is a maximum negative value. The currents and resultant magnetic field are as shown in Fig. 19.1(d).

Similar diagrams to Figs. 19.1(b), (c) and (d) can be produced for all time values and these would show that the magnetic field travels through one revolution for each cycle of the supply voltage applied to the stator windings. By considering the flux values rather than the current values, it is shown below that the rotating magnetic field has a constant value of flux.

The three coils shown in Fig. 19.2(a) are connected in star to a three-phase supply. Let the positive directions of the fluxes produced by currents flowing in the coils be ϕ_A, ϕ_B and ϕ_C, respectively. The directions of ϕ_A, ϕ_B and ϕ_C do not alter, but their magnitudes are proportional to the currents flowing in the coils at any particular time. At time t_1, shown in Fig. 19.2(b), the currents flowing in the coils are: i_B, a maximum positive value, i.e. the flux is towards point P; i_A and i_C, half the maximum value and negative, i.e. the flux is away from point P.

These currents give rise to the magnetic fluxes ϕ_A, ϕ_B and ϕ_C, whose magnitudes and directions are as shown in Fig. 19.2(c). The resultant flux is the phasor sum of ϕ_A, ϕ_B and ϕ_C, shown as Φ in Fig. 19.2(c). At time t_2, the currents flowing are: i_B, 0.866 × maximum positive value, i_C, zero, and i_A, 0.866 × maximum negative value.

The magnetic fluxes and the resultant magnetic flux are as shown in Fig. 19.2(d). At time t_3, i_B is 0.5 × maximum value and is positive, i_A is a maximum negative value, and i_C is 0.5 × maximum value and is positive.

The magnetic fluxes and the resultant magnetic flux are as shown in Fig. 19.2(e). Inspection of Figs. 19.2(c), (d) and (e) shows that the magnitude of the resultant magnetic flux, Φ, in each case is constant and is $1\frac{1}{2}$ × the maxiumum value of ϕ_A, ϕ_B or ϕ_C, but that its direction is changing. The process of determining the resultant flux may be repeated for all values of time and shows that the magnitude of the resultant flux is constant for all values of time and also that it rotates at constant speed, making one revolution for each cycle of the supply voltage.

19.3 Synchronous speed

The rotating magnetic field produced by three phase windings could have been produced by rotating a permanent magnet's north and south pole at synchronous speed (shown as N and S at the ends of the flux phasors in Figs. 19.1(b), (c) and (d)). For this reason, it is called a 2-pole system and an induction motor using three phase windings only is called a 2-pole induction motor.

If six windings displaced from one another by 60° are used, as shown in Fig. 19.3(a), by drawing the current and resultant magnetic field diagrams at various time values, it may be shown that one cycle of the supply current to the stator windings causes the magnetic field to move through half a revolution. The current distribution in the stator windings are shown in Fig. 19.3(a), for the time t shown in Fig. 19.3(b).

It can be seen that for six windings on the stator, the magnetic flux produced is the same as that produced by rotating two permanent magnet north poles and two permanent magnet south poles at synchronous speed. This is called a 4-pole system

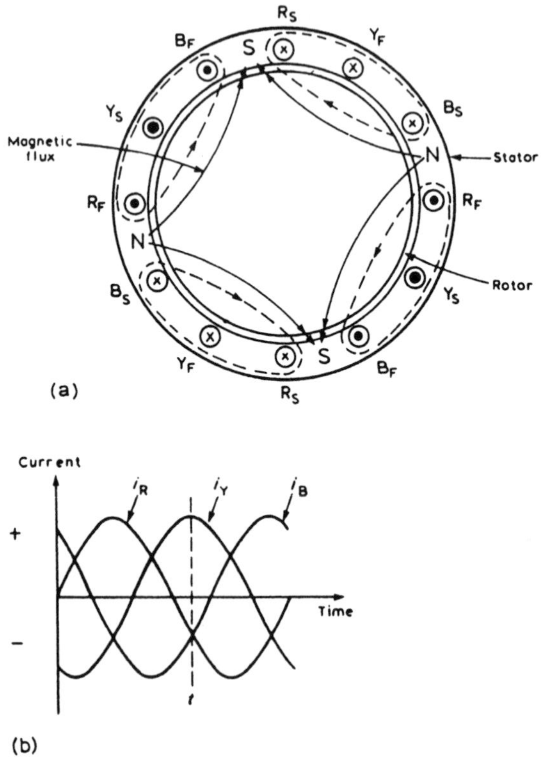

Figure 19.3

and an induction motor using six phase windings is called a 4-pole induction motor. By increasing the number of phase windings the number of poles can be increased to any even number.

In general, if f is the frequency of the currents in the stator windings and the stator is wound to be equivalent to p **pairs** of poles, the speed of revolution of the rotating magnetic field, i.e. the synchronous speed, n_s, is given by

$$n_s = \frac{f}{p} \text{ rev/s}$$

Problem 1. A three-phase 2-pole induction motor is connected to a 50 Hz supply. Determine the synchronous speed of the motor in rev/min.

From above, $n_s = f/p$ rev/s, where n_s is the synchronous speed, f is the frequency in hertz of the supply to the stator and p is the number of **pairs** of poles. Since the motor is connected to a 50 hertz supply, $f = 50$. The motor has a two-pole system, hence p, the number of pairs of poles, is one. Thus, synchronous speed, $n_s = (50/1) = 50$ rev/s $= 50 \times 60$ rev/min $=$ **3000 rev/min**.

Problem 2. A stator winding supplied from a three-phase 60 Hz system is required to produce a magnetic flux rotating at 900 rev/min. Determine the number of poles.

Synchronous speed, $n_s = 900$ rev/min $= 900/60$ rev/s $= 15$ rev/s.

Since $n_2 = (f/p)$ then $p = (f/n_s) = 60/15 = 4$.

Hence the number of pole pairs is 4 and thus **the number of poles is 8.**

Problem 3. A three-phase 2-pole motor is to have a synchronous speed of 6000 rev/min. Calculate the frequency of the supply voltage.

Since $n_2 = (f/p)$ then

$$\textbf{frequency, } f = (n_s)(p)$$

$$= \left(\frac{6000}{60}\right)\left(\frac{2}{2}\right) = \textbf{100 Hz}$$

19.4 Construction of a three-phase induction motor

The stator of a three-phase induction motor is the stationary part corresponding to the yoke of a d.c. machine. It is wound to give a 2-pole, 4-pole, 6-pole, . . . rotating magnetic field, depending on the rotor speed required. The rotor, corresponding to the armature of a d.c. machine, is built up of laminated iron, to reduce eddy currents.

In the type most widely used, known as a **squirrel-cage rotor**, copper or aluminium bars are placed in slots cut in the laminated iron, the ends of the bars being welded or brazed into a heavy conducting ring (see Fig. 19.4(a)). A cross-sectional view of a three-phase induction motor is shown in Fig. 19.4(b). The conductors are placed in slots in the laminated iron rotor core. If the slots are skewed, better starting and quieter running is achieved. This type of rotor has no external connections which means that slip rings and brushes are not needed. The squirrel-cage motor is cheap, reliable and efficient.

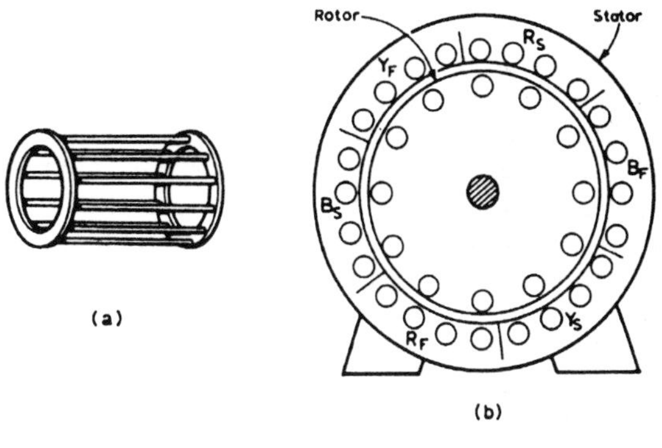

Figure 19.4

Another type of rotor is the **wound rotor**. With this type there are phase windings in slots, similar to those in the stator. The windings may be connected in star or delta and the connections made to three slip rings. The slip rings are used to add external resistance to the rotor circuit, particularly for starting (see section 19.13), but for normal running the slip rings are short-circuited.

The principle of operation is the same for both the squirrel cage and the wound rotor machines.

19.5 Principle of operation of a three-phase induction motor

When a three-phase supply is connected to the stator windings, a rotating magnetic field is produced. As the magnetic flux cuts a bar on the rotor, an e.m.f. is induced in it and since it is joined, via the end conducting rings, to another bar one pole pitch away, a current flows in the bars. The magnetic field associated with this current flowing in the bars interacts with the rotating magnetic field and a force is produced, tending to turn the rotor in the same direction as the rotating magnetic field (see Fig. 19.5). Similar forces are applied to all the conductors on the rotor, so that a torque is produced causing the rotor to rotate.

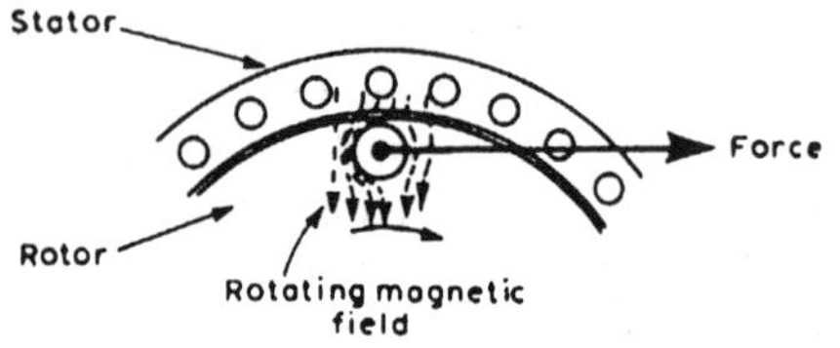

Figure 19.5

19.6 Slip

The force exerted by the rotor bars causes the rotor to turn in the direction of the rotating magnetic field. As the rotor speed increases, the rate at which the rotating magnetic field cuts the rotor bars is less and the frequency of the induced e.m.f.s in the rotor bars is less. If the rotor runs at the same speed as the rotating magnetic field, no e.m.f.s are induced in the rotor, hence there is no force on them and no torque on the rotor. Thus the rotor slows down. For this reason the rotor can never run at synchronous speed.

When there is no-load on the rotor, the resistive forces due to windage and bearing friction are small and the rotor runs very nearly at synchronous speed. As the rotor is loaded, the speed falls and this causes an increase in the frequency of the induced e.m.f.s in the rotor bars and hence the rotor current, force and torque increase. The difference between the rotor speed, n_r, and the synchronous speed, n_s, is called the **slip speed**, i.e.

slip speed = $n_s - n_r$ rev/s

The ratio $(n_s - n_r)/n_s$ is called the **fractional slip** or just the **slip**, s, and is usually expressed as a percentage. Thus

$$\textbf{slip, } s = \left(\frac{n_s - n_r}{n_s}\right) \times 100\%$$

Typical values of slip between no-load and full-load are about 4 to 5% for small motors and 1.5 to 2% for large motors.

> Problem 4. The stator of a 3-phase, 4-pole induction motor is connected to a 50 Hz supply. The rotor runs at 1455 rev/min at full load. Determine (a) the synchronous speed, and (b) the slip at full-load.

(a) The number of pairs of poles, $p = 4/2 = 2$. The supply frequency $f = 50$ Hz. Thus

$$\textbf{synchronous speed, } n_s = \frac{f}{p} = \frac{50}{2} = \textbf{25 rev/s}$$

(b) The rotor speed, $n_r = 1455/60 = 24.25$ rev/s.

$$\textbf{slip, } s = \left(\frac{n_s - n_r}{n_s}\right) \times 100\%$$

$$= \left(\frac{25 - 24.25}{25}\right) \times 100\% = \textbf{3\%}$$

> Problem 5. A 3-phase, 60 Hz induction motor has two poles. If the slip is 2% at a certain load, determine (a) the synchronous speed, (b) the speed of the rotor, and (c) the frequency of the induced e.m.f.s in the rotor.

(a) $f = 60$ Hz, $p = 2/2 = 1$. Hence

$$\textbf{synchronous speed, } n_s = \frac{f}{p} = \frac{60}{1} = 60 \text{ rev/s}$$

or $60 \times 60 =$ **3600 rev/min**

(b) Since slip is given by

$$s = \left(\frac{n_s - n_r}{n_s}\right) \times 100\%$$

$$2 = \left(\frac{60 - n_r}{60}\right) \times 100\%$$

then

$$\frac{2 \times 60}{100} = 60 - n_r$$

i.e.

$$n_r = 60 - \frac{2 \times 60}{100} = 58.8 \text{ rev/s}$$

i.e. the rotor runs at 58.8 × 60 = **3528 rev/min**.

(c) Since the synchronous speed is 60 rev/s and that of the rotor is 58.8 rev/s, the rotating magnetic field cuts the rotor bars at (60 – 58.8), i.e. 1.2 rev/s.
Thus the frequency of the e.m.f.s induced in the rotor bars is 1.2 Hz.

> Problem 6. A three-phase induction motor is supplied from a 50 Hz supply and runs at 1200 rev/min when the slip is 4%. Determine the synchronous speed.

$$\text{Slip, } s = \left(\frac{n_s - n_r}{n_s}\right) \times 100\%$$

Rotor speed, $n_r = 1200/60 = 20$ rev/s, and $s = 4$. Hence

$$4 = \left(\frac{n_s - 20}{n_s}\right) \times 100\%$$

or

$$0.04 = \frac{n_s - 20}{n_s}$$

from which, $n_s(0.04) = n_s - 20$, and $20 = n_s - 0.04n_s = n_s(1 - 0.04)$. Hence

synchronous speed, n_s

$$= \frac{20}{1 - 0.04} = 20.8\dot{3} \text{ rev/s}$$

$$= (20.8\dot{3} \times 60) = \mathbf{1250 \text{ rev/min}}$$

19.7 Rotor e.m.f. and frequency

Rotor e.m.f.

When an induction motor is stationary, the stator and rotor windings form the equivalent of a transformer as shown in Fig. 19.6.

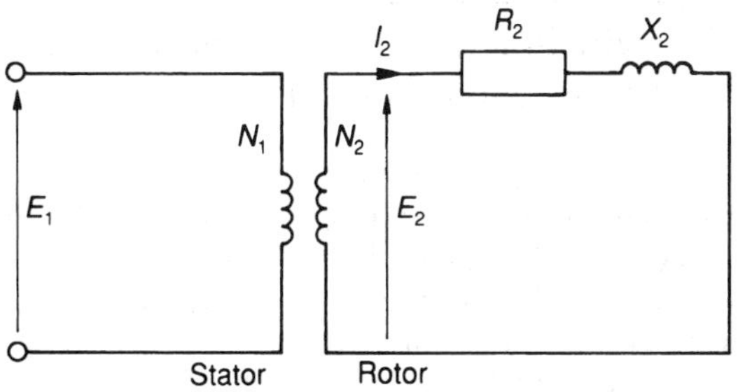

Figure 19.6

The rotor e.m.f. at standstill is given by

$$E_2 = \left(\frac{N_2}{N_1}\right) E_1 \quad (19.1)$$

where E_1 is the supply voltage per phase to the stator.

When an induction motor is running, the induced e.m.f. in the rotor is less since the relative movement between conductors and the rotating field is less. The induced e.m.f. is proportional to this movement, hence it must be proportional to the slip, s.

Hence when running, rotor e.m.f. per phase is given by

$$E_r = sE_2$$

$$= s\left(\frac{N_2}{N_1}\right) E_1 \quad (19.2)$$

Rotor frequency

The rotor e.m.f. is induced by an alternating flux and the rate at which the flux passes the conductors is the slip speed. Thus the frequency of the rotor e.m.f. is given by:

$$f_r = (n_s - n_r)p = \frac{(n_s - n_r)}{n_s}(n_s p)$$

However $(n_s - n_r)/n_s$ is the slip s and $n_s p$ is the supply frequency f, hence

$$\boxed{f_r = sf} \quad (19.3)$$

> Problem 7. The frequency of the supply to the stator of an 8-pole induction motor is 50 Hz and the rotor frequency is 3 Hz. Determine (a) the slip, and (b) the rotor speed.

(a) From equation (19.3), $f_r = sf$, hence $3 = (s)(50)$, from which,

$$\textbf{slip, } s = \frac{3}{50} = \textbf{0.06} \text{ or } \textbf{6\%}$$

(b) Synchronous speed, $n_s = f/p = 50/4 = 12.5$ rev/s or $(12.5 \times 60) = 750$ rev/min.

$$\text{Slip, } s = \left(\frac{n_s - n_r}{n_s}\right)$$

hence

$$0.06 = \left(\frac{12.5 - n_r}{12.5}\right)$$

$$(0.06)(12.5) = 12.5 - n_r$$

and

$$\textbf{rotor speed, } n_r = 12.5 - (0.06)(12.5)$$
$$= \textbf{11.75 rev/s} \text{ or } \textbf{705 rev/min}$$

19.8 Rotor impedance and current

Rotor resistance

The rotor resistance is unaffected by frequency or slip, and hence remains constant.

Rotor reactance

Rotor reactance varies with the frequency of the rotor current. At standstill, reactance per phase, $X_2 = 2\pi fL$. When running, reactance per phase is given by

$$X_r = 2\pi f_r L$$
$$= 2\pi(sf)L \text{ (from equation (19.3))}$$
$$= s(2\pi fL)$$

i.e. $$\boxed{X_r = sX_2} \qquad (19.4)$$

Figure 19.7 represents the rotor circuit when running.

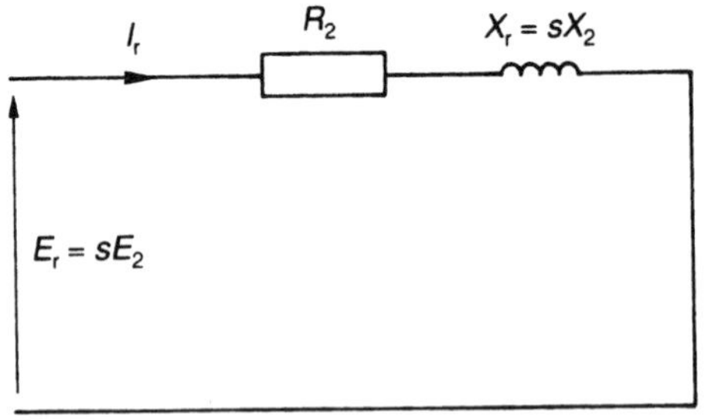

Figure 19.7

Rotor impedance

Rotor impedance per phase,

$$Z_r = \sqrt{[R_2^2 + (sX_2)^2]} \qquad (19.5)$$

At standstill, slip $s = 1$, then

$$Z_2 = \sqrt{[R_2^2 + X_2^2]} \qquad (19.6)$$

Rotor current

From Figs. 19.6 and 19.7, at standstill,

$$\boxed{I_2 = \frac{E_2}{Z_2} = \frac{\left(\frac{N_2}{N_1}\right) E_1}{\sqrt{[R_2^2 + X_2^2]}}} \qquad (19.7)$$

and when running,

$$\boxed{I_r = \frac{E_r}{Z_r} = \frac{s\left(\frac{N_2}{N_1}\right) E_1}{\sqrt{[R_2^2 + (sX_2)^2]}}} \qquad (19.8)$$

19.9 Rotor copper loss

Power $P = 2\pi nT$, where T is the torque in newton metres, hence torque $T = P/2\pi n$.

If P_2 is the power input to the rotor from the rotating field, and P is the mechanical power output (including friction losses) then

$$T = \frac{P_2}{2\pi n_s} = \frac{P_m}{2\pi n_r}$$

from which,

$$\frac{P_2}{n_s} = \frac{P_m}{n_r} \text{ or } \frac{P_m}{P_2} = \frac{n_r}{n_s}$$

Hence

$$1 - \frac{P_m}{P_2} = 1 - \frac{n_r}{n_s}$$

$$\frac{P_2 - P_m}{P_2} = \frac{n_s - n_r}{n_s} = s$$

$P_2 - P_m$ is the electrical or copper loss in the

rotor, i.e. $P_2 - P_m = I_r^2 R_2$. Hence

$$\textbf{slip, } s = \frac{\textbf{rotor copper loss}}{\textbf{rotor input}} = \frac{I_r^2 R_2}{P_2} \qquad (19.9)$$

or power input to the rotor,

$$P_2 = \frac{I_r^2 R_2}{s} \qquad (19.10)$$

19.10 Induction motor losses and efficiency

Figure 19.8 summarizes losses in induction motors.
Motor efficiency,

$$\eta = \frac{\textbf{output power}}{\textbf{input power}} = \frac{P_m}{P_1} \times 100\%$$

Problem 8. The power supplied to a three-phase induction motor is 32 kW and the stator losses are 1200 W. If the slip is 5%, determine (a) the rotor copper loss, (b) the total mechanical power developed by the rotor, (c) the output power of the motor if friction and windage losses are 750 W, and (d) the efficiency of the motor, neglecting rotor iron loss.

(a) Input power to rotor = stator input power − stator losses, i.e.

32 kW − 1.2 kW = 30.8 kW

From equation (19.9),

$$\text{slip} = \frac{\text{rotor copper loss}}{\text{rotor input}}$$

i.e.

$$\frac{5}{100} = \frac{\text{rotor copper loss}}{30.8}$$

from which, **rotor copper loss** = (0.05)(30.8) = **1.54 kW**.

(b) Total mechanical power developed by the rotor = rotor input power − rotor losses, i.e.

30.8 − 1.54 = **29.26 kW**

(c) Output power of motor = power developed by the rotor − friction and windage losses, i.e.

29.26 − 0.75 = **28.51 kW**

(d) Efficiency of induction motor, η

$$= \left(\frac{\text{output power}}{\text{input power}}\right) \times 100\%$$

$$= \left(\frac{28.51}{32}\right) \times 100\% = \mathbf{89.10\%}$$

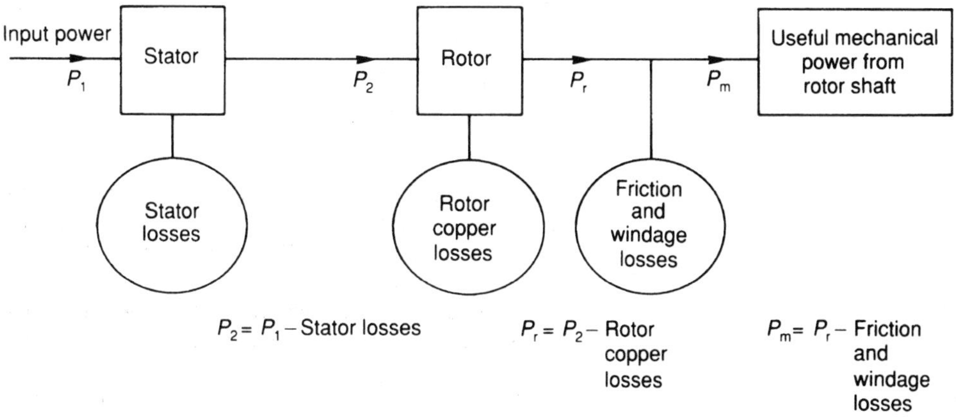

Figure 19.8

Problem 9. The speed of the induction motor of Problem 8 is reduced to 35% of its synchronous speed by using external rotor resistance. If the torque and stator losses are unchanged, determine (a) the rotor copper loss, and (b) the efficiency of the motor.

(a) Slip, $s = \left(\frac{n_s - n_r}{n_s}\right) \times 100\%$

$= \left(\frac{n_s - 0.35n_s}{n_s}\right) \times 100\%$

$= (0.65)(100) = 65\%$

Input power to rotor = 30.8 kW (from Problem 8).

Since s = (rotor copper loss)/(rotor input), then **rotor copper loss** is given by (s)(rotor input)

$= \left(\frac{65}{100}\right)(30.8) = \mathbf{20.02\ kW}$

(b) Power developed by rotor = input power to rotor − rotor copper loss

= 30.8 − 20.02 = 10.78 kW

Output power of motor = power developed by rotor − friction and windage losses

= 10.78 − 0.75 = 10.03 kW

$$\text{Efficiency, } \eta = \frac{\text{output power}}{\text{input power}} \times 100\%$$

$$= \left(\frac{10.03}{32}\right) \times 100\% = \mathbf{31.34\%}$$

19.11 Torque equation for an induction motor

$$\text{Torque } T = \frac{P_2}{2\pi n_s} = \left(\frac{1}{2\pi n_s}\right)\left(\frac{I_r^2 R_2}{s}\right)$$

(from equation (19.10))

From equation (19.8),

$$I_r = \frac{s\left(\frac{N_2}{N_1}\right)E_1}{\sqrt{[R_2^2 + (sX_2)^2]}}$$

Hence torque per phase,

$$T = \left(\frac{1}{2\pi n_s}\right)\left(\frac{s^2\left(\frac{N_2}{N_1}\right)^2 E_1^2}{R_2^2 + (sX_2)^2}\right)\left(\frac{R_2}{s}\right)$$

i.e.

$$T = \left(\frac{1}{2\pi n_s}\right)\left(\frac{s\left(\frac{N_2}{N_1}\right)^2 E_1^2 R_2}{R_2^2 + (sX_2)^2}\right)$$

If these are m phases then

$$\text{torque, } T = \left(\frac{m}{2\pi n_s}\right)\left(\frac{s\left(\frac{N_2}{N_1}\right)^2 E_1^2 R_2}{R_2^2 + (sX_2)^2}\right)$$

i.e.

$$T = \left(\frac{m\left(\frac{N_2}{N_1}\right)^2}{2\pi n_s}\right)\left(\frac{sE_1^2 R_2}{R_2^2 + (sX_2)^2}\right) \quad (19.11)$$

$$= k\left(\frac{sE_1^2 R_2}{R_2^2 + (sX_2)^2}\right)$$

where k is a constant for a particular machine, i.e.

$$\text{torque } T \propto \frac{sE_1^2 R_2}{R_2^2 + (sX_2)^2} \quad (19.12)$$

Under normal conditions, the supply voltage is usually constant, hence equation (19.12) becomes:

$$T \propto \frac{sR_2}{R_2^2 + (sX_2)^2} \propto \frac{R_2}{\frac{R_2^2}{s} + sX_2^2}$$

The torque will be a maximum when the denominator is a minimum and this occurs when $R_2^2/s = sX_2^2$, i.e. when $s = (R_2/X_2)$ or $R_2 = sX_2 = X_r$ (from equation (19.4)).

Thus **maximum torque** occurs when rotor resistance and rotor reactance are equal, i.e. $\boldsymbol{R_2 = X_r}$.

Problems 10 to 13 illustrate some of the characteristics of three-phase induction motors.

Problem 10. A 415 V, three-phase, 50 Hz, 4-pole, star-connected induction motor runs at 24 rev/s on full-load. The rotor resistance and reactance per phase are 0.3 Ω and 3.5 Ω, respectively, and the effective rotor–stator turns ratio is 0.85 : 1. Calculate (a) the synchronous speed, (b) the slip, (c) the full-load torque, (d) the power output if mechanical losses amount to 580 W, (e) the maximum torque, (f) the speed at which maximum torque occurs, and (g) the starting torque.

(a) Synchronous speed, $n_s = f/p = 50/2 =$ **25 rev/s** or $(25 \times 60) =$ **1500 rev/min.**

(b) Slip, $s = \left(\frac{n_s - n_r}{n_s}\right) = \left(\frac{25 - 24}{25}\right) =$ **0.04** or **4%**

(c) Phase voltage, $E_1 = (415/\sqrt{3}) = 239.6$ volts.

Full-load torque, $T =$

$$\left(\frac{m\left(\frac{N_2}{N_1}\right)^2}{2\pi n_s}\right)\left(\frac{sE_1{}^2R_2}{R_2{}^2 + (sX_2)^2}\right)$$

(from equation (19.11))

$$= \left(\frac{3(0.85)^2}{2\pi(25)}\right)\left(\frac{0.04(239.6)^2 0.3}{(0.3)^2 + (0.04 \times 3.5)^2}\right)$$

$$= (0.013\,80)\left(\frac{688.90}{0.1096}\right) = \mathbf{86.74\ N\ m}$$

(d) Output power, including friction losses is given by

$$P_m = 2\pi n_r T$$
$$= 2\pi(24)(86.74)$$
$$= 13\,080 \text{ watts}$$

Power output $= P_m -$ mechanical losses

$13\,080 - 580 = 12\,500$ W $=$ **12.5 kW**

(e) Maximum torque occurs when $R_2 = X_r = 0.35\ \Omega$. Slip, $s = R_2/X_2 = 0.35/3.5 = 0.1$. Hence

maximum torque, $\boldsymbol{T_{max}}$

$$= (0.013\,80)\left(\frac{sE_1{}^2R_2}{R_2{}^2 + (sX_2)^2}\right) \text{ (from part (c))}$$

$$= (0.013\,80)\left(\frac{0.1(239.6)^2 0.35}{0.35^2 + 0.35^2}\right)$$

$$= (0.013\,80)\left(\frac{2009.29}{0.245}\right) = \mathbf{113.18\ N\ m}$$

(f) For maximum torque, slip $s = 0.1$.

Slip, $s = \left(\frac{n_s - n_r}{n_s}\right)$

i.e.

$$0.1 = \left(\frac{25 - n_r}{25}\right)$$

Hence $(0.1)(25) = 25 - n_r$ and $n_r = 25 - (0.1)(25)$. Thus speed at which maximum torque occurs,

$$n_r = 25 - 2.5$$
$$= \mathbf{22.5\ rev/s} \text{ or } \mathbf{1350\ rev/min}$$

(g) At the start, i.e. at standstill, slip $s = 1$. Hence starting torque

$$= \left(\frac{m\left(\frac{N_2}{N_1}\right)^2}{2\pi n_s}\right)\left(\frac{E_1{}^2R_2}{R_2{}^2 + X_2{}^2}\right)$$

(from equation (19.11) with $s = 1$)

$$= (0.013\,80)\left(\frac{(239.6)^2 0.35}{0.35^2 + 3.5^2}\right)$$

$$= (0.013\,80)\left(\frac{20092.86}{12\,.3725}\right)$$

i.e. **starting torque = 22.41 N m.**

(Note that the full load torque (from part (c)) is 86.74 N m but the starting torque is only 22.41 N m.)

Problem 11. Determine for the induction motor in Problem 10 at full-load, (a) the rotor current, (b) the rotor copper loss, and (c) the starting current.

(a) From equation (19.8),

$$\textbf{rotor current, } \boldsymbol{I_r} = \frac{s\left(\frac{N_2}{N_1}\right)^2 E_1}{\sqrt{[R_2{}^2 + (sX_2)^2}}$$

$$= \frac{(0.04)(0.85)^2(239.6)}{\sqrt{[0.35^2 + (0.04 \times 3.5)^2]}}$$

$= \dfrac{6.9244}{0.376\,96} = \mathbf{18.37\ A}$

(b) Rotor copper loss per phase $= I_r^2R_2$

$= (18.37)^2(0.35) = 118.11$ W

Total copper loss (for 3 phases) $= 3 \times 118.11 =$ **354.3 W**.

(c) From equation (19.7), starting current,

$$I_2 = \frac{\left(\dfrac{N_2}{N_1}\right)E_1}{\sqrt{[R_2^2 + X_2^2]}} = \frac{(0.85)(239.6)}{\sqrt{[0.35^2 + 3.5^2]}} = \mathbf{57.90A}$$

(Note that the starting current of 57.90 A is considerably higher than the full-load current of 18.37 A.)

Problem 12. For the induction motor in Problems 10 and 11, if the stator losses are 650 W, determine (a) the power input at full-load, (b) the efficiency of the motor at full-load, and (c) the current taken from the supply at full-load if the motor runs at a power factor of 0.87 lagging.

(a) Output power $P_m = 13.080$ kW from part (d), Problem 10.
Rotor copper loss = 354.3 W = 0.3543 kW from part (b), Problem 11.
Stator input power, $P_1 = P_m$ + rotor copper loss + rotor stator loss

$= 13.080 + 0.3543 + 0.650 =$ **14.084 kW**

(b) Net power output = 12.5 kW from part (d), Problem 10. Hence

$$\text{efficiency, } \eta = \frac{\text{output}}{\text{input}} \times 100\%$$

$$= \left(\frac{12.5}{14.084}\right) \times 100\%$$

$= \mathbf{88.75\%}$

(c) Power input, $P_1 = \sqrt{3}V_LI_L \cos\phi$ (see Chapter 16) from which,

$$\textbf{supply current, } I_L = \frac{P_1}{\sqrt{3}V_L\cos\phi}$$

$$= \frac{14.084 \times 1000}{\sqrt{3}(415)0.87}$$

$= \mathbf{22.52\ A}$

Problem 13. For the induction motor of Problems 10 to 12, determine the resistance of the rotor winding required for maximum starting torque.

From equation (19.4), rotor reactance $X_r = sX_2$. At the moment of starting, slip, $s = 1$. Maximum torque occurs when rotor reactance equals rotor resistance, hence for **maximum torque**, $R_2 = X_r = sX_2 = X_2 =$ **3.5 Ω**.

Thus if the induction motor was a wound rotor type with slip rings then an external star-connected resistance of (3.5 − 0.35) Ω = 3.15 Ω per phase could be added to the rotor resistance to give maximum torque at starting (see section 19.13).

19.12 Induction motor torque – speed characteristics

From Problem 10, parts (c) and (g), it was seen that the normal starting torque may be less than the full-load torque. Also, from Problem 10, parts (e) and (f), it is seen that the speed at which maximum torque occurs is determined by the value of the rotor resistance. At synchronous speed, slip $s = 0$ and torque is zero. From these observations, the torque-speed and torque-slip characteristics of an induction motor are as shown in Fig. 19.9.

The rotor resistance of an induction motor is usually small compared with its reactance (for example, $R_2 = 0.3\ \Omega$ and $X_2 = 3.5\ \Omega$ in the above Problems), so that maximum torque occurs at a high speed, typically about 80% of synchronous speed.

Curve P in Fig. 19.9 is a typical characteristic for an induction motor. The curve P cuts the full-load torque line at X, showing that at full-load the slip is about 4–5%. The normal operating conditions are between 0 and X, thus it can be seen that for normal operation the speed variation with load is quite small – the induction motor is an almost constant-speed machine. Redrawing the speed–torque characteristic between 0 and X gives the characteristic shown in Fig. 19.10, which is similar to a d.c. shunt motor (see Chapter 18).

If maximum torque is required at starting then a high resistance rotor is necessary, which gives characteristic Q in Fig. 19.9. However, as can be seen, the motor has a full-load slip of over 30%, which results in a drop in efficiency. Also such a

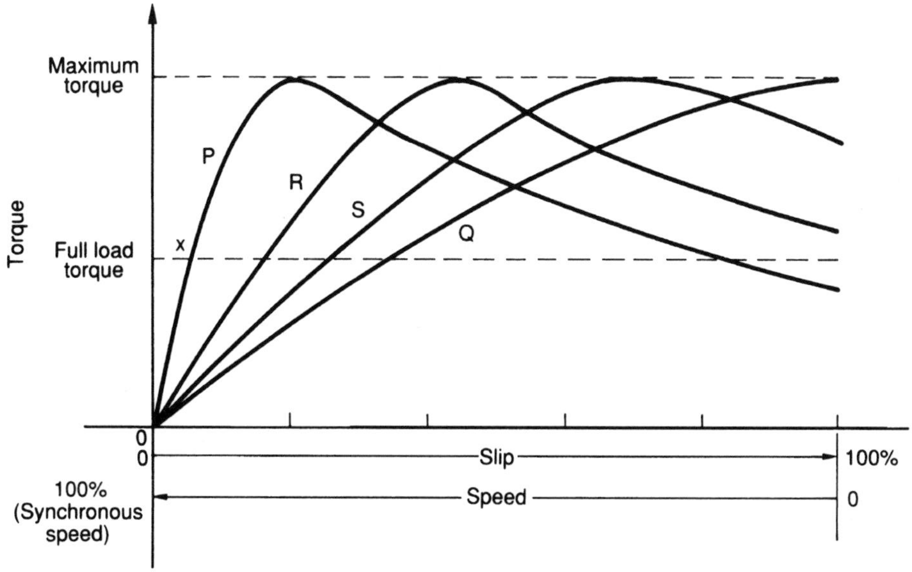

Figure 19.9

motor has a large speed variation with variations of load. Curves R and S of Fig. 19.9 are characteristics for values of rotor resistances between those of P and Q. Better starting torque than for curve P is obtained, but with lower efficiency and with speed variations under operating conditions.

A **squirrel-cage induction motor** would normally follow characteristic P. This type of machine is highly efficient and about constant speed under normal running conditions. However, it has a poor starting torque and must be started off-load or very lightly loaded (see section 19.13). Also, on starting, the current can be four or five times the normal full-load current, due to the motor acting like a transformer with secondary short-circuited. In Problem 11, for example, the current at starting was over three times the full-load current.

A **wound-rotor induction motor** would follow characteristic P when the slip rings are short-circuited, which is the normal running condition. However, the slip rings allow for the addition of resistance to the rotor circuit externally and, as a result, for starting, the motor can have a characteristic similar to curve Q in Fig. 19.9. and the high starting current experienced by the cage induction motor can be overcome.

In general, for three-phase induction motors, the power factor is usually between about 0.8 and 0.9 lagging, and the full-load efficiency is usually about 80–90%.

From equation (19.12), it is seen that torque is proportional to the square of the supply voltage. Any voltage variations therefore would seriously affect the induction motor performance.

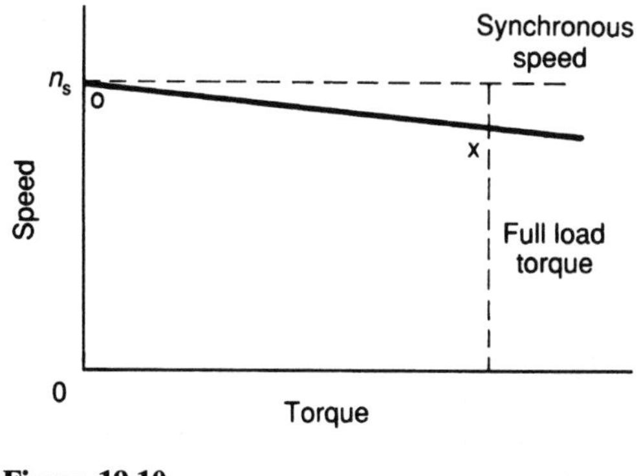

Figure 19.10

19.13 Starting methods for induction motors

Squirrel-cage rotor

(i) **Direct-on-line starting**
With this method, starting current is high and may cause interference with supplies to other consumers.

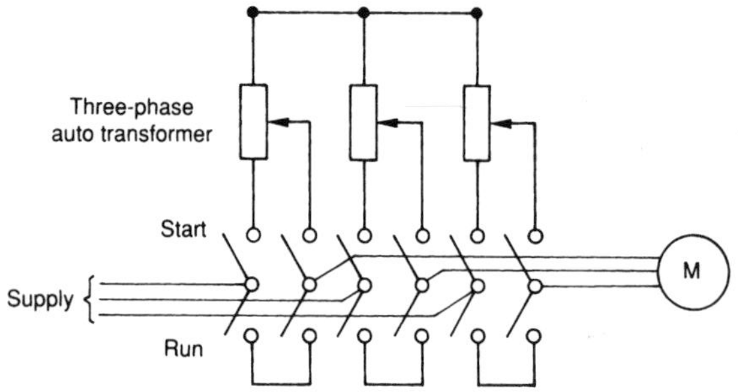

Figure 19.11

(ii) **Auto transformer starting**
With this method, an auto transformer is used to reduce the stator voltage, E_1, and thus the starting current (see equation (19.7)). However, the starting torque is seriously reduced (see equation (19.12)), so the voltage is reduced only sufficiently to give the required reduction of the starting current. A typical arrangement is shown in Fig. 19.11. A double-throw switch connects the auto transformer in circuit for starting, and when the motor is up to speed the switch is moved to the run position which connects the supply directly to the motor.

(iii) **Star–delta starting**
With this method, for starting, the connections to the stator phase winding are star connected, so that the voltage across each phase winding is $1/\sqrt{3}$ (i.e. 0.577) of the line voltage. For running, the windings are switched to delta connection. A typical arrangement is shown in Fig. 19.12. This method of starting is less expensive than by auto transformer.

Wound rotor

When starting on load is necessary, a wound rotor induction motor must be used. This is because maximum torque at starting can be

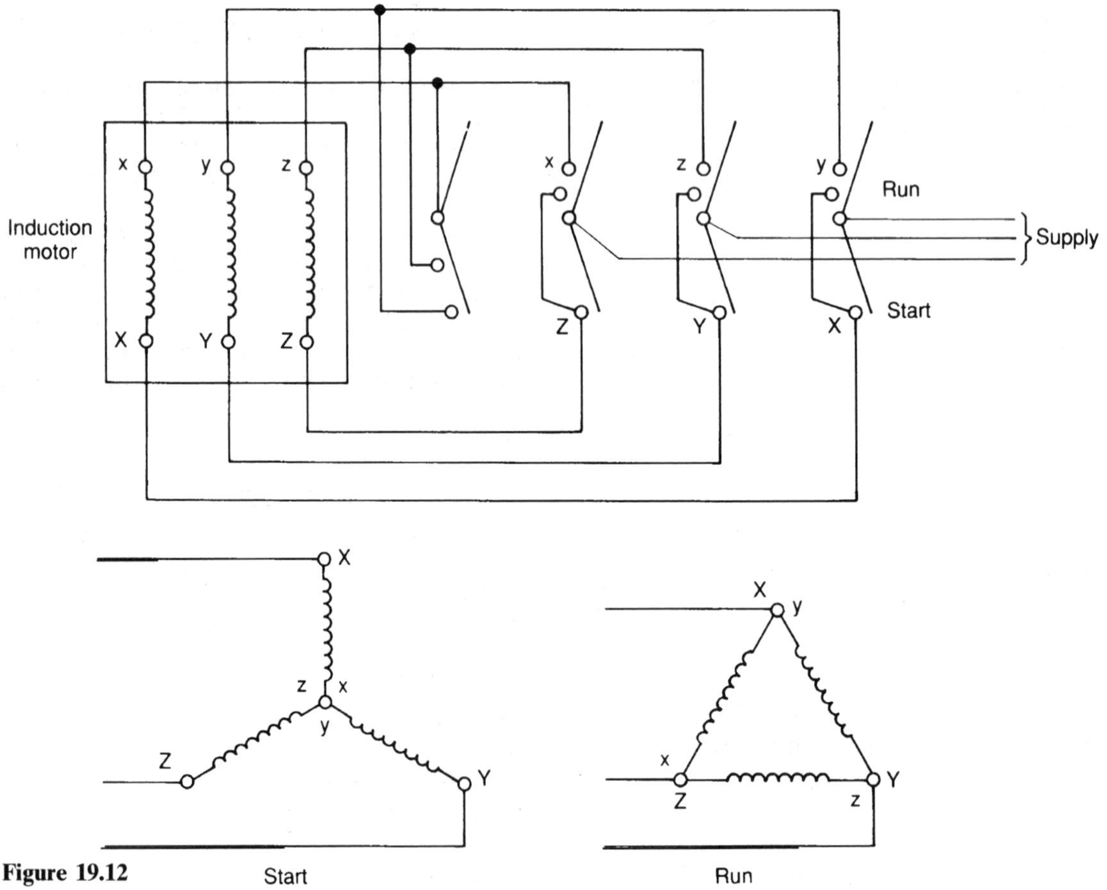

Figure 19.12

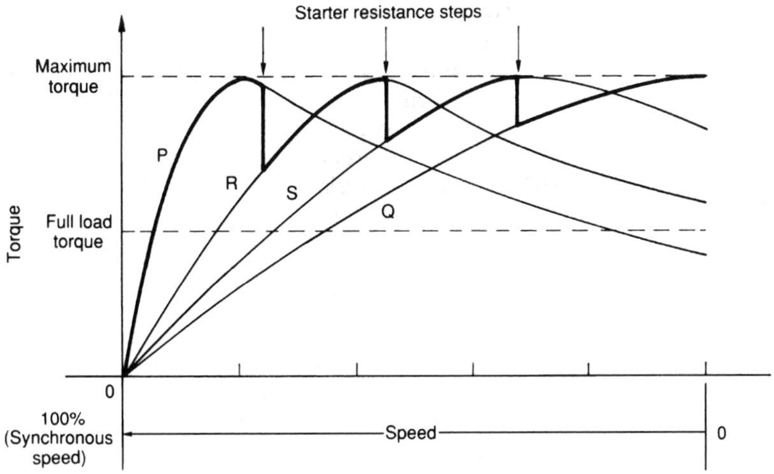

Figure 19.13

obtained by adding external resistance to the rotor circuit via slip rings (see Problem 13). A face-plate type starter is used, and as the resistance is gradually reduced, the machine characteristics at each stage will be similar to Q, S, R and P of Fig. 19.13. At each resistance step, the motor operation will transfer from one characteristic to the next so that the overall starting characteristic will be as shown by the bold line in Fig. 19.13. For very large induction motors, very gradual and smooth starting is achieved by a liquid type resistance.

19.14 Advantages of squirrel-cage induction motors

The advantages of squirrel-cage motors compared with the wound rotor type are that they:

(i) are cheaper and more robust;
(ii) have slightly higher efficiency and power factor;
(iii) are explosion-proof, since the risk of sparking is eliminated by the absence of slip rings and brushes.

19.15 Advantages of wound rotor induction motors

The advantages of the wound rotor motor compared with the cage type are that they:

(i) have a much higher starting torque;
(ii) have a much lower starting current;
(iii) have a means of varying speed by use of external rotor resistance.

19.16 Double cage induction motor

The advantages of squirrel-cage and wound rotor induction motors are combined in the double cage induction motor. This type of induction motor is specially constructed with the rotor having two cages, one inside the other. The outer cage has high resistance conductors so that maximum torque is achieved at or near starting. The inner cage has normal low resistance copper conductors but high reactance since it is embedded deep in the iron core. The torque–speed characteristic of the inner cage is that of a normal

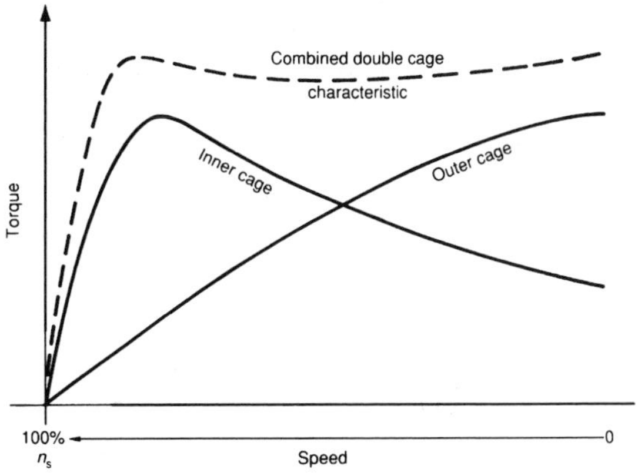

Figure 19.14

induction motor, as shown in Fig. 19.14. At starting, the outer cage produces the torque, but when running the inner cage produces the torque. The combined characteristic of inner and outer cages is shown in Fig. 19.14. The double cage induction motor is highly efficient when running.

19.17 Uses of three-phase induction motors

Three-phase induction motors are widely used in industry and constitute almost all industrial drives where a nearly constant speed is required, from small workshops to the largest industrial enterprises.

Typical applications are with machine tools, pumps and mill motors. The squirrel-cage rotor type is the most widely used of all a.c. motors.

19.18 Multi-choice questions on three-phase induction motors

(Answers on p. 284.)

1. Which of the following statements about a three-phase squirrel-cage induction motor is false?
   (a) It has no external electrical connections to its rotor.
   (b) A three-phase supply is connected to its stator.
   (c) A magnetic flux which alternates is produced.
   (d) It is cheap, robust and requires little or no skilled maintenance.
2. Which of the following statements about a three-phase induction motor is false?
   (a) The speed of rotation of the magnetic field is called the synchronous speed.
   (b) A three-phase supply connected to the rotor produces a rotating magnetic field.
   (c) The rotating magnetic field has a constant speed and constant magnitude.
   (d) It is essentially a constant speed type machine.
3. Which of the following statements is false when referring to a three-phase induction motor?
   (he synchronous speed is half the supply quency when it has four poles.
   2-pole machine, the synchronous is equal to the supply frequency.
   (c) If the number of poles is increased, the synchronous speed is reduced.
   (d) The synchronous speed is inversely proportional to the number of poles.
4. A 4-pole, three-phase induction motor has a synchronous speed of 25 rev/s. The frequency of the supply to the stator is:
   (a) 50 Hz; (b) 100 Hz; (c) 25 Hz; (d) 12.5 Hz.

Questions 5 and 6 refer to a three-phase induction motor. Which statements are false?

5. (a) The slip speed is the synchronous speed minus the rotor speed.
   (b) As the rotor is loaded, the slip decreases.
   (c) The frequency of induced rotor e.m.f.s increases with load on the rotor.
   (d) The torque on the rotor is due to the interaction of magnetic fields.
6. (a) If the rotor is running at synchronous speed, there is no torque on the rotor.
   (b) If the number of poles on the stator is doubled, the synchronous speed is halved.
   (c) At no-load, the rotor speed is very nearly equal to the synchronous speed.
   (d) The direction of rotation of the rotor is opposite to the direction of rotation of the magnetic field to give maximum current induced in the rotor bars.

A three-phase, 4-pole, 50 Hz induction motor runs at 1440 rev/min. In questions 7 to 10, determine the correct answers for the quantities stated, selecting your answer from the list given.

(a) $12\frac{1}{2}$ rev/s	(b) 25 rev/s	(c) 1 rev/s
(d) 50 rev/s	(e) 1%	(f) 4%
(g) 50%	(h) 4 Hz	(i) 50 Hz
(j) 1 Hz		

7. The synchronous speed.
8. The slip speed.
9. The percentage slip.
10. The frequency of induced e.m.f.s in the rotor.
11. The slip speed of an induction motor may be defined as:
    (a) number of pairs of poles ÷ frequency;
    (b) rotor speed – synchronous speed;
    (c) rotor speed + synchronous speed;
    (d) synchronous speed – rotor speed.
12. The slip speed of an induction motor depends upon:
    (a) armature current;
    (b) supply voltage;
    (c) mechanical load;
    (d) eddy currents.

13. The starting torque of a simple squirrel-cage motor is:
    (a) low;
    (b) increases as rotor current rises;
    (c) decreases as rotor current rises;
    (d) high.
14. The slip speed of an induction motor
    (a) is zero until the rotor moves and then rises slightly,
    (b) is 100% until the rotor moves and then decreases slightly,
    (c) is 100% until the rotor moves and then falls to a low value,
    (d) is zero until the rotor moves and then rises to 100%.
15. A four-pole induction motor when supplied from a 50 Hz supply experiences a 5% slip. The rotor speed will be
    (a) 25 rev/s,
    (b) 23.75 rev/s,
    (c) 26.25 rev/s,
    (d) 11.875 rev/s.

19.19 Short answer questions on three-phase induction motors

1. Name three advantages that a three-phase induction motor has when compared with a d.c. motor.
2. Name the principal disadvantage of a three-phase induction motor when compared with a d.c. motor.
3. Write down the two properties of the magnetic field produced by the stator of a three-phase induction motor.
4. The speed at which the magnetic field of a three-phase induction motor rotates is called the .... speed.
5. The synchronous speed of a three-phase induction motor is .... proportional to the supply frequency.
6. The synchronous speed of a three-phase induction motor is .... proportional to the number of pairs of poles.
7. The type of rotor most widely used in a three-phase induction motor is called a .... rotor.
8. The slip of a three-phase induction motor is given by:

   $$s = \frac{\dots\dots\dots}{\dots\dots\dots} \times 100\%$$
9. A typical value for the slip of a small three-phase induction motor at full-load is .... %.
10. As the load on the rotor of a three-phase induction motor increases, the slip ....
11. $$\frac{\text{Rotor copper loss}}{\text{Rotor input power}} = \dots\dots\dots\dots .$$
12. State the losses in an induction motor.
13. Maximum torque occurs when .... = ...
14. Sketch a typical speed–torque characteristic for an induction motor.
15. State two methods of starting squirrel-cage induction motors.
16. Which type of induction motor is used when starting on-load is necessary?
17. Describe briefly a double cage induction motor.
18. State two advantages of cage rotor machines compared with wound rotor machines.
19. State two advantages of wound rotor machines compared with cage rotor machines.
20. Name any three applications of three-phase induction motors.

19.20 Further questions on three-phase induction motors

1. With the aid of diagrams, explain how a rotating magnetic field is produced when a three-phase supply is connected to the stator of an induction motor.
2. Explain briefly what you understand by the term 'a 6-pole induction motor'.
3. The synchronous speed of a 3-phase, 4-pole induction motor is 60 rev/s. Determine the frequency of the supply to the stator windings. [120 Hz]
4. The synchronous speed of a 3-phase induction motor is 25 rev/s and the frequency of the supply to the stator is 50 Hz. Calculate the equivalent number of pairs of poles of the motor. [2]
5. A 6-pole, 3-phase induction motor is connected to a 300 Hz supply. Determine the speed of rotation of the magnetic field produced by the stator. [100 rev/s]
6. A 6-pole, 3-phase induction motor runs at 970 rev/min at a certain load. If the stator is connected to a 50 Hz supply, find the percentage slip at this load. [3%]
7. A 3-phase, 50 Hz induction motor has 8 poles. If the full load slip is $2\frac{1}{2}$%, determine (a) the synchronous speed, (b) the rotor speed, and (c) the frequency of the rotor e.m.f.s.
   [(a) 750 rev/min (b) 731 rev/min (c) 1.25 Hz]

8. A 3-phase induction motor is supplied from a 60 Hz supply and runs at 1710 rev/min when the slip is 5%. Determine the synchronous speed. [1800 rev/min]
9. Explain briefly, with the aid of sketches, the principle of operation of a 3-phase induction motor.
10. Explain how slip-frequency currents are set up in the rotor bars of a 3-phase induction motor and why this frequency varies with load.
11. Explain why a 3-phase induction motor develops no torque when running at synchronous speed. Define the slip of an induction motor and explain why its value depends on the load on the rotor.
12. A 4-pole, 3-phase, 50 Hz induction motor runs at 1440 rev/min at full-load. Calculate (a) the synchronous speed, (b) the slip, and (c) the frequency of the rotor induced e.m.f.s. [(a) 1500 rev/min (b) 4% (c) 2 Hz]
13. A 12-pole, 3-phase, 50 Hz induction motor runs at 475 rev/min. Determine (a) the slip speed, (b) the percentage slip, and (c) the frequency of rotor currents. [(a) 25 rev/min (b) 5% (c) 2.5 Hz]
14. The frequency of the supply to the stator of a 6-pole induction motor is 50 Hz and the rotor frequency is 2%. Determine (a) the slip, and (b) the rotor speed in rev/min. [(a) 0.04 or 4% (b) 960 rev/min]
15. The power supplied to a 3-phase induction motor is 50 kW and the stator losses are 2 kW. If the slip is 4%, determine (a) the rotor copper loss, (b) the total mechanical power developed by the rotor, (c) the output power of the motor if friction and windage losses are 1 kW, and (d) the efficiency of the motor, neglecting rotor iron losses. [(a) 1.92 kW (b) 46.08 kW (c) 45.08 kW (d) 90.16%]
16. By using external rotor resistance, the speed of the induction motor in question 15 is reduced to 40% of its synchronous speed. If the torque and stator losses are unchanged, calculate (a) the rotor copper loss, and (b) the efficiency of the motor. [(a) 28.80 kW (b) 36.40%]
17. A 400 V, 3-phase, 50 Hz, 2-pole, star-connected induction motor runs at 48.5 rev/s on full-load. The rotor resistance and reactance per phase are 0.4 Ω and 4.0 Ω, respectively, and the effective rotor–stator turns ratio is 0.8 : 1. Calculate (a) the synchronous speed, (b) the slip, (c) the full-load torque, (d) the power output if mechanical losses amount to 500 W, (e) the maximum torque, (f) the speed at which maximum torque occurs, and (g) the starting torque. [(a) 50 rev/s or 3000 rev/min (b) 0.03 or 3% (c) 22.43 N m (d) 6.34 kW (e) 40.74 N m (f) 45 rev/s or 2700 rev/min (g) 8.07 N m]
18. For the induction motor in question 17, calculate at full load (a) the rotor current, (b) the rotor copper loss, and (c) the starting current. [(a) 10.62 A (b) 135.3 W (c) 45.96 A]
19. If the stator losses for the induction motor in question 17 are 525 W, calculate at full-load (a) the power input, (b) the efficiency of the motor, and (c) the current taken from the supply if the motor runs at a power factor of 0.84. [(a) 7.49 kW (b) 84.65% (c) 12.87 A]
20. For the induction motor in question 17, determine the resistance of the rotor winding required for maximum starting torque. [4.0 Ω]

Answers to multi-choice questions

PART 1

Chapter 1. Units associated with basic electrical quantities (page 6)

1. (c) 2. (d) 3. (c) 4. (a)
5. (c) 6. (b)

Chapter 2. An introduction to electric circuits (page 15)

1. (b) 2. (b) 3. (c) 4. (b)
5. (d) 6. (d) 7. (b) 8. (c)

Chapter 3. Resistance variation (page 22)

1. (c) 2. (d) 3. (b) 4. (d)
5. (d) 6. (c) 7. (b)

Chapter 4. Chemical effects of electricity (page 30)

1. (a) 2. (b) 3. (c) 4. (b)
5. (d) 6. (d) 7. (b) 8. (c)

Chapter 5. Series and parallel networks (page 42)

1. (a) 2. (c) 3. (c) 4. (c)
5. (a) 6. (b) 7. (d) 8. (b)

Chapter 6. Capacitors and capacitance (page 56

1. (b) 2. (a) 3. (b) 4. (c)
5. (a) 6. (b) 7. (b) 8. (a)
9. (c) 10. (c)

Chapter 7. Magnetic circuits (page 69)

1. (d) 2. (b) 3. (d) 4. (i)
5. (f) 6. (j) 7. (h) 8. (c)
9. (c) 10. (a) and (d); (b) and (f); (c) and (e)

Chapter 8. Electromagnetism (page 81)

1. (d) 2. (c) 3. (d) 4. (a)
5. (b) 6. (c)

Chapter 9. Electromagnetic induction (page 89)

1. (c) 2. (b) 3. (c) 4. (b)
5. (c) 6. (a) 7. (c) 8. (d)
9. (c)

Chapter 10. Electrical measuring instruments and measurements (page 110)

1. (d) 2. (a) or (c) 3. (b)
4. (b) 5. (c) 6. (f) 7. (c)
8. (a) 9. (i) 10. (j) 11. (g)
12. (c) 13. (b) 14. (p) 15. (d)
16. (o) 17. (n) 18. (b) 19. (d)
20. (a) 21. (d) 22. (c)

PART 2

Chapter 11. D.c. circuit theorems (page 136)

1. (d) 2. (c) 3. (b) 4. (c)
5. (a) 6. (d) 7. (c) 8. (a)
9. (c) 10. (c) 11. (b) 12. (d)

Chapter 12. Alternating voltages and currents (page 151)

1. (c)	2. (d)	3. (d)	4. (a)
5. (d)	6. (c)	7. (b)	8. (c)
9. (b)	10. (c)		

Chapter 13. Single-phase series a.c. circuits (page 168)

1. (c)	2. (a)	3. (b)	4. (b)
5. (a)	6. (b)	7. (a)	8. (d)
9. (d)	10. (d)	11. (b)	12. (c)
13. (b)	14. (c)	15. (b)	

Chapter 14. Single-phase parallel a.c. circuits (page 183)

1. (d)	2. (g)	3. (i)	4. (s)
5. (h)	6. (b)	7. (k)	8. (l)
9. (a)	10. (d)	11. (g)	12. (b)
13. (c)	14. (j)	15. (h)	

Chapter 15. D.c. transients (page 198)

1. (c)	2. (b)	3. (b)	4. (g)
5. (g)	6. (e)	7. (l)	8. (c)
9. (a)	10. (d)	11. (g)	12. (b)
13. (c)	14. (j)	15. (h)	

PART 3

Chapter 16. Three-phase systems (page 216)

1. (g)	2. (c)	3. (a)	4. (a)
5. (f)	6. (a)	7. (g)	8. (l)
9. (l)	10. (d)	11. (f)	12. (j)

Chapter 17. Transformers (page 237)

1. (b)	2. (g)	3. (d)	4. (a)
5. (h)	6. (k)	7. (j)	8. (f)
9. (c)	10. (b) and (c)		11. (c)
12. (b)			

Chapter 18. D.c. machines (page 261)

1. (b)	2. (e)	3. (e)	4. (c)
5. (c)	6. (a)	7. (d)	8. (f)
9. (b)	10. (c)	11. (b)	12. (a)
13. (b)	14. (a)	15. (d)	16. (b)
17. (b)	18. (b)	19. (c)	20. (b)

Chapter 19. Three-phase induction motors (page 280)

1. (c)	2. (b)	3. (d)	4. (a)
5. (b)	6. (d)	7. (b)	8. (c)
9. (f)	10. (j)	11. (d)	12. (c)
13. (a)	14. (c)	15. (b)	

Index